Aerodynamics for Engineering Students

Sixth Edition

E.L. Houghton

P.W. Carpenter

Steven H. Collicott

Daniel T. Valentine

ELSEVIER

AMSTERDAM • BOSTON • HEIDELBERG • LONDON
NEW YORK • OXFORD • PARIS • SAN DIEGO
SAN FRANCISCO • SINGAPORE • SYDNEY • TOKYO

Butterworth-Heinemann is an imprint of Elsevier

Butterworth-Heinemann is an imprint of Elsevier
225 Wyman Street, Waltham, MA 02451, USA
The Boulevard, Langford Lane, Kidlington, Oxford, OX5 1GB, UK

MATLAB® is a trademark of The MathWorks, Inc. and is used with permission. The MathWorks does not
warrant the accuracy of the text or exercises in this book. This book's use or discussion of MATLAB® software
or related products does not constitute endorsement or sponsorship by The MathWorks of a particular
pedagogical approach or particular use of the MATLAB® software.

Library of Congress Cataloging-in-Publication Data
Aerodynamics for engineering students / E.L. Houghton ... [et al.]. – 6th ed.
   p. cm.
ISBN: 978-0-08-096632-8 (pbk.)
1. Aerodynamics. 2. Airplanes–Design and construction. I. Houghton, E. L. (Edward Lewis)

TL570.H64 2012
629.132'5–dc23

2011047033

British Library Cataloguing-in-Publication Data
A catalogue record for this book is available from the British Library.

For information on all Butterworth-Heinemann publications
visit our Web site at *www.elsevierdirect.com*

Printed in the United States
12 13 14 15 16 17 18   10 9 8 7 6 5 4 3 2 1

Contents

Preface

This volume is intended for engineering students in introductory aerodynamics courses and as a reference useful for reviewing foundational topics for graduate courses.

The sequence of subject development in this edition begins with definitions and concepts and then moves on to incompressible flow, low speed airfoil and wing theories, compressible flow, high speed wing theories, viscous flow, boundary layers, transition and turbulence, wing design, and concludes with propellers and propulsion.

Reinforcing or teaching first the units, dimensions, and properties of the physical quantities used in aerodynamics addresses concepts that are perhaps both the simplest and the most critical. Common aeronautical definitions are covered before lessons on the aerodynamic forces involved and how the forces drive our definitions of airfoil characteristics. The fundamental fluid dynamics required for the development of aerodynamic studies and the analysis of flows within and around solid boundaries for air at subsonic speeds is explored in depth in the next two chapters. Classical airfoil and wing theories for the estimation of aerodynamic characteristics in these regimes are then developed.

Attention is then turned to the aerodynamics of high speed air flows in Chapters 6 and 7. The laws governing the behavior of the physical properties of air are applied to the transonic and supersonic flow speeds and the aerodynamics of the abrupt changes in the flow characteristics at these speeds, shock waves, are explained. Then compressible flow theories are applied to explain the significant effects on wings in transonic and supersonic flight and to develop appropriate aerodynamic characteristics. Viscosity is a key physical quantity of air and its significance in aerodynamic situations is next considered in depth. The powerful concept of the boundary layer and the development of properties of various flows when adjacent to solid boundaries create a body of reliable methods for estimating the fluid forces due to viscosity. In aerodynamics, these forces are notably skin friction and profile drag. Chapters on wing design and flow control, and propellers and propulsion, respectively, bring together disparate aspects of the previous chapters as appropriate. This permits discussion of some practical and individual applications of aerodynamics.

Obviously aerodynamic design today relies extensively on computational methods. This is reflected in part in this volume by the introduction, where appropriate, of descriptions and discussions of relevant computational techniques. However, this text is aimed at providing the fundamental fluid dynamics or aerodynamics background necessary for students to move successfully into a dedicated course on computation methods or experimental methods. As such, experience in computational techniques or experimental techniques are not required for a complete understanding of the aerodynamics in this book. The authors urge students onward to such advanced courses and exciting careers in aerodynamics.

ADDITIONAL RESOURCES

A set of .m files for the MATLAB routines in the book are available by visiting the book's companion site, www.elsevierdirect.com and searching on 'houghton.' Instructors using the text for a course may access the solutions manual and image bank by visiting www.textbooks.elsevier.com and following the online registration instructions.

ACKNOWLEDGEMENTS

The authors thank the following faculty, who provided feedback on this project through survey responses, review of proposal, and/or review of chapters:

Alina Alexeenko	Purdue University
S. Firasat Ali	Tuskegee University
David Bridges	Mississippi State University
Russell M. Cummings	California Polytechnic State University
Paul Dawson	Boise State University
Simon W. Evans, Ph.D	Worcester Polytechnic Institute
Richard S. Figliola	Clemson University
Timothy W. Fox	California State University Northridge
Ashok Gopalarathnam	North Carolina State University
Dr. Mark W. Johnson	University of Liverpool
Brian Landrum, Ph.D	University of Alabama in Huntsville
Gary L. Solbrekken	University of Missouri
Mohammad E. Taslim	Northeastern University
Valana Wells	Arizona State University

Professors Collicott and Valentine are grateful for the opportunity to continue the work of Professors Houghton and Carpenter and thank Joe Hayton, Publisher, for the invitation to do so. In addition, the professional efforts of Mike Joyce, Editorial Program Manager, Heather Tighe, Production Manager, and Kristen Davis, Designer are instrumental in the creation of this sixth edition.

The products of one's efforts are of course the culmination of all of one's experiences with others. Foremost amongst the people who are to be thanked most warmly for support are our families. Collicott and Valentine thank Jennifer, Sarah, and Rachel and Mary, Clara, and Zach T., respectively, for their love and for the countless joys that they bring to us. Our Professors and students over the decades are major contributors to our aerodynamics knowledge and we are thankful for them. The authors share their deep gratitude for God's boundless love and grace for all.

Basic Concepts and Definitions

"To work intelligently" (Orville and Wilbur Wright)
"one needs to know the effects of variations
incorporated in the surfaces.... The pressures on squares
are different from those on rectangles, circles, triangles, or
ellipses....
The shape of the edge also makes a difference."

from *The Structure of the Plane* – Muriel Rukeyser

LEARNING OBJECTIVES

- Review the fundamental principles of fluid mechanics and thermodynamics required to investigate the aerodynamics of airfoils, wings, and airplanes.

- Recall the concepts of units and dimension and how they are applied to solving and understanding engineering problems.

- Learn about the geometric features of airfoils, wings, and airplanes and how the names for these features are used in aerodynamics communications.

- Explore the aerodynamic forces and moments that act on airfoils, wings, and airplanes and learn how we describe these loads quantitatively in dimensional form and as coefficients.

1.1 INTRODUCTION

The study of aerodynamics requires a number of basic definitions, including an unambiguous nomenclature and an understanding of the relevant physical properties, related mechanics, and appropriate mathematics. Of course, these notions are common to other disciplines, and it is the purpose of this chapter to identify and explain those that are basic and pertinent to aerodynamics and that are to be used in the remainder of the volume.

1.1.1 Basic Concepts

This text is an introductory investigation of aerodynamics for engineering students.[1]
Hence, we are interested in theory to the extent that it can be practically applied
to solve engineering problems related to the design and analysis of aerodynamic
objects.

The design of vehicles such as airplanes has advanced to the level where we
require the wealth of experience gained in the investigation of flight over the past
100 years. We plan to investigate the clever approximations made by the few who
learned how to apply mathematical ideas that led to productive methods and useful
formulas to predict the dynamical behavior of "aerodynamic" shapes. We need to
learn the strengths and, more important, the limitations of the methodologies and
discoveries that came before us.

Although we have extensive archives of recorded experience in aeronautics, there
are still many opportunities for advancement. For example, significant advancements
can be achieved in the state of the art in design analysis. As we develop ideas
related to the physics of flight and the engineering of flight vehicles, we will learn
the strengths and limitations of existing procedures and existing computational tools
(commercially available or otherwise). We will learn how airfoils and wings perform
and how we approach the designs of these objects by analytical procedures.

The fluid of primary interest is air, which is a gas at standard atmospheric conditions. We assume that air's dynamics can be effectively modeled in terms of the
continuum fluid dynamics of an incompressible or simple-compressible fluid. Air is
a fluid whose local thermodynamic state we assume is described either by its mass
density $\rho = $ constant, or by the ideal gas law. In other words, we assume air behaves
as either an incompressible or a simple-compressible medium, respectively. The concepts of a continuum, an incompressible substance, and a simple-compressible gas
will be elaborated on in Chapter 2.

The *equation of state,* known as the ideal gas law, relates two thermodynamic
properties to other properties and, in particular, the pressure. It is

$$p = \rho R T$$

where p is the thermodynamic pressure, ρ is mass density, T is absolute (thermodynamic) temperature, and $R = 287\,\text{J/(kg K)}$ or $R = 1716$ ft-lb $(\text{slug}^\circ\text{R})^{-1}$. Pressure
and temperature are relatively easy to measure. For example, "standard" barometric
pressure at sea level is $p = 101{,}325\,\text{Pascals}$, where a Pascal (Pa) is 1N/m^2. In Imperial units this is 14.675 psi, where psi is lb/in^2 and 1 psi is equal to 6895 Pa (note that

---

[1]It has long been common in engineering schools for an elementary, macroscopic thermodynamics
course to be completed prior to a compressible-flow course. The portions of this text that discuss compressible flow assume that such a course precedes this one, and thus the discussions assume some
elementary experience with concepts such as internal energy and enthalpy.

14.675 psi is equal to 2113.2 lb/ft^2). The standard temperature is 288.15 K (or 15°C, where absolute zero equal to -273.15°C is used). In Imperial units this is 519°R (or 59°F, where absolute zero equal to -459.67°F is used). Substituting into the ideal gas law, we get for the standard density $\rho = 1.225$ kg/m^3 in SI units (and $\rho = 0.00237$ slugs/ft^3 in Imperial units). This is the density of air at sea level given in the table of data for atmospheric air; the table for standard atmospheric conditions is provided in Appendix B.

The thermodynamic properties of pressure, temperature, and density are assumed to be the properties of a mass-point particle of air at a location $\mathbf{x} = (x, y, z)$ in space at a particular instant in time, t. We assume the measurement volume to be sufficiently small to be considered a mathematical point. We also assume that it is sufficiently large so that these properties have meaning from the perspective of equilibrium thermodynamics. And we further assume that the properties are the same as those described in a course on classical equilibrium thermodynamics. Therefore, we assume that *local thermodynamic equilibrium* prevails within the mass-point particle at $\mathbf{x}$ and t regardless of how fast the thermodynamic state changes as the particle moves from one location in space to another. This is an acceptable assumption for our macroscopic purposes because molecular processes are typically faster than changes in the flow field we are interested in from a macroscopic point of view. In addition, we invoke the *continuum hypothesis*, which states that we can define all flow properties as continuous functions of position and time and that these functions are smooth, that is, their derivatives are continuous. This allows us to apply differential integral calculus to solve partial differential equations that successfully model the flow fields of interest in this course. In other words, predictions based on the theory reported in this text have been experimentally verified.

To develop the theory, the fundamental principles of classical mechanics are assumed. They are

- Conservation of mass
- Newton's second law of motion
- First law of thermodynamics
- Second law of thermodynamics

The principle of conservation of mass defines a mass-point particle, which is a fixed-mass particle. Thus the principle also defines mass density ρ, which is mass per unit volume. If a mass-point particle conserves mass, as we have postulated, then density changes can only occur if the volume of the particle changes, because the dimension of mass density is M/L^3, where M is mass and L is length. The SI unit of density is thus kg/m^3.

Newton's second law defines the concept of force in terms of acceleration ("$\mathbf{F} = m\mathbf{a}$"). The acceleration of a mass-point particle is the change in its velocity with respect to a change in time. Let the velocity vector $\mathbf{u} = (u, v, w)$; this is the velocity of a mass-point particle at a point in space, $\mathbf{x} = (x, y, z)$, at a particular instant in time t.

The acceleration of this mass-point particle is

$$\mathbf{a} = \frac{D\mathbf{u}}{Dt} = \frac{\partial \mathbf{u}}{\partial t} + \mathbf{u} \cdot \nabla \mathbf{u}$$

This is known as the substantial derivative of the velocity vector. Since we are interested in the properties at fixed points in space in a coordinate system attached to the object of interest (i.e., the "laboratory" coordinates), there are two parts to mass-point particle acceleration. The first is the local change in velocity with respect to time. The second takes into account the convective acceleration associated with a change in velocity of the mass-point particle from its location upstream of the point of interest to the observation point $\mathbf{x}$ at time t.

We will also be interested in the spatial and temporal changes in any property f of a mass-point particle of fluid. These changes are described by the substantial derivative as follows:

$$\frac{Df}{Dt} = \frac{\partial f}{\partial t} + \mathbf{u} \cdot \nabla f$$

This equation describes the changes in any material property f of a mass point at a particular location in space at a particular instant in time. This is in a laboratory reference frame, the so-called *Eulerian viewpoint*.

The next step in conceptual development of a theory is to connect the changes in flow properties with the forces, moments, and energy exchange that cause these changes to happen. We do this by first adopting the Newtonian simple-compressible viscous fluid model for real fluids (e.g., water and air), which is described in detail in Chapter 2. Moreover, we will apply the simpler, yet quite useful, Euler's perfect fluid model, also described in Chapter 2. It is quite fortunate that the latter model has significant practical use in the design analysis of aerodynamic objects.

Before we proceed to Chapter 2 and look at the development of the equations of motion and the simplifications we will apply to potential flows in Chapters 3, 4 and 5, we review some useful mathematical tools, define the geometry of the wing, and provide an overview of wing performance in the next three sections, respectively.

1.1.2 Measures of Dynamical Properties

The mathematical concepts presented in this section and applied in this text describe the dynamic behavior of a thermo-mechanical fluid. In other words, we neglect electromagnetic, relativistic, and quantum effects on dynamics. Also, as already pointed out, we take the view that the properties are continuous functions of location in space and time. The discussion of units and dimensions here are thus limited to the measures of flow properties of fluids (liquids and gases) near the surface of the Earth under standard conditions.

The units and dimensions of all physical properties and the relevant properties of fluids are recalled, and after a review of the aeronautical definitions of wing and airfoil geometry, the remainder of the chapter discusses aerodynamic force.

The origins of aerodynamic force and how it is manifest on wings and other aeronautical bodies, and the theories that permit its evaluation and design, are to be found in the following chapters. In this chapter the lift, drag, side-wind components, and associated moments of aerodynamic force are conventionally identified, the application of dimensional theory establishing their coefficient form. The significance of the pressure distribution around an aerodynamic body and the estimation of lift, drag, and pitching moment on the body in flight completes the basic concepts and definitions.

1.2 UNITS AND DIMENSIONS

Measurement and calculation require a system of units in which quantities are measured and expressed. Aerospace is a global industry, and to be best prepared for a global career, engineers need to be able to work in both systems in use today. Even when one works for a company with a strict standard for use of one set of units, customers, suppliers, and contractors may be better versed in another, and it is the engineer's job to efficiently reconcile the various documents or specifications without introducing conversion errors. Consider, too, the physics behind the units. That is, one knows that for linear motion, force equals the product of mass and acceleration. The units one uses do not change the physics but change only our quantitative descriptions of the physics. When confused about units, focus on the process or state being described and step through the analysis, tracking units the entire way.

In the United States, "Imperial" or "English" units remain common. Distance (within the scale of an aerodynamic design) is described in inches or feet. Mass is described by either the slug or the pound-mass (lbm). Weight is described by pounds (lb) or by the equivalent unit with a redundant name, the pound-force (lbf). Large distances—for example, the range of an aircraft—are described in miles or nautical miles. Speed is feet per second, miles per hour, or knots, where one knot is one nautical mile per hour. Multimillion dollar aircraft are still marketed and sold using knots and nautical miles (try a web search on "777 range"), so these units are not obsolete.

In other parts of the world, and in K-12 education in the United States, the dominant system of units is the Système International d'Unités, commonly abbreviated as "SI units." It is used throughout this book, except in a very few places as specially noted.

It is essential to distinguish between "dimension" and "unit." For example, the dimension "length" expresses the *qualitative* concept of linear displacement, or distance between two points, as an abstract idea, without reference to actual quantitative measurement. The term "unit" indicates a specified amount of a quantity. Thus a meter is a unit of length, being an actual "amount" of linear displacement, and so is a mile. The meter and mile are different *units*, since each contains a different *amount* of length, but both describe length and therefore are identical *dimensions*.[2]

Expressing this in symbolic form:

x meters $=$ [L] (a quantity of x meters has the dimension of length)
x miles $=$ [L] (a quantity of x miles has the dimension of length)
x meters $\neq x$ miles (x miles and x meters are unequal quantities of length)
$[x$ meters$] = [x$ miles$]$ (the dimension of x meters is the same as the dimension of x miles).

1.2.1 Fundamental Dimensions and Units

There are five fundamental dimensions in terms of which the dimensions of all other physical quantities may be expressed. They are mass [M], length [L], time [T], temperature $[\theta]$, and charge.[3] (Charge is not used in this text so is not discussed further.) A consistent set of units is formed by specifying a unit of particular value for each of these dimensions. In aeronautical engineering the accepted units are, respectively, the kilogram, the meter, the second, and the Kelvin or degree Celsius. These are identical with the units of the same names in common use and are defined by international agreement.

It is convenient and conventional to represent the names of these units by abbreviations:

kg—kilogram, slugs for slugs, and lbm for pound-mass
m—meter and ft for feet
s—second
°C—degree Celsius and °F for degree Fahrenheit
K—Kelvin and R for Rankine (but also for the gas constant)

The degree Celsius is one one-hundredth part of the temperature rise involved when pure water at freezing temperature is heated to boiling temperature at standard pressure. In the Celsius scale, pure water at standard pressure freezes at 0°C (32°F) and boils at 100°C (212°F).

The unit Kelvin (K) is identical in size to the degree Celsius (°C), but the Kelvin scale of temperature is measured from the absolute zero of temperature, which is approximately –273°C. Thus a temperature in K is equal to a temperature in °C plus 273.15. Similarly, degrees Rankine equals °F plus 459.69.

---

[2]Quite often "dimension" appears in the form "a dimension of 8 meters," meaning a specified length. This is thus closely related to the engineer's "unit," and implies linear extension only. Another common example of the use of "dimension" is in "three-dimensional geometry," implying three linear extensions in different directions. References in later chapters to two-dimensional flow, for example, illustrate this. The meaning here must not be confused with either of these uses.

[3]Some authorities express temperature in terms of length and time. This introduces complications that are briefly considered in Section 1.3.8.

1.2.2 Fractions and Multiples

Sometimes, the fundamental units just defined are inconveniently large or inconveniently small for a particular case. If so, the quantity can be expressed as a fraction or multiple of the fundamental unit. Such multiples and fractions are denoted by a prefix appended to the unit symbol. The prefixes most used in aerodynamics are:

M (mega)—1 million
k (kilo)—1 thousand
m (milli)—1-thousandth part
μ (micro)—1-millionth part
n (nano)—1-billionth part

Thus

1 MW = 1,000,000 W
1 mm = 0.001 m
1 μm = 0.001 mm

A prefix attached to a unit makes a new unit so, for example,

$$1\text{mm}^2 = 1(\text{mm})^2 = 10^{-6}\text{m}^2 \left(\text{not } 10^{-3}\text{m}^2\right)$$

For some purposes, the hour or the minute can be used as the unit of time.

For Imperial units, everyday scientific notation is used rather than suffixes or prefixes. One exception is stress or pressure of thousands of pounds per square inch, known as kpsi. Additionally, length may switch from feet to inches or miles. It is common to use fractional inches, but the student engineer needs to be aware that the implied precision in a fraction increases rapidly. For example, $1/2 = 0.5$, but $1/32 = 0.03125$.

1.2.3 Units of Other Physical Quantities

Having defined the four fundamental dimensions and their units, it is possible to establish units of all other physical quantities (see Table 1.1). Speed, for example, is defined as the distance traveled in unit time. It therefore has the dimension LT^{-1} and is measured in meters per second (m s^{-1}). It is sometimes desirable to use kilometers per hour or knots (nautical miles per hour; see Appendix D) as units of speed; care must be exercised to avoid errors of consistency.

To find the dimensions and units of more complex quantities, we use the principle of *dimensional homogeneity*. This simply means that, in any valid physical equation, the dimensions of both sides must be the same. Thus, for example, if $(\text{mass})^n$ appears on the left-hand side of the equation, it must also appear on the right-hand side; similarly for length, time, and temperature.

Thus, to find the dimensions of force, we use Newton's second law of motion

$$\text{Force} = \text{mass} \times \text{acceleration}$$

Table 1.1 Units and Dimensions

Quantity	Dimension	Unit (abbreviation)
Length	L	Meter (m) or feet (ft)
Mass	M	Kilogram (kg) or slug or pound-mass (lbm)
Time	T	Second (s)
Temperature	θ	Degree Celsius (°C) or Fahrenheit (°F) or Kelvin (K) or Rankine (R)
Area	L^2	Square meter (m^2) or square foot (ft^2)
Volume	L^3	Cubic meter (m^3) or cubic foot (ft^3)
Speed	LT^{-1}	Meters per second ($m\,s^{-1}$) or feet per second ($ft\,s^{-1}$)
Acceleration	LT^{-2}	Meters per second per second ($m\,s^{-2}$) or feet per second squared ($ft\,s^{-2}$)
Angle	1	Radian or degree (°) (radian is expressed as a ratio and is therefore dimensionless)
Angular velocity	T^{-1}	Radians per second (s^{-1})
Angular acceleration	T^{-2}	Radians per second per second (s^{-2})
Frequency	T^{-1}	Cycles per second, Hertz (s^{-1}, Hz)
Density	ML^{-3}	Kilograms per cubic meter ($kg\,m^{-3}$) or slugs per cubic foot ($slug\,ft^{-3}$) or pound-mass per cubic foot ($lbm\,ft^{-3}$)
Force	MLT^{-2}	Newton (N) or pound (lb)
Stress	$ML^{-1}T^{-2}$	Newtons per square meter or Pascal ($N\,m^{-2}$ or Pa) or pounds per square inch (psi) or pounds per square foot (psf)
Strain	1	None (expressed as a nondimensional ratio)
Pressure	$ML^{-1}T^{-2}$	Newtons per square meter or Pascal ($N\,m^{-2}$ or Pa) or pounds per square inch (psi) or pounds per square foot (psf)
Energy work	ML^2T^{-2}	Joule (J) or foot-pounds (ft lb)
Power	ML^2T^{-3}	Watt (W) or horsepower (Hp)
Moment	ML^2T^{-2}	Newton meter (Nm) or foot-pounds, (ft lb)
Absolute viscosity	$ML^{-1}T^{-1}$	Kilograms per meter per second or Poiseuilles ($kg\,m^{-1}\,s^{-1}$ or PI) or slugs per foot per second ($slug\,ft^{-1}\,s^{-1}$)
Kinematic viscosity	L^2T^{-1}	Meters squared per second ($m^2\,s^{-1}$) or feet squared per second ($ft^2\,s^{-1}$)
Bulk elasticity	$ML^{-1}T^{-2}$	Newtons per square meter or Pascal ($N\,m^{-2}$ or Pa) or pounds per square inch (psi) or pounds per square foot (psf).

where acceleration is speed ÷ time. Expressed dimensionally, this is

$$\text{Force} = [M] \times \left[\frac{L}{T} \div T\right] = \left[MLT^{-2}\right]$$

Writing in the appropriate units, it is seen that a force is measured in units of $\text{kg}\,\text{m}\,\text{s}^{-2}$. Since, however, the unit of force is given the name Newton (abbreviated usually to N), it follows that

$$1\,\text{N} = 1\,\text{kg}\,\text{m}\,\text{s}^{-2}$$

It should be noted that there can be confusion between the use of m both for "milli" and for "meter." This is avoided by use of a space. Thus ms denotes millisecond while m s denotes the product of meter and second.

The concept of dimension forms the basis of dimensional analysis, which is used to develop important and fundamental physical laws. Its treatment is postponed to Section 1.5.

1.2.4 Imperial Units

Engineers in some parts of the world, the United States in particular, use a set of units based on the Imperial systems[4] in which the fundamental units are

Mass—slug
Length—foot
Time—second
Temperature—degree Fahrenheit or Rankine

AERODYNAMICS AROUND US

Units in Use

Students have long struggled with learning to use units correctly. The danger is that it is simple to create large quantitative errors when even an experienced engineer is hurrying through an analysis task. That the SI system is supposedly "easier" to use than the Imperial system is irrelevant if you are going to commit errors—the only difference between the systems will be how large those errors are. There is no doubt that all humans are fallible. The relevant question is: How can one work with units correctly all of the time? Your pursuit of excellence may be aided by a short review of the basics of unit conversion.

Engineers are wise to remember that the mathematical symbol stating equality between two quantities (the "equals" sign) relates not just numerical equivalence but dimensional equivalence as well. That 5280 feet equals 1 mile is a statement of equivalence between two descriptions of the same distance. The same is true for 100 cm = 1 m. But what about equalities between systems of units, such as how many centimeters equal one inch? We all learn that 2.54 cm = 1 in, but is this

---

[4]Since many valuable texts and papers exist using Imperial units, this book contains, as Appendix D, a table of factors for converting from the Imperial to the SI system.

exact? Yes (see the NIST handbook discussed in a moment for details). Does 1 m = 39.37 in? No. (Can you show this to be the case?)

However, for anything that actually flies, error exists because we can only build and measure with finite precision. Thus many unit conversions—such as 1 m = 39.37 in, 1.15078 mi = 1nm (that is, nautical mile, not nanometers), or 1 slug = 32.174 lb—are sufficiently precise for use throughout aerodynamics. The modern student or engineer can find authoritative values at the web site of the United States' National Institute of Science and Technology (NIST), such as in Appendix C of NIST Handbook 44-2010.

Consider a wing chord of 3.50 meters, which we need to convert to inches. How should we proceed? A quick search on your smartphone tells you 1m $\cong$ 39.37 in. Yet, for the sake of education, let's work through this from the exact relation, 2.54 cm = 1 in. First, convert meters to centimeters using (100 cm/1 m) = 1:

$$3.50 \text{ m} = 3.50 \text{ m} \times 1$$

$$3.50 \text{ m} = 3.50 \text{ m} \times \left(\frac{100 \text{ cm}}{1 \text{ m}} \right)$$

$$3.50 \text{ m} = 350 \text{ cm}$$

Note that the units of meter in the original description, 3.5 m, cancel the units of meter in the denominator of the ratio, leaving units of centimeters on the right. Knowing that 2.54 cm = 1 in, we can form another unit (as in magnitude of one) ratio, 1 = (1 in/2.54 cm). Multiplying 350 cm by 1 does not change the length:

$$3.50 \text{ m} = 350 \text{ cm} \times 1$$

$$3.50 \text{ m} = 350 \text{ cm} \times \left(\frac{1 \text{ in}}{2.54 \text{ cm}} \right)$$

$$3.50 \text{ m} = \frac{350}{2.54} \text{ in} \approx 137.795 \text{ in} \approx 138.0 \text{ in}$$

Remember to review the concept of significant digits, such as in an introductory physics text. Again, a length on the left side of the equality requires a length on the right side, plus proper numerical computation.

Most common among questions about units that confound students each semester involves the relationships between units of mass and force or weight and the physical difference between mass and weight. The latter isn't a question of units at all but a fundamental physics question that requires immediate answer for the student to excel. High school physics teachers teach the difference between mass and weight, but, through no fault of theirs, students generally seem to require several years of thinking about the issue before it is clarified. Consider an astronaut launching into orbit. While on the ground, you know that the astronaut's weight W is computed from his or her mass $M = 80 \text{ kg}$ and the Earth-surface value of the acceleration due to gravity, $g_o = 9.81 \text{ m/s}^2$,

$$W = Mg = 80 \text{ kg} \times 9.8 \text{ m s}^{-2} = 784 \text{ kg m s}^{-2} = 784 \text{ N}$$

where N denotes Newtons. This equation does not equate mass and weight but says that weight is how we describe the effect of Earth's gravity on a mass. Mass is an intrinsic property of the countless subatomic particles that make up the atoms that make up the molecules of the astronaut. When the astronaut is on the ground, her particles all have mass and her mass is the sum of the masses of all those particles. When the astronaut is launched into orbit, she may feel weightless because of the centripetal acceleration of the orbital path, but all of the particles have the same masses they had on the Earth's surface, and the sum of those is still the mass of the astronaut.

Note that Newton's first law, which, as we know, is force equals the product of mass and acceleration (F = ma), relates the units of mass and weight. For example, using the numerical values from the weight example, we know that

$$784\,\mathrm{N} = (80\,\mathrm{kg}) \times \left(9.8\,\mathrm{m\,s^{-2}}\right)$$

Isolate the units on the left and the numbers on the right:

$$\frac{\mathrm{N}}{\mathrm{kg\,m\,s^{-2}}} = \frac{80 \times 9.8}{784}$$

or

$$\frac{\mathrm{N}}{\mathrm{kg\,m\,s^{-2}}} = 1$$

This shows the common definition in a format that students often lose sight of when working on tasks. That is, if

$$\frac{\mathrm{N}}{\mathrm{kg\,m\,s^{-2}}} = 1$$

then any occurrence of Newton may be replaced by the product of kilogram, meter, and inverse seconds squared. Thus, while weight (or force) and mass are different, their units are related, and we can use this relationship in analysis.

This elementary discussion can help the student in working with force (or weight) and mass. Weight is one specific type of force—the action of gravity on mass. Weight is not mass; 1 kilogram weighs 9.8 Newtons, but it does not equal 9.8 Newtons. One slug weighs 32.174 pounds, but does not equal 32.174 pounds because weight (force) and mass are two different things. You cannot equate slugs and pounds any more than you can equate meters and Coulombs.

The kg-m-s and slug-ft-s systems of units are identical in their use in aerodynamics (use an electrodynamics text for lessons when working in that field). In other words,

$$\frac{\mathrm{N}}{\mathrm{kg\,m\,s^{-2}}} = 1$$

and

$$\frac{\mathrm{lb}}{\mathrm{slug\,ft\,s^{-2}}} = 1$$

so

$$\frac{\mathrm{N}}{\mathrm{kg\,m\,s^{-2}}} = \frac{\mathrm{lb}}{\mathrm{slug\,ft\,s^{-2}}}$$

One slug does equal 32.174 pound-mass (lbm) because the pound-mass is a unit of mass. Wherever you see the slug, you can replace it with 32.174 lbm. Thus,

$$\frac{\mathrm{N}}{\mathrm{kg\,m\,s^{-2}}} = \frac{\mathrm{lb}}{\mathrm{slug\,ft\,s^{-2}}} = 1 = \frac{\mathrm{lbf}}{32.174\,\mathrm{lbm\,ft\,s^{-2}}}$$

where pounds are now called pound-force, or lbf, presumably to make a clear distinction between lbf and lbm.

There are good and bad units practices around us. For example, one may see an equation for pressure drop in a system given as "The pressure drop across the device, Δp, is given by $\Delta p = 17.34 Q^2$ for Q in gallons per minute and pressure in psi." The choice in how to present the units information may be fine for a technician who will be sizing specific items or determining if a specific design will meet a specification. Where this statement is poor practice is in engineering reports where you will document and communicate your results to your project team and to the people who will follow you. Consider that today you can fly on commercial airliners whose designs are older than you are (Boeing 737 and 767 in this country and the 727 elsewhere in the world). Engineering knowledge is passed on over the decades in reports, not by word of mouth. A proper way to report the previous pressure loss relation for a device would be to include the units in the equation:

"The pressure drop across the device, Δp, is given by

$$\Delta p = \left(\frac{17.34\,\text{psi}}{\text{gpm}^2}\right) Q^{2}\text{"}$$

Then anyone who uses the equation in their own analysis or computer model will be able to convert rapidly into the units that they wish. It is a good habit in communication to keep the units in the equation, not in an accompanying sentence, especially in this age of computer cut-and-paste.

The authors agree with what is likely running through a student's head, that these different unit systems are a strange way to run an industry. History has led us to where we are. The student aiming for success in the global aerospace industry, in either atmospheric flight or spaceflight, will be wise to practice working problems in all set of units. Focus on the physics involved and write down all the units in your analysis.

1.3 RELEVANT PROPERTIES

Any fluid that we wish to describe exists in some state of matter. For example, if we are working with a flow of nitrogen, is it gaseous nitrogen or liquid nitrogen? For whatever the state the fluid is in, we need a collection of "tools" to use to describe the thermodynamic state of the fluid at a point, over time, and throughout a field. An unambiguous description of the thermodynamic state of the fluid is important of course to a mathematical model of a flow and is vital to effective engineering communication. Thus in this section we develop the tools to use to form these unambiguous descriptions.

1.3.1 Forms of Matter

Matter may exist in three principal forms—solid, liquid, or gas—corresponding in that order to decreasing rigidity of the bonds between the molecules the matter comprises. A special form of a gas, a *plasma*, has properties different from those of a normal gas; although belonging to the third group, it can be regarded justifiably as a separate, distinct form of matter that is relevant to the highest-speed aerodynamics such as flows over spacecraft reentering the atmosphere.

In a solid the intermolecular bonds are very rigid, maintaining the molecules in what is virtually a fixed spatial relationship. Thus a solid has a fixed volume and shape. This is seen clearly in crystals, in which the molecules or atoms are arranged in a definite, uniform pattern, giving all crystals of that substance the same geometric shape.

A liquid has weaker bonds between its molecules. The distances between the molecules are fairly rigidly controlled, but the arrangement in space is free. Therefore, liquid has a closely defined volume but no definite shape, and may accommodate itself to the shape of its container within the limits imposed by its volume.

A gas has very weak bonding between the molecules and therefore has neither definite shape nor definite volume, but rather will fill the vessel containing it.

A plasma is a special form of gas in which the atoms are ionized—that is, they have lost or gained one or more electrons and therefore have an electrical charge. Any electrons that have been stripped from the atoms are wandering free within the plasma and have a negative electrical charge. If the number of ionized atoms and free electrons is such that the total positive and negative charges are approximately equal, so that the gas as a whole has little or no charge, it is termed a plasma. In astronautics plasma is of particular interest for the reentry of rockets, satellites, and space vehicles into the atmosphere.

1.3.2 Fluids

A fluid is a liquid or a gas. Equations of motion for a fluid do not depend on liquid or gas, but the equation of state will differ. The basic feature of a fluid is that it can flow—this is the essence of any definition of it. However, flow applies to substances that are not true fluids—for example a fine powder piled on a sloping surface will flow. For example, flour poured in a column onto a flat surface will form a roughly conical pile, with a large angle of repose, whereas water, which is a true fluid, poured onto a fully wetted surface will spread uniformly over it. Equally, a powder may be heaped in a spoon or bowl, whereas a liquid will always form a level surface. Any definition of a fluid must allow for these facts, so a fluid may be defined as "matter capable of flowing, and either finding its own level (if a liquid), or filling the whole of its container (if a gas)." Once we restrict ourselves to an ideal gas, such as for steady, level atmospheric flight, distinctions between air as a "Newtonian fluid" and fine particulates are clear. A Newtonian fluid is one in which shear stress is proportional to rate of shearing strain; this is never found in particulates.

Experiment shows that an extremely fine powder, in which the particles are not much larger than molecular size, finds its own level and may thus come under the common definition of a liquid. Also, a phenomenon well known in the transport of sands, gravels, and so forth, is that these substances find their own level if they are agitated by vibration or the passage of air jets through the particles. These are special cases, however, and do not detract from the authority of the definition of a fluid as a substance that flows or (tautologically) that possesses fluidity.

1.3.3 Pressure

At any point in a fluid, whether liquid or gas, there is a pressure. If a body is placed in a fluid, its surface is bombarded by a large number of molecules moving at random. Under normal conditions the collisions on a small area of surface are so frequent that they cannot be distinguished as individual impacts but appear as a steady force on the area. The *intensity* of this "molecular bombardment" is its *static pressure*.

Very frequently the static pressure is referred to simply as pressure. The term *static* is rather misleading as it does not imply that the fluid is at rest.

For large bodies moving or at rest in the fluid (e.g., air), the pressure is not uniform over the surface, and this gives rise to *aerodynamic* or *aerostatic force*, respectively.

Since a pressure is force per unit area, it has the dimensions

$$[\text{Force}] \div [\text{area}] = [MLT^{-2}] \div [L^2] = [ML^{-1}T^{-2}]$$

and is expressed in units of Newtons per square meter or in Pascals ($N\,m^{-2}$ or Pa). Pressure is also commonly specified in pounds per square inch (psi) or pounds per square foot (psf). It can also be of use to consider the above equation multiplied by length over length:

$$[\text{Force}] * [\text{Length}]/([\text{Area}] * [\text{Length}]) = [ML^2T^{-2}]/[L^3] = [\text{Energy}]/[\text{Volume}]$$

Thus, besides the most common view of it as a force per area, pressure also has units of energy per volume.

Pressure in Fluid at Rest

Consider a small cubic element containing fluid at rest in a larger bulk of fluid also at rest. The faces of the cube, assumed conceptually to be made of some thin flexible material, are subject to continual bombardment by the molecules of the fluid and thus experience a *force*. The force on any face may be resolved into two components, one acting perpendicular to the face and the other along it (i.e., tangential to it). Consider the tangential components only; there are three significantly different possible arrangements (Fig. 1.1). System (a) would cause the element to rotate, and thus the fluid would not be at rest; system (b) would cause the element to move (upward and to the right for the case shown), and, once again, the fluid would not be at rest. Since a fluid cannot resist shear stress but only rate of change in shear strain (Sections 1.3.6 and 2.7.2), system (c) would cause the element to distort, the degree of distortion increasing with time, and the fluid would not remain at rest. The conclusion is that a fluid at rest cannot sustain tangential stresses.

Pascal's Law

Consider the right prism of length δz in the direction into the page and cross-section ABC, the angle ABC being a right angle (Fig. 1.2). The prism is constructed of material of the same density as a bulk of fluid in which the prism floats at rest with the face BC horizontal.

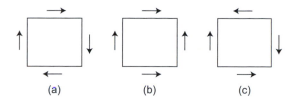

FIGURE 1.1

Fictitious systems of tangential forces in static fluid.

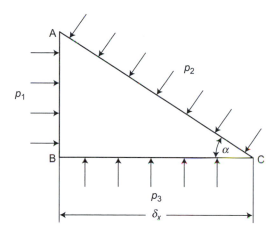

FIGURE 1.2

Prism for Pascal's Law.

Pressures p_1, p_2, and p_3 act on the faces shown and, as just proved, act in the direction perpendicular to the respective face. Other pressures act on the end faces of the prism, but are ignored in the present problem. In addition to these pressures, the weight W of the prism acts vertically downward. Consider the forces acting on the wedge that is in equilibrium and at rest.

Resolving forces horizontally,

$$p_1(\delta x \tan \alpha)\delta y - p_2(\delta x \sec \alpha)\delta y \sin \alpha = 0$$

Dividing by $\delta x\, \delta y \tan \alpha$, this becomes

$$p_1 - p_2 = 0$$

that is,

$$p_1 = p_2 \tag{1.1}$$

Resolving forces vertically,

$$p_3 \delta x \delta y - p_2 (\delta x \sec \alpha) \delta y \cos \alpha - W = 0 \qquad (1.2)$$

Now

$$W = \rho g (\delta x)^2 \tan \alpha \delta y / 2$$

Therefore, substituting this in Eq. (1.2) and dividing by $\delta x\, \delta y$,

$$p_3 - p_2 - \frac{1}{2} \rho g \tan \alpha \delta y = 0$$

If now the prism is imagined to become infinitely small, so that $\delta x \rightarrow 0$, the third term tends to zero, leaving

$$p_3 - p_2 = 0$$

Thus, finally,

$$p_1 = p_2 = p_3 \qquad (1.3)$$

Having become infinitely small, the prism is in effect a point, so this analysis shows that, at a point, the three pressures considered are equal. In addition, the angle α is purely arbitrary and can take any value, while the whole prism can be rotated through a complete circle about a vertical axis without affecting the result. It may be concluded, then, that the pressure acting at a point in a fluid at rest is the same in all directions.

1.3.4 Temperature

In any form of matter the molecules are in motion relative to each other. In gases the motion is random movement of magnitude ranging from approximately 60 nm under normal conditions to some tens of millimeters at very low pressures. The distance of free movement of a molecule of gas is the distance it can travel before colliding with another molecule or the walls of the container. The mean value of this distance for all molecules in a gas is called the length of the mean molecular free path.

By virtue of this motion, the molecules possess kinetic energy, and this energy is sensed as the *temperature* of the solid, liquid, or gas. In the case of a gas in motion it is called the static temperature or, more usually, just the temperature. Temperature has the dimension $[\theta]$ and the units K, °C, °F, or °R (Section 1.1). In practically all calculations in aerodynamics, temperature is measured in K or °R (i.e., from absolute zero).

1.3.5 Density

The density of a material is a measure of the amount of the material contained in a given volume. In a fluid the density may vary from point to point. Consider the fluid contained in a small region of volume δV centered at some point in the fluid, and let the mass of fluid within this spherical region be δm. Then the density of the fluid at the point on which the volume is centered is defined by

$$\text{Density } \rho = \lim_{\delta v \to 0} \frac{\delta m}{\delta V} \qquad (1.4)$$

The dimensions of density are thus ML^{-3}, and density is measured in units of kilogram per cubic meter ($kg\,m^{-3}$). At standard temperature and pressure (288 K, 101, $-325\,N\,m^{-2}$), the density of dry air is $1.2256\,kg\,m^{-3}$.

Difficulties arise in rigorously applying the definition given a real fluid composed of discrete molecules, since the volume, when taken to the limit, either will or will not contain part of a molecule. If it does contain a molecule, the value obtained for the density will be fictitiously high. If it does not contain a molecule, the resultant value will be zero. This difficulty is generally avoided in the range of temperatures and pressures normally encountered in aerodynamics because the molecular nature of a gas may for many purposes—in fact, for nearly every terrestrial flight application—be ignored and the assumption made that the fluid is a continuum—that is, it does not consist of discrete particles. This "continuum assumption" suffices because the mean free path of the molecular motion is much less than the smallest length scale on the vehicle for almost every atmospheric flight regime.

1.3.6 Viscosity

Viscosity is regarded as the tendency of a fluid to resist sliding between layers or, more rigorously (as explained later) a rate of change in shear strain. There is very little resistance to the movement of a knife blade edge-on through air, but to produce the same motion through thick oil requires much more effort. This is because the viscosity of oil is high compared with that of air.

Dynamic Viscosity

The dynamic (more properly, coefficient of dynamic, or absolute) viscosity is a direct measure of the viscosity of a fluid. Consider two parallel flat plates placed a distance h apart, with the space between them filled with fluid. One plate is held fixed, and the other is moved in its own plane at a speed V (see Fig. 1.3). The fluid immediately adjacent to each plate will move with it (i.e., there is no slip). Thus the fluid in contact with the lower plate will be at rest while that in contact with the upper plate will be moving with speed V. Between the plates the speed of the fluid will vary linearly, as shown in Fig. 1.3, in the absence of other influences. As a direct result of viscosity, a force F has to be applied to each plate to maintain the motion, the fluid tending to retard the moving plate and drag the fixed plate to the right. If the area of fluid in contact with each plate is A, the shear stress is F/A. The rate of shear strain caused by the upper plate sliding over the lower is V/h.

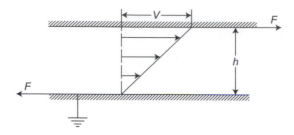

FIGURE 1.3

Simple flow geometry to create a uniform sear.

These quantities are connected by Newton's equation, which serves to define the dynamic viscosity μ:

$$\frac{F}{A} = \mu\left(\frac{V}{h}\right) \tag{1.5}$$

Hence

$$[ML^{-1}T^{-2}] = [\mu][LT^{-1}L^{-1}] = [\mu][T^{-1}]$$

Thus

$$[\mu] = [ML^{-1}T^{-1}]$$

and the units of μ are therefore $kg\,m^{-1}\,s^{-1}$; in the SI system the name Poiseuille (Pl) has been given to this combination of fundamental units. At $0°C$ (273 K) the dynamic viscosity for dry air is $1.714 \times 10^{-5}\,kg\,m^{-1}\,s^{-1}$.

Note that while the relationship of Eq. (1.5) with constant μ applies nicely to aerodynamics, it does not apply to all fluids. For an important class of fluids, which includes blood, some oils, and some paints, μ is not constant but is a function of V/h—that is, the rate at which the fluid is shearing. Numerous classes of "non-Newtonian fluids" are important in fields outside of aerodynamics, and the eager student can explore these best with good knowledge of Newtonian fluid behavior as discussed in this book.

Kinematic Viscosity

The kinematic viscosity (or, more properly, the coefficient of kinematic viscosity) is a convenient form in which the viscosity of a fluid may be expressed. It is formed by combining the density ρ and the dynamic viscosity μ according to the equation

$$v = \frac{\mu}{\rho}$$

and has the dimensions L^2T^{-1} and the units $m^2\,s^{-1}$. It may be regarded as a measure of the relative magnitudes of fluid viscosity and inertia and has the practical advantage, in calculations, of replacing two values representing μ and ρ with a single value.

1.3.7 Speed of Sound and Bulk Elasticity

Bulk elasticity is a measure of how much a fluid (or solid) will be compressed by the application of external pressure. If a certain small volume V of fluid is subjected to a rise in pressure δp, this reduces the volume by an amount $-\delta V$. In other words, it produces a volumetric strain of $-\delta V/V$. Accordingly, bulk elasticity is defined as

$$K = -\frac{\delta p}{\delta V/V} = -V\frac{dp}{dV} \tag{1.6a}$$

The volumetric strain is the ratio of two volumes and is evidently dimensionless, so the dimensions of K are the same as those for pressure: $\mathrm{ML^{-1}T^{-2}}$. The SI unit is $\mathrm{Nm^{-2}}$ (or Pa) and the Imperial unit is psi. When written in terms of density of the air rather than volume, Eq. (1.6a) becomes

$$K = \rho\frac{dp}{d\rho}$$

The propagation of sound waves involves alternating compression and expansion of the medium. Accordingly, bulk elasticity is closely related to the speed of sound a as follows:

$$a = \sqrt{\frac{K}{\rho}} \tag{1.6b}$$

Let the mass of the small volume of fluid be M; then by definition the density $\rho = M/V$. By differentiating this definition, keeping M constant, we obtain

$$d\rho = -\frac{M}{V^2}dV = -\rho\frac{dV}{V}$$

Therefore, combining this with Eqs. (1.6a) and (1.6b), it can be seen that

$$a = \sqrt{\frac{dp}{d\rho}} \tag{1.6c}$$

The propagation of sound in a perfect gas is regarded as a lossless process; that is, no energy is lost and the wave process lacks heat transfer to or from the surrounding fluid. Accordingly (see the passage on *Entropy* to come), the pressure and density are related by Eq. (1.24), so for a perfect gas, for which $P = \rho RT$,

$$a = \sqrt{\frac{\gamma p}{\rho}} = \sqrt{\gamma RT} \tag{1.6d}$$

where γ is the ratio of the specific heats and R is the specific gas constant for that gas. Eq. (1.6d) is the formula normally used to determine the speed of sound in gases for aerodynamics applications.

1.3.8 Thermodynamic Properties

Heat, like work, is a form of energy transfer. Consequently, it has the same dimensions as energy (i.e., ML^2T^{-2}) and is measured in Joules (J) or foot-pounds (ft-lb).

Specific Heat

The specific heat of a material is the amount necessary to raise the temperature of a unit mass of the material by one degree. Thus it has the dimensions $L^2T^{-2}\theta^{-1}$ and is measured in SI units of $J\,kg^{-1}\,K^{-1}$. Imperial units of ft-lb slug^{-1} $^\circ F^{-1}$ or ft-lb slug^{-1} $^\circ R^{-1}$ are most common.

There are countless ways in which gas may be heated. Two important and distinct ways are at constant volume and at constant pressure. These define important thermodynamic properties of the gas.

Specific Heat at Constant Volume

If a unit mass of the gas is enclosed in a cylinder sealed by a piston, and the piston is locked in position, the volume of the gas cannot change. It is assumed that the cylinder and piston do not receive any of the heat. The specific heat of the gas under these conditions is the specific heat at constant volume c_V. For dry air at normal aerodynamic temperatures, $c_V = 718\,J\,kg^{-1}\,K^{-1} = 4290$ ft-lb slug^{-1} $^\circ R^{-1}$.

Internal energy (E) is a measure of the kinetic energy of the molecules that make up the gas, so

$$\text{internal energy per unit mass } E = c_V T$$

or more generally

$$c_V = \left[\frac{\partial E}{\partial T}\right]_\rho \tag{1.7}$$

Specific Heat at Constant Pressure

Assume that the piston just referred to is now freed and acted on by a constant force. The pressure of the gas is that necessary to resist the force and is therefore constant as well. The application of heat to the gas causes its temperature to rise, which leads to an increase in its volume in order to maintain the constant pressure. Thus the gas does mechanical work against the force, so it is necessary to supply the heat required to increase its temperature (as in the case at constant volume) as well as heat equivalent to the mechanical work done against the force. This total amount is called the specific heat at constant pressure c_p and is defined as that amount required to raise the temperature of a unit mass of the gas by one degree, the pressure of the gas being kept constant while heating. Therefore, c_p is always greater than c_V. For dry air at normal aerodynamic temperatures, $c_p = 1005 \ \text{J} \, \text{kg}^{-1} \, \text{K}^{-1} = 6006 \ \text{ft-lb} \, \text{slug}^{-1} \, {}^{\circ}\text{R}^{-1}$.

The sum of internal energy per unit mass and pressure energy per unit mass is known as *enthalpy* (*h* per unit mass) (discussed momentamly). Thus

$$h = c_p T$$

or more generally

$$c_p = \left[\frac{\partial h}{\partial T} \right]_p \qquad (1.8)$$

Ratio of Specific Heats

The ratio of specific heats is a property important in high-speed flows and is defined by the equation

$$\gamma = \frac{c_p}{c_V} \qquad (1.9)$$

(The value of γ for air depends on the temperature, but for much of practical aerodynamics it may be regarded as constant at about 1.403. This value is often in turn approximated to $\gamma = 1.4$, which is in fact the theoretical value for an ideal diatomic gas.)

Enthalpy

The enthalpy *h* of a unit mass of gas is the sum of the internal energy *E* and the pressure energy $p \times 1/\rho$. Thus

$$h = E + p/\rho \qquad (1.10)$$

Enthalpy may be a new term to many students, but it is simply a tool for keeping track of a sum of two energies. It is not an exotic new property, but it is an energy. From the definition of specific heat at constant volume, Eq. (1.7), Eq. (1.10) becomes

$$h = c_V T + p/\rho$$

Again from the definition in Eq. (1.8), Eq. (1.10) gives

$$c_p T = c_V T + p/\rho \qquad (1.11)$$

Now the pressure, density, and temperature are related in the equation of state, which for perfect gases takes the form

$$p/(\rho T) = \text{constant} = R \qquad (1.12)$$

Substituting for p/ρ in Eq. (1.11) yields the relationship

$$c_p - c_V = R \tag{1.13}$$

The specific gas constant R is thus the amount of mechanical work obtained by heating the unit mass of a gas through a unit temperature rise at constant pressure. It follows that R is measured in units of $J\,kg^{-1}\,K^{-1}$. For air over the range of temperatures and pressures normally encountered in aerodynamics, R has the value 287.26 $J\,kg^{-1}\,K^{-1}$, or 1716.6 ft-lb $slug^{-1}\,R^{-1}$.

Introducing the ratio of specific heats (Eq. 1.9), the following expressions are obtained:

$$c_p = \frac{\gamma}{\gamma - 1}R \quad \text{and} \quad c_V = \frac{R}{\gamma - 1} \tag{1.14}$$

Replacing $c_V T$ by $[1/(\gamma - 1)]p/\rho$ in Eq. (1.11) readily gives the enthalpy as

$$c_p T = \frac{\gamma}{\gamma - 1}\frac{p}{\rho} \tag{1.15}$$

It is often convenient to link the enthalpy or total heat to the other energy of motion. This would be kinetic energy $\bar{K}$ per unit mass of gas moving with mean velocity V:

$$\bar{K} = \frac{V^2}{2} \tag{1.16}$$

Thus the total energy flux in the absence of external, tangential surface forces and heat conduction becomes

$$\frac{V^2}{2} + c_p T = c_p T_0 = \text{constant} \tag{1.17}$$

where, with c_p invariant, T_0 is the absolute temperature when the gas is at rest. The quantity $c_p T_0$ is referred to as *total* or *stagnation enthalpy*. This quantity is an important parameter of the equation for the conservation of energy.

Applying the *first law of thermodynamics* to the flow of non-heat-conducting inviscid fluids gives

$$\frac{d(c_V T)}{dt} + p\frac{d(1/\rho)}{dt} = 0 \tag{1.18}$$

Further, if the flow is unidirectional and $c_V T = E$, Eq. (1.18) becomes, on cancelling dt,

$$dE + p\,d\left(\frac{1}{\rho}\right) = 0 \tag{1.19}$$

However, differentiating Eq. (1.10) gives

$$dh = dE + pd\left(\frac{1}{\rho}\right) + \frac{1}{\rho}dp \qquad (1.20)$$

Combining Eqs. (1.19) and (1.20), we get

$$dh = \frac{1}{\rho}dp \qquad (1.21)$$

but

$$dh = c_p dT = \frac{c_p}{R}d\left(\frac{p}{\rho}\right) = \frac{\gamma}{\gamma - 1}\left[\frac{1}{\rho}dp + pd\left(\frac{1}{\rho}\right)\right] \qquad (1.22)$$

which, together with Eq. (1.21), gives the identity

$$\frac{dp}{p} + \gamma\rho d\left(\frac{1}{\rho}\right) = 0 \qquad (1.23)$$

Integrating gives

$$\ln p + \gamma \ln\left(\frac{1}{\rho}\right) = \text{constant}$$

or

$$p = k\rho^\gamma \qquad (1.24)$$

which is the isentropic relationship between pressure and density.

Note that this result is obtained from the equation of state for a perfect gas and from the equation of conservation of energy of the flow of a non-heat-conducting inviscid fluid. Such a flow behaves isentropically and, notwithstanding the apparently restrictive nature of the assumptions made, can be used as a model for a great many aerodynamic applications.

Entropy

Entropy is a function of state that follows from, and indicates the working of, the *second law of thermodynamics*, which is concerned with the *direction* of any process involving heat and energy. Any increase in the entropy of the fluid as it experiences a process is a measure of the energy no longer available to the system. Negative entropy change is possible when work is performed on a system or heat is removed. Zero entropy change indicates an ideal or completely adiabatic and reversible process, and we call such a constant entropy process an isentropic process.

By definition, specific entropy (S)[5] (Joules per kilogram per Kelvin) is given by the integral

$$S = \int \frac{\mathrm{d}Q}{T} \tag{1.25}$$

for any reversible process, with the integration extending from some datum condition; however, as we saw earlier, it is the change in entropy that is important:

$$\mathrm{d}S = \frac{\mathrm{d}Q}{T} \tag{1.26}$$

In this and the previous equation, $\mathrm{d}Q$ is a heat transfer to a unit mass of gas from an external source. This addition will change the internal energy and do work.

Thus, for a reversible process,

$$\mathrm{d}Q = \mathrm{d}E + p\mathrm{d}\left(\frac{1}{\rho}\right)$$

$$\mathrm{d}S = \frac{\mathrm{d}Q}{T} = \frac{c_V \mathrm{d}T}{T} + \frac{p\mathrm{d}(1/\rho)}{T} \tag{1.27}$$

but $p/T = R\rho$; therefore,

$$\mathrm{d}S = \frac{c_V \mathrm{d}T}{T} + \frac{R\mathrm{d}(1/\rho)}{1/\rho} \tag{1.28}$$

Integrating Eq. (1.28) from datum conditions to conditions given by suffix 1,

$$S_1 = c_V \ln \frac{T_1}{T_\mathrm{D}} + R\ln \frac{\rho_\mathrm{D}}{\rho_1}$$

Likewise,

$$S_2 = c_V \ln \frac{T_2}{T_\mathrm{D}} + R\ln \frac{\rho_\mathrm{D}}{\rho_2}$$

The entropy change from condition 1 to condition 2 is given by

$$\Delta S = S_2 - S_1 = c_V \ln \frac{T_2}{T_1} + R\ln \frac{\rho_1}{\rho_2} \tag{1.29}$$

With the use of Eq. (1.14) this is more usually rearranged to a nondimensional form:

$$\frac{\Delta S}{c_V} = \ln \frac{T_2}{T_1} + (\gamma - 1)\ln \frac{\rho_1}{\rho_2} \tag{1.30}$$

---

[5]Note that here the unconventional symbol S is used for specific entropy to avoid confusion with length symbols.

or to the exponential form:

$$e^{\Delta S/c_V} = \frac{T_2}{T_1}\left(\frac{\rho_1}{\rho_2}\right)^{\gamma-1} \tag{1.31}$$

Alternatively, for example, using the equation of state,

$$e^{\Delta S/c_V} = \left(\frac{T_2}{T_1}\right)^{\gamma}\left(\frac{p_1}{p_2}\right)^{\gamma-1} \tag{1.32}$$

These latter expressions are useful in particular problems.

1.4 AERONAUTICAL DEFINITIONS

1.4.1 Airfoil Geometry

If a horizontal wing is cut by a vertical plane parallel to the centerline, the shape of the resulting section is usually like that Fig. 1.4. This is an airfoil section, which for subsonic use almost always has a rounded leading edge (early stealth designs being

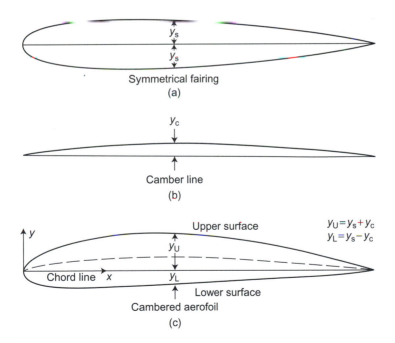

FIGURE 1.4

Airfoil (wing section) geometry and definitions.

the primary exceptions). The thickness increases smoothly to a maximum that usually occurs between one-quarter and halfway along the profile and thereafter tapers off toward the rear of the section.

If the leading edge is rounded, it is described by a planar curve and therefore has a definite radius of curvature. It is here that the curvature of the airfoil shape is the greatest aside from the trailing edge. The trailing edge may be sharp or may also have a very small radius of curvature or bluntness.

Consider a circle that is larger than and contains the airfoil. As its diameter is reduced, the circle will, for some diameter, contact the airfoil at two points only. These are the leading and trailing edges, and the diameter that connects them is the chord line. The length of the chord line is the airfoil chord, denoted c.

The point where the chord line intersects the front (or nose) of the section is used as the origin of a pair of axes: the x-axis is the chord line; the y-axis is perpendicular to the chord line, positive in the upward direction. The shape of the section is then usually given as a table of values of x and corresponding values of y. These section ordinates are usually expressed as percentages of the chord.

Camber

At any distance along the chord from the nose, a point may be marked midway between the upper and lower surfaces. The locus of all such points, usually curved, is the median line of the section and is called the camber line (here the word "line" is sloppy; it does not mean that the camber curve is straight, but it is used throughout the industry). The maximum height of the camber line above the chord line is denoted δ, and the quantity δ/c is called the maximum camber of the section. Airfoil sections have cambers usually in the range from 0% (a symmetrical section) to 5%, although much larger cambers are used in cascades (e.g., turbine blades).

Seldom can a camber line be expressed in simple geometric or algebraic forms, although a few simple curves, such as circular arcs or parabolas, have been used.

Thickness Distribution

Having found the median, or camber, line, the distances from it to the upper and lower surfaces may be measured at any value of x. These distances are, by the definition of the camber line, equal. They may be measured at all points along the chord and then plotted against x from a straight line. The result is a symmetrical shape, called the thickness distribution or symmetrical fairing.

An important parameter of the thickness distribution is the maximum thickness t, which, when expressed as a fraction of the chord, is called the thickness-to-chord ratio and commonly expressed as a percentage. Current values vary tremendously as aircraft now fly in many scales, from extreme low-Reynolds-number micro-air vehicles to massive airliners, along with super-cruise fighters and hypersonic flight test vehicles. However, airfoils with greater than about 18% thickness are rare.

The position along the chord at which maximum thickness occurs is another important characteristic of the thickness distribution. Values usually lie between 30% and 60% of the chord from the leading edge. For some older sections the value is about 25% of the chord, whereas for some more extreme sections it is more than 60% of the chord behind the leading edge.

Any airfoil section may be regarded as a thickness distribution plotted around a camber line. American and British conventions differ in the exact derivation of an airfoil section from a given camber line and thickness distribution. The British convention is to plot the camber line and then plot the thickness ordinates from this, perpendicular to the chord line. Thus the thickness distribution is in effect sheared until its median line, initially straight, has been distorted to coincide with the given camber line. The American convention is to plot the thickness ordinates perpendicular to the curved camber line, so the thickness distribution is regarded as bent until its median line coincides with the given camber line.

Since the camber-line curvature is generally very small, the difference in airfoil section shape given by these two conventions is also very small.

1.4.2 Wing Geometry

The *planform* of a wing is its shape seen on a plan (top) view of the aircraft. Fig. 1.5 illustrates this and defines the symbols for the various planform-geometry parameters. Note that the root ends of the leading and trailing edges have been connected across the fuselage by straight lines. An alternative to this is to produce the leading and trailing edges, if straight, to the aircraft centerline.

Wingspan

The wingspan is the dimension b, the distance between the two wingtips. The distance $b/2$ from each tip to the centerline is the wing semi-span.

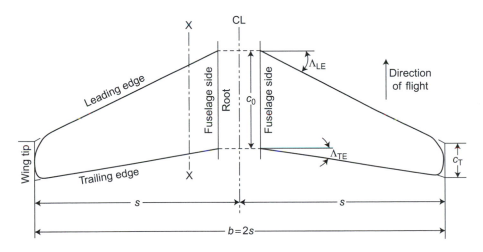

FIGURE 1.5

Wing planform geometry.

Chords

The two lengths c_T and c_0 are the tip and root chords, respectively; with the alternative convention, the root chord is the distance between the intersections with the fuselage centerline of the leading and trailing edges produced. The ratio c_T/c_0 is the taper ratio λ. Sometimes the reciprocal of this, c_0/c_T, is taken as the taper ratio. For most wings, $c_T/c_0 < 1$.

Wing Area

The plan-area of the wing including its continuation in the fuselage is the gross wing area S_G. The unqualified term wing area S usually means this gross wing area. The plan-area of the exposed wing (i.e., excluding the continuation in the fuselage) is the net wing area S_N.

Mean Chords

A useful parameter is the standard mean chord or the geometric mean chord, denoted $\bar{c}$ and defined by $\bar{c} = S_G/b$ or S_N/b. It should be clear whether S_G or S_N is used. The definition may also be written as

$$\bar{c} = \frac{\int_{-b/2}^{b/2} c\,dy}{\int_{-b/2}^{b/2} dy}$$

where y is distance measured from the centerline toward the starboard tip (right-hand to the pilot). "Standard mean chord" is often abbreviated as "SMC."

Another mean chord is the aerodynamic mean chord (AMC) which is denoted $\bar{c}_A$ or $\bar{\bar{c}}$ and is defined by

$$\bar{\bar{c}} = \frac{\int_{-b/2}^{b/2} c^2\,dy}{\int_{-b/2}^{b/2} c\,dy}$$

Aspect Ratio

Aspect ratio is a measure of the narrowness of the wing planform. It is denoted AR and is given by

$$AR = \frac{\text{span}}{\text{area}} = \frac{b^2}{c}$$

Note that only for a rectangular wing does $AR = b/c$.

Wing Sweep

The sweep angle of a wing is that between a line drawn along the span at a constant fraction of the chord from the leading edge, and a line perpendicular to the centerline. It is usually denoted Λ. Sweep-back is commonly measured on the leading edge

FIGURE 1.6

Dihedral angle.

(Λ_{LE}), on the quarter-chord line (i.e., the line one-quarter of the chord behind the leading edge ($\Lambda_{1/4}$)), or on the trailing edge (Λ_{TE}).

Dihedral Angle

If an airplane is viewed from directly ahead, it is seen that the wings are generally not in a single geometric plane but instead inclined to each other at a small angle. Imagine lines drawn on the wings along the locus of the intersections between the chord lines and the section noses, as in Fig. 1.6. Then the angle 2Γ is the dihedral angle of the wings. If the wings are inclined upward, they are said to have *dihedral*; if inclined downward, they have *anhedral*.

Incidence, Twist, Wash-out, and Wash-in

When an airplane is in flight, the chord lines of the various wing sections are not normally parallel to the direction of flight. The angle between the chord line of a given airfoil section and the direction of flight or of the undisturbed stream is the geometric angle of attack α.

Carrying this concept of incidence to the twist of a wing, it may be said that, if the geometric angles of attack of all sections are not the same, the wing is twisted. If the angle of attack increases towards the tip, the wing has *wash-in*; if it decreases towards the tip, the wing has *wash-out*.

1.5 DIMENSIONAL ANALYSIS

1.5.1 Fundamental Principles

The theory of dimensional homogeneity has more uses than those described in Section 1.2.3. By predicting how one variable may depend on a number of others, that variable may be used to direct an experiment, analyze experimental results, derive analytical results, or minimize computational effort. For example, when fluid flows past a circular cylinder the axis of which is perpendicular to the stream, eddies are formed behind the cylinder at a frequency that depends on factors such as the size of the cylinder, the speed of the stream, and so forth.

In an experiment to investigate the variation of eddy frequency, the obvious procedure is to take several sizes of cylinder, place them in streams of various fluids at a number of different speeds, and count the frequency of the eddies in each case. No matter how detailed, the results apply directly only to the cases tested, and it is necessary to find some pattern underlying them. A theoretical guide is helpful in achieving this, and it is here that dimensional analysis is of use.

Following the methods set forth by Buckingham in the early twentieth century (and easily available online these days), any collection of N quantities that are collectively described by k dimensions (mass, length, time, etc.) can be formed into $(N - k)$ nondimensional parameters that fully describe the problem. In the previous problem the frequency of eddies n will depend primarily on

- The size of the cylinder, represented by its diameter d
- The speed of the stream V
- The density of the fluid ρ
- The kinematic viscosity of the fluid ν

It should be noted that either μ or ν may be used to represent fluid viscosity.

Another factor should be the geometric shape of the body. Since the problem here is concerned only with long circular cylinders with their axes perpendicular to the stream, this factor is common to all readings and may be ignored in this analysis. It is also assumed that the speed is low compared to the speed of sound in the fluid, so compressibility (represented by the modulus of bulk elasticity) may be ignored as well. Gravitational effects are also excluded.

Then

$$n = f(d, V, \rho, \nu)$$

and, assuming that this function may be put in the form

$$n = \sum C d^a V^b \rho^e \nu^f \qquad (1.33)$$

where C is a constant and a, b, e, and f are some unknown indices. Putting Eq. (1.33) in dimensional form leads to

$$[T^{-1}] = [L^a (LT^{-1})^b (ML^{-3})^e (L^2 T^{-1})^f] \qquad (1.34)$$

where each factor has been replaced by its dimensions. Now the dimensions of both sides must be the same, and therefore the indices of M, L, and T on the two sides of the equation may be equated as follows:

$$\text{Mass (M)} \qquad 0 = e \qquad (1.35a)$$

$$\text{Length (L)} \qquad 0 = a + b - 3e + 2f \qquad (1.35b)$$

$$\text{Time (T)} \qquad -1 = -b - f \qquad (1.35c)$$

Here are three equations in four unknowns. One unknown must therefore be left undetermined: f, the index of ν, is selected for this role, and the equations are solved for a, b, and e in terms of f. The solution is therefore

$$b = 1 - f \tag{1.35d}$$

$$e = 0 \tag{1.35e}$$

$$a = -1 - f \tag{1.35f}$$

Substituting these values in Eq. (1.33),

$$n = \sum C d^{-1-f} V^{1-f} \rho^0 v^f \tag{1.36}$$

Rearranging Eq. (1.36), it becomes

$$n = \sum C \frac{V}{d} \left(\frac{Vd}{v} \right)^{-f} \tag{1.37}$$

or, alternatively,

$$\left(\frac{nd}{V} \right) = g \left(\frac{Vd}{v} \right) \tag{1.38}$$

where g represents some function that, as it includes the undetermined constant C and index f, is unknown from the present analysis.

Although it may not appear so at first sight, Eq. (1.38) is extremely valuable, as it shows that the values of nd/V should depend only on the corresponding value of Vd/v, regardless of the actual values of the original variables. This means that if, for each observation, the values of nd/V and Vd/v are calculated and plotted as a graph, all the results should lie on a single curve representing the unknown function g. An engineer wishing to estimate the eddy frequency for some given cylinder, fluid, and speed need only calculate the value of Vd/v, read from the curve the corresponding value of nd/V, and convert this to eddy frequency n. Thus the results of the series of observations are now in a usable form.

Consider for a moment the two compound variables just derived:

- nd/V. The dimensions of this are given by

$$\frac{nd}{V} = [T^{-1} \times L \times (LT^{-1})^{-1}] = [L^0 T^0] = [1]$$

- Vd/v. The dimensions of this are given by

$$\frac{Vd}{v} = [(LT^{-1})^{-1} \times L \times (L^2 T^{-1})^{-1}] = [1]$$

Thus the analysis has collapsed the five original variables n, d, V, p, and v into two compound variables, both of which are nondimensional. This has two advantages: (1) the values obtained for these two quantities are independent of the consistent system of units used; and (2) the influence of four variables on a fifth term can be shown on a single graph instead of an extensive range of graphs.

It can now be seen why the index f was left unresolved. The variables with indices that were resolved appear in both dimensionless groups, although in the group nd/V the density ρ is to the power zero. These repeated variables have been combined in turn with each of the other variables to form dimensionless groups.

There are certain problems—for example, the frequency of vibration of a stretched string, in which all the indices may be determined, leaving only the constant C undetermined. It is, however, usual to have more indices than equations, requiring one index or more to be left undetermined as before.

It must be noted that, while dimensional analysis will show which factors are not relevant to a given problem, it cannot indicate which relevant factors, if any, have been left out. It is therefore advisable to include all factors likely to have any bearing on a given problem, leaving out only those that, on *a priori* considerations, can be shown to have little or no relevance.

1.5.2 Dimensional Analysis Applied to Aerodynamic Force

In discussing aerodynamic force, it is necessary to know how the dependent variables, aerodynamic force and moment, vary with the independent variables thought to be relevant. Assume, then, that the aerodynamic force, or one of its components, is denoted F and, when fully immersed, depends on the following quantities: fluid density ρ, fluid kinematic viscosity v, stream speed V, and fluid bulk elasticity K. Force and moment will also depend on the shape and size of the body and its orientation to the stream. If, however, attention is confined to geometrically similar bodies (e.g., spheres, or models of a given airplane to different scales), the effects of shape as such are eliminated and the size of the body can be represented by a single typical dimension—for example, the sphere diameter, or the wingspan of the model airplane, denoted D. Then, following the method just given,

$$F = \mathrm{f}(V, D, \rho, v, K)$$

$$= \sum C V^a D^b \rho^c v^d K^e \tag{1.39}$$

In dimensional form this becomes

$$\left[\frac{ML}{T^2}\right] = \left[\left(\frac{L}{T}\right)^a (L)^b \left(\frac{M}{L^3}\right)^c \left(\frac{L^2}{T}\right)^d \left(\frac{M}{LT^2}\right)^e\right]$$

Equating indices of mass, length, and time separately leads to the three equations

(Mass) $\qquad 1 = c + e$ $\hfill$ (1.40a)

(Length) $\qquad 1 = a + b - 3c + 2d - e$ $\hfill$ (1.40b)

(Time) $\qquad -2 = -a - d - 2e$ $\hfill$ (1.40c)

With five unknowns and three equations, it is impossible to determine completely all unknowns, and so two must be left undetermined. These are d and e. The variables whose indices are solved here represent the most important characteristic of the body (the diameter), the most important characteristic of the fluid (the density), and the speed. These variables are known as repeated variables because they appear in each dimensionless group formed.

Equations (1.40a) through (1.40c) may then be solved for a, b, and c in terms of d and e, giving

$$a = 2 - d - 2e$$
$$b = 2 - d$$
$$c = 1 - e$$

Substituting these in Eq. (1.39) gives

$$F = V^{2-d-2e} D^{2-d} \rho^{1-e} \nu^d K^e$$
$$= \rho V^2 D^2 \left(\frac{\nu}{VD}\right)^d \left(\frac{K}{\rho V^2}\right)^e \qquad (1.41)$$

The speed of sound is given by Eqs. (1.6b) and (1.6d):

$$a^2 = \frac{\gamma p}{\rho} = \frac{K}{\rho}$$

Then

$$\frac{K}{\rho V^2} = \frac{\rho a^2}{\rho V^2} = \left(\frac{a}{V}\right)^2$$

and V/a is the Mach number M of the free stream. Therefore, Eq. (1.41) may be written as

$$F = \rho V^2 D^2 \mathrm{g}\left(\frac{VD}{\nu}\right) \mathrm{h}(M) \qquad (1.42)$$

where $g(VD/\nu)$ and $h(M)$ are undetermined functions of the stated compound variables. Thus it can be concluded that the aerodynamic forces acting on a family of geometrically similar bodies (the similarity including the orientation to the stream) obey the law

$$\frac{F}{\rho V^2 D^2} = \text{function}\left\{\frac{VD}{\nu}; M\right\} \qquad (1.43)$$

This relationship is sometimes known as *Rayleigh's equation*.

The term VD/ν may also be written, from the definition of ν, as $\rho VD/\mu$, as earlier in the problem relating to eddy frequency in the flow behind a circular cylinder. It is a very important parameter in fluid flows and is called the *Reynolds number*.

Now consider any parameter representing the geometry of the flow around the bodies at any point relative to them. If this parameter is expressed in a suitable nondimensional form, it can easily be shown by dimensional analysis that it is a function of the Reynolds number and the Mach number only. If, therefore, the values of Re (a common symbol for Reynolds number) and M are the same for a number of flows around geometrically similar bodies, it follows that all of the flows are geometrically similar in all respects, differing only in geometric scale and/or speed. This is true even though some of the fluids may be gaseous and others liquid. Flows that obey these conditions are said to be dynamically similar, and the concept of dynamic similarity is essential in wind-tunnel experiments.

It has been found, for most flows of aeronautical interest, that the effects of compressibility can be disregarded for Mach numbers less than 0.3 to 0.5, and in cases where this limit is not exceeded, a Reynolds number may be used as the only criterion of dynamic similarity.

Example 1.1

An aircraft and some scale models of it are tested under various conditions, given in the table. Which cases are dynamically similar to the aircraft in flight, given as case A?

	Case A	Case B	Case C	Case D	Case E	Case F
Span (m)	15	3	3	1.5	1.5	3
Relative density	0.533	1	3	1	10	10
Temperature (°C)	−24.6	+15	+15	+15	+15	+15
Speed (TAS) (m s^{-1})	100	100	100	75	54	54

Case A represents the full-size aircraft at 6000 m. The other cases represent models under test in various types of wind tunnel. Cases C, E, and F, where the relative density is greater than unity, represent a special type of tunnel, the compressed-air tunnel, which may be operated at static pressures in excess of atmospheric pressure.

From the figures just given Reynolds numbers $VD\rho/\mu$ may be calculated for each case. These are found to be

Case A	$Re = 5.52 \times 10^7$	Case D	$Re = 7.75 \times 10^6$
Case B	$Re = 1.84 \times 10^7$	Case E	$Re = 5.55 \times 10^7$
Case C	$Re = 5.56 \times 10^7$	Case F	$Re = 1.11 \times 10^8$

It is seen that the values of Re for cases C and E are very close to that for the full-size aircraft. Cases A, C, and E are therefore dynamically similar, and the flow patterns in these three cases are geometrically similar. In addition, the ratios of the local velocity to the free-stream velocity at any point on the three bodies are the same for these three cases. Hence, from Bernoulli's equation, the pressure coefficients are similarly the same in these three cases, and thus the forces on the bodies are simply and directly related. Cases B and D have Reynolds numbers considerably less than that for A, and are therefore said to represent a "smaller aerodynamic scale." The flows around these models, and the forces acting on them, are not simply or directly related to the force or flow pattern on the full-size aircraft. In case F the value of Re is larger than that for any other case and it has the largest aerodynamic scale of the six.

Example 1.2

An airplane approaches to land at a speed of 40 m s^{-1} at sea level. A 1/5th-scale model is tested under dynamically similar conditions in a compressed-air tunnel (CAT) working at 10 atmospheres pressure and 15°C. It is found that the load on the horizontal stabilizer is subject to impulsive fluctuations at a frequency of 20 cycles per second, owing to eddies being shed from the wing-fuselage junction. If the natural frequency of flexural vibration of the horizontal stabilizer is 8.5 cycles per second, can this represent a dangerous condition?

For dynamic similarity, the Reynolds numbers must be equal. Since the temperature of the atmosphere equals that in the tunnel, 15°C, the value of μ is the same in both the model and the full-scale case. Thus, for similarity

$$V_f d_f \rho_f = V_m d_m \rho_m$$

In this case, then, since

$$V_f = 40\,\mathrm{m\,s}^{-1}$$

$$40 \times 1 \times 1 = V_m \times \frac{1}{5} \times 10 = 2V_m$$

giving

$$V_m = 20\,\mathrm{m\,s}^{-1}$$

Now Eq. (1.38) covers this case of eddy shedding and is

$$\frac{nd}{V} = g(Re)$$

For dynamic similarity,

$$\left(\frac{nd}{V}\right)_f = \left(\frac{nd}{V}\right)_m$$

Therefore,

$$\frac{n_f \times 1}{40} = \frac{20 \times \frac{1}{5}}{20}$$

giving

$$n_f = 8 \text{ cycles per second}$$

This is very close to the given natural frequency of the horizontal stabilizer, and there is thus a considerable danger that the eddies might excite the natural frequencies of the horizontal stabilizer structure, possibly leading to its structural failure. Thus the shedding of eddies at this frequency is very dangerous to the aircraft.

Example 1.3

An aircraft flies at a Mach number of 0.85 at 18,300 m, where the pressure is $7160 \, \mathrm{N\,m^{-2}}$ and the temperature is $-56.5°C$. A model of 1/10th scale is to be tested in a high-speed wind tunnel. Calculate the total pressure of the tunnel stream necessary to give dynamic similarity if the total temperature is $50°C$. It may be assumed that the dynamic viscosity is related to the temperature as follows:

$$\frac{\mu}{\mu_0} = \left(\frac{T}{T_0}\right)^{3/4}$$

where $T_0 = 273°C$ and $\mu_0 = 1.71 \times 10^{-5} \, \mathrm{kg \ m^{-1} s^{-1}}$

• For the full-scale aircraft

$$M = 0.85, \ a = 20.05(273 - 56.5)^{1/2} = 297 \, \mathrm{m\,s^{-1}}$$

$$V = 0.85 \times 297 = 252 \, \mathrm{m\,s^{-1}}$$

$$\rho = \frac{p}{RT} = \frac{7160}{287.3 \times 216.5} = 0.1151 \, \mathrm{kg\,m^{-3}}$$

$$\frac{\mu_0}{\mu} = \left(\frac{273}{216.5}\right)^{3/4} = 1.19$$

$$\mu = \frac{1.71}{1.19} \times 10^{-5} = 1.44 \times 10^{-5} \, \mathrm{kg\,m^{-1}\,s^{-1}}$$

Consider a dimension that, on the aircraft, has a length of 10 m. Then, basing the Reynolds number on this dimension,

$$Re_f = \frac{Vd\rho}{\mu} = \frac{252 \times 10 \times 0.1151}{1.44 \times 10^{-5}} = 20.2 \times 10^6$$

- For the model

Total temperature $T_s = 273 + 50 = 323$ K
Therefore at $M = 0.85$

$$\frac{T_s}{T} = 1 + \frac{1}{5}(0.85)^2 = 1.1445$$

$$T = 282 \text{ K}$$

so

$$a = 20.05 \times (282)^{1/2} = 337 \,\text{m s}^{-1}$$

$$V = 0.85 \times 337 = 287 \,\text{m s}^{-1}$$

$$\frac{\mu}{\mu_0} = \left(\frac{282}{273}\right)^{3/4} = 1.0246$$

giving

$$\mu = 1.71 \times 1.0246 \times 10^{-5} = 1.751 \times 10^{-5} \,\text{kg m}^{-1}\,\text{s}^{-1}$$

For dynamic similarity the Reynolds numbers must be equal:

$$\frac{287 \times 1 \times \rho}{1.75 \times 10^{-5}} = 20.2 \times 10^6$$

giving

$$\rho = 1.23 \,\text{kg m}^{-3}$$

Thus the static pressure required in the test section is

$$p = \rho RT = 1.23 \times 287.3 \times 282 = 99\,500 \,\text{N m}^{-2}$$

The total pressure p_s is given by

$$\frac{p_s}{p} = \left(1 + \frac{1}{5}M^2\right)^{3.5} = (1.1445)^{3.5} = 1.605$$

$$p_s = 99\,500 \times 1.605 = 160\,000 \,\text{N m}^{-2}$$

If the total pressure available in the tunnel is less than the value just given, it is not possible to achieve equality of the Mach and Reynolds numbers. Either the Mach number may be achieved at a lower value of Re, or Re may be made equal at a lower Mach number. In such a case it is normally preferable to make the Mach number correct since, provided the Reynolds number in the tunnel is not too low, the effects of compressibility are more important than the effects of aerodynamic scale at Mach numbers of this order. Moreover, techniques are available that can mitigate errors due to unequal aerodynamic scales.

In particular, the position at which laminar-turbulent transition (see Section 8.1 and 8.9) of the boundary layer occurs at full scale can be fixed on the model by roughening the model surface. This can be done by gluing on a line of grit, the proper sizing of which should be determined by investigating the technical literature for similar geometries and Reynolds numbers.

1.6 BASIC AERODYNAMICS

We now have a method to succinctly describe the thermodynamic state of the air in a manner which is relevant to the topic of flight. Now we need to have another set of well-defined tools that we can use to describe the effect of the air on the aircraft. This effect is primarily a force, or a force per area. Only in higher speed flight do we typically become interested in the effects of temperature of the air on a vehicle. Such "aerothermodynamics" topics are best served by more advanced texts than this one. To develop the necessary tools for creating concise and accurate descriptions of the aerodynamic loads (effects) on a vehicle, a number of forces and moments are first defined. Then the force and moment coefficients are defined, followed by common definitions of airfoil or wing characteristics. We follow what is standard practice in much of the world.

1.6.1 Aerodynamic Force and Moment

Air flowing past an airplane, or any other body, must be diverted from its original path; such deflections lead to changes in air speed. Bernoulli's equation shows that the pressure exerted by the air on the airplane is altered from that of the undisturbed stream. Also, the viscosity of the air leads to frictional forces tending to resist the air's flow. As a result of these processes, the airplane experiences an aerodynamic force and moment. It is conventional and convenient to separate aerodynamic force and moment into three components each, as follows.

Lift, L (+Z Direction)

Lift is the component of force acting upward, perpendicular to the direction of flight or of the undisturbed stream. The word "upward" is used in the same sense that the pilot's head is above his feet. Figure 1.7 illustrates the meaning in various attitudes of flight. The arrow V represents the direction of flight, the arrow L represents the lift acting upward, and the arrow W is the weight of the aircraft and shows the downward vertical. Comparison of (a) and (c) shows that this upward is not fixed relative to the aircraft, while (a), (b), and (d) show that the meaning is not fixed relative to the Earth. As a general rule, if it is remembered that the lift is always a component perpendicular to the flight direction, the exact direction in which the lift acts will be obvious, particularly after reference to Fig. 1.7. This may not apply to certain guided missiles that have no obvious top or bottom, so the exact meaning of "up" must then be defined with care.

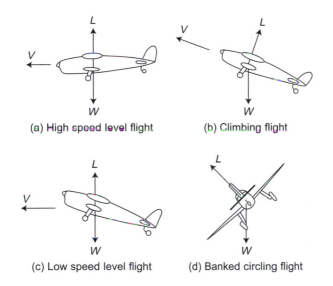

(a) High speed level flight   (b) Climbing flight

(c) Low speed level flight   (d) Banked circling flight

FIGURE 1.7

Direction of lift force. Note that lift is always normal to flight velocity; drag force is always parallel to velocity.

Drag, D(–X)

Drag is the component of force acting in the opposite direction to the line of flight, or in the same direction as the motion of the undisturbed stream. It is the force that resists the motion of the aircraft. There is no ambiguity regarding its direction or sense.

Crosswind Force, Y

Crosswind is the component of force mutually perpendicular to the lift and the drag—that is, in a spanwise direction. It is reckoned positive when acting toward the starboard wingtip (right-hand to the pilot).

Pitching Moment, M

Pitching is the moment acting in the plane containing the lift and the drag—that is, in the vertical plane when the aircraft is flying horizontally. It is defined positive when it tends to increase the angle of attack or raise the nose of the aircraft upward (using this word in the sense discussed earlier).

Rolling Moment, L_R

Rolling is the moment tending to make the aircraft roll about the flight direction—that is, tending to depress one wingtip and raise the other. It is positive when it tends to depress the starboard wingtip.

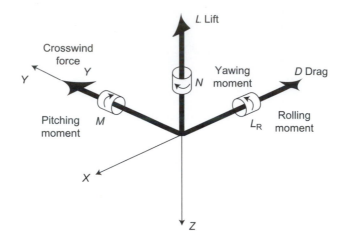

FIGURE 1.8

Systems of force and moment components. Broad arrows represent forces used in elementary work; line arrows, the system in control and stability studies. Moments are common to both systems.

Yawing Moment, N

Yawing is the moment that tends to rotate the aircraft about the lift direction—that is, to swing the nose to one side or the other of the flight direction. This is comparable to the turning of an automobile. It is positive when it swings, or tends to swing, the nose to the right (starboard).

The relationships of these components is shown in Fig. 1.8. In each case the arrow shows the direction of the positive force or moment. All three forces are mutually perpendicular, and each acts about the line of one of the forces.

The system of forces and moments just described is conventionally used for performance analysis and other simple problems. For aircraft stability and control studies, however, it is more convenient to use a slightly different system of forces.

1.6.2 **Force and Moment Coefficients**

The nondimensional quantity called a force coefficient, $F/(\rho V^2 S)$ (compare Eq. (1.43), where F is an aerodynamic force and S is an area), is similar to the type often developed and used in aerodynamics. It is not, however, used in precisely this form. In place of ρV^2 it is conventional for incompressible flow to use $\frac{1}{2}\rho V^2$, the dynamic pressure of the free-stream flow. The actual physical area of the body, such as the planform area of the wing, or the maximum cross-sectional area of a fuselage is usually used for S. Thus the aerodynamic force coefficient is usually defined as follows:

$$C_F = \frac{F}{\frac{1}{2}\rho V^2 S} \tag{1.44a}$$

The two most important force coefficients are lift and drag, defined by

$$\text{Lift coefficient } C_L = \text{lift}/\tfrac{1}{2}\rho V^2 S \tag{1.44b}$$

$$\text{Drag coefficient } C_D = \text{drag}/\tfrac{1}{2}\rho V^2 S \tag{1.44c}$$

When the body in question is a wing, the area S is almost invariably the planform area as defined in Section 1.4.1. For the drag of a body such as a fuselage, sphere, or cylinder, S is usually the projected frontal area, the maximum cross-sectional area, or the (volume)$^{2/3}$. The area used for definition of the lift and drag coefficients of such a body is thus seen to be variable from case to case and therefore needs to be stated for each one.

The impression is sometimes that lift and drag coefficients cannot exceed unity. This is not true; with modern developments some wings can produce lift coefficients of 10 or more based on their plan-area.

Aerodynamic moments also can be expressed in the form of nondimensional coefficients. Since a moment is the product of a force and a length, it follows that a nondimensional form for a moment is $Q/\rho V^2 Sl$, where Q is any aerodynamic moment and l is a reference length. Here again it is conventional to replace ρV^2 with $\tfrac{1}{2}\rho V^2$. In the case of the pitching moment of a wing, the area is the plan-area S and the length is the mean wing chord $\bar{c}$ or $\bar{c}_A$ (see Section 1.4.1). Then the pitching moment coefficient C_M is defined by

$$C_M = \frac{M}{\tfrac{1}{2}\rho V^2 S\bar{c}} \tag{1.45}$$

1.6.3 Pressure Distribution on an Airfoil

The pressure on the surface of an airfoil in flight is not uniform. Figure 1.9 shows typical pressure distributions for a given section at various angles of incidence. It is convenient to deal with nondimensional pressure differences using p_∞, the pressure far upstream, as the datum. Thus the coefficient of pressure is introduced as

$$C_p = \frac{p - p_\infty}{\tfrac{1}{2}\rho V^2}$$

Looking at the sketch for zero incidence ($\alpha = 0$), we see that there are small regions at the nose and tail where C_p is positive but that over most of the section it is negative. At the trailing edge the pressure coefficient comes close to $+1$ but does not actually reach it (more will be said on this point later). The reduced pressure on the upper surface is tending to draw the section upward while that on the lower surface has the opposite effect.

With the pressure distribution as sketched, the effect on the upper surface is larger, and there is a resultant upward force on the section—that is, the lift.

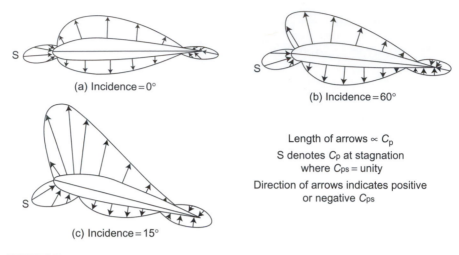

(a) Incidence = 0°

(b) Incidence = 60°

Length of arrows ∝ C_p

S denotes C_p at stagnation
where C_{ps} = unity

Direction of arrows indicates positive
or negative C_{ps}

(c) Incidence = 15°

FIGURE 1.9

Typical pressure distributions on an airfoil section.

As incidence is increased from zero, the following airfoil behaviors are noted:

- The low pressure on the upper surface becomes lower and covers a larger extent of the airfoil until, at large incidence, it actually encroaches on a small part of the front lower surface.
- The stagnation point moves progressively further back on the lower surface, and the increased pressure on the lower surface covers a greater proportion of it. The pressure reduction on the lower surface is simultaneously decreased in both intensity and extent.

The large negative values of C_p reached on the upper surface at high incidences (e.g., 15 degrees) are also noteworthy. In some cases values of –6 or –7 are found. This corresponds to local flow speeds of nearly three times the speed of the undisturbed stream.

From the foregoing, the following conclusions may be drawn:

- At low incidences the lift is generated by the difference between the pressure reductions on the upper and lower surfaces.
- At higher incidences the lift is partly due to pressure reduction on the upper surface and partly due to pressure increase on the lower surface.

At angles of incidence around 18 or 20 degrees the pressure reduction on the upper surface suddenly collapses and what little lift remains is due principally to the pressure increase on the lower surface. A picture drawn for one small negative incidence (for this airfoil section, about −4 degrees) would show equal suction effects on the upper and lower surfaces, and the section would give no lift. At more negative incidences the lift would be negative.

The relationship between the pressure distribution and the drag of an airfoil section is discussed later (Section 1.6.5).

1.6.4 Pitching Moment

The pitching moment on a wing may be estimated experimentally by two principal methods: direct measurement on a balance or pressure plotting, as described in Section 1.6.6. With a computational model, the moment contributions of all points on the surface pressure distribution are integrated to produce the total moment. In any case, the pitching moment coefficient is measured about some definite point on the airfoil chord, while for some particular purpose it may be desirable to know what it is about some other point. To convert from one reference point to the other is a simple application of the parallel axis theorem from mechanics or statics.

Suppose, for example, that lift and drag are known, as is the pitching moment M_a about a point distance a from the leading edge. We want to find the pitching moment M_x about a different point distance x behind the leading edge. The situation is as shown in Fig. 1.10. Figure 1.10(a) represents the known conditions and Figure 1.10(b), the unknown conditions. These represent two ways of looking at the same physical system and must therefore give identical effects on the airfoil.

Obviously, $L = L$ and $D = D$. Taking moments in each case about the leading edge,

$$M_{\mathrm{LE}} = M_a - La\cos\alpha - Da\sin\alpha = M_x - Lx\cos\alpha - Dx\sin\alpha$$

Then

$$M_x = M_a - (L\cos\alpha - D\sin\alpha)(a - x)$$

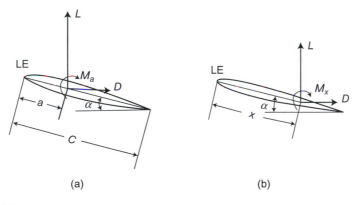

(a)   (b)

FIGURE 1.10

Pitching moment definitions. Lift and drag are unchanged between (a) and (b).

Converting to coefficient form by dividing by $\frac{1}{2}\rho V^2 Sc$ gives

$$C_{M_x} = C_{M_a} - (C_L \cos\alpha - C_D \sin\alpha)\left(\frac{a}{c} - \frac{x}{c}\right) \qquad (1.46)$$

With this equation it is easy to calculate C_{M_x}, for any value of x/c. As a particular case, if the known pitching moment coefficient is that about the leading edge $C_{M_{LE}}$, then $a = 0$ and Eq. (1.46) becomes

$$C_{M_x} = C_{M_{LE}} + \frac{x}{c}(C_L \cos\alpha + C_D \sin\alpha) \qquad (1.47)$$

Aerodynamic Center

If the pitching moment coefficient at each point along the chord is calculated for each of several values of C_L, one very special point is found for which C_M is virtually constant, independent of the lift coefficient. This point is the aerodynamic center. For incidences up to 10 degrees or so it is a fixed point close to, but not generally on, the chord line, between 23% and 25% of the chord behind the leading edge.

For a flat or curved plate in inviscid, incompressible flow the aerodynamic center is theoretically exactly one-quarter of the chord behind the leading edge; however, thickness of the section and viscosity of the fluid tend to place it a few percent further forward as indicated earlier, while compressibility tends to move it backward. For a thin airfoil (or infinite aspect ratio wing) in supersonic flow, the aerodynamic center is theoretically at 50% of the chord.

Knowledge of how the pitching moment coefficient about a point distance a behind the leading edge varies with C_L may be used to find the position of the aerodynamic center behind the leading edge and the value of the pitching moment coefficient there $C_{M_{AC}}$. Let the position of the aerodynamic center be a distance x_{AC} behind the leading edge. Then, with Eq. (1.46) slightly rearranged,

$$C_{M_a} = C_{M_{AC}} - (C_L \cos\alpha + C_D \sin\alpha)\left(\frac{x_{AC}}{c} - \frac{a}{c}\right)$$

Now, at moderate incidences between, say, 3 and 7 degrees:

$$C_L = O[20C_D] \quad \text{and} \quad \cos\alpha = O[10\sin\alpha]$$

where $O[\ ]$ means "of the order of"; that is, C_L is of the order of 20 times C_D. Then

$$C_L \cos\alpha = O[200 C_D \sin\alpha]$$

and therefore $C_D \sin\alpha$ can be neglected compared with $C_L \cos\alpha$. With this approximation and the further approximation $\cos\alpha = 1$,

$$C_{M_a} = C_{M_{AC}} - C_L\left(\frac{x_{AC}}{c} - \frac{a}{c}\right) \qquad (1.48)$$

Differentiating Eq. (1.48) with respect to C_L gives

$$\frac{d}{dC_L}(C_{M_a}) = \frac{d}{dC_L}(C_{M_{AC}}) - \left(\frac{x_{AC}}{c} - \frac{a}{c}\right)$$

But the aerodynamic center is, by definition, that point about which C_M is independent of C_L, and therefore the first term on the right-hand side is identically zero. Thus

$$\frac{d}{dC_L}(C_{M_a}) = 0 - \left(\frac{x_{AC}}{c} - \frac{a}{c}\right) = \frac{a}{c} - \frac{x_{AC}}{c} \qquad (1.49)$$

$$\frac{x_{AC}}{c} = \frac{a}{c} - \frac{d}{dC_L}(C_{M_a}) \qquad (1.50)$$

If, then, C_{M_a} is plotted against C_L and the slope of the resulting line is measured, subtracting this value from a/c gives the aerodynamic center position x_{AC}/c.

In addition if, in Eq. (1.48), C_L is made zero, that equation becomes

$$C_{M_a} = C_{M_{AC}} \qquad (1.51)$$

That is, the pitching moment coefficient about an axis at zero lift is equal to the constant pitching moment coefficient about the aerodynamic center. Because of this association with zero lift, $C_{M_{AC}}$ is often denoted C_{M_0}.

Example 1.4

For a particular airfoil section, the pitching moment coefficient about an axis a third of the chord behind the leading edge varies with the lift coefficient in the following manner:

C_L	0.2	0.4	0.6	0.8
C_M	-0.02	0.00	$+0.02$	$+0.04$

Find the aerodynamic center and the value of C_{M_0}.

It is seen that C_M varies linearly with C_L, the value of dC_M/dC_L being

$$\frac{0.04 - (-0.02)}{0.80 - 0.20} = +\frac{0.06}{0.60} = +0.10$$

Therefore, from Eq. (1.50), with $a/c = 1/3$,

$$\frac{x_{AC}}{c} = \frac{1}{3} - 0.10 = 0.233$$

The aerodynamic center is thus at 23.3% of the chord behind the leading edge. Plotting C_M against C_L gives the value of C_{M_0}, the value of C_M when $C_L = 0$, as -0.04.

A particular case is one in which the known values of C_M are those about the leading edge, namely $C_{M_{LE}}$. In this case $a = 0$ and therefore

$$\frac{x_{AC}}{c} = -\frac{d}{dC_L}(C_{M_{LE}}) \tag{1.52}$$

Taking this equation with the statement made earlier about the normal position of the aerodynamic center implies that, for all airfoils at low Mach numbers,

$$\frac{d}{dC_L}(C_{M_{LE}}) \simeq -\frac{1}{4} \tag{1.53}$$

Center of Pressure

The aerodynamic forces on an airfoil section may be represented by lift, drag, and pitching moment. At each value of the lift coefficient there will be one particular point about which the pitching moment coefficient is zero, and the aerodynamic effects on the airfoil section may be represented by the lift and the drag alone acting at that point. This special point is termed the center of pressure.

Whereas the aerodynamic center is a fixed point that always lies within the profile of a normal airfoil section, the center of pressure moves with change in lift coefficient and is not necessarily within the airfoil profile. Figure 1.11 shows the forces on the airfoil regarded as either of the following:

- Lift, drag, and moment acting at the aerodynamic center.
- Lift and drag only acting at the center of pressure, a fraction k_{CP} of the chord behind the leading edge.

Then, taking moments about the leading edge,

$$M_{LE} = M_{AC} - (L\cos\alpha - D\sin\alpha)x_{AC} = -(L\cos\alpha + D\sin\alpha)k_{CP}c$$

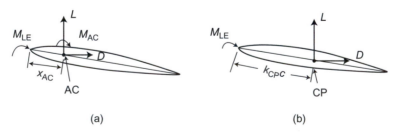

(a)   (b)

FIGURE 1.11

Determination of center-of-pressure position.

Dividing by $\frac{1}{2}\rho V^2 Sc$, this becomes

$$C_{M_{AC}} - (C_L\cos\alpha - C_D\sin\alpha)\frac{x_{AC}}{c} = -(C_L\cos\alpha + C_D\sin\alpha)k_{CP}$$

giving

$$k_{CP} = \frac{x_{AC}}{c} - \frac{C_{M_{AC}}}{C_L\cos\alpha + C_D\sin\alpha} \tag{1.54}$$

Again making the approximations that $\cos\alpha \approx 1$ and $C_D\sin\alpha \ll 1$, Eq. (1.54) becomes

$$k_{CP} = \frac{x_{AC}}{c} - \frac{C_{M_{AC}}}{C_L} \tag{1.55}$$

At first sight this suggests that k_{CP} is always less than x_{AC}/c. However, $C_{M_{AC}}$ is almost invariably negative, so in fact k_{CP} is numerically greater than x_{AC}/c and the center of pressure is behind the aerodynamic center.

Example 1.5

For the airfoil section from Example 1.4, plot a curve showing the approximate variation of the position of the center of pressure with the lift coefficient, for lift coefficients between zero and unity. For this case

$$k_{CP} \simeq 0.233 - (-0.04 / C_L)$$

$$\simeq 0.233 + (-0.04 / C_L)$$

the corresponding curve is given in Fig. 1.12, which shows that k_{CP} tends asymptotically to x_{AC} as C_L increases, and tends to infinity behind the airfoil as C_L tends to zero. For values of C_L less than 0.05, the center of pressure is actually behind the airfoil.

For a symmetrical section (zero camber) and for some special camber lines, the pitching moment coefficient about the aerodynamic center is zero. It then follows, from Eq. (1.55), that $k_{CP} = x_{ac}/c$ (i.e., the center of pressure and the aerodynamic center coincide) and that, for moderate incidences, the center of pressure is therefore stationary at about the quarter-chord point.

1.6.5 Types of Drag

Attempts have been made to rationalize the definitions and terminology associated with drag. [1] On the whole, the new terms have not been widely adopted. Here we will use the widely accepted traditional terms and indicate alternatives in parentheses.

Total Drag

Total drag is formally defined as the force corresponding to the rate of decrease in momentum in the direction of the undisturbed external flow around the body. This

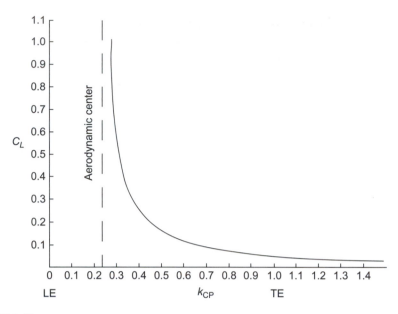

FIGURE 1.12

Center-of-pressure position for Example 1.5.

decrease is calculated between stations at infinite distances upstream and downstream of the body, so it is the total force or drag in the direction of the undisturbed flow. It is also the total force resisting the motion of the body through the surrounding fluid.

There are a number of separate contributions to total drag. As a first step it may be divided by physical effect into *pressure drag* and *skin-friction drag*.

Skin-Friction Drag (or Surface-Friction Drag)

Skin-friction drag is generated by the resolved components of the traction due to shear stresses acting on the body surface. This traction is due directly to viscosity and acts tangentially at all points on the body surface. At each point it has a component aligned with but opposing the undisturbed flow (i.e., opposite to the direction of flight). The total effect of these components, integrated over the entire exposed surface of the body, is the skin-friction drag. Skin-friction drag cannot exist in an invisicid flow.

Pressure Drag

Pressure drag is generated by the resolved components of the forces due to pressure acting normal to the surface at all points. It is computed as the integral of the flight-path direction component of the pressure forces acting on all points on the body. Pressure distribution, and thus pressure drag, has several distinct contributions:

- *Induced* drag (sometimes known as "drag due to lift" or "vortex drag").

- *Wave* drag, when there exists a supersonic region in the flow regardless of the flight Mach number being less than or greater than 1.
- *Form* drag (sometimes known as boundary-layer pressure drag).

Induced Drag (or Vortex Drag)

Induced drag is discussed in more detail in Section 1.6.7 and Section 5.6. For now it may be noted that it depends on lift, does not depend directly on viscous effects, and can be both understood and estimated by assuming inviscid flow.

Wave Drag

Wave drag is associated with the formation of shock waves in high-speed flight. It is described in more detail in Chapter 6.

Form Drag (or Boundary-Layer Pressure Drag)

Form drag is caused by differences between the pressure distribution over a body in viscous flow and that in an ideal inviscid flow (Fig. 1.13). If the flow is inviscid, it can be shown that the flow speed at the trailing edge is zero, implying that the pressure coefficient is +1. But in a real flow (see Fig. 1.13(a)) the body plus the boundary-layer displacement thickness has a finite width at the trailing edge, so the flow speed does not fall to zero and therefore the pressure coefficient is less than +1. The variation in coefficient of pressure due to real flow around an airfoil is shown in Fig. 1.13(b). This combines to generate a net drag as follows.

The relatively high pressures around the nose of the airfoil tend to push it backward. The region of the suction pressures that follows, extending up to the point of maximum thickness, acts to generate a thrust pulling the airfoil forward. The region of suction pressures downstream of the point of maximum thickness generates a retarding force on the airfoil, whereas the relatively high-pressure region around the trailing edge generates a thrust. In an inviscid flow, these various contributions cancel out exactly and the net drag is zero. In a real viscous flow, this exact cancellation does not occur. The pressure distribution ahead of the point of maximum thickness is little altered by real-flow effects. The drag generated by the suction pressures downstream

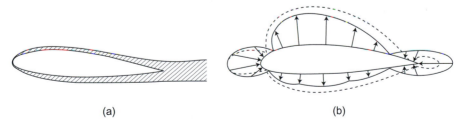

(a) (b)

FIGURE 1.13

(a) Displacement thickness of the boundary layer (hatched area) representing an effective change in airfoil shape (boundary-layer thickness is greatly exaggerated). (b) Pressure distribution on an airfoil section in viscous flow (dotted line) and inviscid flow (solid line).

of the point of maximum thickness is slightly reduced in a real flow. However, this effect is greatly outweighed by a substantial reduction in the thrust generated by the high-pressure region around the trailing edge. Thus the exact cancellation of the pressure forces found in an inviscid flow is destroyed in a real flow, resulting in an overall rearward force. This force is the *form* drag.

We reemphasize that both form and skin-friction drag depend on viscosity for their existence and cannot exist in an inviscid flow.

Profile Drag (or Boundary-Layer Drag)

The profile drag is the sum of the skin-friction and form drags. (See also the formal definition given for the previous item.)

Comparison of Drags for Various Body Types

Normal Flat Plate (Fig. 1.14)   In the case of a flat plate set broadside to a uniform flow, the drag is entirely form drag, coming mostly from the large negative pressure coefficients over the rear face. Although viscous tractions exist, they act along the surface of the plate and therefore have no rearward component to produce skin-friction drag.

Parallel Flat Plate (Fig. 1.15)   In this case, the drag is entirely skin-friction drag. Whatever the distribution of pressure may be, it can have no rearward component, and therefore the form drag must be zero.

Circular Cylinder (Fig. 1.16)   Figure 1.16 is a sketch of the distribution of pressure around a circular cylinder in inviscid flow (solid lines) and in a viscous fluid (dotted lines). The perfect symmetry in the inviscid case shows that there is no resultant force on the cylinder. The drastic modification of the pressure distribution due to viscosity is apparent, the result being a large form drag. In this case, only some 5% of the drag is skin-friction drag, the remaining 95% being form drag, although these proportions depend on the Reynolds number.

Airfoil or Streamlined Strut   The pressure distributions for this case are given in Fig. 1.13(b). The effect of viscosity on the pressure distribution is much less than for the circular cylinder, and the form drag is much lower as a result. The percentage of the total drag represented by skin-friction drag depends on the Reynolds number, the

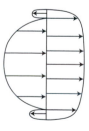

FIGURE 1.14

Pressure on a normal flat plate, flow from left to right.

FIGURE 1.15

Viscous tractions on a tangential flat plate.

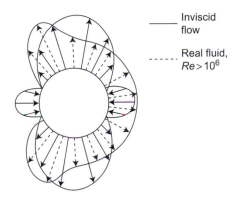

_____ Inviscid
flow

------ Real fluid,
$Re > 10^6$

FIGURE 1.16

Pressure on a circular cylinder with its axis normal to the stream (see also Fig. 3.23).

thickness/chord ratio, and a number of other factors, but between 40% and 80% is fairly typical.

The Wake

Behind any body moving in air is a wake. Although the wake in air is not normally visible, it may be felt, as when, for example, a bus passes by. The total drag of a body appears as a loss of momentum and an increase of energy in the wake. The loss of momentum appears as a reduction of average flow speed, while the increase of energy is seen as violent eddying (or vorticity). The size and intensity of the wake is therefore an indication of the body's profile drag. Figure 1.17 shows comparative widths of the wakes behind a few bodies.

1.6.6 Estimation of Lift, Drag, and Pitching Moment Coefficients from the Pressure Distribution

Let Fig. 1.18 represent an airfoil at an angle of attack α to a fluid flow traveling from left to right at speed V. The axes Ox and Oz are respectively aligned along and perpendicular to the chord line. The chord length is denoted c.

Taking the airfoil to be a wing section of constant chord and unit spanwise length, we consider the forces acting on a small element of the upper airfoil surface as having length δs. The inward force perpendicular to the surface is given by $p_u \delta s$. This force may be resolved into components δX and δZ in the x and z directions. It can be seen that

$$\delta Z_u = -p_u \cos \varepsilon \qquad (1.56)$$

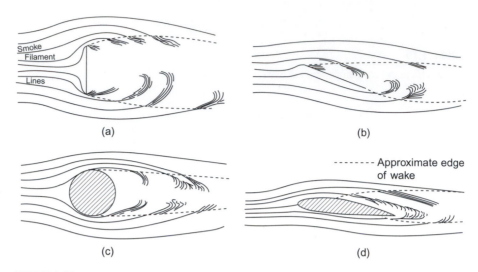

FIGURE 1.17

Behavior of smoke filaments in flows past various bodies, showing wakes. (a) Normal flat plate. In this case the wake oscillates up and down at several cycles per second. Half a cycle later the picture would be reversed, with the upper filaments curving back, as the lower filaments curve in this sketch. (b) Flat plate at fairly high incidence. (c) Circular cylinder at low Re. (For a pattern at higher Re, see Fig. 7.14.) (d) Airfoil section at moderate incidence and low Re.

and, from the geometry,

$$\delta s \, \cos \varepsilon = \delta x \tag{1.57}$$

so that

$$\delta Z_u = -p_u \delta x \quad \text{per unit span}$$

Similarly, for the lower surface,

$$\delta Z_\ell = -p_\ell \delta x \quad \text{per unit span}$$

We now add these two contributions and integrate with respect to x between $x = 0$ and $x = c$ to get

$$Z = -\int_0^c p_u dx + \int_0^c p_\ell dx$$

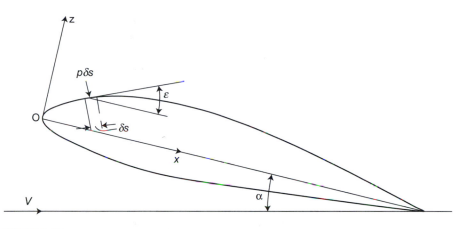

FIGURE 1.18

Normal pressure force on an element of the airfoil surface.

However, we can always subtract a constant pressure from both p_u and p_ℓ without altering the value of Z, so we can write

$$Z = -\int_0^c (p_u - p_\infty)\mathrm{d}x + \int_0^c (p_\ell - p_\infty)\mathrm{d}x \qquad (1.58)$$

where p_∞ is the pressure in the free stream (we could equally well use any other constant pressure, say the stagnation pressure in the free stream).

Equation (1.58) can readily be converted into coefficient form. Recalling that the airfoil section is of unit $S = 1 \times c = c$, we obtain

$$C_Z \equiv \frac{Z}{\frac{1}{2}\rho V^2 c} = -\frac{1}{\frac{1}{2}\rho V^2 c}\int_0^c [(p_u - p_\infty) - (p_\ell - p_\infty)]\,\mathrm{d}x$$

Remembering that $(1/c)\mathrm{d}x = \mathrm{d}(x/c)$ and that the definition of pressure coefficient is

$$C_\mathrm{p} = \frac{p - p_\infty}{\frac{1}{2}\rho V^2}$$

we see that

$$C_Z = -\int_0^1 (C_{p_u} - C_{p_\ell})\mathrm{d}(x/c) \qquad (1.59a)$$

or, simply,

$$C_Z = \oint_C C_p \cos \varepsilon \, d(s/c) = \oint_C C_p d(x/c), \qquad (1.59b)$$

where the contour integral is evaluated by following a counter-clockwise direction around the contour C of the airfoil.

Similar arguments lead to the following relations for X:

$$\delta X_u = p_u \delta s \sin \varepsilon, \quad \delta X_\ell = p_\ell \delta s \sin \varepsilon, \quad \delta s \sin \varepsilon = \delta z,$$

giving

$$C_X = \oint_c C_p \sin \varepsilon \, d(s/c) = \oint_c C_p d(z/c) = \int_{z_{m_\ell}}^{z_{m_u}} C_p d\left(\frac{z}{c}\right) \qquad (1.60)$$

where z_{m_u} and z_{m_ℓ} are, respectively, the maximum and minimum values of z, and ΔC_p is the difference between the values of C_p acting on the fore and rear points of an airfoil for a fixed value of z.

The pitching moment can also be calculated from the pressure distribution. For simplicity, we will calculate the pitching moment about the leading edge. The contribution of force δZ acting on a slice of airfoil of length δx is given by

$$\delta M = (p_u - p_\ell)x\delta x = [(p_u - p_\infty) - (p_\ell - p_\infty)]x\delta x$$

so, remembering that the coefficient of pitching moment is defined as

$$C_M = \frac{M}{\frac{1}{2}\rho V^2 Sc} = \frac{M}{\frac{1}{2}\rho V^2 c^2} \quad \begin{array}{l}\text{in this case, for an airfoil}\\ \text{where } M \text{ is the moment per span,}\end{array}$$

the coefficient of pitching moment due to the Z force is given by

$$C_{MZ} = -\oint_C C_p \frac{x}{c} d\left(\frac{x}{c}\right) = \int_0^c [C_{p_u} - C_{p_\ell}]\frac{x}{c} d\left(\frac{x}{c}\right) \qquad (1.61)$$

Similarly, the much smaller contribution due to the X force may be obtained as

$$C_{MX} = -\oint_c C_p \sin \varepsilon \frac{z}{c} d\left(\frac{s}{c}\right) = \int_{z_{m_\ell}}^{z_{m_u}} \Delta C_p \frac{z}{c} d\left(\frac{z}{c}\right) \qquad (1.62)$$

The integrations just given are usually performed using a computer or graphically.

The force coefficients C_X and C_Z are parallel and perpendicular to the chord line, whereas the more usual coefficients C_L and C_D are defined with reference to the

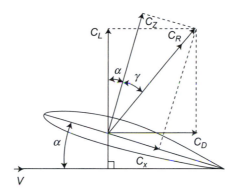

FIGURE 1.19

Definitions: axial, normal, lift, and drag force directions.

direction of the free-stream airflow. The conversion from one pair of coefficients to the other may be carried out with reference to Fig. 1.19, in which C_R, the coefficient of the resultant aerodynamic force, acts at an angle γ to C_Z. C_R is the result both of C_X and C_Z and of C_L and C_D; therefore, from Fig. 1.19, it follows that

$$C_L = C_R \cos(\gamma + \alpha) = C_R \cos\gamma \cos\alpha - C_R \sin\gamma \sin\alpha$$

But $C_R \cos\gamma = C_Z$ and $C_R \sin\gamma = C_X$, so

$$C_L = C_Z \cos\alpha - C_X \sin\alpha \qquad (1.63)$$

Similarly,

$$C_D = C_R \sin(\alpha + \gamma) = C_Z \sin\alpha + C_X \cos\alpha \qquad (1.64)$$

The total pitching moment coefficient is

$$C_M = C_{M_z} + C_{M_x} \qquad (1.65)$$

In Fig. 1.20 are the graphs necessary to evaluate the aerodynamic coefficients for the midsection of a three-dimensional wing with an elliptic-Zhukovsky profile.

1.6.7 Induced Drag

Consider what is happening at some point y along the wingspan (Fig. 1.21). Each of the trailing vortices produces a downward component of velocity w at y, known as downwash or induced velocity. This causes the flow over that section of the wing to inclined slightly downward from the direction of the undisturbed stream V (Fig. 1.22) by the angle ε, the induced angle of attack, or the downwash angle. The local flow is also at a slightly different speed q.

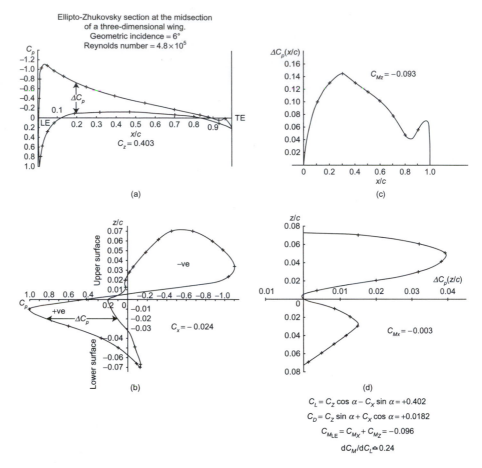

FIGURE 1.20

Example of pressure distribution on an airfoil surface.

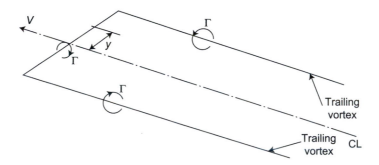

FIGURE 1.21

Simplified horseshoe vortex system. (This geometry is used extensively in Chapter 5 as the basis for numerous wing models.)

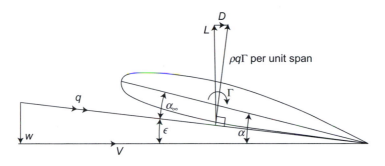

FIGURE 1.22

Flow conditions and forces at a section of a three-dimensional lifting wing.

If the angle between the airfoil chord line and the direction of the undisturbed stream, the geometric angle of attack, is α, the angle between the chord line and the actual flow at that section of the wing is equal to $\alpha - \varepsilon$; this is called the effective incidence α_∞. Effective incidence determines the lift coefficient at that section of the wing, and thus the wing is lifting less strongly than the geometric angle of attack would suggest. Since the circulation and therefore w and ε increase with the lift coefficient, it follows that the lift of a three-dimensional wing increases less rapidly with incidence than does that for a two-dimensional wing, which has no trailing vortices.

Now the circulation around this section of the wing will have a value Γ appropriate to α_∞, and the lift force corresponding to this circulation will be $\rho q \Gamma$ per unit length, according to the Kutta-Joukowski theorem developed in Chapter 3, acting perpendicular to the direction of q as shown—that is, inclined backward from the vertical by the angle ε. This force therefore has a component perpendicular to the undisturbed stream V, that by definition is called the lift; it is of magnitude

$$l = \rho q \Gamma \cos \varepsilon = \rho q \Gamma \frac{V}{q} = \rho V \Gamma \text{ per unit length}$$

There is also a rearward component of magnitude

$$d = \rho q \Gamma \sin \varepsilon = \rho q \Gamma \frac{w}{q} = \rho w \Gamma \text{ per unit length}$$

which must be reckoned a drag and is, in fact, the induced drag. Thus the induced drag arises essentially from the downward velocity induced over the wing by the wingtip vortices.

The further apart the wingtip vortices, the less their effectiveness in producing induced incidence and drag. It is therefore to be expected that these induced quantities will depend on the wing aspect ratio (*AR*). Some results obtained in Chapter 5 are

$$\frac{dC_L}{d\alpha} = a = \frac{a_\infty}{1 + a_\infty/\pi(AR)}$$

where a_∞ is the lift-curve slope for the two-dimensional wing, and the trailing vortex drag coefficient C_{D_v} is given by

$$C_{D_v} = \frac{D_V}{\frac{1}{2}\rho V^2 S} = \frac{C_L^2}{\pi(AR)}(1 + \delta)$$

where δ is a small positive number, constant for a given wing.

1.6.8 Lift-Dependent Drag

It has been seen that the induced drag coefficient is proportional to C_L^2, and it may exist in an inviscid fluid. On a complete aircraft, interference at wing-fuselage, wing/engine-nacelle, and other such junctions leads to modification of the boundary layers over the isolated wing, fuselage, and so forth. This interference, which is actually part of the profile drag, usually varies with the lift coefficient in such a manner that it may be treated as of the form $(a + bC_L^2)$. The part of this profile drag coefficient that is represented by the term (bC_L^2) may be added to the induced drag. The sum so obtained is known as the lift-dependent drag coefficient, which is actually defined as "the difference between the drag at a given lift coefficient and the drag at some datum lift coefficient."

If this datum lift coefficient is taken to be zero, the total drag coefficient of a complete airplane may be taken, to a good approximation in most cases, as

$$C_D = C_{D_0} + kC_L^2$$

where C_{D_0} is the drag coefficient at zero lift, and kC_L^2 is the lift-dependent drag coefficient, denoted by C_{D_L}.

1.6.9 Airfoil Characteristics

We now have the tools to describe quantitatively and unambiguously the integrated, or "overall," effects of the air pressure and shear on the vehicle. It is now instructive to use our descriptors to discuss how some of these aerodynamic loads typically vary with angle of attack, aspect ratio, or other parameters. The reader is cautioned that as the quantitative powers of aerodynamics are applied to topics ranging from insect-like micro-air-vehicles to the Airbus 380 and even to ballistic entry to the Martian atmosphere, one should expect any general discussion to be improper for some extreme cases. Thus, there are jobs for engineers.

1.6 Basic Aerodynamics   **59**

Lift Coefficient: Incidence

Lift coefficient is illustrated in Fig. 1.23 for a two-dimensional (infinite-span) wing. Considering first the full curve (a), which is for a moderately thick (13%) section of zero camber, it is seen to consist of a straight line passing through the origin, curving over at the higher values of C_L, reaching a maximum value of $C_{L_{max}}$ at an incidence of α_s, known as the stalling point. After the stalling point, the lift coefficient decreases, tending to level off at some lower value for higher incidences. The slope of the straight portion of the curve is the two-dimensional lift-curve slope, $(dC_L/d\alpha)_\infty$, or a_∞. Its theoretical value for a thin section (strictly a curved or flat plate) is 2π per radian (see Section 4.4.1). For a section of finite thickness in air, a more accurate empirical value is

$$\left(\frac{dC_L}{d\alpha}\right)_\infty = 1.8\pi\left(1 + 0.8\frac{t}{c}\right) \tag{1.66}$$

The value of $C_{L_{max}}$ is a very important airfoil characteristic because it determines the minimum speed at which an airplane can fly. A typical value for the type of airfoil section mentioned is about 1.5. The corresponding value of α_s is around 18 degrees.

Curves (b) and (c) in Fig. 1.23 are for sections that have the same thickness distribution but are cambered, (c) more so than (b). The effect of camber is merely to reduce the incidence at which a given lift coefficient is produced: to shift the whole lift curve somewhat to the left, with negligible change in the value of the lift-curve slope or in the shape of the curve. This shift is measured by the incidence at which the lift coefficient is zero, or the no-lift incidence, denoted α_0. A typical value is -3 degrees. The same reduction occurs in α_s. Thus a cambered section has the same value of $C_{L_{max}}$ as does its thickness distribution, but this occurs at a smaller incidence.

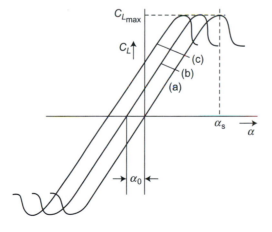

FIGURE 1.23

Typical lift curves for sections of moderate thickness and various cambers.

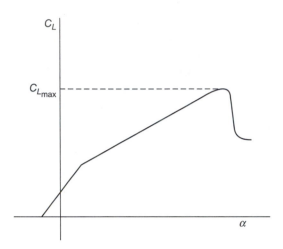

FIGURE 1.24

Lift curve for a thin airfoil section with a small nose radius of curvature.

Modern thin, sharp-nosed sections display a slightly different characteristic from just described, as shown in Fig. 1.24. In this case, the lift curve has two approximately straight portions, of different slopes. The slope of the lower portion is almost the same as that for a thicker section, but, at a moderate incidence, it takes a different, smaller value, leading to a smaller value of $C_{L_{\max}}$, typically on the order of unity. This change in the lift-curve slope is due to a change in the type of flow near the nose of the airfoil.

Effect of Aspect Ratio on the C_L versus α Curve

The induced angle of attack ε is given by

$$\varepsilon = \frac{kC_L}{\pi A}$$

where A is the aspect ratio and thus

$$\alpha_\infty = \alpha - \frac{kC_L}{\pi A}$$

Considering a number of wings of the same symmetrical section but of different aspect ratios, the expression just given leads to a family of C_L, α curves, as in Fig. 1.25, since the actual lift coefficient at a given section of the wing is equal to that for a two-dimensional wing at an incidence of α_∞.

For highly swept wings of very low aspect ratio (less than 3 or so), the lift-curve slope becomes very small, leading to values of $C_{L_{\max}}$ of about 1.0, at stalling incidences of around 45 degrees. This is reflected in the extreme nose-up landing attitudes of many aircraft designed with wings of this type.

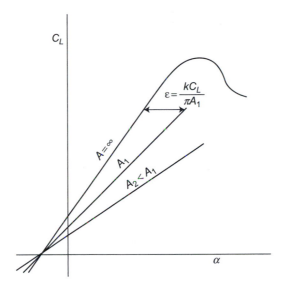

FIGURE 1.25

Influence of wing aspect ratio on lift curve.

Effect of Reynolds Number on the C_L versus α Curve

Reduction of the Reynolds number moves the transition point of the boundary layer rearward on the upper surface of the wing. At low Re values this may permit a laminar boundary layer to extend into the adverse pressure gradient region of the airfoil. Because a laminar boundary layer is much less able than a turbulent boundary layer to overcome an adverse pressure gradient, the flow separates from the surface at a lower angle of attack. This causes a reduction of $C_{L_{\max}}$, which is a problem in model testing because it is always difficult to match full-scale and model Reynolds numbers. Transition can be fixed artificially on the model by roughening its surface with appropriately sized grit at the calculated full-scale point.

Drag Coefficient versus Lift Coefficient

For a two-dimensional wing at low Mach numbers, the drag contains no induced or wave drag, and the drag coefficient is C_{D_0}. There are two distinct variations of C_D with C_L, both illustrated in Fig. 1.26.

In the figure, curve (a) represents a typical conventional airfoil with C_{D_0} fairly constant over the working range of the lift coefficient, increasing rapidly toward the two extreme values of C_L. Curve (b) represents the type of variation found for low-drag airfoil sections. Over much of the C_L range the drag coefficient is rather larger than for the conventional airfoil type, but, within a restricted range of the lift coefficient (C_{L_1} to C_{L_2}), it is considerably less. This range of C_L is known as the favorable range for the section, and the low drag coefficient is due to the design of the airfoil section, which permits a comparatively large extent of laminar boundary layer. It is for this reason that airfoils of this type are also known as laminar-flow sections.

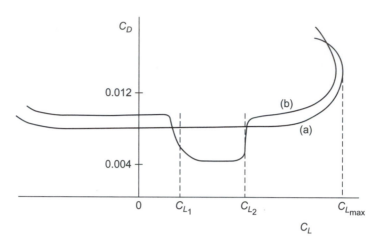

FIGURE 1.26

Typical variation of sectional drag coefficient with lift coefficient.

The width and depth of this favorable range or, more graphically, low-drag bucket, is determined by the shape of the thickness distribution. The central value of the lift coefficient is known as the optimum or ideal $C_{L_{opt}}$ or C_{L_i}. Its is decided by the shape of the camber line and the degree of camber, and thus the favorable range may be placed where desired by suitable camber-line design. It may be placed to cover the most common range of the lift coefficient for a particular airplane; for example, C_{L_2} may be slightly larger than the lift coefficient used on the climb, and C_{L_1} may be slightly less than the cruising lift coefficient. In such a case the airplane will have the benefit of a low drag coefficient value for the wing throughout most of the flight, with obvious benefits in performance and economy. Unfortunately, it is not possible to have large areas of laminar flow on swept wings at high Reynolds numbers. To maintain natural laminar flow, sweep-back angles are limited to about 15 degrees.

The effect of a finite aspect ratio is to give rise to induced drag, and this drag coefficient is proportional to C_L^2 and must be added to the curves of Fig. 1.26.

Drag Coefficient versus (Lift Coefficient)²

Since

$$C_{Dv} = \frac{C_L^2}{\pi A}(1 + \delta)$$

it follows that a curve of C_{Dv} against C_L^2 will be a straight line of slope $(1 + \delta)/\pi A$. If the curve C_{D_0} against C_L^2 from Fig. 1.26 is added to the induced drag coefficient—that is, to the straight line—the result is the total drag coefficient variation with C_L^2, as shown in Fig. 1.27, for the two types of section considered in Fig. 1.26. Taking an idealized case in which C_{D_0} is independent of the lift coefficient, the C_{Dv} versus $(C_L)^2$ curve for a family of wings of various aspect ratios is as shown in Fig. 1.28.

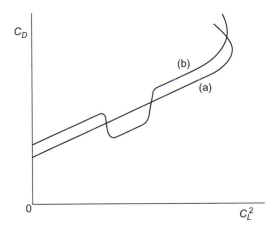

FIGURE 1.27

Variation of total wing drag coefficient with (lift coefficient)^2. Note the "drag bucket" in the curve labeled (b).

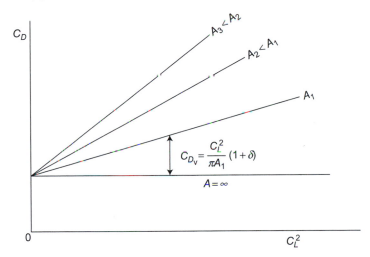

FIGURE 1.28

Idealized variation of total wing drag coefficient with (lift coefficient)^2 for a family of three-dimensional wings of various aspect ratios.

Pitching Moment Coefficient

In Section 1.6.4 it was shown that

$$\frac{dC_M}{dC_L} = \text{constant}$$

the value of the constant depending on the point of the airfoil section about which C_M is measured. Thus a curve of C_M against C_L is theoretically as shown in Fig. 1.29.

In the figure, line (a), for which $dC_M/dC_L = -\frac{1}{4}$, is for the case where C_M is measured about the leading edge. Line (c), for which the slope is zero, is for the case where C_M is measured about the aerodynamic center. Line (b) is obtained if C_M is measured about a point between the leading edge and the aerodynamic center, while for line (d) the reference point is behind the aerodynamic center. These curves are straight only for moderate values of C_L. As the lift coefficient approaches $C_{L\max}$, the C_M against C_L curve departs from the straight line.

The two possibilities are sketched in Fig. 1.30, where for curve (a) the pitching moment coefficient becomes more negative near the stall, thus tending to decrease the incidence and unstall the wing. This is known as a stable break. Curve (b), on the other hand, shows that, near the stall, the pitching moment coefficient becomes less

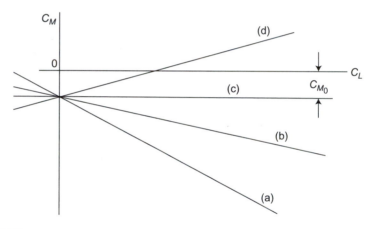

FIGURE 1.29

Variation of C_M with C_L for an airfoil section for four different reference points.

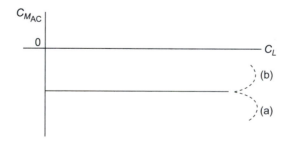

FIGURE 1.30

Behavior of the pitching moment coefficient in the region of the stalling point, showing stable and unstable breaks.

negative. The tendency, then, is for the incidence to increase, aggravating the stall. Such a characteristic is an unstable break and is commonly found with highly swept wings, although measures can be taken to counteract this undesirable behavior.

1.7 EXERCISES

1.1 Verify the dimensions and units given in Table 1.1.

1.2 The constant of gravitation G is defined by

$$F = G\frac{mM}{r^2} F = G\frac{m_1 m_2}{r^2}$$

where F is the gravitational force between two masses m_1 and m_2 whose centers of mass are a distance r apart. Find the dimensions of G and its units in both SI and Imperial units.
(*Answer*: MT^2L^{-3}, $kg\,m^{-3}\,s^2$, $slug\,ft^{-3}\,s^2$)

1.3 Assuming the period of oscillation of a simple pendulum τ to depend on the mass of the object, the length of the pendulum l, and the acceleration due to gravity g, use the theory of dimensional analysis to show that the mass of the object is not, in fact, relevant. Then find a suitable expression for the period of oscillation in terms of the other variables.
(*Answer*: $\tau = c\sqrt{l/g}$, where c is a constant)

1.4 A thin flat disc of diameter D is rotated about a spindle through its center, at a speed of ω radians per second, in a fluid of density ρ and kinematic viscosity ν. Show that the power P needed to rotate the disc may be expressed as

(a) $P = \rho\omega^3 D^5 f\left(\frac{\nu}{\omega D^2}\right)$

(b) $P = \frac{\rho\nu^3}{D}h\left(\frac{\omega D^2}{\nu}\right)$

Note: For (a) solve in terms of the index of ν and for (b) solve in terms of the index of ω.

Further, show that $\omega D^2/\nu$, $PD/\rho\nu^3$, and $P/\rho\omega^3 D^5$ are all nondimensional quantities.

1.5 Spheres of various diameters D and densities σ are allowed to fall freely under gravity through various fluids (represented by their densities ρ and kinematic viscosities ν) and their terminal velocities V are measured. Find a rational expression connecting V with the other variables, and hence suggest a suitable graph in which the results can be presented.

Note: There will be five unknown indices, and therefore two must remain undetermined, which will give two unknown functions on the right-hand side. Make the unknown indices those of σ and ν.

(*Answer:* $V = \sqrt{Dg}\, f\left(\frac{\sigma}{\rho}\right) h\left(\frac{D}{v}\sqrt{Dg}\right)$; therefore, plot curves of $\frac{V}{\sqrt{Dg}}$ against $\left(\frac{D}{v}\right)\sqrt{Dg}$ for various values of σ/ρ.)

1.6 An airplane weighs 60,000 N and has a wingspan of 17 m. A 1/10th-scale model is tested, flaps down, in a compressed-air tunnel at 15 atmospheres pressure and 15°C at various speeds. The maximum lift on the model is measured at the various speeds, with the results as given:

Speed (m s^{-1})	20	21	22	23	24
Maximum lift (N)	2960	3460	4000	4580	5200

Estimate the minimum flying speed of the aircraft at sea level (i.e., the speed at which the maximum lift of the aircraft is equal to its weight).
(*Answer:* 33 m s^{-1})

1.7 The pressure distribution over a section of a two-dimensional wing at 4 degrees of incidence may be approximated as follows: Upper surface: C_p constant at –0.8 from the leading edge to 60% chord, then increasing linearly to +0.1 at the trailing edge. Lower surface: C_p constant at –0.4 from the leading edge to 60% chord, then increasing linearly to +0.1 at the trailing edge. Estimate the lift coefficient and the pitching moment coefficient about the leading edge due to lift.
(*Answer:* 0.3192; –0.13)

1.8 Static pressure is measured at a number of points on the surface of a long circular cylinder of 150 mm in diameter with its axis perpendicular to a stream of standard density at 30 m s^{-1}. The pressure points are defined by the angle θ, which is the angle subtended at the center by the arc between the pressure point and the front stagnation point. In the table following, values are given for p–p_0, where p is the pressure on the surface of the cylinder and p_0 is the undisturbed pressure of the free stream, for various angles θ, all pressures being in N m^{-2}. The readings are identical for the upper and lower halves of the cylinder. Estimate the form pressure drag per meter run and the corresponding drag coefficient.

θ	(degrees)	0	10	20	30	40	50	60
$p - p_0$	(N m^{-2})	+569	+502	+301	–57	–392	–597	–721
θ	(degrees)	70	80	90	100	110	120	
$p - p_0$	(N m^{-2})	–726	–707	–660	–626	–588	–569	

For values of θ between 120 and 180 degrees, $p - p_0$ is constant at –569 N m^{-2}.
(*Answer:* $C_D = 0.875$, $D = 7.25$ N m^{-1})

1.9 A sailplane has a wing of 18 m span and aspect ratio of 16. The fuselage is 0.6 m wide at the wing root, and the wing taper ratio is 0.3 with square-cut wingtips. At a true air speed of 115 km h^{-1} at an altitude where the relative density is 0.7, the lift and drag are 3500 N and 145 N, respectively. The wing pitching moment coefficient about the quarter-chord point is –0.03 based on the gross wing area and the aerodynamic mean chord. Calculate the lift and drag coefficients based on the gross wing area, and the pitching moment about the quarter-chord point.
(*Answer*: $C_L = 0.396$, $C_D = 0.0169$, $M = -322$ N m, since $\bar{c}_A \approx 1.245$ m)

1.10 Describe qualitatively the results expected from the pressure plotting of a conventional, symmetrical, low-speed two-dimensional airfoil. Indicate the changes expected with incidence and discuss the processes for determining the resultant forces. Are any further tests needed to determine the overall forces of lift and drag? Include in the discussion the order of magnitude expected for the various distributions and forces described.

1.11 Show that for geometrically similar aerodynamic systems the nondimensional force coefficients of lift and drag depend only on Reynolds number and Mach number. Discuss briefly the importance of this theorem in wind-tunnel testing and simple performance theory.

1.12 "A pint's a pound the world around" is the old rhyme describing the weight of water. Given that a pint is an eighth of a gallon, which is 231.8 cubic inches, and the density of freshwater (not seawater) is 1.93 slugs per cubic foot, solve for the percentage error in the old rhyme.

1.13 In high school physics, W=mg shows that 1 slug weighs 32.174 pounds. Yet 1 slug $= 1\frac{\text{pound } s^2}{\text{ft}}$. Similarly, 1 kg weighs 9.8 N. What does 1 kg equal? Note the difference between "weighs" and "equals" that we often lose sight of during homework assignments and exams. Note too that the working slug-ft-sec system is identical to the mks units. It is the lbm-ft-sec system that is unusual.

1.14 The plot at the top of the following page shows data for publicly available maximum thrust versus maximum takeoff weight for a wide range of twin-engine business jets. Planes from the "very light jet" market are at the lower left, and the 737 business jet is at the upper right. Show that the maximum lift-to-drag ratio for all of these aircraft is approximately 6, assuming that they can take off with one engine out. You should assume that the takeoff roll is horizontal and that enough lift is generated to begin accelerating vertically. Similarly, assume that the takeoff roll at liftoff has negligible acceleration—that is, nearly constant speed.

1.15 One important performance measure of a commercial airliner is "dollars per seat-mile," or the cost to fly one passenger one mile. The smaller this number, the better the aircraft performs. This number is not just a function of the shape

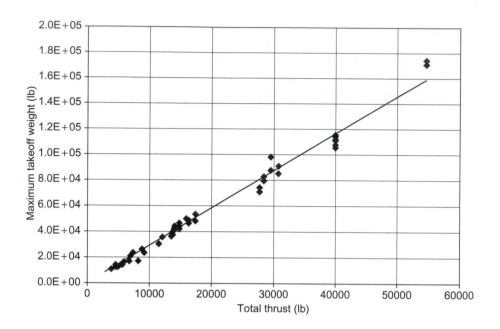

of the wing and the engine; it depends on how heavily the aircraft is loaded. If you can sail a small balsa wood or foam glider from a repeatable height (ladder, balcony, hilltop, etc.), you can optimize the "dollars per seat-mile" for it. Maximizing dollars per seat-mile for the glider is a task of minimizing the denominator: seats times miles, or the product of payload and distance. Load your glider with varying payloads and record the distance it travels (you must retain good trim, so tape on weights at the glider's center of mass). What payload provides you the maximum payload-range product? The minimum?

Fundamental Equations of Fluid Mechanics

LEARNING OBJECTIVES

- Learn the physics and mathematics of one-dimensional fluid motion. Recognize that many of the physical phenomena evident in all stages of aerodynamics are most readily approached by considering the one-dimensional mode, without prejudice to the wider analysis of two- and three-dimensional motions.

- Learn the mathematical formulas for the laws governing changes in the physical properties of air. These laws are applied to accelerating gas as it moves out of the low-speed (incompressible) regime and into the transonic and supersonic regimes, where abrupt changes in properties are manifest.

- Learn that, even for incompressible flows, the fundamental formulas cannot be solved. Hence, learn that solving aeronautical engineering problems is an art of approximation.

- Learn that the one practical approximation appropriate for the design and analysis of airfoils and wings is that of the outer-potential-flow/boundary-layer theory.

- Learn that a class of methods playing an ever increasing role in engineering analysis of designs is commercially available computational fluid dynamics (CFD). These tools are approximate; therefore, you must learn their strengths and limitations (as with any other engineering method).

2.1 INTRODUCTION

The equations describing the motion of air were known in the nineteenth century through the work of Navier at the beginning of the century and the work of Stokes at the end. There were a number of other notable works published between the papers of Navier and Stokes on the problem of flow of real fluids. However, it was the work of Bernoulli and Euler in the eighteenth century and work of Prandtl in the first decade of the twentieth century that led the way to the procedures described in this text as applied to the airfoil and the wing. Prandtl's breakthrough led to the productive application of ideal-fluid potential-flow theory to airfoil and wing design. Prandtl demonstrated that for air moving over airfoil shapes at speeds of interest the viscous effects are confined to relatively thin regions adjacent to the boundaries of the flow. Hence, for relatively thin airfoils the boundary layer can be neglected altogether and the pressure distribution can be reasonably predicted by applying ideal potential-flow theory. Since we cannot solve the Navier-Stokes equations for most practical flow problems, and since there are still some open questions about their solutions, it is quite fortunate that the aeronautical engineer can apply potential flow theory productively in the design of aerodynamic shapes. The foundations of this theory are described in this chapter.

The engineering science of aerodynamics is built on a foundation comprising three sets of tools. These are the interconnected theoretical, experimental, and computational tools of classical continuum mechanics of fluids. This foundation was introduced at the beginning of Chapter 1. We pointed out there that the two fluids of interest are the simple-compressible Newtonian viscous fluid model of "real" flows and Euler's inviscid perfect fluid model. The mathematical statements of the fundamental principles and the constitutive models of material behavior necessary to formulate a complete set of equations are developed next. The approach will begin with control-volume analysis. For further more detailed accounts of mathematical modeling of fluid dynamics the interested reader should consult Shapiro (1953), Batchelor (1967), and Bird, Stewart, and Lightfoot (2002).

2.1.1 Selection of Coordinates

Consider an airplane in steady flight. To an observer on the ground, it is flying into air substantially at rest, assuming no wind, and any air movement is caused by the airplane's motion through it. The pilot might think that he or she is stationary, that a stream of air is flowing past him or her, and that the airplane modifies the motion of the air. Of course, both viewpoints are mathematically and physically correct. Both observers may use the same equations to study the mutual effects of the air and the airplane, and they will both arrive at the same answers for, say, the forces exerted by the air on the airplane. However, if indeed steady-state conditions occur, from the pilot's point of view the time-dependent terms in the equations of motion will be zero. This is an important observation and, hence, a simplification that makes it easier to solve the flow problem as compared with the ground-based observer where the flow field always appears time-dependent. Because of this it is convenient to regard most problems in aerodynamics as cases of air flowing past a body at rest, with consequent simplification of the mathematics. It is also from this point of view that an important *design condition* is typically established, that is, the desired performance at steady cruise speed at constant altitude through quiescent air.

Types of Flow

The flow around a body may be steady or unsteady. A steady flow is one in which the flow parameters, (e.g., speed, direction, pressure) may vary from point to point in the flow but at any point are constant with respect to time. That is, measurements of the flow parameters at a given point in the flow at various times remain the same. In an unsteady flow the flow parameters at any point vary with time.

2.1.2 **A Comparison of Steady and Unsteady Flow**

Figure 2.1(a) shows a section of a stationary wing with air flowing past. The velocity of the air along way from the wing is constant at V, as shown. The flow parameters are measured at some fixed point relative to the wing (e.g., at $P(x,y)$). The flow perturbations produced at P by the body will be the same at all times; that is, the flow is steady relative to a set of axes fixed in the body.

Figure 2.1(b) represents the same wing moving at the same speed V through air, which, a long way from the body, is at rest. The flow parameters are measured at a point $P'(x',y')$ fixed relative to the stationary air. The wing thus moves past P'. At times t_1, when the wing is at A_1, P' is a fairly large distance ahead of the wing and the perturbations at P' are small. Later, at time t_2, the wing is at A_2, directly beneath P', and the perturbations are much greater. Later still, at time t_3, P' is far behind the wing, which is now at A_3, and the perturbations are again small. Thus, the perturbation at P' started from a small value, increased to a maximum, and finally decreased to a small value. The perturbation at the fixed point P' is therefore not constant with respect to time, and so the flow, referred to axes fixed in the fluid, is

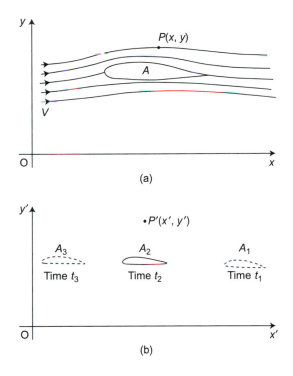

(a)

(b)

FIGURE 2.1

(a) Air moves at speed v past axes fixed relative to the airfoil. (b) Airfoil moves at speed v through air initially at rest. Axes Ox' and Oy' are fixed relative to undisturbed air at rest.

not steady. Thus, changing the axes of reference from a set fixed relative to the air flow to a different set fixed relative to the body causes changes in the flow from unsteady to steady. This produces the mathematical simplification mentioned earlier by eliminating time from the equations. Since the flow relative to the air flow can, by a change of axes, be made steady, it is sometimes known as "quasi-steady."

True Unsteady Flow

An example of true unsteady flow is the wake behind a bluff body—for example, a circular cylinder (Fig. 2.2). The air is flowing from left to right, and the system of eddies or vortices behind the cylinder is moving in the same direction at a somewhat lower speed. This region of slower moving fluid is the "wake." Consider a point P, fixed relative to the cylinder, in the wake. Sometimes the point will be immersed in an eddy and sometimes not. Thus the flow parameters will be changing rapidly at P, and the flow there is unsteady. Moreover, it is impossible to find a set of axes relative to which the flow is steady. At a point Q well outside the wake, the fluctuations are so small that they may be ignored and the flow at Q may, with little error, be regarded as steady. Thus, even though the flow in some region may be unsteady, there may be another region where the unsteadiness is negligibly small, so that the flow there may be regarded as steady with sufficient accuracy for all practical purposes.

Three concepts are useful in describing fluid flows:

- A streamline, defined as "an imaginary line drawn in the fluid such that there is no flow across it at any point" or as "a line that is always in the same direction as the local velocity vector." It follows that since this is identical to the condition at a solid boundary:
    - Any streamline may be replaced by a solid boundary without modifying the flow. (This is strictly true only if viscous effects are ignored.)
    - Any solid boundary is itself a streamline of the flow around it.
- A filament (or streak) line, defined as the line taken up by successive particles of fluid passing through some given point. A fine filament of smoke injected into the flow through a nozzle traces out a filament line. The lines shown in Fig. 2.2 are examples.
- A path line or particle path, defined as the path traced out by any one particle of the fluid in motion. In unsteady flow, these three are in general different, while

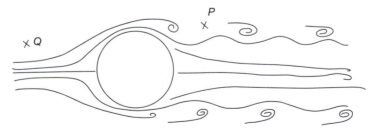

FIGURE 2.2

True unsteady flow.

in steady flow all three they are identical. Also, in steady flow it is convenient to define a stream tube as an imaginary bundle of adjacent streamlines.

2.2 ONE-DIMENSIONAL FLOW: THE BASIC EQUATIONS

In all real flow situations the physical laws of classical mechanics apply. These refer to the conservation of both mass and energy. The momentum equation, based on Newton's second law, also applies. The equation of state completes the set that needs to be solved if some or all of the parameters controlling the flow are unknown. If a real flow can be "modeled" by a similar but simplified system the degree of complexity in handling the resulting equations may be considerably reduced.

Historically, the lack of mathematical tools available to the engineer required that many simplifying assumptions be made. The simplifications used depend on the particular problem but are not arbitrary. In fact, judgment is required to decide which parameters in a flow process may be reasonably ignored, at least to a first approximation. For example, in much of aerodynamics the gas (air) is considered to behave as an incompressible fluid (see Section 2.3.4), and an even wider assumption is that the air flow is unaffected by its viscosity. This last assumption would appear at first to be utterly inappropriate since viscosity plays an important role in the mechanism by which aerodynamic force is transmitted from the air flow to the body and vice versa. Nevertheless, the science of aerodynamics has progressed far on this assumption, and much of the aeronautical technology now available follows from theories based on it.

Other examples will be invoked from time to time and it is salutary, and good engineering practice, to acknowledge those "simplifying" assumptions made in order to arrive at an understanding of, or a solution to, a physical problem.

2.2.1 One-Dimensional Flow: The Basic Equations of Conservation

A prime simplification of the algebra involved without any loss of physical significance may be made by examining the changes in the flow properties along a stream tube that is essentially straight or for which the cross-section changes slowly (i.e., a so-called quasi-one-dimensional flow).

Conservation of Mass

This law states that in normally perceived engineering situations matter cannot be created or destroyed. For steady flow in the stream tube shown in Fig. 2.3, let the flow properties at stations 1 and 2 be a distance s apart, as shown. If the values for the flow velocity v and the density ρ at station 1 are the same across the tube, which is a reasonable assumption if the tube is thin, then the quantity flowing into the volume comprising the element of stream tube is

$$\text{velocity} \times \text{area} = v_1 A_1$$

The mass flowing in through station 1 is

$$\rho_1 v_1 A_1 \tag{2.1}$$

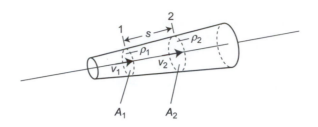

FIGURE 2.3

Stream tube for conservation of mass.

Similarly the mass outflow at station 2, on the same assumptions, is

$$\rho_2 v_2 A_2 \tag{2.2}$$

These two quantities, Eqs. (2.1) and (2.2), must be the same if the tube does not leak or gain fluid and if matter is to be conserved. Thus

$$\rho_1 v_1 A_1 = \rho_2 v_2 A_2 \tag{2.3}$$

or in a general form:

$$\rho v A = \text{constant} \tag{2.4}$$

Momentum Equation

The momentum equation requires that the time rate of momentum change in a given direction be equal to the sum of the forces acting in that direction. This is known as Newton's second law of motion and in the model used here the forces concerned are gravitational (body) and surface.

Consider a fluid in steady flow, and take any small stream tube as in Fig. 2.4. The distance s is measured along the tube's axis from some arbitrary origin. A is the cross-sectional area of the stream tube at distance s from the arbitrary origin.

p, ρ, and v represent pressure, density, and flow speed, respectively. A, p, ρ, and v vary with s (i.e., with position along the stream tube) but not with time since the motion is steady. Now consider the small element of fluid shown in Fig. 2.5, which is immersed in fluid of varying pressure. The element is the right frustrum of a cone of length δs, area A at the upstream section, area $A + \delta A$ on the downstream section. The pressure acting on one face of the element is p, and on the other face is $p + (dp/ds)\delta s$. Around the curved surface the pressure may be taken to be the mean value $p + (dp/ds)\delta s/2$. In addition, the weight W of the fluid in the element acts vertically as shown. Shear forces on the surface due to viscosity would add another force, which is ignored here.

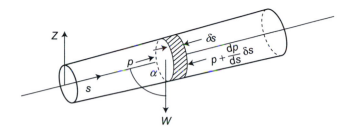

FIGURE 2.4

Stream tube and element for the momentum equation.

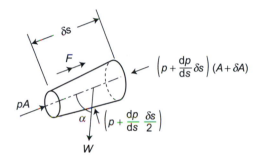

FIGURE 2.5

Forces on the element.

As a result of these pressures and the weight, there is a resultant force F acting along the axis of the cylinder where F is given by

$$F = pA - \left(p + \frac{dp}{ds}\delta s\right)(A + \delta A) + \left(p + \frac{dp}{ds}\frac{\delta s}{2}\right)\delta A - W\cos\alpha \qquad (2.5)$$

where α is the angle between the axis of the stream tube and the vertical.

From Eq. (2.5) it is seen that on neglecting quantities of small order such as $(dp/ds)\delta s\delta A$ and cancelling, we get

$$F = -\frac{dp}{ds}A\delta s - \rho g A(\delta s)\cos\alpha \qquad (2.6)$$

since the gravitational force on the fluid in the element is $\rho g A\,\delta s$, volume $\times$ density $\times g$.

Now, Newton's second law of motion (force $=$ mass $\times$ acceleration) applied to the element of Fig. 2.5 gives

$$-\rho g A\delta s\cos\alpha - \frac{dp}{ds}A\,\delta s = \rho A\,\delta s\frac{dv}{dt} \qquad (2.7)$$

where t represents time. Dividing by $A\delta s$ this becomes

$$-\rho g \cos\alpha - \frac{dp}{ds} = \rho \frac{dv}{dt}$$

But

$$\frac{dv}{dt} = \frac{dv}{ds}\frac{ds}{dt} = v\frac{dv}{ds}$$

and therefore

$$\rho v\frac{dv}{ds} + \frac{dp}{ds} + \rho g \cos\alpha = 0$$

or

$$v\frac{dv}{ds} + \frac{1}{\rho}\frac{dp}{ds} + g\cos\alpha = 0$$

Integrating along the stream tube, this becomes

$$\int \frac{dp}{\rho} + \int v\,dv + g\int \cos\alpha\,ds = \text{constant}$$

but since

$$\int \cos\alpha\,ds = \text{increase in vertical coordinate } z$$

and

$$\int v\,dv = \frac{1}{2}v^2$$

then

$$\int \frac{dp}{\rho} + \frac{1}{2}v^2 + gz = \text{constant} \qquad (2.8)$$

This result is known as *Bernoulli's equation* and is discussed momentarily.

The Conservation of Energy

Conservation of energy implies that changes in energy, heat transferred, and work done by a system in steady operation are in balance. In seeking an equation to represent the conservation of energy in the steady flow of a fluid, it is useful to consider a length of stream tube—for example, between stations 1 and 2 (Fig. 2.6), as constituting the *control surface* of an "open thermodynamic system" or *control volume.* For the flow at stations 1 and 2, let the fluid properties be as shown.

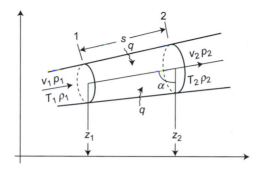

FIGURE 2.6

Control volume for the energy equation.

A unit mass of fluid entering the system through the inlet section will possess internal energy $c_V T_1$, kinetic energy $v_1^2/2$, and potential energy gz_1:

$$\left(c_V T_1 + \frac{v_1^2}{2} + gz_1 \right)$$

(2.9a)

Likewise on exit from the system across station 2, a unit mass will possess energy

$$\left(c_V T_2 + \frac{v_2^2}{2} + gz_2 \right)$$

(2.9b)

Now, to enter the system a unit mass must possess a volume $1/\rho_1$, which must push against the pressure p_1 and utilize energy to the value of $p_1 \times 1/\rho_1$ pressure $\times$ (specific) volume. At exit, p_2/ρ_2 is utilized in a similar manner.

Between the inlet and exit stations, the system accepts or rejects heat q per unit mass. As all the quantities are flowing steadily, the energy entering plus the heat transfer must equal the energy leaving.[1] Thus, with a positive heat transfer it follows, from conservation of energy, that

$$c_V T_1 + \frac{v_1^2}{2} + \frac{p_1}{\rho_1} + gz_1 + q = c_V T_2 + \frac{v_2^2}{2} + \frac{p_2}{\rho_2} + gz_2$$

---

[1]It should be noted that in a general system the fluid would also do work, which should be considered in the equation, but it is disregarded here for the particular case of flow in a stream tube.

However, enthalpy per unit mass of fluid is $c_V T + p/\rho = c_p T$. Thus

$$\left(c_p T_2 + \frac{v_2^2}{2} + g z_2 \right) - \left(c_p T_1 + \frac{v_1^2}{2} + g z_1 \right) = q$$

or in differential form

$$\frac{d}{ds}\left(c_p T + \frac{v^2}{2} + gs\cos\alpha \right) = \frac{dq}{ds} \qquad (2.10)$$

For an adiabatic (no heat transfer) horizontal flow system, Eq. (2.10) becomes zero and thus (neglecting the change in gravitational potential energy)

$$c_p T + \frac{v^2}{2} = \text{constant} \qquad (2.11)$$

Equation of State

The equation of state for a perfect gas is

$$p/(\rho T) = R$$

Substituting for p/ρ in Eq. (1.11) yields Eqs. (1.13) and (1.14):

$$c_p - c_V = R, \quad c_p = \frac{\gamma}{\gamma - 1} R \quad c_V = \frac{1}{\gamma - 1} R$$

The first law of thermodynamics requires that the gain in internal energy of a mass of gas plus the work done by the mass be equal to the heat supplied; that is, for a unit mass of gas with no heat transfer

$$E + \int p \, d\left(\frac{1}{\rho} \right) = \text{constant}$$

or

$$dE + p \, d\left(\frac{1}{\rho} \right) = 0 \qquad (2.12)$$

Differentiating Eq. (1.10) for enthalpy gives

$$dh = dE + p \, d\left(\frac{1}{\rho} \right) + \frac{1}{\rho} dp = 0 \qquad (2.13)$$

and combining Eqs. (2.12) and (2.13) yields

$$dh = \frac{1}{\rho} dp \qquad (2.14)$$

But

$$dh = c_p dT = \frac{c_p}{R} d\left(\frac{p}{\rho}\right) = \frac{\gamma}{\gamma - 1}\left[\frac{1}{\rho}dp + pd\left(\frac{1}{\rho}\right)\right] \qquad (2.15)$$

Therefore, from Eqs. (2.14) and (2.15)

$$\frac{dp}{p} + \gamma\rho d\left(\frac{1}{\rho}\right) = 0$$

which on integrating gives

$$\ln p + \gamma \ln\left(\frac{1}{\rho}\right) = \text{constant}$$

or

$$p = k\rho^{\gamma}$$

where k is a constant. This is the isentropic relationship between pressure and density and has been replicated for convenience from Eq. (1.24).

Momentum Equation for an Incompressible Fluid

Provided velocity and pressure changes are small, density changes will be very small, and it is permissible to assume that the density ρ is constant throughout the flow. With this assumption, Eq. (2.8) may be integrated as

$$\int dp + \frac{1}{2}\rho v^2 + \rho gz = \text{constant}$$

Performing this integration between two conditions represented by suffices 1 and 2 gives

$$(p_2 - p_1) + \frac{1}{2}\rho\left(v_2^2 - v_1^2\right) + \rho g(z_2 - z_1) = 0$$

That is,

$$p_1 + \frac{1}{2}\rho v_1^2 + \rho gz_1 = p_2 + \frac{1}{2}\rho v_2^2 + \rho gz_2$$

In the foregoing analysis 1 and 2 were completely arbitrary choices, and therefore the same equation must apply to conditions at any other points. Thus

$$p + \frac{1}{2}\rho v^2 + \rho gz = \text{constant} \qquad (2.16)$$

This is *Bernoulli's equation* for an incompressible fluid—that is, a fluid that cannot be compressed or expanded and for which the density is invariable. Note that Eq. (2.16) can be applied more generally to two- and three-dimensional steady flows, provided that viscous effects are neglected. In the more general case, however, it is important to note that Bernoulli's equation can only be applied along a streamline, and in certain cases the constant may vary from streamline to streamline.

2.2.2 Comments on the Momentum and Energy Equations

Referring to Eq. (2.8), which expresses the momentum principle in algebraic form,

$$\int \frac{\mathrm{d}p}{\rho} + \frac{1}{2}v^2 + gz = \text{constant}$$

where the first term is the internal energy of unit mass of the air, $\frac{1}{2}v^2$ is the kinetic energy per unit mass, and gz is the potential energy per unit mass. Thus, Bernoulli's equation in this form is really a statement of the principle of conservation of energy in the absence of heat exchanged and work done. As a corollary, this equation applies only to flows where there is no mechanism for the dissipation of energy into some form not included in the above three terms. In aerodynamics a common form of energy dissipation is that due to viscosity. Thus, the equation cannot be strictly applied in this form to a flow where the effects of viscosity are appreciable, such as that in a boundary layer.

AERODYNAMICS AROUND US

Bernoulli's Equation and a Spring-Mass System

You have seen that Bernoulli's equation in aerodynamics is written $P_\circ = P + \frac{1}{2}\rho v^2$, but at first you are likely unsure of just what the dynamic pressure $q = \frac{1}{2}\rho v^2$ is. Static pressure you experience clearly in your ears as you ascend or descend a thousand feet in air or underwater. Total, or stagnation, pressure you might consider as the increased static pressure you feel when your hand is held out the car window with palm facing forward. Dynamic pressure seems less physical to most students.

Recall an elementary physics problem: a mass on a frictionless surface attached to a spring. When the mass is pulled to extend the spring, energy is stored in the spring. If you release the mass, it moves toward the spring, converting spring energy into kinetic energy. The sum of spring energy and kinetic energy is constant; if the energy is not in the motion, it is in the spring. Pretty simple.

Consider pressure, which you are likely accustomed to thinking of as a force per area. But the units of force per area are also units of energy per volume. So Bernoulli's equation is also a description of how three energies (per volume) are related. What three energies? Because density is mass per volume, dynamic pressure is kinetic energy per volume. This is added to another energy per volume, static pressure, to equal stagnation pressure, which is a constant in many flows. Thus, static pressure is like spring energy per volume. In this view, Bernoulli's equation tells us that an incompressible steady flow is like a spring-mass system; if the energy isn't in the spring, (static pressure) then it is in the motion (kinetic energy). There is no other place for the energy to go in such a system.

Bernoulli's law simply tells us the motion and compression trade-off back and forth in a steady incompressible flow, like a mass and spring system, except of course, the spring can be in tension and the air cannot.

2.3 MEASUREMENT OF AIR SPEED

2.3.1 Pitôt-Static Tube

Consider an instrument of the form sketched in Fig. 2.7, called a Pitôt-static tube. It consists of two concentric tubes A and B. The mouth of A is open and faces directly into the airstream, while the end of B is attached to the outer edge of A, causing it to be sealed off. Some very fine holes are drilled in the wall of B, as at C, allowing B to communicate with the surrounding air. The right-hand ends of A and B are connected to opposite sides of a manometer. The instrument is placed into a stream of air, with the mouth of A pointing directly upstream, the stream being of speed v and of static pressure p. The air flowing past the holes at C will be moving at a speed little different from v, and its pressure will therefore be equal to p and will be communicated to the interior of tube B through holes C. The pressure in B is, therefore, the static pressure of the stream.

Air entering the mouth of A will, on the other hand, be brought to rest (in the ultimate analysis by the fluid in the manometer). Its pressure will therefore be equal to the total head of the stream. As a result, a pressure difference exists between the air in A and that in B, and this may be measured on the manometer. Denote the pressure in A by p_A, that in B by p_B, and the difference between them by Δp. Then

$$\Delta p = p_A - p_B \tag{2.17}$$

However, by Bernoulli's equation (for incompressible flow)

$$p_A + \frac{1}{2}\rho(0)^2 = p_B + \frac{1}{2}\rho v^2$$

and therefore

$$p_A - p_B = \frac{1}{2}\rho v^2 \tag{2.18}$$

or

$$\Delta p = \frac{1}{2}\rho v^2$$

whence

$$v = \sqrt{2\Delta p / \rho} \tag{2.19}$$

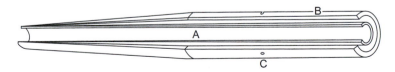

FIGURE 2.7

Simple Pitôt-static tube.

The value of ρ, which is constant in incompressible flow, may be calculated from the ambient pressure and the temperature. This, together with the measured value of Δp, permits calculation of the speed v.[2]

The quantity $\frac{1}{2}\rho v^2$ is the *dynamic pressure* of the flow. Since $p_A =$ total pressure $= p_0$ (i.e., the pressure of the air at rest, also referred to as the stagnation pressure), and $p_B =$ static pressure $= p$, then

$$p_0 - p = \frac{1}{2}\rho v^2 \tag{2.20}$$

which may be expressed in words as

$$\text{stagnation pressure} - \text{static pressure} = \text{dynamic pressure}$$

It should be noted that this equation applies at all speeds, but the dynamic pressure is equal to $\frac{1}{2}\rho v^2$ only in incompressible flow. Note also that

$$\frac{1}{2}\rho v^2 = \left[ML^{-3}L^2T^{-2} \right] = \left[ML^{-1}T^{-2} \right]$$

$$= \text{units of pressure}$$

as is of course essential.

Defining the *stagnation pressure coefficient* as

$$C_{p0} = \frac{p_0 - p}{\frac{1}{2}\rho v^2} \tag{2.21}$$

it follows immediately from Eq. (2.20) that for incompressible flow

$$C_{p0} = 1 \text{ (always)} \tag{2.22}$$

2.3.2 Pressure Coefficient

In Chapter 1 it was seen that it is often convenient to express variables in a nondimensional coefficient form. The coefficient of pressure was introduced in Section 1.6.3. The stagnation pressure coefficient has already been defined as

$$C_{p0} = \frac{p_0 - p}{\frac{1}{2}\rho v^2}$$

---

[2]Note that, notwithstanding the formal restriction of Bernoulli's equation to inviscid flows, the Pitôt-static tube is commonly used to determine the local velocity in wakes and boundary layers with no apparent loss of accuracy.

An alternative form of the pressure coefficient can be defined as follows:

$$C_p = \frac{p - p_\infty}{\frac{1}{2}\rho v^2} \tag{2.23}$$

where

$C_p =$ pressure coefficient

$p =$ static pressure at some point in the flow where the velocity is q

$p_\infty =$ static pressure of the undisturbed flow

$\rho =$ density of the undisturbed flow

$v =$ speed of the undisturbed flow

Now, in incompressible flow,

$$p + \frac{1}{2}\rho q^2 = p_\infty + \frac{1}{2}\rho v^2$$

Then

$$p - p_\infty = \frac{1}{2}\rho\left(v^2 - q^2\right)$$

and therefore

$$C_p = 1 - \left(\frac{q}{v}\right)^2 \tag{2.24}$$

Then

1. If C_p is positive, $p > p_\infty$ and $q < v$.
2. If C_p is zero, $p = p_\infty$ and $q = v$.
3. If C_p is negative, $p < p_\infty$ and $q > v$.

2.3.3 Air-Speed Indicator: Indicated and Equivalent Air Speeds

A Pitôt-static tube is commonly used to measure air speed both in the laboratory and on aircraft. There are, however, differences in the requirements for the two applications. In the laboratory, liquid manometers provide a simple and direct method for measuring pressure. These would be completely unsuitable for use on an aircraft where a pressure transducer is used that converts the pressure measurement into an electrical signal. Pressure transducers are also becoming more common in laboratory measurements.

When the measured pressure difference is converted into air speed, the correct value for the air density should, of course, be used in Eq. (2.19). This is easy enough in the laboratory, although for accurate results the variation of density with the ambient atmospheric pressure should be taken into account.

There is a relationship between true air speed (TAS) and equivalent air speed (EAS) that is important when using a Pitot-static tube on an aircraft. This relationship is derived next. Using the correct value of density ρ Eq. (2.19) shows that the relationship between the measured pressure difference and true air speed is

$$\Delta p = \frac{1}{2}\rho v^2 \tag{2.25}$$

whereas if the standard value of density, $\rho_0 = 1.226$ kg/m^3, is used, we find

$$\Delta p = \frac{1}{2}\rho_0 v_E^2 \tag{2.26}$$

where v_E is the equivalent air speed. But the values of Δp in Eqs. (2.25) and (2.26) are the same, and therefore

$$\frac{1}{2}\rho_0 v_E^2 = \frac{1}{2}\rho v^2 \tag{2.27}$$

or

$$v_E = v\sqrt{\rho / \rho_0} \tag{2.28}$$

If the relative density $\sigma = \rho/\rho_0$ is introduced, Eq. (2.28) can be written as

$$v_E = v\sqrt{\sigma} \tag{2.29}$$

The term *indicated air speed* (IAS) is used for the measurement made with an actual (imperfect) air-speed indicator. Owing to instrument error, IAS will normally differ from EAS.

The following definitions may therefore be stated: IAS is the uncorrected reading shown by an actual air-speed indicator. EAS is the uncorrected reading that would be shown by a hypothetical, error-free air-speed indicator. *True air speed* (TAS) is the actual speed of the aircraft relative to the air. Only when $\sigma = 1$ will TAS and EAS be equal. Normally EAS is less than the TAS.

Formerly, the aircraft navigator needed to calculate the TAS from the IAS. But in modern aircraft, the conversion is done electronically. The calibration of the air-speed indicator also makes an approximate correction for compressibility.

2.3.4 Incompressibility Assumption

As a first step in calculating the stagnation pressure coefficient in compressible flow, we use Eq. (1.6d) to rewrite the dynamic pressure as follows:

$$\frac{1}{2}\rho v^2 = \frac{1}{2}\left(\frac{\rho}{\gamma p}\right)\gamma p v^2 = \frac{1}{2}\gamma p \frac{v^2}{a^2} = \frac{1}{2}\gamma p M^2 \tag{2.30}$$

where M is Mach number.

When the ratio of the specific heats is given the value 1.4 (the approximate value for air), the stagnation pressure coefficient then becomes

$$C_{p0} = \frac{p_0 - p}{0.7pM^2} = \frac{1}{0.7M^2}\left(\frac{p_0}{p} - 1\right) \tag{2.31}$$

Now

$$\frac{p_0}{p} = \left[1 + \frac{1}{5}M^2\right]^{7/2} \qquad \text{(Eq. (6.18a))}$$

Expanding this by the binomial theorem gives

$$\frac{p_0}{p} = 1 + \frac{7}{2}\left(\frac{1}{5}M^2\right) + \frac{7}{2}\frac{5}{2}\frac{1}{2!}\left(\frac{1}{5}M^2\right)^2 + \frac{7}{2}\frac{5}{2}\frac{3}{2}\frac{1}{3!}\left(\frac{1}{5}M^2\right)^3$$
$$+ \frac{7}{2}\frac{5}{2}\frac{3}{2}\frac{1}{2}\frac{1}{4!}\left(\frac{1}{5}M^2\right)^4 + \cdots$$
$$= 1 + \frac{7M^2}{10} + \frac{7M^4}{40} + \frac{7M^6}{400} + \frac{7M^8}{16000} + \cdots$$

Then

$$C_{p0} = \frac{10}{7M^2}\left(\frac{p_0}{p} - 1\right)$$
$$= \frac{10}{7M^2}\left[\frac{7M^2}{10} + \frac{7M^4}{40} + \frac{7M^6}{400} + \frac{7M^8}{16000} + \cdots\right] \tag{2.32}$$
$$= 1 + \frac{M^2}{4} + \frac{M^4}{40} + \frac{M^6}{1600} + \cdots$$

It can be seen that this will become unity, the incompressible value, at $M = 0$. This is the practical meaning of the incompressibility assumption—that is, that any velocity changes are small compared with the speed of sound in the fluid. The result given in Eq. (2.32) is the correct one, which applies at all Mach numbers less than unity. At supersonic speeds, shock waves may be formed, in which case the physics of the flow are completely altered.

Table 2.1 shows the variation of C_{p0} with Mach number. It is seen that the error in assuming $C_{p0} = 1$ is only 2% at $M = 0.3$ but rises rapidly at higher Mach numbers, being slightly more than 6% at $M = 0.5$ and 27.6% at $M = 1.0$.

It is often convenient to regard the effects of compressibility as negligible if the flow speed nowhere exceeds about $100\,\text{ms}^{-1}$. However, it must be remembered that this is an entirely arbitrary limit. Compressibility applies at all flow speeds, and therefore ignoring it always introduces an error. It is thus necessary to consider, for each problem, whether or not the error can be tolerated.

Table 2.1 Variation of Stagnation Pressure Coefficient with Mach Numbers Less Than Unity

M	0	0.2	0.4	0.6	0.7	0.8	0.9	1.0
C_{p0}	1	1.01	1.04	1.09	1.13	1.16	1.217	1.276

In the following examples the speed of sound is required. It can be written as follows:

$$a = \sqrt{\gamma R T}$$

For air, with $\gamma = 1.4$ and $R = 287.3\,\mathrm{J\,kg^{-1}K^{-1}}$, this becomes

$$a = 20.05\sqrt{T}\,\mathrm{m\,s^{-1}} \tag{2.33}$$

where T is the temperature in K.

Example 2.1

The air-speed indicator fitted to a particular airplane has no instrument errors and is calibrated assuming incompressible flow in standard conditions. While flying at sea level in the ISA, the indicated air speed is 950 $\mathrm{km\,h^{-1}}$. What is the true air speed? Note that 950 $\mathrm{km\,h^{-1}} = 264\ \mathrm{m\,s^{-1}}$ and that this is the speed corresponding to the pressure difference applied to the instrument based on the stated calibration. This pressure difference can therefore be calculated by

$$p_0 - p = \Delta p = \frac{1}{2}\rho_0 v_E^2$$

and therefore

$$p_0 - p = \frac{1}{2} \times 1.226(264)^2 = 42670\ \mathrm{N\,m^{-2}}$$

Now

$$\frac{p_0}{p} = \left[1 + \frac{1}{5}M^2\right]^{3.5}$$

In standard conditions $p = 101325\ \mathrm{N\,m^{-2}}$. Therefore

$$\frac{p_0}{p} = \frac{42670}{101325} + 1 = 1.421$$

So

$$1 + \frac{1}{5}M^2 = (1.421)^{2/7} = 1.106$$

$$\frac{1}{5}M^2 = 0.106$$

$$M^2 = 0.530$$

$$M = 0.728$$

The speed of sound at standard conditions is

$$a = 20.05(288)^{\frac{1}{2}} = 340.3\,\mathrm{m\,s^{-1}}$$

Therefore, true air speed $= Ma = 0.728 \times 340.3$, that is,

$$Ma = 248\,\mathrm{m\,s^{-1}} = 891\,\mathrm{km/h^{-1}}$$

In this example, $\sigma = 1$, and therefore there is no effect due to density; that is, the difference is due entirely to compressibility. Thus it is seen that neglecting compressibility in the calibration has led the air-speed indicator to overestimate the true air speed by $59\,\mathrm{km\,h^{-1}}$.

2.4 TWO-DIMENSIONAL FLOW

Consider flow in two dimensions only. It is the same as that between two planes set parallel and a little distance apart. The fluid can then flow in any direction between and parallel to the planes but not at right angles to them. This means that in the subsequent mathematics there are only two space variables: x and y in Cartesian (or rectangular) coordinates or r and θ in polar coordinates. For convenience, a unit length of the flow field is assumed in the z direction perpendicular to x and y. This simplifies the treatment of two-dimensional flow problems, but care must be taken in the matter of units.

In practice, if two-dimensional flow is to be simulated experimentally, the method of constraining the flow between two close parallel plates is often used (e.g., small smoke tunnels and some high-speed tunnels).

To summarize, two-dimensional flow is fluid motion where the velocity at all points is parallel to a given plane.

We have already seen how the principle of conservation of mass and the momentum equation can be applied to one-dimensional flows to obtain the continuity and momentum equations (see Section 2.2). We will now derive the governing equations for two-dimensional flow by applying conservation of mass and the momentum equation to an infinitesimal rectangular control volume (see Fig. 2.8).

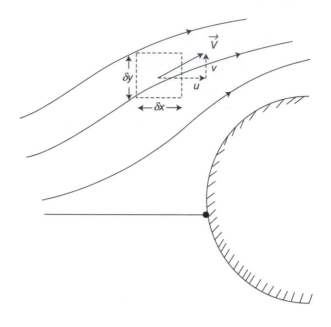

FIGURE 2.8

Infinitesimal control volume in a typical two-dimensional flow field.

2.4.1 Component Velocities

In general the local velocity in a flow is inclined to the reference axes Ox, Oy, and it is usual to resolve the velocity vector $\vec{v}$ (magnitude q) into two components mutually at right angles.

In a Cartesian coordinate system let a particle move from point $P(x, y)$ to point $Q(x + \delta x, y + \delta y)$, a distance of δs in time δt (Fig. 2.9). Then the velocity of the particle is

$$\lim_{\delta \to 0} \frac{\delta s}{\delta t} = \frac{ds}{dt} = q$$

Going from P to Q, the particle moves horizontally through δx, giving the horizontal velocity $u = dx/dt$ positive to the right. Similarly going from P to Q, the particle moves vertically through δy with the vertical velocity $v = dy/dt$ (upward positive). By geometry

$$(\delta s)^2 = (\delta x)^2 + (\delta y)^2$$

Thus

$$q^2 = u^2 + v^2$$

and the direction of q relative to the x-axis is $\alpha = \tan^{-1}(v/u)$.

In a polar coordinate system (Fig. 2.10) the particle moves distance δs from $P(r, \theta)$ to $Q(r + \delta r, \theta + \delta \theta)$ in time δt. The component velocities are

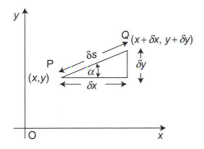

FIGURE 2.9

Infinitesimal motion of a particle from P to Q in Cartesian coordinates.

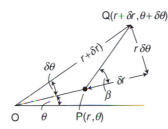

FIGURE 2.10

Infinitesimal motion of a particle from P to Q in polar coordinates.

$$\text{radially (outwards positive) } q_n = \frac{dr}{dt}$$

$$\text{tangentially (counterclockwise positive) } q_t = r\frac{d\theta}{dt}$$

Again

$$(\delta s)^2 = (\delta r)^2 + (r\delta\theta)^2$$

Thus

$$q^2 = q_n^2 + q_t^2$$

and the direction of q relative to the radius vector is given by

$$\beta = \tan^{-1}\frac{q_t}{q_n}$$

Fluid Acceleration

The equation of acceleration of a fluid mass is rather different from that of a vehicle, for example, and a note on fluid acceleration follows. Let a fluid particle move from P to Q in time δt in a two-dimensional flow (Fig. 2.11). At the point $P(x, y)$ the velocity components are u and v. At the adjacent point $Q(x + \delta x, y + \delta y)$ the velocity components are $u + \delta u$ and $v + \delta v$. That is, in general the velocity component has

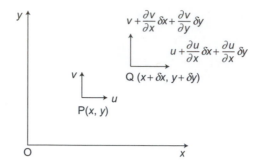

FIGURE 2.11

Velocity vector components at P and Q in Cartesian coordinates.

changed in each direction by an increment δu or δv. This incremental change is the result of a spatial displacement, and as u and v are functions of x and y the velocity components at Q are

$$u + \delta u = u + \frac{\partial u}{\partial x}\delta x + \frac{\partial u}{\partial y}\delta y \quad \text{and} \quad v + \delta v = v + \frac{\partial v}{\partial x}\delta x + \frac{\partial v}{\partial y}\delta y \qquad (2.34)$$

The component of acceleration in the Ox direction is thus

$$\frac{d(u + \delta u)}{dt} = \frac{\partial u}{\partial t} + \frac{\partial u}{\partial x}\frac{dx}{dt} + \frac{\partial u}{\partial y}\frac{dy}{dt}$$
$$= \frac{\partial u}{\partial t} + u\frac{\partial u}{\partial x} + v\frac{\partial u}{\partial y} \qquad (2.35)$$

and in the Oy direction

$$\frac{d(v + \delta v)}{dt} = \frac{\partial u}{\partial t} + u\frac{\partial v}{\partial x} + v\frac{\partial v}{\partial y} \qquad (2.36)$$

The change in other flow variables, such as pressure, between points P and Q may be dealt with in a similar way. Thus, if the pressure takes the value p at P, at Q it takes the value

$$p + \delta p = p + \frac{\partial p}{\partial x}\delta x + \frac{\partial p}{\partial y}\delta y \qquad (2.37)$$

The total time rate of change in velocity vector components, i.e., the acceleration [Eqs. (2.35) and (2.36)], and the total time rate of change in pressure [Eq. (2.37)], i.e., a flow property, was introduced in Section 1.1.1 on Basic Concepts.

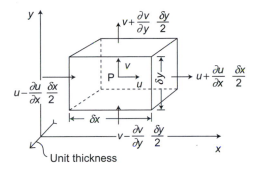

FIGURE 2.12

Rectangular space of volume $\delta x \times \delta y \times 1$ at point $P(x,y)$, where the velocity components are u and v and the density is ρ.

2.4.2 Equation of Continuity or Conservation of Mass

Consider a typical elemental control volume like the one illustrated in Fig. 2.8. This is a small rectangular region of space of sides δx, δy, and unity, centered at the point $P(x, y)$ in a fluid motion, which is referred to the axes Ox, Oy. At $P(x,y)$ the local velocity components are u and v and the density ρ, where each of these three quantities is a function of x, y, and t (Fig. 2.12). Dealing with the flow into the box in the Ox direction, the amount of mass flowing into the region of space per second through the left-hand vertical face is mass flow per unit area $\times$ area:

$$\left(\rho u - \frac{\partial(\rho u)}{\partial x} \frac{\delta x}{2} \right) \delta y \times 1 \tag{2.38}$$

The amount of mass leaving the box per second through the right-hand vertical face is

$$\left(\rho u + \frac{\partial(\rho u)}{\partial x} \frac{\delta x}{2} \right) \delta y \times 1 \tag{2.39}$$

The accumulation of mass per second in the box due to the horizontal flow is the difference of Eqs. (2.38) and (2.39):

$$- \frac{\partial(\rho u)}{\partial x} \delta x \delta y \tag{2.40}$$

Similarly, the accumulation per second in the Oy direction is

$$- \frac{\partial(\rho v)}{\partial y} \delta x \delta y \tag{2.41}$$

so that the total accumulation per second is

$$-\left(\frac{\partial(\rho u)}{\partial x} + \frac{\partial(\rho v)}{\partial y}\right)\delta x \delta y \tag{2.42}$$

As mass cannot be destroyed or created, Eq. (2.42) must represent the rate of change in mass of the fluid in the box and can also be written as

$$\frac{\partial(\rho \times \text{volume})}{\partial t}$$

but with the elementary box having constant volume ($\delta x \delta y \times 1$), this becomes

$$\frac{\partial \rho}{\partial t}\delta x \delta y \times 1 \tag{2.43}$$

Equating (2.42) and (2.43) gives the general equation of continuity, thus

$$\frac{\partial \rho}{\partial t} + \frac{\partial(\rho u)}{\partial x} + \frac{\partial(\rho v)}{\partial y} = 0 \tag{2.44}$$

This can be expanded to

$$\frac{\partial \rho}{\partial t} + u\frac{\partial \rho}{\partial x} + v\frac{\partial \rho}{\partial y} + \rho\left(\frac{\partial u}{\partial x} + \frac{\partial v}{\partial y}\right) = 0 \tag{2.45}$$

and if the fluid is incompressible and the flow steady, the first three terms are all zero since the density cannot change and the equation reduces for incompressible flow to

$$\frac{\partial u}{\partial x} + \frac{\partial v}{\partial y} = 0 \tag{2.46}$$

This equation is fundamental and important, and it should be noted that it expresses a physical reality. For example, in the case given by Eq. (2.46)

$$\frac{\partial u}{\partial x} = -\frac{\partial v}{\partial y}$$

This reflects the fact that if the flow velocity increases in the x direction it must decrease in the y direction.

For three-dimensional flows Eqs. (2.45) and (2.46) are written in the forms

$$\frac{\partial \rho}{\partial t} + u\frac{\partial \rho}{\partial x} + v\frac{\partial \rho}{\partial y} + w\frac{\partial \rho}{\partial z} + \rho\left(\frac{\partial u}{\partial x} + \frac{\partial v}{\partial y} + \frac{\partial w}{\partial z}\right) = 0 \tag{2.47a}$$

$$\frac{\partial u}{\partial x} + \frac{\partial v}{\partial y} + \frac{\partial w}{\partial z} = 0 \tag{2.47b}$$

2.4.3 Equation of Continuity in Polar Coordinates

A corresponding equation can be found in the polar coordinates r and θ, where the velocity components are q_n and q_t radially and tangentially. By carrying out a similar development for the accumulation of fluid in an elemental box of space, the equation of continuity corresponding to Eq. (2.44) can be found as follows. Taking the element to be at $P(r,\theta)$ where the mass flow is ρq per unit length (Fig. 2.13), the accumulation per second radially is

$$\left(\rho q_n - \frac{\partial(\rho q_n)}{\partial r}\frac{\delta r}{2}\right)\left(r - \frac{\delta r}{2}\right)\delta\theta - \left(\rho q_n + \frac{\partial(\rho q_n)}{\partial r}\frac{\delta r}{2}\right)\left(r + \frac{\delta r}{2}\right)\delta\theta$$
$$= -\rho q_n \delta r \delta\theta - \frac{\partial(\rho q_n)}{\partial r}r\delta r\delta\theta \qquad (2.48)$$

and the accumulation per second tangentially is

$$\left(\rho q_t - \frac{\partial(\rho q_t)}{\partial \theta}\frac{\delta\theta}{2}\right)\delta r - \left(\rho q_t + \frac{\partial(\rho q_t)}{\partial \theta}\frac{\delta\theta}{2}\right)\delta r = -\frac{\partial(\rho q_t)}{\partial \theta}\delta r \delta\theta \qquad (2.49)$$

The total accumulation per second is

$$-\left(\frac{\rho q_n}{r} + \frac{\partial(\rho q_n)}{\partial r} + \frac{1}{r}\frac{\partial(\rho q_t)}{\partial \theta}\right)r\delta r\delta\theta \qquad (2.50)$$

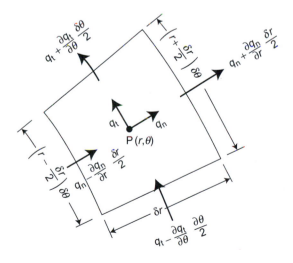

FIGURE 2.13

Rectangular element at $P(r,\theta)$ in a system of polar coordinates.

and this by the previous argument equals the rate of change in mass within the region of space

$$\frac{\partial(\rho r \delta r \delta \theta)}{\partial t} \tag{2.51}$$

Equations (2.50) and (2.51) give

$$\frac{\rho q_n}{r} + \frac{\partial \rho}{\partial t} + \frac{\partial(\rho q_n)}{\partial r} + \frac{1}{r}\frac{\partial(\rho q_t)}{\partial \theta} = 0 \tag{2.52}$$

Hence for steady flow

$$\frac{\partial(\rho r q_n)}{\partial r} + \frac{\partial(\rho q_t)}{\partial \theta} = 0 \tag{2.53}$$

and the incompressible equation in this form becomes

$$\frac{q_n}{r} + \frac{\partial q_n}{\partial r} + \frac{1}{r}\frac{\partial q_t}{\partial \theta} = 0 \tag{2.54}$$

2.5 STREAM FUNCTION AND STREAMLINE

Streamlines are lines tangent to the velocity vectors that describe the velocity field. The concepts of stream function and streamlines are examined in the next two subsections, respectively. In the third subsection the relationship between the velocity components and the stream function is discussed.

2.5.1 Stream Function ψ

Imagine being on the banks of a shallow river of a constant depth of 1 m at a position O (Fig. 2.14) with a friend directly opposite at A, 40 m away. Mathematically the bank can be represented by the Ox-axis, and the line joining you to your friend at A the Oy-axis in the two-coordinate system. Now if the stream speed is $2\,\mathrm{m\,s^{-1}}$ the amount of water passing between you and your friend is $40 \times 1 \times 2 = 80\mathrm{m^3\,s^{-1}}$, and this is the amount of water flowing past any point anywhere along the river and can be measured at a weir downstream. Suppose you now throw a buoyant rope to your friend who catches the end but allows the slack to fall in the river and float into a curve as shown. The amount of water flowing under the line is still $80\ \mathrm{m^3\,s^{-1}}$, no matter what shape the rope takes, and is unaffected by the configuration of the rope.

Suppose your friend moves along to point B somewhere downstream, still holding his end of the line but with sufficient rope paid out as he goes. The volume of water passing under the rope is still only $80\ \mathrm{m^3\,s^{-1}}$ providing he has not stepped over a tributary stream or an irrigation drain in the bank. It follows that, if no water can enter or leave the stream, the quantity flowing past the line will be the same as before and furthermore will be unaffected by the shape of the line between O and B. The amount or quantity of fluid passing such a line per second is called the *stream function* or *current function* and is denoted ψ.

Consider now a pair of coordinate axes set in a two-dimensional air stream that is moving generally from left to right (Fig. 2.15). The axes are arbitrary space references and in no way interrupt the fluid streaming past. Similarly the line joining O to a point P in the flow in no way interrupts the flow since it is as imaginary as the reference axes Ox and Oy. An algebraic expression can be found for the line in x and y.

Let the flow past the line at any point Q on it be at velocity q over a small length δs of line where the direction of q makes angle β to the tangent of the curve at Q. The component of the velocity q perpendicular to the element δs is $q \sin \beta$ and therefore, assuming the depth of stream flow to be unity, the amount of fluid crossing the element of line δs is $q \sin \beta \times \delta s \times 1$ per second. Adding up all such quantities crossing similar elements along the line from O to P, the total amount of flow past the line (sometimes called flux) is

$$\int_{OP} q \sin \beta \, ds$$

which is the line integral of the normal velocity component from O to P.

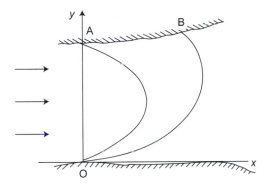

FIGURE 2.14

Shape of "rope" lines in a two-dimensional flow.

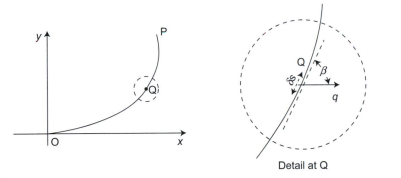

Detail at Q

FIGURE 2.15

Infinitesimal element of "rope" line.

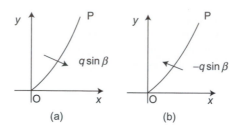

FIGURE 2.16

(a) Flow normal to OP. (b) Opposite direction of flow normal to OP.

If this quantity of fluid flowing between O and P remains the same irrespective of the path of integration, (i.e., independent of the curve of the rope), then $\int_{OP} q\sin\beta\, ds$ is called the stream function of P with respect to O and

$$\psi_P = \int_{OP} q\sin\beta\, ds$$

Note: it is implicit that $\psi = 0$ at point O.

Sign Convention for Stream Functions

It is necessary here to consider a sign convention since quantities of fluid are being considered. When integrating the cross-wise component of flow along a curve, the component can go either from left to right, or vice versa, across the path of integration (Fig. 2.16). Integrating the normal flow components from O to P, the flow components are, looking in the direction of integration, either from left to right or from right to left. The former is considered positive flow while the latter is negative flow. The convention is therefore the following:

Flow *across* the path of integration is *positive* if, when looking in the direction of integration, it crosses the path from left to right.

2.5.2 Streamline

From the statement above, ψ_P is the flow across the line OP. Suppose there is a point P_1 close to P that has a stream function equal in value to that of point P (Fig. 2.17). Then the flow across any line OP_1 equals that across OP, and the amount of fluid flowing into area OPP_1O across OP equals the amount flowing out across OP_1. Therefore, no fluid crosses line PP_1 and the velocity of flow must be along, or tangential to, PP_1.

All other points P_2, P_3, and so forth, which have a stream function equal in value to that of P have, by definition, the same flow across any lines joining them to O, so by the same argument the velocity of the flow in the region of P_1, P_2, P_3, and so forth, must be along PP_1, P_2, P_3, and so forth, and no fluid crosses the line $PP_1, P_2, \ldots, P_n$. Since $\psi_{P_1} = \psi_{P_2} = \psi_{P_3} = \psi_P = $ constant, the line $PP_1, P_2, \ldots P_n$, etc. is a line of

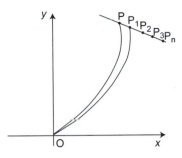

FIGURE 2.17

Streamline emanating from P.

constant ψ and is called a streamline. It follows further that since no flow can cross the line PP_n the velocity along the line must always be in the direction tangential to it. This leads to the two common definitions of a streamline, each of which indirectly has the other's meaning. They are

A *streamline* is a line of constant ψ.
A *streamline* is a line of fluid particles, the velocity of each particle being tangential to the line.

It should be noted that the velocity can change in magnitude along a streamline but by definition the direction is always that of the tangent to the line.

2.5.3 Velocity Components in Terms of ψ

In this subsection we examine the velocity components in terms of the stream function in Cartesian coordinates and in polar coordinates.

In Cartesian coordinates: Let point $P(x, y)$ be on the streamline AB in Fig. 2.18(a) of constant ψ, and let point $Q(x + \delta x, y + \delta y)$ be on the streamline CD of constant $\psi + \delta\psi$. Thus, the volume rate of fluid flowing across any path between P and Q is equal to the change in stream function between P and Q.

The most convenient path along which to integrate in this case is PRQ, point R being given by the coordinates $(x + \delta x, y)$. Then the flow across $PR = -v\delta x$ (since the flow is from right to left and thus by convention negative), and that across $RQ = u\delta y$. Therefore, total flow across the line PRQ is

$$\delta\psi = u\delta y - v\delta x \qquad (2.55)$$

Now ψ is a function of two independent variables x and y in steady motion, and thus

$$\delta\psi = \frac{\partial\psi}{\partial x}\delta x + \frac{\partial\psi}{\partial y}\delta y \qquad (2.56)$$

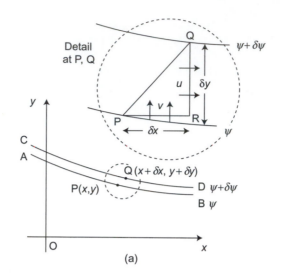

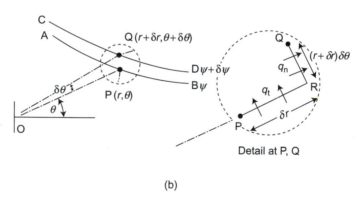

FIGURE 2.18

(a) Flow between streamlines in Cartesian coordinates. (b) Flow between streamlines in polar coordinates.

$\partial\psi/\partial x$ and $\partial\psi/\partial y$ being the partial derivatives with respect to x and y, respectively. Then, equating terms

$$u = \partial\psi \,/\, \partial y \qquad\qquad (2.56a)$$

and

$$v = -\partial\psi \,/\, \partial x \qquad\qquad (2.56b)$$

These are the components of velocity at a point x,y in a flow given by stream function ψ.

In polar coordinates: Let the point $P(r, \theta)$ be on the streamline AB (Fig. 2.18(b)) of constant ψ, and let point $Q(r + \delta r, \theta + \delta\theta)$ be on the streamline CD of constant $\psi + \delta\psi$. The velocity components are q_n and q_t, radially and tangentially, respectively. Here the most convenient path of integration is PRQ, where OP is added to R so that $PR = \delta r$; that is, R is given by ordinates $(r + \delta r, \theta)$. Then

$$\delta\psi = -q_t\delta r + q_n(r + \delta r)\delta\theta$$
$$= -q_t\delta r + q_n r\delta\theta + q_n\delta r\delta\theta$$

To the first order of small quantities:

$$\delta\psi = -q_t\delta r + q_n r\delta\theta \tag{2.57}$$

But here ψ is a function of (r, θ), and again

$$\delta\psi = \frac{\partial\psi}{\partial r}\delta r + \frac{\partial\psi}{\partial\theta}\delta\theta$$

Equating terms in this equation with the terms in Eq. (2.57), we get

$$q_t = -\frac{\partial\psi}{\partial r} \tag{2.58a}$$

$$q_n = \frac{1}{r}\frac{\partial\psi}{\partial\theta} \tag{2.58b}$$

these being velocity components at a point r, θ in a flow given by stream function ψ.

In general terms the velocity q in any direction s is found by differentiating the stream function ψ partially with respect to the direction n normal to q, where n is taken in the counterclockwise sense looking along q (Fig. 2.19):

$$q = \frac{\partial\psi}{\partial n}$$

FIGURE 2.19

Relationship between velocity in the s direction along a streamline and the derivative of the stream function perpendicular to the streamline.

2.6 MOMENTUM EQUATION

The momentum equation for two- or three-dimensional flow embodies Newton's second law of motion (mass times acceleration = force, or rate of change in momentum = force) to an infinitesimal control volume in a fluid flow (see Fig. 2.8). This law takes the form of a set of partial differential equations. Physically it states that the rate of increase in momentum within the control volume plus the net rate at which momentum flows out of the control volume equals the force acting on the fluid within the control volume.

There are two distinct classes of force that act on the fluid particles within the control volume. *Body forces* act on all the fluid within the control volume. Here the only body force of interest is the force of gravity or weight of the fluid. *Surface forces* act only on the control surface; their effect on the fluid inside the control volume cancels out. They are always expressed in terms of stress (force per unit area). Two main types of surface force are involved.

Pressure force. Pressure p is a stress that *always* acts perpendicular to the control surface and in the opposite direction to the unit normal (see Fig. 1.3). In other words it always tends to compress the fluid in the control volume. Although p can vary from point to point in the flow field, it is invariant with direction at a particular point. In other words, irrespective of the orientation of the infinitesimal control volume, the pressure force on any face will be $-p\delta A$, where δA is the area of the face (see Fig. 1.3). As is evident from Bernoulli's Eq. (2.16), the pressure depends on the flow speed.

Viscous forces. In general the viscous force acts at an angle to any particular face of the infinitesimal control volume, so in general it will have two components in two-dimensional flow (three for three-dimensional flow) acting on each face (one due to direct stress acting perpendicularly to the face and one shear stress (two for three-dimensional flow) acting tangentially to the face. As an example, let us consider the stresses acting on two faces of a square infinitesimal control volume (Fig. 2.20). For the top face the unit normal would be $\mathbf{j}$ (unit vector in the y direction) and the unit tangential vector would be $\mathbf{i}$ (unit vector in the x direction). In this case, then, the viscous force acting on this face and the side face is given by

$$\left(\sigma_{yx}\mathbf{i} + \sigma_{yy}\mathbf{j}\right)\delta x \times 1, \quad \left(\sigma_{xx}\mathbf{i} + \sigma_{xy}\mathbf{j}\right)\delta y \times 1$$

respectively. Note that, as in Section 2.4, we are assuming unit length in the z direction. The viscous shear stress is what is termed a second-order *tensor*—that is, it is a quantity that is characterized by a magnitude and *two* directions (compare a vector or first-order tensor that is characterized by a magnitude and one direction). The stress tensor can be expressed in terms of four components (nine for three-dimensional flow) in matrix form as

$$\begin{pmatrix} \sigma_{xx} & \sigma_{xy} \\ \sigma_{yx} & \sigma_{yy} \end{pmatrix}$$

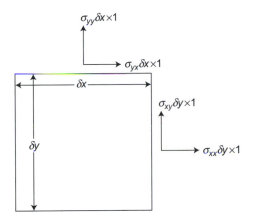

FIGURE 2.20

Surface forces acting on an infinitesimal control volume.

Owing to symmetry $\sigma_{xy} = \sigma_{yx}$. Just as the components of a vector change when the coordinate system is changed, so do the components of the stress tensor. In many engineering applications the direct viscous stresses (σ_{xx}, σ_{yy}) are negligible compared with the shear stresses. The viscous stress is generated by fluid motion and cannot exist in a still fluid.

Other surface forces (e.g., surface tension), can be important in some engineering applications.

When the momentum equation is applied to an infinitesimal control volume (c.v.), it can be written in the form

$$\underbrace{\text{Rate of increase of momentum within the c.v.}}_{\text{(i)}}$$

$$+ \underbrace{\text{Net rate at which momentum leaves the c.v.}}_{\text{(ii)}} \qquad (2.59)$$

$$= \underbrace{\text{Body force}}_{\text{(iii)}} + \underbrace{\text{pressure force}}_{\text{(iv)}} + \underbrace{\text{viscous force}}_{\text{(v)}}$$

We will now consider the evaluation of each of terms (i) through (v) in turn for the case of two-dimensional incompressible flow.

Term (i) is dealt with in a similar way to Eq. (2.43), once it is recalled that momentum is (mass) × (velocity), so term (i) is given by

$$\frac{\partial}{\partial t}(\rho \times \text{volume} \times \vec{v}) = \frac{\partial \rho \vec{v}}{\partial t} \delta x \delta y \times 1 = \left(\frac{\partial \rho u}{\partial t}, \frac{\partial \rho v}{\partial t} \right) \delta x \delta y \times 1 \qquad (2.60)$$

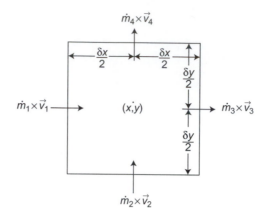

FIGURE 2.21

Momentum transport across each face of the control surface of the infinitesimal control volume.

To evaluate term (ii) we will make use of Fig. 2.21 (compare Fig. 2.12). Note that the rate at which momentum crosses any face of the control volume is the rate at which mass crosses the face times the velocity. So if we denote the rate at which mass crosses a face by $\dot{m}$, term (ii) is given by

$$\dot{m}_3 \times \vec{v}_3 - \dot{m}_1 \times \vec{v}_1 + \dot{m}_4 \times \vec{v}_4 - \dot{m}_2 \times \vec{v}_2 \tag{2.61}$$

But $\dot{m}_3$ and $\dot{m}_1$ are given by Eqs. (2.38) and (2.39), respectively, and $\dot{m}_2$ and $\dot{m}_4$ by similar expressions. In a similar way it can be seen that, recalling $\vec{v} = (u, v)$

$$\vec{v}_1 = (u, v) - \left(\frac{\partial u}{\partial x}, \frac{\partial v}{\partial x}\right)\frac{\delta x}{2}, \quad \vec{v}_3 = (u, v) + \left(\frac{\partial u}{\partial x}, \frac{\partial v}{\partial x}\right)\frac{\delta x}{2}$$

$$\vec{v}_2 = (u, v) - \left(\frac{\partial u}{\partial y}, \frac{\partial v}{\partial y}\right)\frac{\delta y}{2}, \quad \vec{v}_4 = (u, v) + \left(\frac{\partial u}{\partial y} + \frac{\partial v}{\partial y}\right)\frac{\delta y}{2}$$

So the x component of Eq. (2.61) becomes

$$\rho\left(u + \frac{\partial u}{\partial x}\frac{\delta x}{2}\right)\delta y \times 1\left(u + \frac{\partial u}{\partial x}\frac{\delta x}{2}\right) - \rho\left(u - \frac{\partial u}{\partial x}\frac{\delta x}{2}\right)\delta y \times 1\left(u - \frac{\partial u}{\partial x}\frac{\delta x}{2}\right)$$

$$+ \rho\left(v + \frac{\partial v}{\partial y}\frac{\delta y}{2}\right)\delta x \times 1\left(u + \frac{\partial u}{\partial y}\frac{\delta y}{2}\right) - \rho\left(v - \frac{\partial v}{\partial y}\frac{\delta y}{2}\right)\delta x \times 1\left(u - \frac{\partial u}{\partial y}\frac{\delta y}{2}\right)$$

Cancelling like terms and neglecting higher-order terms simplifies this expression to

$$\rho\left(2u\frac{\partial u}{\partial x}+v\frac{\partial u}{\partial y}+u\frac{\partial v}{\partial y}\right)\delta x\delta y\times 1$$

This can be rearranged as

$$\rho\left(u\frac{\partial u}{\partial x}+v\frac{\partial u}{\partial y}+u\underbrace{\left\{\frac{\partial u}{\partial x}+\frac{\partial v}{\partial y}\right\}}_{=0\ \text{Eq. (2.46)}}\right)\delta x\delta y\times 1 \qquad (2.62\text{a})$$

In the exact same way the y component of Eq. (2.61) can be shown to be

$$\rho\left(u\frac{\partial v}{\partial x}+v\frac{\partial v}{\partial y}\right)\delta x\delta y\times 1 \qquad (2.62\text{b})$$

Term (iii), the body force acting on the control volume, is simply given by the weight of the fluid—the mass of the fluid multiplied by the acceleration (vector) due to gravity. Thus

$$\rho\delta x\delta y\times 1\times\vec{g}=(\rho g_x,\rho g_y)\delta x\delta y\times 1 \qquad (2.63)$$

Normally, of course, gravity acts vertically downwards, so $g_x=0$ and $g_y=-g$.

The evaluation of term (iv), the net pressure force acting on the control volume, is illustrated in Fig. 2.22. In the x direction the net pressure force is given by

$$\left(p-\frac{\partial p}{\partial x}\frac{\delta x}{2}\right)\delta y\times 1-\left(p+\frac{\partial p}{\partial x}\frac{\delta x}{2}\right)\delta y\times 1=-\frac{\partial p}{\partial x}\delta x\delta y\times 1 \qquad (2.64\text{a})$$

Similarly, the y component of the net pressure force is given by

$$-\frac{\partial p}{\partial y}\delta x\delta y\times 1 \qquad (2.64\text{b})$$

The evaluation of the x component of term (v), the net viscous force, is illustrated in Fig. 2.23. In a similar way as for Eqs. (2.64a,b), we obtain the net viscous force in the x and y directions respectively as

$$\left(\frac{\partial\sigma_{xx}}{\partial x}+\frac{\partial\sigma_{xy}}{\partial y}\right)\delta x\delta y\times 1 \qquad (2.65\text{a})$$

$$\left(\frac{\partial\sigma_{yx}}{\partial x}+\frac{\partial\sigma_{yy}}{\partial y}\right)\delta x\delta y\times 1 \qquad (2.65\text{b})$$

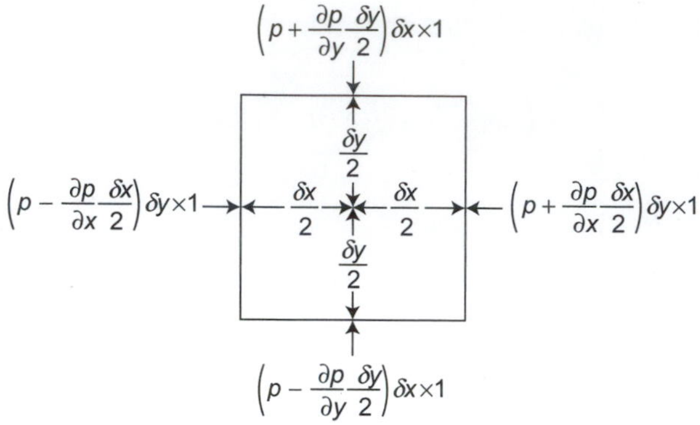

$$\left(p + \frac{\partial p}{\partial y}\frac{\delta y}{2}\right)\delta x \times 1$$

FIGURE 2.22

Pressure forces acting on the infinitesimal control volume.

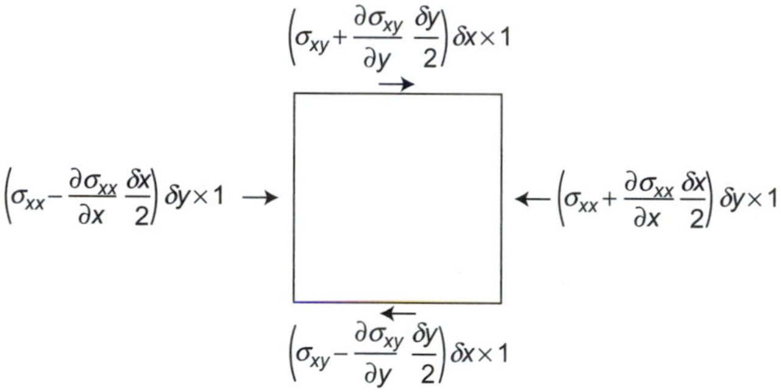

FIGURE 2.23

x component of forces due to viscous stress acting on the infinitesimal control volume.

We now substitute Eqs. (2.61) through (2.65) into Eq. (2.59) and cancel the common factor $\delta x \delta y \times 1$ to obtain

$$\rho\left(\frac{\partial u}{\partial t} + u\frac{\partial u}{\partial x} + v\frac{\partial u}{\partial y}\right) = \rho g_x - \frac{\partial p}{\partial x} + \frac{\partial \sigma_{xx}}{\partial x} + \frac{\partial \sigma_{xy}}{\partial y} \qquad (2.66a)$$

$$\rho\left(\frac{\partial v}{\partial t} + u\frac{\partial v}{\partial x} + v\frac{\partial v}{\partial y}\right) = \rho g_y - \frac{\partial p}{\partial y} + \frac{\partial \sigma_{yx}}{\partial x} + \frac{\partial \sigma_{yy}}{\partial y} \qquad (2.66b)$$

These are the momentum equations in the form of partial differential equations.

For three-dimensional flows the momentum equations can be written in the form

$$\rho\left(\frac{\partial u}{\partial t} + u\frac{\partial u}{\partial x} + v\frac{\partial u}{\partial y} + w\frac{\partial u}{\partial z}\right) = \rho g_x - \frac{\partial p}{\partial x} + \frac{\partial \sigma_{xx}}{\partial x} + \frac{\partial \sigma_{xy}}{\partial y} + \frac{\partial \sigma_{xz}}{\partial z} \qquad (2.67a)$$

$$\rho\left(\frac{\partial v}{\partial t}+u\frac{\partial v}{\partial x}+v\frac{\partial v}{\partial y}+w\frac{\partial v}{\partial z}\right)=\rho g_y-\frac{\partial p}{\partial y}+\frac{\partial \sigma_{yx}}{\partial x}+\frac{\partial \sigma_{yy}}{\partial y}+\frac{\partial \sigma_{yz}}{\partial z} \qquad (2.67b)$$

$$\rho\left(\frac{\partial w}{\partial t}+u\frac{\partial w}{\partial x}+v\frac{\partial w}{\partial y}+w\frac{\partial w}{\partial z}\right)=\rho g_z-\frac{\partial p}{\partial z}+\frac{\partial \sigma_{zx}}{\partial x}+\frac{\partial \sigma_{zy}}{\partial y}+\frac{\partial \sigma_{zz}}{\partial z} \qquad (2.67c)$$

where g_x, g_y, g_z are the components of the acceleration $\mathbf{g}$ due to gravity, the body force per unit volume being given by $\rho\mathbf{g}$.

The only approximation made to derive Eqs. (2.66) and (2.67) is the *continuum model*. In other words, we ignore the fact that matter consists of myriad molecules and treat it as continuous. Although we have made use of the incompressible form of the continuity Eq. (2.46) to simplify Eqs. (2.58a,b), Eqs. (2.62) and (2.63) apply equally well to compressible flow. In order to show this to be true, it is necessary to allow density to vary in the derivation of term (i) and to simplify it using the compressible form of the continuity Eq. (2.45).

2.6.1 Euler Equations

For some applications in aerodynamics it can be an acceptable approximation to neglect the viscous stresses. In this case Eqs. (2.66a,b) simplify to

$$\rho\left(\frac{\partial u}{\partial t}+u\frac{\partial u}{\partial x}+v\frac{\partial u}{\partial y}\right)=\rho g_x-\frac{\partial p}{\partial x} \qquad (2.68a)$$

$$\rho\left(\frac{\partial v}{\partial t}+u\frac{\partial v}{\partial x}+v\frac{\partial v}{\partial y}\right)=\rho g_y-\frac{\partial p}{\partial y} \qquad (2.68b)$$

These are known as the *Euler* equations. In principle, Eqs. (2.68a,b), together with the continuity Eq. (2.46), can be solved to give the velocity components u and v and pressure p. However, in general, this is difficult because Eqs. (2.68a,b) can be regarded as the governing equations for u and v, but p does not appear explicitly in the continuity equation. Except for special cases, solution of the Euler equations can only be achieved numerically using a computer. A very special and comparatively simple case is *irrotational* flow (see Section 2.7.6). For this case the Euler equations reduce to a single simpler equation, the *Laplace* equation. This equation is much more amenable to analytical solution and is the subject of Chapter 3.

2.7 RATES OF STRAIN, ROTATIONAL FLOW, AND VORTICITY

As they stand, the momentum Eqs. (2.66) and (2.67), together with the continuity Eqs. (2.46) or (2.47) cannot be solved, even in principle, for flow velocity and pressure. Before this can become possible it is necessary to link the viscous stresses to the velocity field through a *constitutive* equation. Air and all other homogeneous gases and liquid are closely approximated by the *Newtonian* fluid model. This means that the viscous stress is proportional to the rate of strain. In a moment, we consider the distortion experienced by an infinitesimal fluid element as it travels through the flow field. In this way we can derive the rate of strain in terms of velocity gradients. The important flow properties, vorticity and circulation, will also emerge as part of this process.

2.7.1 Distortion of Fluid Element in Flow Field

Figure 2.24 shows how a fluid element is transformed as it moves through a flow field. In general the transformation comprises the following operations:

1. *Translation*: movement from one position to another.

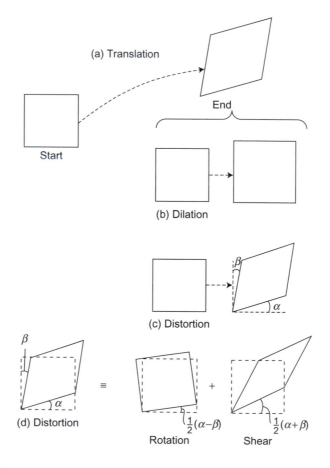

FIGURE 2.24

Transformation of a fluid element as it moves through the flow field.

2. *Dilation/Compression*: the shape remains invariant, but volume reduces or increases. For *incompressible* flow the volume remains invariant from one position to another.
3. *Distortion*: change of shape, keeping the volume invariant. Distortion can be decomposed into counterclockwise *rotation* through angle $(\alpha - \beta)/2$ and a *shear* of angle $(\alpha + \beta)/2$.

The angles α and β are the shear strains.

2.7.2 **Rate of Shear Strain**

Consider Fig. 2.25. This shows an elemental control volume $ABCD$ that initially at time $t = t_i$ is square. After an interval of time δt has elapsed, $ABCD$ has moved and distorted into $A'B'C'D'$. The velocities at $t = t_i$ at A, B, and C are given by

$$u_A = u - \frac{\partial u}{\partial x}\frac{\delta x}{2} - \frac{\partial u}{\partial y}\frac{\delta y}{2}, \quad v_A = v - \frac{\partial v}{\partial x}\frac{\delta x}{2} - \frac{\partial v}{\partial y}\frac{\delta y}{2} \qquad (2.69a)$$

$$u_B = u - \frac{\partial u}{\partial x}\frac{\delta x}{2} + \frac{\partial u}{\partial y}\frac{\delta y}{2}, \quad v_B = v - \frac{\partial v}{\partial x}\frac{\delta x}{2} + \frac{\partial v}{\partial y}\frac{\delta y}{2} \qquad (2.69b)$$

$$u_C = u + \frac{\partial u}{\partial x}\frac{\delta x}{2} - \frac{\partial u}{\partial y}\frac{\delta y}{2}, \quad v_C = v + \frac{\partial v}{\partial x}\frac{\delta x}{2} - \frac{\partial v}{\partial y}\frac{\delta y}{2} \qquad (2.69c)$$

$$x_{A'} = u_A \delta t, \quad y_{A'} = v_A \delta t, \text{ etc.} \qquad (2.70)$$

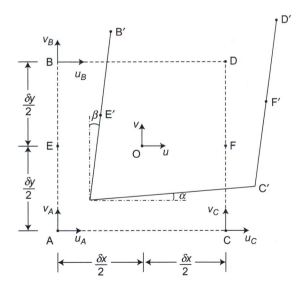

FIGURE 2.25

Distortion of an elemental control volume.

Therefore, if we neglect the higher-order terms,

$$
\alpha = \frac{y_{C'} - y_{A'}}{\delta x} = (v_C - v_A)\frac{\delta t}{\delta x} = \left\{ v + \frac{\partial v}{\partial x}\frac{\delta x}{2} - \frac{\partial v}{\partial y}\frac{\delta y}{2} \right.
$$

$$
\left. - \left(v - \frac{\partial v}{\partial x}\frac{\delta x}{2} - \frac{\partial v}{\partial y}\frac{\delta y}{2} \right) \right\}\frac{\delta t}{\delta x} = \frac{\partial v}{\partial x}\delta t \quad (2.71a)
$$

$$
\beta = \frac{x_{B'} - x_{A'}}{\delta y} = (u_B - u_A)\frac{\delta t}{\delta y} = \left\{ u - \frac{\partial u}{\partial x}\frac{\delta x}{2} + \frac{\partial u}{\partial y}\frac{\delta y}{2} \right.
$$

$$
\left. - \left(u - \frac{\partial u}{\partial x}\frac{\delta x}{2} - \frac{\partial u}{\partial y}\frac{\delta y}{2} \right) \right\}\frac{\delta t}{\delta y} = \frac{\partial u}{\partial y}\delta t \quad (2.71b)
$$

The rate of shear strain in the xy plane is given by

$$
\frac{d\gamma_{xy}}{dt} = \frac{d}{dt}\left(\frac{\alpha + \beta}{2} \right) = \left(\frac{\partial v}{\partial x}\delta t + \frac{\partial u}{\partial y}\delta t \right)\frac{1}{2\delta t} = \frac{1}{2}\left(\frac{\partial v}{\partial x} + \frac{\partial u}{\partial y} \right) \quad (2.72a)
$$

In much the same way, for three-dimensional flows it can be shown that there are two other components of the rate of shear strain:

$$
\frac{d\gamma_{xz}}{dt} = \frac{1}{2}\left(\frac{\partial w}{\partial x} + \frac{\partial u}{\partial z} \right), \quad \frac{d\gamma_{yz}}{dt} = \frac{1}{2}\left(\frac{\partial v}{\partial z} + \frac{\partial w}{\partial y} \right) \quad (2.72\,b,c)
$$

2.7.3 Rate of Direct Strain

Following an analogous process, we can also calculate the direct strains and their corresponding rates of strain. For example,

$$
\varepsilon_{xx} = \frac{x_{F'} - x_{E'}}{x_F - x_E} = \frac{(u_{F'} - u_{E'})\delta t}{\delta x} = \left\{ u + \frac{\partial u}{\partial x}\frac{\delta x}{2} - \left(u - \frac{\partial u}{\partial x}\frac{\delta x}{2} \right) \right\}\frac{\delta t}{\delta x} = \frac{\partial u}{\partial x}\delta t \quad (2.73)
$$

The other direct strains are obtained in a similar way; thus the rates of direct strain are given by

$$
\frac{d\varepsilon_{xx}}{dt} = \frac{\partial u}{\partial x}, \quad \frac{d\varepsilon_{yy}}{dt} = \frac{\partial v}{\partial y}, \quad \frac{d\varepsilon_{zz}}{dt} = \frac{\partial w}{\partial z} \quad (2.74a,b,c)
$$

In this way we can introduce a rate of strain tensor analogous to the stress tensor (see Section 2.6) and for which components in two-dimensional flow can be represented in matrix form as follows:

$$\begin{pmatrix} \dot{\varepsilon}_{xx} & \dot{\gamma}_{xy} \\ \dot{\gamma}_{yx} & \dot{\varepsilon}_{yy} \end{pmatrix} \tag{2.75}$$

where $(\dot{\gamma})$ is used to denote a time derivative.

2.7.4 Vorticity

The instantaneous rate of rotation of a fluid element is given by $(\dot{\alpha} - \dot{\beta})/2$ as was just shown. This corresponds to a fundamental property of fluid flow called *vorticity* that, using Eq. (2.71), in two-dimensional flow is defined as

$$\zeta = \frac{d\alpha}{dt} - \frac{d\beta}{dt} = \frac{\partial v}{\partial x} - \frac{\partial u}{\partial y} \tag{2.76}$$

In three-dimensional flow vorticity is a vector given by

$$\boldsymbol{\Omega} = (\xi, \eta, \zeta) = \left(\frac{\partial w}{\partial y} - \frac{\partial v}{\partial z}, \frac{\partial u}{\partial z} - \frac{\partial w}{\partial x}, \frac{\partial v}{\partial x} - \frac{\partial u}{\partial y} \right) \tag{2.77a,b,c}$$

It can be seen that the three components of vorticity are twice the instantaneous rates of rotation of the fluid element about the three coordinate axes. Mathematically vorticity is given by the following vector operation:

$$\boldsymbol{\Omega} = \nabla \times \mathbf{v} \tag{2.78}$$

Vortex lines can be defined analogously to streamlines as tangential to the vorticity vector at all points in the flow field. Similarly the concept of *vortex tube* is analogous to that of stream tube. Physically we can think of flow structures like vortices as comprising bundles of vortex tubes. In many respects vorticity and vortex lines are even more fundamental to understanding the flow physics than are velocity and streamlines.

2.7.5 Vorticity in Polar Coordinates

Referring to Section 2.4.3, where polar coordinates were introduced, the corresponding definition of vorticity in polar coordinates is

$$\zeta = \frac{q_t}{r} + \frac{\partial q_t}{\partial r} - \frac{1}{r} \frac{\partial q_n}{\partial \theta} \tag{2.79}$$

Note that, consistent with vorticity's physical interpretation as rate of rotation, its units are radians per second.

2.7.6 Rotational and Irrotational Flow

It will be made clear in Section 2.8 that the generation of shear strain in a fluid element, as it travels through the flow field, is closely linked with the effects of viscosity. It is also plain from its definition in Eq. (2.76) that vorticity is related to rate of shear strain. Thus, in aerodynamics vorticity is associated with the effects of viscosity.[3] Accordingly, when the effects of viscosity can be neglected, vorticity is usually zero. This means that the individual fluid elements do not rotate, or distort, as they move through the flow field. For incompressible flow, then, this corresponds to the state of pure translation that is illustrated in Fig. 2.26. Such a flow is termed *irrotational*. Mathematically, it is characterized by the existence of a velocity potential and is therefore also called *potential* flow. This is the subject of Chapter 3. The converse of irrotational flow is *rotational* flow.

2.7.7 Circulation

The total amount of vorticity passing through any plane region within a flow field is called *circulation*, Γ. This is illustrated in Fig. 2.27, which shows a bundle of vortex tubes passing through a plane region of area A located in the flow field. The perimeter of the region is denoted C. At a typical point P on the perimeter, the velocity vector is designated $\mathbf{q}$ or, equivalently, $\vec{q}$. At P, the infinitesimal portion of C has length δs and points in the tangential direction defined by the unit vector $\mathbf{t}$ (or $\vec{t}$). It is important to understand that the region of area A and its perimeter C have no physical existence. Like the control volumes used for the conservation of mass and the momentum equation, they are purely theoretical constructs.

Mathematically the total strength of the vortex tubes can be expressed as an integral over the area A; thus

$$\Gamma = \iint\limits_A \mathbf{n} \cdot \boldsymbol{\Omega}\,dA \tag{2.80}$$

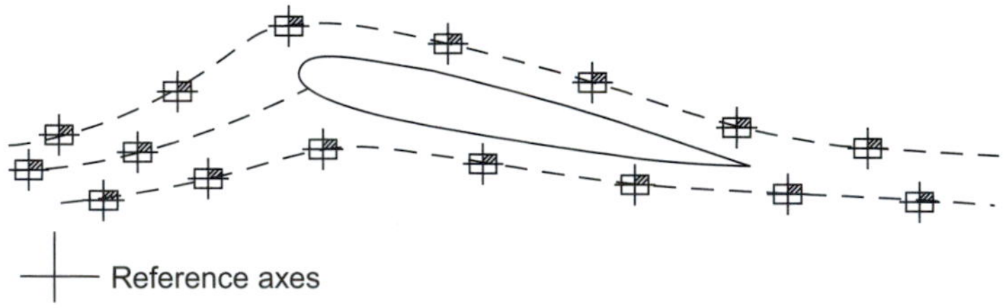

FIGURE 2.26

Irrotational flow of elemental fluid particles.

---

[3] Vorticity can also be created by other agencies, such as spatially varying body forces in the flow field. This could correspond to the presence of particles in the flow field, for example.

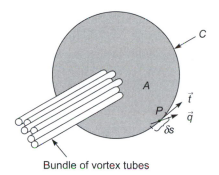

Bundle of vortex tubes

FIGURE 2.27

Bundling of vortex tubes.

where **n** is the unit normal to the area A. In two-dimensional flow the vorticity is in the z direction perpendicular to the two-dimensional flow field in the (x, y) plane. Thus $\mathbf{n} = \mathbf{k}$ (i.e., the unit vector in the z direction) and $\mathbf{\Omega} = \zeta\mathbf{k}$, so that Eq. (2.80) simplifies to

$$\Gamma = \iint_A \zeta\, dA \tag{2.81}$$

Circulation can be regarded as a measure of the combined strength of the total number of vortex lines passing through A. It is a measure of the *vorticity flux* carried through A by these vortex lines. The relationship between circulation and vorticity is broadly similar to that between momentum and velocity or that between internal energy and temperature. Thus circulation is the property of the region A bounded by control surface C, whereas vorticity is a flow variable, like velocity, defined at a point. Strictly, it makes no more sense to speak of conservation, generation, or transport of vorticity than its does to speak of conservation, generation, or transport of velocity. These terms should logically be applied to circulation just as they are to momentum rather than velocity. However, human affairs frequently defy logic, and aerodynamics is no exception. We have become used to speaking about vorticity in terms of conservation and the like. It would be considered pedantic to insist on circulation in this context, even though this would be strictly correct. Our only reason for making such fine distinctions here is to elucidate the meaning and significance of circulation. Henceforth we will adhere to the common usage of the terms *vorticity* and *circulation*.

In two-dimensional flow, in the absence of the effects of viscosity, circulation is conserved. This can be expressed mathematically as follows:

$$\frac{\partial \zeta}{\partial t} + u\frac{\partial \zeta}{\partial x} + v\frac{\partial \zeta}{\partial y} = 0 \tag{2.82}$$

In view of what was written in Section 2.7.6 about the link between vorticity and viscous effects, it may seem somewhat illogical to neglect such effects in Eq. (2.82). Nevertheless, this equation often provides a useful approximation.

Circulation can also be evaluated by means of an integration around the perimeter C. This can be shown elegantly by applying *Stokes's theorem* to Eq. (2.81):

$$\Gamma = \iint_A \mathbf{n} \cdot \boldsymbol{\Omega} \, dA = \iint_A \mathbf{n} \cdot \nabla \times \mathbf{q} \, dA = \oint_C \mathbf{q} \cdot \mathbf{t} \, ds \qquad (2.83)$$

This commonly serves as the definition of circulation in most aerodynamics texts.

The concept of circulation is central to the theory of lift. This will become clear in Chapters 4 and 5.

Example 2.2

For the rectangular region of a two-dimensional flow field depicted in Fig. 2.28, starting with the definition of circulation given in Eq. (2.81), show that it can also be evaluated by means of the integral around the closed circuit appearing as the last term in Eq. (2.83).

From Eqs. (2.76) and (2.81) it follows that

$$\Gamma = \int_{y_1}^{y_2} \int_{x_1}^{x_2} \left(\frac{\partial v}{\partial x} - \frac{\partial u}{\partial y} \right) dx \, dy = \int_{y_1}^{y_2} \underbrace{\int_{x_1}^{x_2} \frac{\partial v}{\partial x} \, dx}_{v(x_2,y) - v(x_1,y)} \, dy - \int_{x_1}^{x_2} \underbrace{\int_{y_1}^{y_2} \frac{\partial u}{\partial y} \, dy}_{u(x,y_2) - u(x,y_1)} \, dx$$

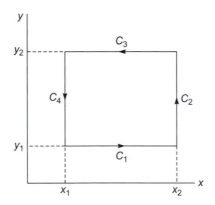

FIGURE 2.28

Contour for the circulation integral.

Therefore

$$\Gamma = \int_{y_1}^{y_2} v(x_2, y)dy - \int_{y_1}^{y_2} v(x_1, y) - \int_{x_1}^{x_2} u(x, y_2)dx + \int_{x_1}^{x_2} u(x, y_1)dx$$

$$= \int_{x_1}^{x_2} u(x, y_1)dx + \int_{y_1}^{y_2} v(x_2, y) + \int_{x2}^{x_1} u(x, y_2)dx + \int_{y_2}^{y_1} v(x_1, y)$$

(2.84)

But along the lines: C_1, $\mathbf{q} = u\mathbf{i}$, $\mathbf{t} = \mathbf{i}$, $ds = dx$; C_2, $\mathbf{q} = v\mathbf{j}$, $\mathbf{t} = \mathbf{j}$, $ds = dy$; C_3, $\mathbf{q} = u\mathbf{i}$, $\mathbf{t} = -\mathbf{i}$, $ds = -dx$; and C_4, $\mathbf{q} = u\mathbf{j}$, $\mathbf{t} = -\mathbf{j}$, $ds = -dy$. It therefore follows that Eq. (2.84) is equivalent to

$$\Gamma = \oint_C \mathbf{q} \cdot \mathbf{t} ds$$

2.8 NAVIER-STOKES EQUATIONS

In this section the Navier-Stokes equations are presented. They are the equations that describe the motion of "real" viscous Newtonian fluids.

2.8.1 Relationship between Rates of Strain and Viscous Stresses

In solid mechanics the fundamental theoretical model linking the stress and strain fields is Hooke's law, which states that

$$\text{Stress} \propto \text{Strain} \tag{2.85}$$

The equivalent in fluid mechanics is the model of the *Newtonian fluid*, for which it is assumed that

$$\text{Stress} \propto \text{Rate of strain} \tag{2.86}$$

However, there is a major difference in status between the two models. At best Hooke's law is a reasonable approximation for describing small deformations of some solids, particularly structural steel, whereas Newtonian fluid is a very accurate model for the behavior of almost all homogeneous fluids, in particular water and air. It does not give good results for pseudo-fluids formed from suspensions of particles in homogeneous fluids (e.g., blood, toothpaste, slurries). Various models are required to describe such non-Newtonian fluids, which are of little interest in aerodynamics and will be considered no further here.

For two-dimensional flows, the constitutive law (2.86) can be written

$$\begin{pmatrix} \sigma_{xx} & \sigma_{xy} \\ \sigma_{yx} & \sigma_{yy} \end{pmatrix} = 2\mu \begin{pmatrix} \dot{\varepsilon}_{xx} & \dot{\gamma}_{xy} \\ \dot{\gamma}_{yx} & \dot{\varepsilon}_{yy} \end{pmatrix} \tag{2.87}$$

where ($\cdot$) denotes time derivatives. The factor 2 is merely used for convenience so as to cancel out the factor 1/2 in the expression (2.72a) for the rate of shear strain. Equation (2.87) is sufficient in the case of an incompressible fluid. For a compressible fluid, however, we should also allow for the possibility of direct stress being generated by the rate of change in volume, or *dilation*. Thus we need to add the following to the right-hand side of (2.87):

$$\lambda \begin{pmatrix} \dot{\varepsilon}_{xx} + \dot{\varepsilon}_{yy} & 0 \\ 0 & \dot{\varepsilon}_{xx} + \dot{\varepsilon}_{yy} \end{pmatrix} \tag{2.88}$$

μ and λ are called the first and second coefficients of viscosity. More frequently μ is just termed the *dynamic viscosity* in contrast to the *kinematic viscosity* $v \equiv \mu/\rho$. If it is required that the actual pressure $p - \frac{1}{3}(\sigma_{xx} + \sigma_{yy}) + \sigma_{zz}$ in a viscous fluid be identical to the thermodynamic pressure p, then it is easy to show that

$$3\lambda + 2\mu = 0 \quad \text{or} \quad \lambda = -\frac{2}{3}\mu$$

This is often called the *Stokes hypothesis*. In effect, it assumes that the bulk viscosity μ' linking the average viscous direct stress to the rate of volumetric strain is zero, i.e.,

$$\mu' = \lambda + \frac{2}{3}\mu \approx 0 \tag{2.89}$$

This is still a rather controversial question. Bulk viscosity is of no importance in the great majority of engineering applications, but can be important for describing the propagation of sound waves in liquids and sometimes in gases. Here, for the most part, we will assume incompressible flow, so that

$$\dot{\varepsilon}_{xx} + \dot{\varepsilon}_{yy} = \frac{\partial u}{\partial x} + \frac{\partial v}{\partial y} = 0$$

and Eq. (2.87) will, accordingly, be valid.

AERODYNAMICS AROUND US

Observing Air Flows

So many features of the flows we study in this text are visible to you, in spite of how transparent air is. That air is transparent may be best understood by observing stars at night—if you view a star at the horizon you are peering through approximately 400 miles of air. Yet there are clues that provide insight into how this transparent material is moving around us.

Clouds, of course, indicate the direction and speed of the air at that altitude. When the winds aloft differ in direction from the wind at ground level, you should see this as a skewed velocity profile in a very thick boundary layer. That is, the direction of the wind differs from its direction near the bottom of the boundary layer (your position) up to the altitude of the clouds. This same change in direction, and likely in magnitude as well occurs in boundary layers near the tips on straight wings and in boundary layers all over swept wings. Picture in your mind a series of velocity vectors stacked up from you to the cloud, with tails on a vertical line, the length of the vectors increasing with height, and the arrowheads slowly pointing in different directions from wind direction where you are to the direction at the clouds.

The authors are fortunate to teach where snow falls. When a day brings the large snowflakes of children's storybooks, almost any wind at all carries (convects) these flakes with it, and by watching the paths of the snowflakes one sees the pathlines in an unsteady air flow. If you find a fairly steady flow, you can see streamlines. One flow feature that is easily visible on snowy days is flow separation, which is the name given to the condition where streamlines that are following the shape of a solid surface depart from the surface. Flow separation is present to a greater or lesser extent on all components of vehicles that move through air or water. Snow blowing from west to east across the flat roof of a building continues west to east when the edge of the roof is reached. The flow does not turn the corner and flow down the wall to the ground (cross-stream pressure gradients are required to turn a flow, as we will see in Chapter 4). Then a vortex forms and is visible in the snow because of the "shear layer" that the air encounters after separating from the building. Suppose too that your face feels chilled, so you stand in a corner to avoid the wind. Here the air speed is slower and the snowflakes move slower. In wind tunnel research we would say that the snowflakes are "markers" for the flow that we use for "flow visualization."

Water can help as well; paddling a canoe will show you flow separation vortices behind on both sides of the paddle. A starting vortex created by the acceleration of a lifting surface is easily viewed in calm water with a canoe paddle or even with your hand if you sit by a calm pool. Though the descriptions are too long to go into here, note that there are times when shallow water flow in fountains can be seen to be governed by partial differential equations that are parabolic, like boundary-layer flows, or hyperbolic, like supersonic flows. Put your smallest finger into the flow and see what the response is; does the perturbation you make in the flow affect only the downstream flow, or does it move (propagate) upstream? Students who have raced small sailboats know additional clues to air motion. "Roughened" patches of the lake's surface moving downwind are higher-wind regions because the greater air speed over the water causes more or larger ripples in the surface than slow air does. As sailors know, a high pressure center nearby causes clockwise wind shifts and a low pressure center causes counterclockwise shifts. Even fluid forces can be sensed nicely on a windsurfer; to travel in a straight line you must balance the aerodynamic force created by the sail with the center of mass of the system, which you do by tipping the rig fore and aft. In a sailboat you can also trim the craft to have zero load on your tiller (neutral helm) by adjusting the rake of the mast and the centerboard.

2.8.2 Derivation of the Navier-Stokes Equations

Restricting our derivation to two-dimensional flow, Eq. (2.87), with Eqs. (2.72a) and (2.73), gives

$$\sigma_{xx} = 2\mu\frac{\partial u}{\partial x}, \quad \sigma_{yy} = 2\mu\frac{\partial v}{\partial y}, \quad \sigma_{xy} = \sigma_{yx} = \mu\left(\frac{\partial u}{\partial y} + \frac{\partial v}{\partial x}\right) \qquad (2.90)$$

Thus the right-hand side of the momentum Eq. (2.66a) becomes

$$gx - \frac{\partial p}{\partial x} + 2\mu\frac{\partial}{\partial x}\left(\frac{\partial u}{\partial x}\right) + \mu\left(\frac{\partial u}{\partial y} + \frac{\partial v}{\partial x}\right)$$

$$= gx - \frac{\partial p}{\partial x} + \mu\left(\frac{\partial^2 u}{\partial x^2} + \frac{\partial^2 u}{\partial y^2}\right) + \mu\frac{\partial}{\partial x}\underbrace{\left(\frac{\partial u}{\partial x} + \frac{\partial v}{\partial y}\right)}_{=0,\ \text{Eq. (2.46)}} \tag{2.91}$$

The right-hand side of Eq. (2.66b) can be dealt with in a similar way, so the momentum Eqs. (2.66a,b) can be written in the form

$$\rho\left(\frac{\partial u}{\partial t} + u\frac{\partial u}{\partial x} + v\frac{\partial u}{\partial y}\right) = \rho g_x - \frac{\partial p}{\partial x} + \mu\left(\frac{\partial^2 u}{\partial x^2} + \frac{\partial^2 u}{\partial y^2}\right) \tag{2.92a}$$

$$\rho\left(\frac{\partial v}{\partial t} + u\frac{\partial v}{\partial x} + v\frac{\partial v}{\partial y}\right) = \rho g_y - \frac{\partial p}{\partial y} + \mu\left(\frac{\partial^2 v}{\partial x^2} + \frac{\partial^2 v}{\partial y^2}\right) \tag{2.92b}$$

This form is known as the *Navier-Stokes equations* for two-dimensional flow. With the inclusion of the continuity equation

$$\frac{\partial u}{\partial x} + \frac{\partial v}{\partial y} = 0 \tag{2.93}$$

we now have three governing equations for three unknown flow variables u, v, p.

The Navier-Stokes equations for three-dimensional incompressible flows are given here:

$$\frac{\partial u}{\partial x} + \frac{\partial v}{\partial y} + \frac{\partial w}{\partial z} = 0 \tag{2.94}$$

$$\rho\left(\frac{\partial u}{\partial t} + u\frac{\partial u}{\partial x} + v\frac{\partial u}{\partial y} + w\frac{\partial u}{\partial z}\right) = \rho g_x - \frac{\partial p}{\partial x} + \mu\left(\frac{\partial^2 u}{\partial x^2} + \frac{\partial^2 u}{\partial y^2} + \frac{\partial^2 u}{\partial z^2}\right) \tag{2.95a}$$

$$\rho\left(\frac{\partial v}{\partial t} + u\frac{\partial v}{\partial x} + v\frac{\partial v}{\partial y} + w\frac{\partial v}{\partial z}\right) = \rho g_y - \frac{\partial p}{\partial y} + \mu\left(\frac{\partial^2 v}{\partial x^2} + \frac{\partial^2 v}{\partial y^2} + \frac{\partial^2 v}{\partial z^2}\right) \tag{2.95b}$$

$$\rho\left(\frac{\partial w}{\partial t} + u\frac{\partial w}{\partial x} + v\frac{\partial w}{\partial y} + w\frac{\partial w}{\partial z}\right) = \rho g_z - \frac{\partial p}{\partial z} + \mu\left(\frac{\partial^2 w}{\partial x^2} + \frac{\partial^2 w}{\partial y^2} + \frac{\partial^2 w}{\partial z^2}\right) \tag{2.95c}$$

2.9 PROPERTIES OF THE NAVIER-STOKES EQUATIONS

At first sight the Navier-Stokes equations, especially the three-dimensional version, Eqs. (2.95), may appear rather formidable. It is important to recall that they are nothing more than the application of Newton's second law of motion to fluid flow. For example, the left-hand side of Eq. (2.95a) represents the total rate of change of the x component of momentum per unit volume. Indeed it is often written as

$$\rho \frac{Du}{Dt} \quad \text{where} \quad \frac{D}{Dt} \equiv \frac{\partial}{\partial t} + u\frac{\partial}{\partial x} + v\frac{\partial}{\partial y} + w\frac{\partial}{\partial z} \tag{2.96}$$

is called the *total* or *material* derivative. It represents the total rate of change with time following the fluid motion. The left-hand sides of Eqs. (2.95b,c) can be written in a similar form. The three terms on the right-hand side represent the x components of body force, pressure force, and viscous force, respectively, acting on a unit volume of fluid.

The compressible versions of the Navier-Stokes equations plus the continuity equation encompass almost the whole of aerodynamics, although, to be sure, applications involving combustion or rarified flow require additional chemical and physical principles. Why, then, do we need the rest of the book, not to mention the remaining vast, ever growing literature devoted to aerodynamics? Given the power of modern computers, could we not merely solve the Navier-Stokes equations numerically for any aerodynamics application of interest? The short answer is no! Moreover, there is no prospect of this ever being possible. To explain in full detail is rather difficult in an introductory course. We will nevertheless attempt to give a brief idea of the nature of the problem.

Let us begin by noting that the Navier-Stokes equations are a set of partial differential equations. Few analytical solutions exist that are useful in aerodynamics. (The most useful examples will be described in Section 2.10.) Accordingly, it is essential to seek approximate solutions. Nowadays it is often possible to obtain very accurate numerical solutions with computers. In many respects these can be regarded almost as exact, although one must never forget that computer-generated solutions are subject to error. However, for most practical problems of interest to the aeronautical engineer it is not possible to obtain numerical solutions to the Navier-Stokes equations because the spatial and temporal scales of motion that must be resolved in a highly unsteady turbulent flow cannot be obtained with the computer technology available today and/or the foreseeable future. Here are two main sources of difficulty. First, the equations are nonlinear. The nonlinearity arises from the left-hand sides—the terms representing the rate of change of momentum or the so-called *inertial* terms. To appreciate why these terms are nonlinear, simply note that if you take a term on the right-hand side of the equations (e.g., the pressure terms), when the flow variable (e.g., pressure) is doubled the term is also doubled. This is also true for the viscous terms. Thus these terms are proportional to the unknown flow variables; that is, they are linear. Now consider a typical inertial term, say $u\partial u/\partial x$, which is plainly proportional to u^2 and not u and is therefore nonlinear.

The second source of difficulty is more subtle. It involves the complex effects of viscosity. In order to understand it, it is necessary to make the Navier-Stokes equations nondimensional. The motivation for working with nondimensional variables and equations is that it makes the theory scale-invariant and accordingly more universal (see Section 1.4). In order to fix ideas, let us consider the air flowing at speed U_∞ toward a body, a circular cylinder or wing, say, of length L (see Fig. 2.29). The space variables x, y, and z can be made nondimensional by dividing by L. L/U_∞ can be used as the reference time to make time nondimensional. Thus we introduce the nondimensional coordinates

$$X = x/L, \quad Y = y/L, \quad Z = z/L, \quad \text{and} \quad T = tU/L \tag{2.97}$$

U_∞ can be used as the reference flow speed to make the velocity components dimensionless, and ρU_∞^2 (compare Bernoulli's Eq. (2.16)) can be used as the reference pressure. (For incompressible flow, at least, only pressure difference is of significance and not the absolute value of the pressure.) This allows us to introduce the following nondimensional flow variables:

$$U = u/U_\infty, \quad V = v/U_\infty, \quad W = w/U_\infty, \quad \text{and} \quad P = p/\left(\rho U_\infty^2\right) \tag{2.98}$$

If, by writing $x = XL$, and so forth, the nondimensional variables given in Eqs. (2.97) and (2.98) are substituted into Eqs. (2.94) and (2.95) with the body-force terms omitted, we obtain the Navier-Stokes equations in the form

$$\frac{\partial U}{\partial X} + \frac{\partial V}{\partial Y} + \frac{\partial W}{\partial Z} = 0 \tag{2.99}$$

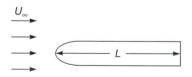

FIGURE 2.29

Uniform flow approaching a body.

$$\frac{DU}{DT} = -\frac{\partial P}{\partial X} + \frac{1}{Re}\left(\frac{\partial^2 U}{\partial X^2} + \frac{\partial^2 U}{\partial Y^2} + \frac{\partial^2 U}{\partial Z^2}\right) \tag{2.100a}$$

$$\frac{DV}{DT} = -\frac{\partial P}{\partial Y} + \frac{1}{Re}\left(\frac{\partial^2 V}{\partial X^2} + \frac{\partial^2 V}{\partial Y^2} + \frac{\partial^2 V}{\partial Z^2}\right) \tag{2.100b}$$

$$\frac{DW}{DT} = -\frac{\partial P}{\partial Z} + \frac{1}{Re}\left(\frac{\partial^2 W}{\partial X^2} + \frac{\partial^2 W}{\partial Y^2} + \frac{\partial^2 W}{\partial Z^2}\right) \tag{2.100c}$$

where the shorthand notation (2.96) for the material derivative has been used. A feature of Eqs. (2.100) is the appearance of the dimensionless quantity known as the *Reynolds number*:

$$Re \equiv \frac{\rho U_\infty L}{\mu} \tag{2.101}$$

From the manner in which it has emerged from making the Navier-Stokes equations dimensionless, it is evident that the Reynolds number (see also Section 1.4) represents the ratio of the inertial to the viscous terms (i.e., the ratio of rate of change in momentum to the viscous force). It would be difficult to overstate the significance of the Reynolds number for aerodynamics.

It should now be clear from Eqs. (2.99) and (2.100) that if one were to calculate the nondimensional flow field for a given shape—a circular cylinder, for example—the overall flow pattern obtained would depend on the Reynolds number and, in the case of unsteady flows, on the dimensionless time T. The flow around a circular cylinder is a good example for illustrating just how much the flow pattern can change over a wide range of Reynolds numbers (see Section 8.4 and Fig. 8.13 in particular). Incidentally, the simple dimensional analysis carried out above shows that it is not always necessary to solve equations in order to extract useful information from them.

For high-speed flows where compressibility becomes important, the absolute value of pressure is significant. As explained in Section 2.3.4 (see also Section 1.4), this leads to the appearance of the Mach number M (the ratio of the flow speed to the speed of sound) in the stagnation pressure coefficient. Thus, when compressibility becomes important (see Section 2.3.4), the Mach number becomes a second dimensionless quantity characterizing the flow field.

The Navier-Stokes equations are deceptively simple in form, but at high Reynolds numbers the resulting flow fields can be exceedingly complex even for simple geometries. This is basically a consequence of the behavior of the regions of vortical flow at high Reynolds numbers. Vorticity can be created only in a viscous flow and can be regarded as a marker for regions where the effects of viscosity are important in some sense.

For engineering applications of aerodynamics the Reynolds numbers are very large, with values well in excess of 10^6 commonplace. Accordingly, one would expect that to a good approximation the viscous terms on the right-hand side of the dimensionless Navier-Stokes Eqs. (2.100) could be dropped. In general, however, this view would be mistaken and one never achieves a flow field similar to the inviscid field no matter how high the Reynolds number. The reason is that the regions of nonzero vorticity where viscous effects cannot be neglected become confined to exceedingly thin boundary layers adjacent to the body surface. As $Re \to \infty$, the boundary-layer thickness $\delta \to 0$. If the boundary layers remained attached to the surface they have little effect beyond giving rise to skin-friction drag. But in all real flows the boundary layers separate from the surface of the body, either because of the effects of an adverse pressure gradient or because they reach the rear of the body or its trailing edge. When these thin regions of vortical flow separate, they form complex unsteady vortex-like structures in the wake. These structures take their most extreme form in turbulent flow, which is characterized by vortical structures with a wide range of length and time scales.

As we have seen from the discussion just given, it is not necessary to solve the Navier-Stokes equations in order to obtain useful information from them. This is also illustrated by following example.

Example 2.3

Aerodynamic modeling: Let us suppose that we are carrying out tests on a model in a wind tunnel in order to study and determine the aerodynamic forces exerted on a motor vehicle traveling at normal motorway speeds. In this case the speeds are sufficiently low to ensure that the effects of compressibility are negligible. Thus for a fixed geometry the flow field will be characterized only by Reynolds number.[4] In this case we can use U_∞, the speed at which the vehicle travels (the air speed in the wind-tunnel working section) as the reference flow speed, and L can be the width or length of the vehicle. So the Reynolds number $Re = \rho U_\infty L / \mu$. For a fixed geometry it is clear from Eqs. (2.99) and (2.100) that the nondimensional flow variables, $U, V, W,$ and P are functions only of the dimensionless coordinates X, Y, Z, T and the dimensionless quantity Re. In a steady flow the aerodynamic force, being an overall characteristic of the flow field, will not depend on $X, Y, Z,$ or T. It will, in fact, depend only on Re. Thus if we make an aerodynamic force, drag (D), say, dimensionless by introducing a force (i.e., drag) coefficient defined as

$$C_D = \frac{D}{\frac{1}{2}\rho U_\infty^2 L^2} \tag{2.102}$$

(see Section 1.5.2 and noting that here we have used L^2 in place of area S), it should be clear that

$$C_D = F(Re) \quad \text{(i.e., a function of } Re \text{ only)} \tag{2.103}$$

[4]In fact, this statement is somewhat of an oversimplification. Technically the turbulence characteristics of the oncoming flow also influence the details of the flow field.

If we wish the model tests to produce useful information about general characteristics of the prototype's flow field, particularly estimates for its aerodynamic drag, it is necessary for the model and prototype to be *dynamically similar*—that is, for the forces to be scale-invariant. It can be seen from Eq. (2.103) that this can only be achieved provided

$$Re_m = Re_p \tag{2.104}$$

where subscripts m and p denote model and prototype respectively.

It is not usually practicable to use any other fluid but air for the model tests. In standard wind tunnels the air properties are not greatly different from those experienced by the prototype. Accordingly, Eq. (2.104) implies that

$$U_m = \frac{L_p}{L_m} U_p \tag{2.105}$$

Thus, if we use a 1/5-scale model, Eq. (2.105) implies that $U_m = 5U_p$, so a prototype speed of 100 km/hr (30 m/s) implies a model speed of 500 km/hr (150 m/s). At such a speed compressibility effects are no longer negligible. This illustrative example suggests that, in practice, it is rarely possible to achieve dynamic similarity in aerodynamic model tests using standard wind tunnels. In fact, dynamic similarity can usually be achieved in aerodynamics only by using very large and expensive facilities where the dynamic similarity is achieved by compressing the air (thereby increasing its density) and using large models.

In this example we briefly revisited the material covered in Section 1.4. The objective was to show how dimensional analysis of the Navier-Stokes equations (effectively the exact governing equations of the flow field) can establish more rigorously the concepts introduced in Section 1.4.

2.10 EXACT SOLUTIONS OF THE NAVIER-STOKES EQUATIONS

Few physically realizable exact solutions of the Navier-Stokes equations exist. Even fewer are of much interest in engineering. Here we will present the two simplest solutions, Couette flow (simple shear flow) and plane Poiseuille flow (channel flow). These are useful for engineering applications, although not for the aerodynamics of wings and bodies. The third exact solution represents flow in the vicinity of a stagnation point. This is important for calculating the flow around wings and bodies. It also illustrates a common and, at first sight, puzzling feature, which is that if the dimensionless Navier-Stokes equations can be reduced to an ordinary differential equation, this is regarded as tantamount to an exact solution because the essentials of the flow field can be represented in terms of one or two curves plotted on a single graph. Also, numerical solutions to ordinary differential equations can be obtained to any desired accuracy.

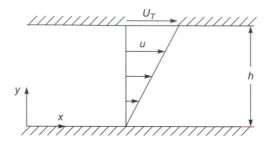

FIGURE 2.30

Velocity profile for the planar Couette flow.

2.10.1 Couette Flow: Simple Shear Flow

This is the simplest exact solution. It corresponds to the flow field created between two infinite, plane, parallel surfaces—the upper one moving tangentially at speed U_T, the lower one being stationary (see Fig. 2.30). Since the flow is steady and two-dimensional, derivatives with respect to z and t are zero, and $w = 0$. The streamlines are parallel to the x-axis, so $v = 0$. Therefore Eq. (2.93) implies $\partial u / \partial x = 0$ (i.e., u is a function only of y). There is no external pressure field, so Eq. (2.92a) reduces to

$$\mu \frac{\partial^2 u}{\partial y^2} = 0 \quad \text{implying} \quad u = C_1 y + C_2 \tag{2.106}$$

where C_1 and C_2 are constants of integration, $u = 0$ and U_T when $y = 0$ and h, respectively. Thus Eq. (2.106) becomes

$$u = U_T \frac{y}{h} = \frac{\tau}{\mu} y \tag{2.107}$$

where τ is the constant viscous shear stress.

   This solution approximates well the flow between two concentric cylinders, with the inner one rotating at fixed speed, provided the clearance is small compared with the cylinder's radius R. This is the basis of a viscometer—an instrument for measuring viscosity—since the torque required to rotate the cylinder at constant speed ω is proportional to τ, which is given by $\mu \omega R / h$. Thus if the torque and rotational speed are measured, the viscosity can be determined.

2.10.2 Plane Poiseuille Flow: Pressure-Driven Channel Flow

This also corresponds to the flow between two infinite, plane, parallel surfaces (see Fig. 2.31). Unlike Couette flow, both surfaces are stationary and flow is produced by the application of pressure. Thus all the arguments used in Section 2.10.1 to simplify the Navier-Stokes equations still hold. The only difference is that the

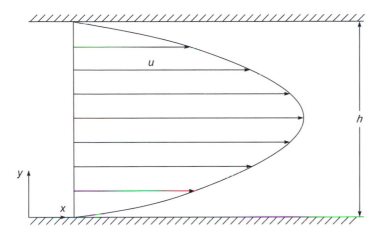

FIGURE 2.31

Velocity profile for the planar Poiseuille flow.

pressure term in Eq. (2.95a) is retained so that it simplifies to

$$-\frac{dp}{dx} + \mu \frac{\partial^2 u}{\partial y^2} = 0, \quad \text{implying} \quad u = \frac{1}{\mu}\frac{dp}{dx}\frac{y^2}{2} + C_1 y + C_2 \qquad (2.108)$$

The no-slip condition implies that $u = 0$ at $y = 0$ and h, so Eq. (2.108) becomes

$$u = -\frac{h^2}{2\mu}\frac{dp}{dx}\frac{y}{h}\left(1 - \frac{y}{h}\right) \qquad (2.109)$$

Thus the velocity profile is parabolic in shape.

The true Poiseuille flow is found in capillaries with round sections. A very similar solution can be found for this case in a similar way to Eq. (2.109) that again has a parabolic velocity profile. From this solution, Poiseuille's law can be derived linking the flow rate through a capillary of diameter d to the pressure gradient:

$$Q = -\frac{\pi d^4}{128\mu}\frac{dp}{dx} \qquad (2.110)$$

Poiseuille was a French physician who derived his law in 1841 in the course of his studies on blood flow. His law is the basis of another type of viscometer by which the flow rate driven through a capillary by a known pressure difference is measured. The value of viscosity can be determined from this measurement using Eq. (2.110).

2.10.3 Hiemenz Flow: Two-Dimensional Stagnation-Point Flow

The simplest example of this type of flow, illustrated in Fig. 2.32, is generated by uniform flow impinging perpendicularly on an infinite plane. The flow divides equally about a stagnation point (strictly, a line). The velocity field for the corresponding inviscid potential flow (see Chapter 3) is

$$u = ax \quad v = -ay, \quad \text{where } a \text{ is a constant} \tag{2.111}$$

The real viscous flow must satisfy the no-slip condition at the wall, as shown in Fig. 2.32, but the potential flow may offer some hints on seeking the full viscous solution.

This special solution is of particular interest for aerodynamics. All two-dimensional stagnation flows behave in a similar way near the stagnation point. This point can therefore be used as the starting solution for boundary-layer calculations in the case of two-dimensional bodies with rounded noses or leading edges (see Example 2.4). There is also an equivalent axisymmetric stagnation flow.

The approach used to find a solution to the two-dimensional Navier-Stokes equations (Eqs. 2.92 and 2.93) is to reduce them to an ordinary differential equation. This is done by assuming that, when appropriately scaled, the nondimensional velocity profile remains the same shape throughout the flow field. Thus the nature of the flow field suggests that the normal velocity component is independent of x, so that

$$v = -f(y) \tag{2.112}$$

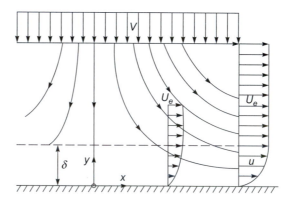

FIGURE 2.32

Stagnation-zone flow field.

where $f(y)$ is a function of y that has to be determined. Substitution of Eq. (2.112) into the continuity Eq. (2.93) gives

$$\frac{\partial u}{\partial x} = f'(y); \quad \text{integrate to get} \quad u = xf'(y) \tag{2.113}$$

where $(\ )'$ denotes differentiation with respect to y. The constant of integration in Eq. (2.113) is equivalent to zero, as $u = v = 0$ at $x = 0$ (the stagnation point), and is therefore omitted.

For a potential flow the Bernoulli equation gives

$$p + \frac{1}{2}\rho \left(\underbrace{u^2 + v^2}_{a^2 x^2 + a^2 y^2} \right) = p_0 \tag{2.114}$$

So for the full viscous solution we will try the form

$$p_0 - p = \frac{1}{2}\rho a^2 [x^2 + F(y)] \tag{2.115}$$

where $F(y)$ is another function of y. If the assumptions (Eqs. 2.112 and 2.115) are incorrect, we will fail in our objective of reducing the Navier-Stokes equations to ordinary differential equations.

We substitute Eqs. (2.112), (2.113), and (2.115) into Eqs. (2.92a, b) to get

$$\underbrace{\rho u \frac{\partial u}{\partial x}}_{\rho x f'^2} + \underbrace{\rho v \frac{\partial u}{\partial y}}_{\rho x f f''} = \underbrace{-\frac{\partial p}{\partial x}}_{-\rho a^2 x} + \mu \left(\underbrace{\frac{\partial^2 u}{\partial x^2}}_{0} + \underbrace{\frac{\partial^2 u}{\partial y^2}}_{\mu x f'''} \right) \tag{2.116}$$

$$\underbrace{\rho u \frac{\partial v}{\partial x}}_{0} + \underbrace{\rho v \frac{\partial v}{\partial y}}_{-\rho f f'} = \underbrace{-\frac{\partial p}{\partial y}}_{-\rho a^2 F'/2} + \mu \left(\underbrace{\frac{\partial^2 v}{\partial x^2}}_{0} + \underbrace{\frac{\partial^2 v}{\partial y^2}}_{\mu f''} \right) \tag{2.117}$$

Simplifying these two equations gives

$$f'^2 - f f'' = a^2 + v f''' \tag{2.118}$$

$$f f' = \frac{1}{2}a^2 F' - v f'' \tag{2.119}$$

where the definition of kinematic viscosity ($v = \mu/\rho$) has been used. Evidently the assumptions made above are acceptable since we have succeeded in reducing the Navier-Stokes equations to ordinary differential equations. Also note that the second equation (Eq. 2.119) is only required to determine the pressure field; Eq. (2.118) on its own can be solved for f, thus determining the velocity field.

The boundary conditions at the wall are straightforward:

$$u = v = 0 \quad \text{at} \quad y = 0, \quad \text{implying} \quad f = f' = 0 \quad \text{at} \quad y = 0 \tag{2.120}$$

As $y \rightarrow \infty$ the velocity will tend to its form in the corresponding potential flow. Thus

$$u \rightarrow ax \quad \text{as} \quad y \rightarrow \infty, \quad \text{implying} \quad f' = a \quad \text{as} \quad y \rightarrow \infty \tag{2.121}$$

In its present form Eq. (2.118) contains both a and v, so f depends on these parameters as well as being a function of y. It is desirable to derive a universal form of Eq. (2.118) so that we only need to solve it once. We attempt to achieve this by scaling the variables $f(y)$ and y:

$$f(y) = \beta \phi(\eta), \quad \eta = \alpha y \tag{2.122}$$

where α and β are constants to be determined by substituting Eq. (2.122) into Eq. (2.118). Noting that

$$f' = \frac{df}{dy} = \frac{d\eta}{dy} \beta \frac{d\phi}{d\eta} = \alpha\beta\phi'$$

Eq. (2.118) thereby becomes

$$\alpha^2 \beta^2 \phi'^2 - \alpha^2 \beta^2 \phi\phi'' = a^2 + v\alpha^3 \beta\phi''' \tag{2.123}$$

thus providing

$$\alpha^2 \beta^2 = a^2 = v\alpha^3 \beta, \quad \text{implying} \quad \alpha = \sqrt{a/v}, \quad \beta = \sqrt{av} \tag{2.124}$$

They can be cancelled as common factors, and Eq. (2.124) reduces to the universal form:

$$\phi''' + \phi\phi'' - \phi'^2 + 1 = 0 \tag{2.125}$$

with boundary conditions

$$\phi(0) = \phi'(0) = 0, \quad \phi'(\infty) = 1$$

In fact, $\varphi' = u/U_e$ where $U_e = ax$, the velocity in the corresponding potential flow found when $\eta \rightarrow \infty$. It is plotted in Fig. 2.33. We can regard the point at which $\varphi' = 0.99$ as marking the edge of the viscous region. This occurs at $\eta \approx 2.4$. This viscous region can be regarded as the boundary layer in the vicinity of the stagnation point (note, though, that no approximation was made to obtain the solution). Its thickness

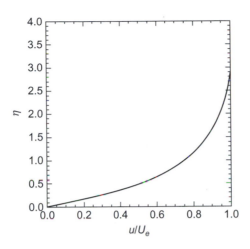

FIGURE 2.33

Hiemenz-flow boundary layer.

does not vary with x and is given by

$$\delta \approx 2.4\sqrt{v/a} \qquad (2.126)$$

Example 2.4

We will estimate the boundary-layer thickness in the stagnation zone of (i) a circular cylinder of 120-mm diameter in a wind tunnel at a flow speed of 20 m/s, and (ii) the leading edge of a Boeing 747 wing with a leading edge radius of 150 mm at a flight speed of 250 m/s.

For a circular cylinder the potential-flow solution for the tangential velocity at the surface is given by $2U_\infty \sin\varphi$ (see Eq. (3.44)). Therefore in case (i) in the stagnation zone, $x = R\sin\phi \approx R\phi$, so the velocity tangential to the cylinder is

$$U_e \approx 2U_\infty \phi = 2\frac{U_\infty}{R}\underbrace{R\phi}_{x}$$

Therefore, as shown in Fig E2.4, if we draw an analogy with the analysis in Section 2.10.3, $a = 2U_\infty/R = 2 \times 20/0.06 = 666.7$ sec^{-1}. Thus from Eq. (2.126), given that for air the kinematic viscosity is $v \approx 15 \times 10^{-6}$ m^2/s,

$$\delta \approx 2.4\sqrt{\frac{v}{a}} = 2.4\sqrt{\frac{15 \times 10^{-6}}{666.7}} = 360\,\mu\text{m}$$

For the aircraft wing in case (ii) we regard the leading edge as analogous locally to a circular cylinder and follow the same procedure as for case (i). Thus $R = 150$ mm $= 0.15$ m

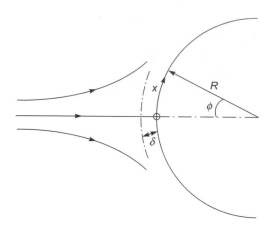

FIGURE E2.4

Stagnation-point flow of Example 2.4.

and $U_\infty = 250$ m/s, so in the stagnation zone $a = 2U_\infty/R = 2 \times 250/0.15 = 3330$ sec^{-1} and

$$\delta \approx 2.4\sqrt{\frac{v}{a}} = 2.4\sqrt{\frac{15 \times 10^{-6}}{3330}} = 160\,\mu m$$

These results underline just how thin the boundary layer is!

2.11 PRANDTL'S BOUNDARY-LAYER EQUATIONS

We will examine the boundary-layer theory in more detail in Chapter 8. However, it is useful at this point to derive Prandtl's boundary-layer equations and to examine the boundary-layer thickness predicted by Blasius for the laminar flow over a flat plate, and by Prandtl for the turbulent boundary layer past a flat plate, to gain insight into the boundary-layer concept put forth by Prandtl in 1904. This section is intended to set the stage for the investigation of potential-flow theory of airfoils and wings in the next two chapters by recognizing two facts of flows at high Reynolds numbers. The first is that viscous effects are confined to thin layers adjacent the boundary of an airfoil; thus the flow outside the boundary layer and wake is essentially an inviscid potential flow. The second is that the real flow leaves the trailing edge of a well-designed airfoil smoothly. In other words, the points of separation depicted in the figure are located at the trailing edge. Hence, in the potential-flow model we apply the insight known as the *Kutta condition*, which states that, although there are many potential-flow solutions of the velocity field for the flow around an airfoil

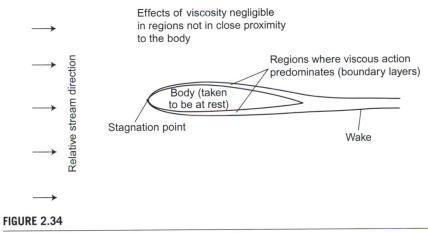

Effects of viscosity negligible in regions not in close proximity to the body

Regions where viscous action predominates (boundary layers)

Body (taken to be at rest)

Relative stream direction

Stagnation point

Wake

FIGURE 2.34

Real fluid flow about an airfoil. *The thickness of the boundary layers and wake is greatly exaggerated.*

(a relatively thin object with a well rounded leading edge and a sharp trailing edge), there is only one that avoids the infinite velocity around the trailing edge. How we construct a solution that satisfies this condition is investigated in Chapters 4 and 5 as part of the vortex theory of airfoils and wings.

To appreciate the concept of the boundary layer, consider the flow of a fluid past a body of reasonably slender form as illustrated in Fig. 2.34. How thick are the viscous boundary layers? In aerodynamics, almost invariably, the fluid viscosity is relatively small (i.e., the Reynolds number is high), so, unless the transverse velocity gradients are appreciable, the shearing stresses developed (given by Newton's equation $\tau = \mu(\partial u/\partial y)$) will be very small. Studies of flows, such as that indicated in Fig. 2.34, show that the transverse velocity gradients are usually negligible throughout the flow field except for thin layers of fluid immediately adjacent to the solid boundaries. Within these boundary layers, however, large shearing velocities are produced with consequent shearing stresses of appreciable magnitude. Consideration of the intermolecular forces between solids and fluids leads to the assumption that at the boundary between a solid and a fluid (other than a rarefied gas) there is a condition of no slip. In other words, the relative velocity of the fluid tangential to the surface is everywhere zero. Since the mainstream velocity at a small distance from the surface may be considerable, it is evident that appreciable shearing velocity gradients may exist within this boundary region.

Prandtl pointed out that these boundary layers are usually very thin, provided that the body is of streamline form, at a moderate angle of incidence to the flow, and that the flow Reynolds number is sufficiently large so that, as a first approximation, the layers might be ignored in order to estimate the pressure field produced about the body. For airfoil shapes, this pressure field is, in fact, only slightly modified by the boundary-layer flow, since almost the entire lifting force is produced by normal pressures at the airfoil surface. It is possible to develop theories for the evaluation of the lift force by consideration of the flow field outside the boundary layers, where the flow is essentially inviscid. Herein lies the importance of the inviscid-flow methods considered in subsequent chapters. We will find in Section 4.1, however, that no drag force, other than induced drag, ever results from the inviscid theories. The drag force is mainly due to shearing stresses at the body surface, and it is in the estimation of these that the study of boundary-layer behavior is essential.

The enormous simplification in the study of the whole problem, which follows from Prandtl's boundary-layer concept, is that the equations of viscous motion need be considered only in the limited regions of the boundary layers, where appreciable simplifying assumptions can reasonably be made. This was the major single impetus to the rapid advance in aerodynamic theory that took place in the first half of the twentieth century. However, in spite of this simplification, the prediction of boundary-layer behavior is by no means simple. Modern methods of computational fluid dynamics provide powerful tools for this; however, they are only accessible to specialists; it still remains essential to study boundary layers in a more fundamental way to gain insight into their behavior and influence on the flow field as a whole.

To begin with, we will consider the general physical behavior of boundary layers. We will then simplify the Navier-Stokes equations to obtain the boundary-layer equations in a similar fashion as did Prandtl more than a century ago. We will examine the boundary layer on a flat plate (or an infintiely thin symmetric airfoil). We will end this section by examining the prediction of the boundary-layer thickness for a turbulent boundary layer due to Prandtl (also discussed in more detail in Chapter 8).

2.11.1 Development of the Boundary Layer

For the flow around a body with a sharp leading edge (like a flat plate parallel to the onset flow), the boundary layer on any surface will grow from zero thickness at the leading edge of the body. For a typical airfoil shape, with a bluff nose, boundary layers will develop on the top and bottom surfaces from the front stagnation point, but will not have zero thickness there (see Section 2.10.3).

On proceeding downstream along a surface, large shearing gradients and stresses will develop adjacent to the surface because of the relatively large velocities in the mainstream and the condition of no slip at the surface. This shearing action is greatest at the body surface and retards the layers of fluid immediately adjacent to it. These layers, since they are now moving more slowly than those above them, will then influence the latter and so retard them. In this way, as the fluid near the surface passes downstream, the retarding action penetrates farther and farther away from the surface and the boundary layer of retarded or "tired" fluid grows in thickness.

Velocity Profile

Further thought about the thickening process will make it evident that the increase in velocity that takes place along a normal to the surface must be continuous. Let y be the perpendicular distance from the surface at any point, and let u be the corresponding velocity parallel to the surface. If u were to increase discontinuously with y at any point, then at that point $\partial u / \partial y$ would be infinite. This would imply an infinite shearing stress (since the shear stress $\tau = \mu(\partial u / \partial y)$), which is obviously untenable.

Consider again a small element of fluid (Fig. 2.35) of unit depth normal to the flow plane, having a unit length in the direction of motion and a thickness δy normal to the flow direction. The shearing stress on the lower face AB will be $\tau = \mu(\partial u / \partial y)$ while that on the upper face CD will be $\tau + (\partial \tau / \partial y)\delta y$, in the directions shown, assuming u to increase with y. Thus the resultant shearing force in the x direction will be $[\tau + (\partial \tau / \partial y)\delta y] - \tau = (\partial \tau / \partial y)\delta y$ (since the area parallel to the x direction is unity), but $\tau = \mu(\partial u / \partial y)$ so that the net shear force on the element equals $\mu(\partial^2 u / \partial y^2)\delta y$. Unless μ is zero, it follows that $\partial^2 u / \partial y^2$ cannot be infinite and therefore the rate of change in the velocity gradient in the boundary layer must also be continuous.

Also shown in Fig. 2.35 are the streamwise pressure forces acting on the fluid element. It can be seen that the net pressure force is $-(\partial p / \partial x)\delta x$. Actually, owing to the very small total thickness of the boundary layer, the pressure hardly varies at all normal to the surface. Consequently, the net transverse pressure force is zero to a very good approximation and Fig. 2.35 shows all the significant fluid forces. The effects of streamwise pressure change are discussed in Section 2.11.6. At this stage it is assumed that $\partial p / \partial x = 0$.

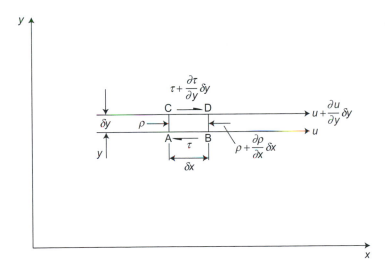

FIGURE 2.35

Elemental fluid particle in a parallel shear flow.

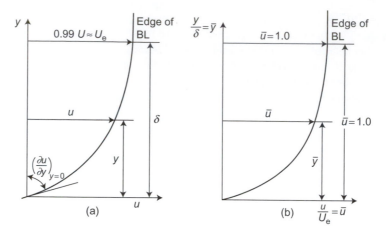

FIGURE 2.36

Illustration of the shape of the boundary layer in (a) dimensional and
(b) dimensionless form.

If the velocity u is plotted against the distance y, it is now clear that a smooth curve of the general form shown in Fig. 2.36(a) must develop. Note that at the surface the curve is not tangential to the u axis as this would imply an infinite gradient $\partial u / \partial y$, and therefore an infinite shearing stress, at the surface. It is also evident that as the shearing gradient decreases, the retarding action decreases, so at some distance from the surface, when $\partial u / \partial y$ becomes very small, the shear stress becomes negligible, although theoretically a small gradient must exist out to $y = \infty$.

2.11.2 Boundary-Layer Thickness

In order to make the idea of a boundary layer realistic, an arbitrary decision must be made as to its extent, and the usual convention is that the boundary layer extends to an unknown distance from the surface such that the velocity u at that distance is 99% of the local mainstream velocity Ue that would exist at the surface in the absence of the boundary layer. Thus δ is the physical thickness of the boundary layer so far as it needs to be considered, and when defined specifically as above it is usually designated the 99%, or general, thickness. Further thickness definitions are given in Section 2.12.2.

2.11.3 Nondimensional Profile

To compare boundary-layer profiles of different thicknesses, it is convenient to express the profile shape nondimensionally. This may be done by writing $\bar{u} = u / U_e$ and $\bar{y} = y / \delta$ so that the profile shape is given by $\bar{u} = f(\bar{y})$. Over the range $y = 0$ to $y = \delta$, the velocity parameter $\bar{u}$ varies from 0 to 0.99. For convenience when using $\bar{u}$ values as integration limits, negligible error is introduced by using $\bar{u} = 1.0$ at the outer boundary and considerable arithmetical simplification is achieved. The velocity profile is then plotted as in Fig. 2.36(b).

2.11.4 Laminar and Turbulent Flows

Closer experimental study of boundary-layer flows discloses that, like flows in pipes, there can be two different regimes: laminar flow and turbulent flow. In *laminar flow* the layers of fluid slide smoothly over one another and there is little interchange of fluid mass between adjacent layers. The shearing tractions that develop because of the velocity gradients are thus due entirely to the viscosity of the fluid. In other words, the momentum exchanges between adjacent layers are on a molecular scale only.

In *turbulent flow* considerable seemingly random motion exists, in the form of velocity fluctuations both along the mean direction of flow and perpendicular to it. As a result of the perpendicular fluctuations, there are appreciable transports of mass between adjacent layers. Owing to these fluctuations the velocity profile varies with time. However, a time-averaged, or mean, velocity profile can be defined. As there is a mean velocity gradient in the flow, there will be corresponding interchanges of streamwise momentum between the adjacent layers that will result in shearing stresses between them. These shearing stresses may well be of much greater magnitude than those that develop as the result of purely viscous action, and the velocity profile shape in a turbulent boundary layer is very largely controlled by these *Reynolds stresses*, as they are termed (more details are provided in Chapter 8).

As a consequence of the essential differences between laminar and turbulent flow shearing stresses, the velocity profiles in the two types of layer are also different. Figure 2.37 shows a typical laminar-layer profile and a typical turbulent-layer profile plotted to the same nondimensional scale. These profiles are typical of those on a flat plate where there is no streamwise pressure gradient.

In the laminar boundary layer, energy from the mainstream is transmitted towards the slower-moving fluid near the surface through viscosity alone, and only a relatively small penetration results. Consequently, an appreciable proportion of the boundary-layer flow has a considerably reduced velocity. Throughout the boundary layer, the shearing stress τ is given by $\tau = \mu(\partial u/\partial y)$ and the wall shearing stress is thus, say, $\tau_w = \mu(\partial u/\partial y)_{y=0} = \mu(\partial u/\partial y)_w$.

In the turbulent boundary layer, as has already been noted, large Reynolds stresses are set up because mass interchanges in a direction perpendicular to the surface, so energy from the mainstream may easily penetrate to fluid layers quite close to the surface. This fact results in the turbulent boundary away from the immediate influence of the wall having a velocity that is not much less than that of the mainstream. However, in layers that are very close to the surface (at this stage of the discussion considered smooth), the velocity fluctuations perpendicular to the wall are evidently damped out; thus in a very limited region immediately adjacent to the surface, the flow is close to purely viscous.

In this *viscous sublayer* the shearing action becomes, once again, purely viscous and the velocity within it falls very sharply, and almost linearly, to zero at the surface.

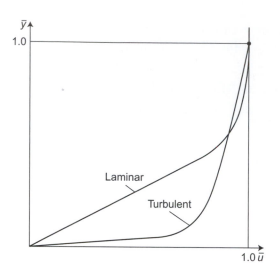

FIGURE 2.37

Comparison of the dimensionless velocity profiles for laminar and turbulent boundary layers.

Since at the surface the wall shearing stress now depends on viscosity only (i.e., $\tau_w = \mu(\partial u/\partial y)_w$), it will be clear that the surface friction stress under a turbulent layer will be far greater than that under a laminar layer of the same thickness, since $(\partial u/\partial y)_w$ is much greater. It should be noted, however, that the viscous shear-stress relation is only employed in the viscous sublayer very close to the surface and not throughout the turbulent boundary layer.

Clearly, from the preceding discussion, the viscous shearing stress at the surface, and thus the surface friction stress, depends only on the slope of the velocity profile at the surface, whatever the boundary-layer type. Therefore, accurate estimation of the profile, in either case, will enable correct predictions of skin-friction drag to be made.

2.11.5 Growth along a Flat Surface

If the boundary layer that develops on the surface of a flat plate held edgeways on the free stream is studied, it is found that, in general, a laminar boundary layer starts to develop from the leading edge. This laminar boundary layer grows in thickness, in accordance with the argument of Section 2.11, from zero at the leading edge to some point on the surface, where a rapid transition to turbulence occurs (see Fig. 2.38(a)). This transition is accompanied by a corresponding rapid thickening of the layer. Beyond this transition region, the turbulent boundary layer then continues to thicken steadily as it proceeds towards the trailing edge. Because of the greater shear stresses in the turbulent boundary layer, its thickness is greater than that

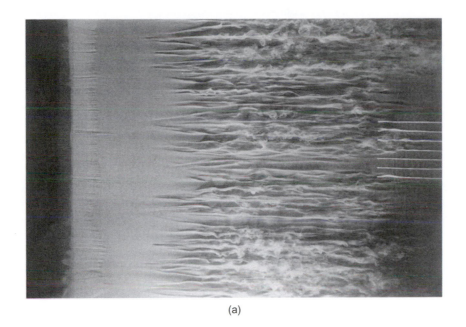

(a)

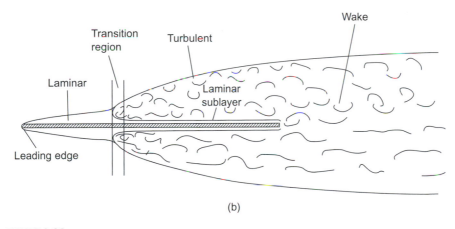

(b)

FIGURE 2.38

(a) Laminar-turbulent transition in a flat-plate boundary layer: This is a planform view of a dye sheet emitted upstream parallel to the wall into water flowing from left to right. Successive stages of transition are revealed (i.e., laminar flow on the upstream side), then the two-dimensional tollmien-schlichting waves, followed by the formation of turbulent spots, and finally fully developed turbulent flow. The reynolds number based on distance along the wall is about 75,000. (*Source: The photograph taken by H. Werlé, ONERA, France.*) (b) Note that the scale normal to surface of plate is greatly exaggerated.

for a laminar boundary layer. But, away from the immediate vicinity of the transition region, the actual rate of growth along the plate is lower for turbulent boundary layers. At the trailing edge the boundary layer along the upper surface joins the boundary layer along the lower surface to form a wake of retarded velocity, which also tends to thicken slowly as it flows downstream (see Fig. 2.38(b)).

On a flat plate the laminar profile has a constant shape at each point along the surface, although, of course, the thickness changes so that one nondimensional relationship for $\bar{u} = f(\bar{y})$ is sufficient. A similar argument applies to a reasonable approximation of the turbulent layer. This constancy of profile shape means that flat-plate boundary-layer studies are greatly simplified, and much work has been undertaken to study them both theoretically and experimentally.

However, in most aerodynamic problems the surface is usually of a streamline form such as with a wing or fuselage. The major difference, which affects the boundary-layer flow in these cases, is that mainstream velocity, and hence pressure, in a streamwise direction is no longer constant. The effect of a pressure gradient along the flow can be initially discussed purely qualitatively in order to ascertain how the boundary layer is likely to react.

2.11.6 Effects of an External Pressure Gradient

In the previous section, it was noted that in most practical aerodynamic applications mainstream velocity and pressure change in the streamwise direction. This has a profound effect on the development of the boundary layer. It can be seen from Fig. 2.39 that the net streamwise force acting on a small fluid element within the boundary layer is

$$\frac{\partial \tau}{\partial y}\delta y - \frac{\partial p}{\partial x}\delta x$$

When the pressure decreases (and, correspondingly, when the velocity along the edge of the boundary layer increases) with passage along the surface, the *external* pressure gradient is said to be *favorable*. This is because $\partial p/\partial x < 0$; therefore, noting that $\partial \tau/\partial y < 0$, it can be seen that the streamwise pressure forces help to counter the effects, discussed earlier, of the shearing action and shear stress at the wall. Consequently, the flow is not decelerated so markedly at the wall, leading to a fuller velocity profile and the boundary layer grows more slowly along the surface than does a flat plate.

The converse case is when the pressure increases and mainstream velocity decreases along the surface. The external pressure gradient is now said to be *unfavorable* or *adverse*. This is because the pressure forces now reinforce the effects of the shearing action and shear stress at the wall. The flow thus decelerates more markedly near the wall and the boundary layer grows more rapidly than in the case of the flat plate. Under these circumstances the velocity profile is much less full than for a flat plate and develops a point of inflection (Fig. 2.39). In fact, as indicated in Fig. 2.39,

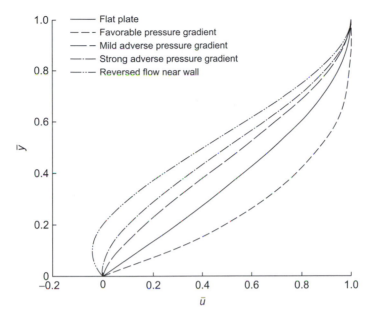

FIGURE 2.39

Effect of external pressure gradient on the velocity profile in the boundary layer.

if the adverse pressure gradient is sufficiently strong or prolonged, the flow near the wall so greatly decelerates that it begins to reverse direction. Flow reversal indicates that the boundary layer has *separated* from the surface. Boundary-layer separation can have profound consequences for the entire flow field and is discussed in more detail in Section 2.13.

2.12 BOUNDARY-LAYER EQUATIONS

To fix ideas it is helpful to think about the flow over a flat plate. This is a particularly simple flow, although, like much else in aerodynamics, the more one studies the details the less simple it becomes. If we consider the case of an infinite Reynolds number—that is, ignore viscous effects completely—the flow becomes exceedingly simple. The streamlines are everywhere parallel to the flat plate and the velocity is uniform and equal to U_∞, the value in the free stream infinitely far from the plate. Of course, there is no drag, since the shear stress at the wall is equivalently zero. (This is a special case of d'Alembert's paradox, which states that no force is generated by irrotational flow around any body irrespective of its shape.) Experiments on flat plates would confirm that the potential (i.e., inviscid) flow solution is indeed a good approximation at a high Reynolds number. It is found that the higher the Reynolds number, the closer the streamlines to being everywhere parallel with the plate. Furthermore, the nondimensional drag, or drag coefficient (see Section 1.4.5), becomes smaller and smaller the higher the Reynolds number indicating that the drag tends to zero as the Reynolds number tends to infinity.

But even though the drag is very small at a high Reynolds number, it is evidently important in applications of aerodynamics to estimate its value. So, how may we use this excellent infinite-Reynolds-number approximation (i.e., potential flow) to do this? Prandtl's boundary-layer concept and theory show us how. In essence, Prandtl assumed that potential flow is a good approximation everywhere except in a thin boundary layer adjacent to the surface. Because the boundary layer is very thin, it hardly affects the flow outside it. Accordingly, it may be assumed that the flow velocity at the edge of the boundary layer is given to a good approximation by the potential-flow solution for the flow velocity along the surface itself. For the flat plate, then, the velocity at the edge of the boundary layer is U_∞. In the more general case of the flow over a streamlined body, like the one depicted in Fig. 2.34, the velocity at the edge of the boundary layer varies and is denoted U_e. Prandtl went on to show, as explained next, how the Navier-Stokes equations may be simplified for application in this thin boundary layer.

2.12.1 Derivation of the Laminar Boundary-Layer Equations

At high Reynolds numbers the boundary-layer thickness can be expected to be very small compared with the length of the plate or streamlined body. (In aeronautical examples, such as the wing of a large aircraft, δ/L is typically around 0.01 and would be even smaller if the boundary layer were laminar rather than turbulent.) We will assume that in the hypothetical case of $Re_L \to \infty$ (where $Re_L = \rho U_\infty L/\mu$), $\delta \to 0$. Thus if we introduce the small parameter

$$\varepsilon = \frac{1}{Re_L} \qquad (2.127)$$

we expect that $\delta \to 0$ as $\varepsilon \to 0$, so that

$$\frac{\delta}{L} \propto \varepsilon^n \qquad (2.128)$$

where n is a positive exponent that is to be determined.

Suppose that we wish to estimate the magnitude of the velocity gradient within the laminar boundary layer. By considering the changes across the boundary layer along line AB in Fig. 2.40, it is evident that a rough approximation can be obtained by writing

$$\frac{\partial u}{\partial y} \simeq \frac{U_\infty}{\delta} = \frac{U_\infty}{L}\frac{1}{\varepsilon^n}$$

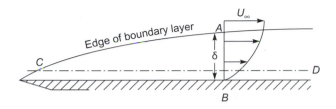

FIGURE 2.40

Scales of motion in boundary layers.

Although this is plainly very rough, it does have the merit of remaining valid as the Reynolds number becomes very high. This is recognized by a special symbol for the rough approximation and writing

$$\frac{\partial u}{\partial y} = O\left(\frac{U_\infty}{L}\frac{1}{\varepsilon^n}\right)$$

For the more general case of a streamlined body (e.g., Fig. 2.34), we use x to denote the distance along the surface from the leading edge (strictly, from the fore stagnation point) and y to denote the distance along the local normal to the surface. Since the boundary layer is very thin and its thickness is much smaller than the local radius of curvature of the surface, we can use the Cartesian form, Eqs. (2.92a,b) and (2.93), of the Navier-Stokes equations. In this more general case, the velocity varies along the edge of the boundary layer, and we denote it by U_e; thus

$$\frac{\partial u}{\partial y} = O\left(\frac{U_e}{\delta}\right) = O\left(\frac{U_e}{L}\frac{1}{\varepsilon^n}\right) \tag{2.129}$$

where U_e replaces U_∞, so that Eq. (2.129) applies to the more general case of a boundary layer around a streamlined body. Engineers think of $O(U_e/\delta)$ as meaning order of magnitude of U_e/δ, or very roughly a similar magnitude to U_e/δ. To mathematicans $F = O(1/\varepsilon^n)$ means that $F \propto 1/\varepsilon^n$ as $\varepsilon \to 0$. It should be noted that the order-of-magnitude estimate is the same irrespective of whether the term is negative or positive.

Estimating $\partial u/\partial y$ is fairly straightforward, but what about $\partial u/\partial x$? To estimate this quantity, consider the changes along the line CD in Fig. 2.40. Evidently, $u = U_\infty$ at C and $u \to 0$ as D becomes more distant from the leading edge of the plate. So the total change in u is approximately $U_\infty - 0$ and takes place over a distance $\Delta x \approx L$. Thus for the general case where the flow velocity varies along the edge of the boundary layer, we deduce that

$$\frac{\partial u}{\partial x} = O\left(\frac{U_e}{L}\right) \tag{2.130}$$

Finally, in order to estimate second derivatives like $\partial^2 u/\partial y^2$, we again consider the changes along the vertical line AB in Fig. 2.40. At B the estimate (2.129) holds for $\partial u/\partial y$, whereas at A $\partial u/\partial y \approx 0$. Therefore, the total change in $\partial u/\partial y$ across the boundary layer is approximately $(U_\infty/\delta) - 0$ and occurs over a distance δ. So, making use of Eq. (2.127), in the general case we obtain

$$\frac{\partial^2 u}{\partial y^2} = O\left(\frac{U_e}{\delta^2}\right) = O\left(\frac{U_e}{L^2}\frac{1}{\varepsilon^{2n}}\right) \tag{2.131}$$

In a similar way we deduce that

$$\frac{\partial^2 u}{\partial x^2} = O\left(\frac{U_e}{L^2}\right) \tag{2.132}$$

We can now use the order-of-magnitude estimates, Eqs. (2.129) through (2.132), to estimate the order of magnitude of each of the terms in the Navier-Stokes equations. We begin with the continuity Eq. (2.93).

$$\underbrace{\frac{\partial u}{\partial x}}_{O(U_e/L)} + \underbrace{\frac{\partial v}{\partial y}}_{O(v/(\varepsilon^n L))} = 0 \tag{2.133}$$

If both terms are the same order of magnitude, we can deduce from Eq. (2.133) that

$$v = O\left(U_e\frac{\delta}{L}\right) = O\left(U_e\varepsilon^n\right) \tag{2.134}$$

One might question the assumption that two terms are the same order of magnitude. But the slope of the streamlines in the boundary layer is equal to v/u by definition and will also be given approximately by δ/L, so Eq. (2.134) is evidently correct.

We will now use Eqs. (2.129) through (2.132) and (2.134) to estimate the orders of magnitude of the terms in the Navier-Stokes equations (Eqs. 2.92a and 2.92b). We will assume steady flow, ignore the body-force terms, and divide throughout by ρ (noting that the kinematic viscosity $v = \mu/\rho$); thus

$$\underbrace{u\frac{\partial u}{\partial x}}_{O(U_e^2/L)} + \underbrace{v\frac{\partial u}{\partial y}}_{O(U_e^2/L)} = \underbrace{-\frac{1}{\rho}\frac{\partial p}{\partial x}}_{O(\text{unknown})} + \underbrace{v\left(\frac{\partial^2 u}{\partial x^2}\right)}_{O(vU_e/L^2)=O(\varepsilon U_e^2/L)}$$

$$+ \underbrace{v\left(\frac{\partial^2 u}{\partial y^2}\right)}_{O(vU_e/(L^2\varepsilon^{2n}))=O(\varepsilon^{1-2n}U_e^2/L)} \tag{2.135}$$

$$u\frac{\partial v}{\partial x} + v\frac{\partial v}{\partial y} = -\frac{1}{\rho}\frac{\partial p}{\partial y} + \nu\left(\frac{\partial^2 v}{\partial x^2}\right)$$

$\underbrace{\phantom{u\frac{\partial v}{\partial x}}}_{O(\varepsilon^n U_e^2/L)}$ $\underbrace{\phantom{v\frac{\partial v}{\partial y}}}_{O(\varepsilon^n U_e^2/L)}$ $\underbrace{\phantom{\frac{1}{\rho}\frac{\partial p}{\partial y}}}_{O(\text{unknown})}$ $\underbrace{\phantom{\nu\left(\frac{\partial^2 v}{\partial x^2}\right)}}_{O(\nu U_e \varepsilon^n/L^2)=O(\varepsilon^{1+n}U_e^2/L)}$

$$+ \quad \nu\left(\frac{\partial^2 v}{\partial y^2}\right)$$

$\underbrace{\phantom{\nu\left(\frac{\partial^2 v}{\partial y^2}\right)}}_{O(\nu U_e/(L^2\varepsilon^n))=O(\varepsilon^{1-n}U_e^2/L)}$

(2.136)

Now $\varepsilon = 1/Re_L$ is a very small quantity so that $O(\varepsilon U_e^2/L)$ is negligible compared with $O(U_e^2/L)$. It therefore follows that the second term on the right-hand side of Eq. (2.135) can be dropped in comparison with the terms on the left-hand side. What about the third term on the right-hand side of Eq. (2.135)? If $2n = 1$, it will be the same order of magnitude as those on the left-hand side. If $2n < 1$, the remaining viscous term will be negligible as compared with the terms on the left-hand side. This cannot be so because we know that the viscous effects are not negligible within the boundary layer. On the other hand, if $2n > 1$, the terms on the left-hand side of Eq. (2.135) will be negligible in comparison with the remaining viscous term. So, for the flat plate for which $\partial p/\partial x \equiv 0$, Eq. (2.135) reduces to

$$\frac{\partial^2 u}{\partial y^2} = 0$$

This can be readily integrated to give

$$u = f(x)y + g(x)$$

Note that, as the second derivative of velocity with respect to y is a partial derivative, arbitrary functions of x, $f(x)$, and $g(x)$ take the place of constants of integration. In order to satisfy the no-slip condition ($u = 0$) at the surface, $y = 0$, $g(x) = 0$, so that $u \propto y$. This conclusion does not conform to the required smooth velocity profile depicted in Figs 2.36 and 2.40. We therefore conclude that the only possibility that fits the physical requirements is

$$2n = 1 \quad \text{implying} \quad \delta \propto \varepsilon^{1/2}\left(= \frac{1}{\sqrt{Re_L}}\right) \tag{2.137}$$

and Eq. (2.135) simplifies to

$$u\frac{\partial u}{\partial x} + v\frac{\partial u}{\partial y} = -\frac{1}{\rho}\frac{\partial p}{\partial x} + \nu\left(\frac{\partial^2 u}{\partial y^2}\right) \tag{2.138}$$

It is now plain that all terms in Eq. (2.136) must be $O(\varepsilon^{1/2}U_e^2/L)$ or even smaller and are therefore negligibly small compared to the terms retained in Eq. (2.138). We therefore conclude that Eq. (2.136) simplifies drastically to

$$\frac{\partial p}{\partial y} \approx 0 \tag{2.139}$$

In other words, the pressure does not change across the boundary layer. (In fact, this can be deduced from the fact that the boundary layer is very thin so that the streamlines are almost parallel with the surface.) This implies that p depends only on x and can be determined in advance from the potential-flow solution. Thus Eq. (2.138) simplifies further to

$$u\frac{\partial u}{\partial x} + v\frac{\partial u}{\partial y} = \underbrace{-\frac{1}{\rho}\frac{dp}{dx}}_{\text{a known function of } x} + v\left(\frac{\partial^2 u}{\partial y^2}\right) \tag{2.140}$$

Equations (2.140) and (2.133) together are usually known as the (Prandtl) *boundary-layer equations*.

To sum up, then, the velocity profiles in the boundary layer can be obtained as follows:

1. Determine the potential flow around the body using the methods described in Chapter 3.
2. From this potential-flow solution determine the pressure and the velocity along the surface.
3. Solve Eqs. (2.133) and (2.140) subject to the boundary conditions

$$u = v = 0 \quad \text{at} \quad y = 0; \quad u = U_e \quad \text{at} \quad y = \delta \text{ (or } \infty) \tag{2.141}$$

The boundary condition $u = 0$ is usually referred to as the *no-slip condition* because it implies that the fluid adjacent to the surface must stick to it. Explanations can be offered for why this should be so, but fundamentally it is an empirical observation. The second boundary condition $v = 0$ is referred to as the *no-penetration condition* because it states that fluid cannot pass into the wall. Plainly, this condition will not hold when the surface is porous, as with boundary-layer suction. The third boundary condition (Eq. 2.141) is applied at the boundary-layer edge, where it requires the flow velocity to be equal to the potential-flow solution. For the approximate methods described in Chapter 8, one usually applies this condition at $y = \delta$. For accurate solutions of the boundary-layer equations, however, no clear edge can be defined; the velocity profile is such that u approaches ever closer to U_e the larger y becomes. Thus for accurate solutions one usually chooses to apply the boundary condition at $y = \infty$, although it is commonly necessary to choose a large finite value of y for seeking computational solutions.

2.12.2 Boundary-Layer Thickness for Laminar and Turbulent Flows

In this section we will examine the thickness of the boundary layer along a flat plate. This flow is illustrated in Figs. 2.38 and 2.40. The boundary-layer thickness is defined as the value of $y = \delta$, which is the distance perpendicular to the plate where the velocity is equal to 99% of the free-stream value. The exact solution (described in more detail in Chapter 8) is due to Blasius. The following formula is in excellent agreement with Blasius's numerical solution for the boundary-layer thickness for laminar boundary layers on flat plates:

$$\delta = 4.64x/(Re_x)^{1/2} \tag{2.142}$$

This result illustrates the fact that a laminar boundary layer grows as $Re_x^{-1/2}$, where the Reynolds number is based on distance from the leading edge, $Re_x = U_\infty x/\nu$. A formula for the boundary-layer thickness for turbulent boundary layers on flat plates, as determined by the approximate method based on a momentum-integral analysis (described in Chapter 8) is as follows:

$$\delta = 0.383 \frac{x}{(Re_x)^{1/5}} \tag{2.143}$$

We see that a turbulent boundary layer grows as $Re_x^{-1/5}$. Fig. 2.41 illustrates the growth of the laminar and turbulent boundary layers for a free stream speed of 60 m/s. Note that at 1 m from the leading edge, the turbulent boundary layer is 1.8% of the distance from the leading edge (or origin of this boundary layer) and the laminar boundary layer is approximately 0.2% of this distance. The transition from laminar to turbulent boundary layer flow occurs at a critical distance from the leading edge in the range of $5 \times 10^5 < Re_{xc} < 3 \times 10^6$, where $Re_{xc} = U_\infty x_c/\nu$ and x_c is the distance from the leading edge where the laminar boundary layer becomes unstable and transitions to a turbulent boundary layer.

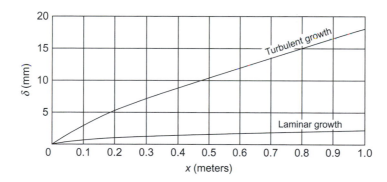

FIGURE 2.41

Boundary-layer growth on a flat plate at a free-stream speed of 60 m s^{-1}.

The main purpose of this section is to point out that viscous effects are confined to thin boundary layers in high-Reynolds-number flows of practical interest to the aeronautical engineer. In addition, as put forth by Prandtl, the flow outside the boundary layer behaves essentially as if it were inviscid. Hence, potential-flow theory is applied to determine the flow field outside the boundary layer and, hence, the pressure gradients in the direction of flow to which the boundary layer is exposed.

2.12.3 Boundary-Layer/Potential-Flow Model of Airfoils and Wings

Recall from the last section that the thickness of a boundary layer is inversely proportional to the Reynolds number (this is the case whether the flow is laminar or turbulent in the boundary layer). Hence, boundary layers are indeed thin for high-Reynolds-number flows, as Prandtl demonstrated from experimental observations and theoretical arguments more than a century ago. Across the thin viscous layer the pressure is constant and it is equal to the value of potential-flow theory to a very good approximation. Therefore, to predict the pressure distribution for an attached flow over an airfoil, potential-flow theory is useful, because a number of shapes can be examined in the course of design rather quickly as compared with attempting to solve approximations of the Navier-Stokes equations with the "advanced computational techniques" of computational fluid dynamics (CFD).

2.13 EXERCISES

2.1 *Continuity equation for axisymmetric flow*

(a) Consider an axisymmetric flow field expressed in terms of the cylindrical coordinate system (r, φ, z), where all flow variables are independent of the azimuthal angle φ—for example, the axial flow over a body of revolution. If the velocity components (u, w) correspond to the coordinate directions (r, z), respectively, show that the continuity equation is given by

$$\frac{\partial u}{\partial r} + \frac{u}{r} + \frac{\partial w}{\partial z} = 0$$

(b) Show that the continuity equation can be automatically satisfied by a stream function ψ of a form such that

$$u = \frac{1}{r}\frac{\partial \psi}{\partial z}, \qquad w = -\frac{1}{r}\frac{\partial \psi}{\partial r}$$

2.2 *Continuity equation for two-dimensional flow in polar coordinates*

(a) Consider a two-dimensional flow field expressed in terms of the cylindrical coordinate system (r, φ, z), where all flow variables are independent of the azimuthal angle φ—for example, the flow over a circular cylinder. If the velocity components (u, v) correspond to the coordinate directions (r, φ), respectively, show that the continuity equation is given by

$$\frac{\partial u}{\partial r} + \frac{u}{r} + \frac{1}{r}\frac{\partial v}{\partial \phi} = 0$$

(b) Show that the continuity equation can be automatically satisfied by a stream function ψ of a form such that

$$u = \frac{1}{r}\frac{\partial \psi}{\partial \phi}, \qquad v = -\frac{\partial \psi}{\partial r}$$

2.3 *Transport equation for contaminant in two-dimensional flow field*
In many engineering applications one is interested in the transport of a contaminant by the fluid flow. The contaminant could be anything from a polluting chemical to particulate matter. To derive the governing equation one needs to recognize that, provided that the contaminant is not being created within the flow field, the mass of contaminant is conserved. The contaminant matter can be transported by two distinct physical mechanisms, *convection* and *molecular diffusion*. Let C be the concentration of contaminant (i.e., mass per unit volume of fluid); then the rate of transport of contamination per unit area is given by

$$-D\nabla C = -D\left(\mathbf{i}\frac{\partial C}{\partial x} + \mathbf{j}\frac{\partial C}{\partial y}\right)$$

where $\mathbf{i}$ and $\mathbf{j}$ are the unit vectors in the x and y directions respectively, and D is the diffusion coefficient (units m^2/s, the same as kinematic viscosity).

Note that diffusion transports the contaminant down the concentration gradient (i.e., the transport is from a higher to a lower concentration); hence the minus sign. It is analogous to thermal conduction.

(a) Consider an infinitesimal rectangular control volume. Assume that no contaminant is produced within it and that the contaminant is sufficiently dilute to leave the fluid flow unchanged. By considering a mass balance for the control volume, show that the transport equation for a contaminant in a two-dimensional flow field is given by

$$\frac{\partial C}{\partial t} + u\frac{\partial C}{\partial x} + v\frac{\partial C}{\partial y} - D\left(\frac{\partial^2 C}{\partial x^2} + \frac{\partial^2 C}{\partial y^2}\right) = 0$$

(b) Why is it necessary to assume a dilute suspension of contaminant? What form would the transport equation take if this assumption were not made? Finally, how could the equation be modified to take account of the contaminant being produced by a chemical reaction at the rate of $\dot{m}_c$ per unit volume.

2.4   *Euler equations for axisymmetric flow*

(a) For the flow field and coordinate system of Exercise 2.1, show that the Euler equations (inviscid momentum equations) take the form

$$\rho\left(\frac{\partial u}{\partial t}+u\frac{\partial u}{\partial r}+w\frac{\partial u}{\partial z}\right)=\rho g_r-\frac{\partial p}{\partial r}$$

$$\rho\left(\frac{\partial w}{\partial t}+u\frac{\partial w}{\partial r}+w\frac{\partial w}{\partial z}\right)=\rho g_z-\frac{\partial p}{\partial z}$$

2.5   *Navier-Stokes equations for two-dimensional axisymmetric flow*

(a) Show that the strain rates and vorticity for an axisymmetric viscous flow like that described in Exercise 2.1 are given by

$$\dot{\varepsilon}_{rr}=\frac{\partial u}{\partial r};\qquad \dot{\varepsilon}_{zz}=\frac{\partial w}{\partial z};\qquad \dot{\varepsilon}_{\phi\phi}=\frac{u}{r}$$

$$\dot{\gamma}_{rz}=\frac{1}{2}\left(\frac{\partial w}{\partial r}+\frac{\partial u}{\partial z}\right);\qquad \eta=\frac{\partial w}{\partial r}-\frac{\partial u}{\partial z}$$

Hint: Note that the azimuthal strain rate is not zero. The easiest way to determine it is to recognize that $\dot{\varepsilon}_{rr}+\dot{\varepsilon}_{\phi\phi}+\dot{\varepsilon}_{zz}=0$ must be equivalent to the continuity equation.

(b) Show that the Navier-Stokes equations for axisymmetric flow are given by

$$\rho\left(\frac{\partial u}{\partial t}+u\frac{\partial u}{\partial r}+w\frac{\partial u}{\partial z}\right)=\rho g_r-\frac{\partial p}{\partial r}+\mu\left(\frac{\partial^2 u}{\partial r^2}+\frac{1}{r}\frac{\partial u}{\partial r}-\frac{u}{r^2}+\frac{\partial^2 u}{\partial z^2}\right)$$

$$\rho\left(\frac{\partial w}{\partial t}+u\frac{\partial w}{\partial r}+w\frac{\partial w}{\partial z}\right)=\rho g_z-\frac{\partial p}{\partial z}+\mu\left(\frac{\partial^2 w}{\partial r^2}+\frac{1}{r}\frac{\partial w}{\partial r}+\frac{\partial^2 w}{\partial z^2}\right)$$

2.6   *Euler equations for two-dimensional flow in polar coordinates*

(a) For the two-dimensional flow described in Exercise 2.2, show that the Euler equations (inviscid momentum equations) take the form

$$\rho\left(\frac{\partial u}{\partial t}+u\frac{\partial u}{\partial r}+\frac{v}{r}\frac{\partial u}{\partial \phi}-\frac{v^2}{r}\right)=\rho g_r-\frac{\partial p}{\partial r}$$

$$\rho\left(\frac{\partial v}{\partial t}+u\frac{\partial v}{\partial r}+\frac{v}{r}\frac{\partial v}{\partial \phi}-\frac{uv}{r}\right)=\rho g_{phi}-\frac{1}{r}\frac{\partial p}{\partial \phi}$$

Hints: (i) The momentum components perpendicular to and entering and leaving the side faces of the elemental control volume have small components in the radial direction that must be taken into account; likewise (ii) the pressure forces acting on these faces have small radial components.

2.7 Show that the strain rates and vorticity for the flow and coordinate system of Exercise 2.6 are given by

$$\dot{\varepsilon}_{rr} = \frac{\partial u}{\partial r}; \qquad \dot{\varepsilon}_{\varphi\varphi} = \frac{1}{r}\frac{\partial v}{\partial \varphi} + \frac{u}{r}$$

$$\dot{\gamma}_{r\varphi} = \frac{1}{2}\left(\frac{\partial v}{\partial r} - \frac{v}{r} + \frac{1}{r}\frac{\partial u}{\partial \varphi}\right); \qquad \zeta = \frac{1}{r}\frac{\partial u}{\partial \varphi} - \frac{\partial v}{\partial r} + \frac{v}{r}$$

Hint: (i) The angle of distortion (β) of the side face must be defined relative to the line joining the origin O to the center of the infinitesimal control volume.

2.8 The flow in the narrow gap (of width h) between two concentric cylinders of length L, with the inner one of radius R rotating at angular speed ω, can be approximated by the Couette solution to the Navier-Stokes equations. Show that the torque T and power P required to rotate the shaft at a rotational speed of ω rad/s are given by

$$T = \frac{2\pi\mu\omega R^3 L}{h}, \qquad P = \frac{2\pi\mu\omega^2 R^3 L}{h}$$

2.9 *Axisymmetric stagnation-point flow*
Carry out a similar analysis to that described in Section 2.10.3 using the axisymmetric form of the Navier-Stokes equations given in Exercise 2.5 for axisymmetric stagnation-point flow, and show that the equivalent to Eq. (2.118) is

$$\phi''' + 2\phi\phi'' - \phi'^2 + 1 = 0$$

where φ' denotes differentiation with respect to the independent variable $\zeta = \sqrt{a/vz}$ and φ is defined in exactly the same way as for the two-dimensional case.

Potential Flow

LEARNING OBJECTIVES

- Learn to calculate the air flow and pressure distribution around various body shapes.

- Learn more about the classical assumption of irrotational flow and its meaning that the vorticity is everywhere zero. Note that this also implies inviscid flow. Irrotational flows are potential fields.

- Learn about the potential function, known as the velocity potential. Learn that the velocity components can be determined from the velocity potential.

- Learn that the equations of motion for irrotational flow reduce to a single partial differential equation for velocity potential known as the Laplace equation.

- Examine the classical analytical techniques described for obtaining two-dimensional and axisymmetric solutions to the Laplace equation for aerodynamic applications.

- Learn how computational tools are applied for predicting the potential flows around arbitrary two-dimensional geometries.

3.1 TWO-DIMENSIONAL FLOWS

The problem of interest in this section is illustrated in Fig. 3.1, which illustrates the two-dimensional steady motion of a body moving at constant speed U in a quiescent fluid. The quiescent fluid is incompressible with uniform density ρ and uniform pressure p_∞. The figure shows this from the point of view of a coordinate system attached to the body. Hence, the body appears stationary in a uniform *onset* flow. In this coordinate system, we can assume steady-state conditions. Note that the x direction of the coordinate system illustrated in the figure was selected, for convenience, to be in the direction of the far-field velocity vector; the far-field velocity vector is $\mathbf{U}_\infty = U\mathbf{i}$.

The aerodynamic problem is to determine the velocity, $\mathbf{u} = (u, v)$, and pressure p for the flow illustrated in Fig. 3.1 as functions of the coordinates (x, y). Since the flow of interest is a two-dimensional incompressible flow, the continuity equation is

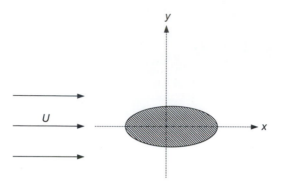

FIGURE 3.1

Arbitrary body in a uniform stream.

(as discussed in Section 2.4.2)

$$\frac{\partial u}{\partial x} + \frac{\partial v}{\partial y} = 0 \tag{3.1}$$

We next assume that the fluid is inviscid (Euler's "perfect fluid"). We also assume steady flow and neglect body forces (e.g., gravity). In this case, the Euler equations (described in Section 2.6.1) reduce to the following form:

$$u\frac{\partial u}{\partial x} + v\frac{\partial u}{\partial y} = -\frac{1}{\rho}\frac{\partial p}{\partial x} \tag{3.2}$$

$$u\frac{\partial v}{\partial x} + v\frac{\partial v}{\partial y} = -\frac{1}{\rho}\frac{\partial p}{\partial y} \tag{3.3}$$

Equations (3.1), (3.2), and (3.3) are in three unknowns. The unknowns are u, v, and p. Thus, they represent a complete system of equations that describes the flows of interest. We can simplify this system of equations as follows.

Equation (3.2) can be rearranged to read

$$\frac{\partial u^2/2}{\partial x} + v\frac{\partial u}{\partial y} = -\frac{\partial p/\rho}{\partial x}$$

Adding and subtracting a term (the second and third term in the following), we get

$$\frac{\partial}{\partial x}\left(\frac{u^2}{2} + \frac{v^2}{2}\right) - \frac{\partial}{\partial x}\left(\frac{v^2}{2}\right) + v\frac{\partial u}{\partial y} = -\frac{\partial}{\partial x}\frac{p}{\rho}$$

Or, after rearranging terms,

$$\frac{\partial}{\partial x}\left(\frac{u^2}{2} + \frac{v^2}{2} + \frac{p}{\rho}\right) = v\left(\frac{\partial v}{\partial x} - \frac{\partial u}{\partial y}\right) \tag{3.4}$$

Similarly, we can operate on and rearrange Eq. (3.3) to get

$$\frac{\partial}{\partial y}\left(\frac{u^2}{2} + \frac{v^2}{2} + \frac{p}{\rho}\right) = -u\left(\frac{\partial v}{\partial x} - \frac{\partial u}{\partial y}\right) \tag{3.5}$$

The equation for the z component of the vorticity (the only finite component of the vorticity vector in two-dimensional flow) is (as described in Section 2.7.4)

$$\frac{\partial v}{\partial x} - \frac{\partial u}{\partial y} = \zeta$$

For irrotational flows, the condition for irrotationality is $\zeta = 0$; hence

$$\frac{\partial v}{\partial x} - \frac{\partial u}{\partial y} = 0 \tag{3.6}$$

Applying Eq. (3.6) to Eqs. (3.4) and (3.5) and integrating the latter pair, we get

$$\frac{u^2}{2} + \frac{v^2}{2} + \frac{p}{\rho} = H \tag{3.7}$$

where H is a constant. This is Bernoulli's equation. If we evaluate H far upstream where the flow is uniform at speed U and at pressure p_∞, we get

$$\frac{u^2}{2} + \frac{v^2}{2} + \frac{p}{\rho} = \frac{U^2}{2} + \frac{p_\infty}{\rho}$$

Rearranging this equation,

$$C_p = \frac{p - p_\infty}{\frac{1}{2}\rho U^2} = 1 - \left(\frac{u^2 + v^2}{U^2}\right) \tag{3.8}$$

where C_p is a dimensionless pressure coefficient.

Equation (3.6), the irrotationality condition, allows us to define a scalar function ϕ as follows:

$$u = \frac{\partial \phi}{\partial x} \quad \text{and} \quad v = \frac{\partial \phi}{\partial y} \tag{3.9}$$

Substituting this into Eq. (3.6), we get

$$\frac{\partial^2 \phi}{\partial x \partial y} - \frac{\partial^2 \phi}{\partial y \partial x} = \frac{\partial^2 \phi}{\partial x \partial y} - \frac{\partial^2 \phi}{\partial x \partial y} = 0$$

which illustrates the fact that we can write the velocity vector $\mathbf{u} = (u, v)$ as the gradient of a scalar function ϕ. That is, if $\mathbf{u} = \nabla\phi$, then $\nabla \times \mathbf{u} = 0$ (in two dimensions, these are Eqs. (3.9) and (3.6), respectively).

Substituting Eq. (3.9) into the continuity equation, Eq. (3.1), we get

$$\frac{\partial^2 \phi}{\partial x^2} + \frac{\partial^2 \phi}{\partial y^2} = 0 \qquad (3.10)$$

This is the Laplace equation for ϕ; hence ϕ is a *harmonic potential* function that we call the *velocity potential*. (Further discussion of this function and its relationship to the streamlines is given in Section 3.1.1.)

Equation (3.1), the continuity equation, also allows us to define the stream function ψ as follows:

$$u = \frac{\partial \psi}{\partial y}, \quad v = -\frac{\partial \psi}{\partial x} \qquad (3.11)$$

Substituting this into Eq. (3.6), the irrotationality condition, we get

$$\frac{\partial^2 \psi}{\partial x^2} + \frac{\partial^2 \psi}{\partial y^2} = 0 \qquad (3.12)$$

This indicates that the *stream function* ψ (already described in Section 2.5) is also a harmonic potential. In fact, since

$$\frac{\partial \phi}{\partial x} = \frac{\partial \psi}{\partial y} \quad \text{and} \quad \frac{\partial \phi}{\partial y} = -\frac{\partial \psi}{\partial x}$$

lines of the constant value of ϕ are perpendicular (or orthogonal) to lines of the constant value of ψ. The former are *equipotential lines*, and the latter are *streamlines*. These relationships are the *Cauchy-Riemann equations*, and they play an important role in the mathematical justification for the practical procedures we apply herein to solve potential-flow problems.

Equation (3.10) or (3.12) indicates that any solution to the Laplace equation is a possible potential flow. Since the Laplace equation is a linear partial differential equation, if, for example, two solutions are summed (*linearly superimposed*), a third solution is obtained. In other words, we can construct solutions by the superposition of elementary solutions to solve particular potential-flow problems. Before we examine elementary solutions to the Laplace equation, let us summarize the procedure to solve potential-flow problems in general.

Note that any streamline can be a boundary of a body on which we wish to predict the pressure distribution. Streamlines are lines tangent to the velocity vectors; therefore, in the direction normal (or perpendicular) to a streamline, the normal component of the velocity is zero:

$$\mathbf{n} \cdot \nabla \phi = \frac{\partial \phi}{\partial n} = 0 \ \text{ on } \ S_B \qquad (3.13)$$

where S_B is the streamline that represents the body surface or the wall of interest in an investigation, and $\mathbf{n}$ is the unit vector perpendicular to the streamline. An example of S_B is the boundary surrounding the cross-hatched region (or body) in Fig. 3.1.

To solve boundary-value problems associated with solving Eq. (3.10), the Laplace equation for ϕ, we need boundary conditions on all components of the boundary that completely surrounds the flow field of interest. In addition to a condition on S_B (i.e., Eq. 3.13), we need to specify the far-field condition, which for the problem illustrated in Fig. 3.1, is

$$\phi_\infty = Ux \qquad (3.14)$$

which is the potential for a uniform stream in the x direction. This completes the specification of the boundary-value problem illustrated in Fig. 3.1. How we construct solutions to this class of problems is discussed in this chapter and in the two subsequent chapters.

The steps to solve aerodynamics problems can be summarized as follows. The first step is to construct the flow field by either solving for ϕ or for ψ. Once one of these fields (or functions) is known, then the velocity field is computed by solving Eq. (3.9) or (3.11) given ϕ or ψ, respectively. Knowing the velocity field, we compute the pressure field by applying Eq. (3.8). Finally, we integrate the pressure distribution on S_B appropriately to determine the force and moment acting on S_B due to pressure. Before we begin to deal with solving particular aerodynamic problems in Section 3.2, in the next three subsections we examine in more detail the velocity potential.

3.1.1 The Velocity Potential

The stream function (see Section 2.5) at a point has been defined as the quantity of fluid moving across some convenient imaginary line in the flow pattern, and lines of constant stream function (amount of flow or flux) may be plotted to give a picture of the flow pattern (see Section 2.5). Another mathematical definition, giving a different pattern of curves, can be obtained for the same flow system. In this case, an expression giving the amount of flow along the convenient imaginary line is found.

In a general two-dimensional fluid flow, consider any (imaginary) line OP joining the origin of a pair of axes to the point P(x, y). Again, the axes and this line do not impede the flow and are used only to form a reference datum. At a point Q on the line, let the local velocity q meet the line OP in β (Fig. 3.2(a)). Then the component of velocity parallel to δs is $q \cos \beta$. The amount of fluid flowing along δs is $q \cos \beta \delta s$. The total amount of fluid flowing along the line toward P is the sum of all such amounts

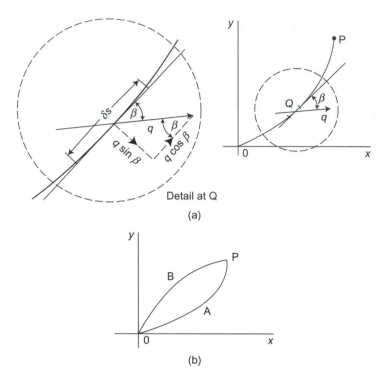

FIGURE 3.2

(a) Relationship between velocity vector **q** and line OP at its segment δs. (b) Two lines connecting points OP.

and is given mathematically as the integral $\int q \cos \beta ds$. This function is called the velocity potential of P with respect to O and is denoted ϕ.

Now OQP can be any line between O and P and a necessary condition for $\int q \cos \beta ds$ to be the velocity potential ϕ is that the value of ϕ is unique for the point P, irrespective of the path of integration. Then the velocity potential is

$$\phi = \int_{OP} q \cos \beta ds \qquad (3.15)$$

If this were not the case, and integrating the tangential flow component from O to P via A (Fig. 3.2(b)) did not produce the same magnitude of ϕ as integrating from O to P via some other path such as B, there would be some flow components circulating in the circuit OAPBO. This in turn would imply that the fluid within the circuit possesses vorticity. The existence of a velocity potential must therefore imply zero vorticity in the flow or, in other words, a flow without circulation (see Section 2.7.7)—that is, an *irrotational* flow. Such flows are also called *potential* flows.

The sign convention for the velocity potential selected in this text is based on the following notion. The tangential flow along a curve is the product of the local velocity component and the elementary length of the curve. Now, if the velocity component is in the direction of integration, it is considered a *positive* increment of the velocity potential.

3.1.2 The Equipotential

Consider a point P having a velocity potential ϕ (ϕ is the integral of the flow component along OP), and let another point P_1 close to P have the same velocity potential ϕ. This means that the integral of flow along OP_1 equals the integral of flow along OP (Fig. 3.3). But, by definition, OPP_1 is another path of integration from O to P_1. Therefore,

$$\phi = \int_{OP} q\cos\beta\,ds = \int_{OP_1} q\cos\beta\,ds = \int_{OPP_1} q\cos\beta\,ds$$

where q is the magnitude of $\mathbf{q}$; however, since the integral along OP equals that along OP_1, there can be no flow along the remaining portions of the path of the third integral, that is, along PP_1. Similarly for other points such as P_2 and P_3 having the same velocity potential, there can be no flow along the line joining P_1 to P_2.

The line joining P, P_1, P_2, and P_3 joins points having the same velocity potential and is called an *equipotential* or a line of constant velocity potential (i.e. a line of constant ϕ). The significant characteristic of an equipotential is that there is no flow along such a line.

The flow in the region of points P and P_1 should be investigated more closely. From the previous discussion, there can be no flow along the line PP_1, but there is fluid flowing in this region so it must be flowing in such a way that there is no component of velocity in the direction PP_1. Thus the flow can only be at right angles to PP_1, which means that the flow in the region PP_1 must be normal to PP_1. Now, the streamline in this region, the line to which the flow is tangential, must also be at right angles to PP_1, which is itself the local equipotential.

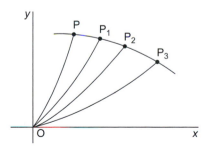

FIGURE 3.3

Equipotential perpendicular to the flow direction.

This relation applies at all points in a homogeneous continuous fluid and can be stated thus: Streamlines and equipotentials meet orthogonally (i.e., always at right angles. It follows from this statement that, for a given streamline pattern, there is a unique equipotential pattern for which the equipotentials are everywhere normal to the streamlines.

3.1.3 Velocity Components in Terms of ϕ

In Cartesian coordinates: Let a point P(x,y) be on an equipotential ϕ and a neighboring point Q$(x+\delta x, y+\delta y)$ be on the equipotential $\phi+\delta\phi$ (Fig. 3.4). Then, by definition, the increase in velocity potential from P to Q is the line integral of the tangential velocity component along any path between P and Q. Taking PRQ as the most convenient path, where the local velocity components are u and v,

$$\delta\phi = u\delta x + v\delta y$$

but

$$\delta\phi = \frac{\partial\phi}{\partial x}\delta x + \frac{\partial\phi}{\partial y}\delta y$$

Thus, equating terms,

$$u = \frac{\partial\phi}{\partial x} \quad \text{and} \quad v = \frac{\partial\phi}{\partial y} \tag{3.16}$$

In polar coordinates: Let a point P(r,θ) be on an equipotential ϕ and a neighboring point Q$(r+\delta r, \theta+\delta\theta)$ be on an equipotential $\phi+\delta\phi$ (Fig. 3.5). By definition, the increase $\delta\phi$ is the line integral of the *tangential* component of velocity along any path. For convenience, choose PRQ where point R is $(r+\delta r, \theta)$. Then, integrating

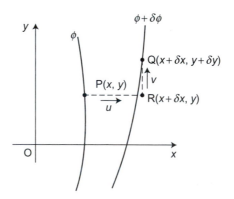

FIGURE 3.4

Relationship between equipotentials and velocity components in Cartesian coordinates.

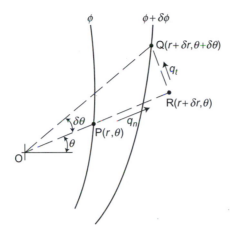

FIGURE 3.5

Relationship between equipotentials and velocity components in polar coordinates.

along PR and RQ where the velocities are q_n and q_t, respectively, and are both in the direction of integration,

$$\delta\phi = q_n \delta r + q_t (r + \delta r) \delta\theta \approx q_n \delta r + q_t r \delta\theta$$

where the right-hand side is a first-order approximation; it neglects terms that are products of small quantities. But, since ϕ is a function of two independent variables,

$$\delta\phi = \frac{\partial\phi}{\partial r}\delta r + \frac{\partial\phi}{\partial\theta}\delta\theta$$

and

$$q_n = \frac{\partial\phi}{\partial r} \quad \text{and} \quad q_t = \frac{1}{r}\frac{\partial\phi}{\partial\theta} \tag{3.17}$$

Again, in general, the velocity q in any direction s is found by differentiating the velocity potential ϕ partially with respect to the direction s of q:

$$q = \frac{\partial\phi}{\partial s}$$

3.2 STANDARD FLOWS IN TERMS OF ψ AND ϕ

There are four basic two-dimensional flow fields, from combinations of which all other steady-flow conditions may be modeled. The four flows are *uniform parallel* stream, source (sink), doublet, and point vortex. Examples of how these are applied to construct useful flows are presented in this section.

3.2.1 Uniform Flow

In this subsection we examine a number of ways to represent a uniform flow. We first examine the velocity potential and stream function that represents a uniform stream in the x direction (this is the uniform stream upstream of the body illustrated in Fig. 3.1). Then we examine the same for a uniform stream in an arbitrary direction.

Flow of Constant Velocity Parallel to the x-Axis from Left to Right

The potentials for a uniform flow in the x direction are

$$\phi = Ux, \quad \psi = Uy \tag{3.18}$$

Thus, applying Eq. (3.9) or (3.11), we get

$$\mathbf{u} = (u, v) = (U, 0) \tag{3.19}$$

Note that, if ψ is constant, streamlines are parallel to the x-axis, as illustrated in Figure 3.6. Similarly, if ϕ is constant, equipotential lines are parallel to the y-axis. The pattern of the potentials for a uniform flow in the x direction is shown in the figure. The MATLAB script used to generate this figure is given in Table 3.1.

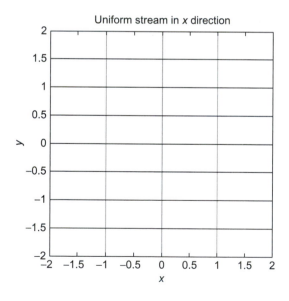

FIGURE 3.6

Vertical lines are equipotentials; horizontal lines are streamlines. The flow in this case is from left to right at uniform speed U.

Table 3.1 MATLAB Script Applied to Plot Figure 3.6

```
U = 1;      % Uniform stream
% GRID:
x = -2:.02:2;
y = -2:.02:2;
for m = 1:length(x)
    for n = 1:length(y)
        xx(m,n) = x(m); yy(m,n) = y(n);
        % Velocity potential function:
        phi_UniFlow(m,n) = U * x(m);
        % Stream function:
        psi_UniFlow(m,n) = U * y(n);
        end
end
% Plots
% Uniform stream in x direction:
contour(xx,yy,psi_UniFlow,[-2:.5:2],'k'),hold on
contour(xx,yy,phi_UniFlow,[-10:1:5],'r')
axis image,hold off
title('Uniform stream in x direction')
xlabel('x'),ylabel('y')
```

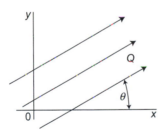

FIGURE 3.7

Uniform stream at an arbitrary angle θ from the x-axis.

Flow of Constant Velocity in Any Direction

Consider the flow streaming past the x- and y-axes at some velocity **Q**, making an angle θ with the x-axis, as shown in Fig. 3.7. Velocity **Q** can be resolved into two components U and V parallel to the x- and y-axes, respectively, where $\mathbf{Q} = U\mathbf{i} + V\mathbf{j}$, $Q^2 = U^2 + V^2$, and $\tan\theta = V/U$. The potentials for this flow field are

$$\phi = Ux + Vy \quad \text{and} \quad \psi = -Vx + Uy \tag{3.20}$$

where U and V are constants. Thus the lines of constant ψ are streamlines defined by the straight lines

$$-Vx + Uy = \text{constant}$$

Assigning a different value of ψ for every streamline plotted leads to a set of parallel lines in the direction of $\mathbf{Q}$, as shown in Fig. 3.7.

Example 3.1

Interpret the flow given by the stream function (units: $m^2\,s^{-1}$)

$$\psi = 6x + 12y$$

This represents a uniform stream with velocity vector $\mathbf{U} = (U, V)$. The constant component of the velocity in the x direction is $u = \partial\psi/\partial y = +12\ ms^{-1}$. The constant component of the velocity in the y direction is $v = -\partial\psi/\partial x = -6\ m\,s^{-1}$. Therefore, the flow equation represents uniform flow inclined to the x-axis by angle θ where $\tan\theta = -6/12$ (i.e., inclined downward).

The speed of flow is given by

$$Q = \sqrt{6^2 + 12^2} = \sqrt{180}\ \ ms^{-1}$$

3.2.2 Two-Dimensional Flow from a Source (or towards a Sink)

Let us find cylindrically symmetric solutions to the Laplace equation, $\phi = \Phi(r)$. In this case, the Laplace equation in polar coordinates is useful:

$$\frac{1}{r}\frac{\partial}{\partial r}\left(r\frac{\partial\phi}{\partial r}\right) + \frac{1}{r^2}\frac{\partial^2\phi}{\partial\theta^2} = 0$$

For cylindrically symmetric solutions this reduces to

$$\frac{1}{r}\frac{\partial}{\partial r}\left(r\frac{\partial\Phi}{\partial r}\right) = 0$$

Integrating twice, we get

$$\Phi = C_1 \ln r + C_2$$

where C_1 and C_2 are arbitrary constants. For example, if we set $C_1 = 1$ and $C_2 = 0$, then

$$\Phi_o = \ln r$$

which is often called the *fundamental* *solution* to the two-dimensional Laplace equation. This formula satisfies the Laplace equation everywhere except at $r = 0$, where it is logarithmically singular. The source flow field it induces is examined next.

A source (sink) of strength $m(-m)$ is a point at which fluid is appearing (or disappearing) at a uniform rate of $m(-m)$ m^2 s^{-1}. Let us assume this source is located at $r = 0$; hence, the constant $C_1 = m/(2\pi)$ and the constant $C_2 = 0$. In this case, the velocity potential is

$$\phi = \frac{m}{2\pi} \ln r$$

We can interpret the flow field for this potential as follows. Consider the analogy of a small hole in a large flat plate through which fluid is welling (the source). If there is no obstruction and the plate is perfectly flat and level, the fluid puddle will get larger and larger, all the while remaining circular in shape. The path that any particle of fluid will trace out as it emerges from the hole and travels outward is a purely radial one; it cannot go sideways because its fellow particles are also moving outward.

Also, its velocity must lessen as it goes outwards. Fluid issues from the hole at a rate of m m^2 s^{-1}. The velocity of flow over a circular boundary of 1-m radius is $m/2\pi$ m s^{-1}. Over a circular boundary of 2-m radius it is $m/(2\pi \times 2)$ (i.e., half as much, and over a circle of diameter $2r$ the velocity is $m/2\pi r$ m s^{-1}. Therefore, the velocity of flow is inversely proportional to the distance of the particle from the source.

All the previous applies to a sink except that fluid is being drained away through the hole and is moving toward the sink radially, increasing in speed as the sink is approached. Thus the particles all move radially, and the streamlines must be radial lines with their origin at the source (or sink).

The stream function that describes the source at the origin of the coordinate system is the stream function

$$\psi = \frac{m}{2\pi} \tan^{-1} \frac{y}{x} \quad \text{or} \quad \psi = \frac{m}{2\pi} \theta \qquad (3.21)$$

where $\theta = \tan^{-1}(y/x)$ (see Fig. 3.8). Again, we place the source (for convenience) at the origin of a system of axes, to which the point P is at (x, y) or (r, θ). The velocity potential for this flow is

$$\phi = \frac{m}{4\pi} \ln \left(x^2 + y^2 \right) \quad \text{or} \quad \phi = \frac{m}{2\pi} \ln r \qquad (3.22)$$

where $x = r\cos\theta$, $y = r\sin\theta$ and, hence, $r^2 = x^2 + y^2$ were used.

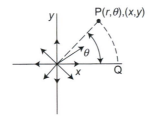

FIGURE 3.8

Source flow and polar coordinates.

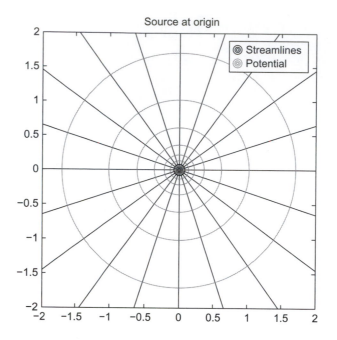

FIGURE 3.9

The potentials of a source: the radial lines emanating from the origin are the streamlines; the concentric circles surrounding the origin are lines of constant velocity potential.

Fig. 3.9 is an illustration of the pattern of the potentials for the source. The MATLAB file used to generate this figure is given in Table 3.2.

The components of the velocity vector due to the source at $(x, y) = (0, 0)$ are

$$u = \frac{\partial \phi}{\partial x} = \frac{m}{2\pi} \frac{x}{x^2 + y^2}, \quad v = \frac{\partial \phi}{\partial y} = \frac{m}{2\pi} \frac{y}{x^2 + y^2} \tag{3.23}$$

or

$$u_r = \frac{\partial \phi}{\partial r} = \frac{1}{r}\frac{\partial \psi}{\partial \theta} = \frac{m}{2\pi r}, \quad u_\theta = \frac{1}{r}\frac{\partial \phi}{\partial \theta} = -\frac{\partial \psi}{\partial r} = 0 \tag{3.24}$$

where u_r and u_θ are the radial and circular components of the velocity in polar coordinates.

3.2.3 Doublet Located at (x, y)=(0, 0)

The doublet given in this section points in the $-x$ direction. The potentials that describe this flow field are

$$\phi = \mu \frac{x}{x^2 + y^2} \quad \text{and} \quad \psi = -\mu \frac{y}{x^2 + y^2} \tag{3.25}$$

Table 3.2 MATLAB Script to Plot Figure 3.9

```
sig = 1; % Source strength
% GRID:
x = -2:.02:2;
y = -2:.02:2;
for m = 1:length(x)
    for n = 1:length(y)
        xx(m,n) = x(m); yy(m,n) = y(n);
    % Velocity potential function:
    phi_Source(m,n) = (sig/4/pi) * log(x(m)^2+(y(n)+.01)^2);
    % Stream function:
    psi_Source(m,n) = (sig/2/pi) * atan2(y(n),x(m));
    end
end
% Plots
% Source at origin of coordinate systex:
contour(xx,yy,psi_Source,[-.5:.05:.5],'k'),hold on
contour(xx,yy,phi_Source,10,'r')
legend('streamlines' ,'potential')
title(' Source at origin')
axis image,hold off
```

The components of the corresponding velocity field are

$$u = \mu \frac{y^2 - x^2}{\left(x^2 + y^2\right)^2}, \quad v = -\mu \frac{2xy}{\left(x^2 + y^2\right)^2} \tag{3.26}$$

The potentials for this flow field are illustrated in Fig. 3.10. The MATLAB script used to generate this figure is given in Table 3.3.

3.2.4 Line (Point) Vortex

The flow induced by a line (point) vortex described next can be interpreted as the flow induced by a straight-line vortex that is infinitely long in the z direction. In the (x, y) plane, it is a point. The potentials for this solution of the Laplace equation, if the point vortex is located at $(x, y) = (0, 0)$, are

$$\phi = -\frac{\Gamma}{2\pi} \tan^{-1} \frac{y}{x}, \quad \psi = -\frac{\Gamma}{4\pi} \ln\left(x^2 + y^2\right) \tag{3.27}$$

The corresponding components of the velocity field are

$$u = \frac{\partial \phi}{\partial x} = -\frac{\Gamma}{2\pi} \frac{y}{x^2 + y^2}, \quad v = \frac{\partial \phi}{\partial y} = \frac{\Gamma}{2\pi} \frac{x}{x^2 + y^2} \tag{3.28}$$

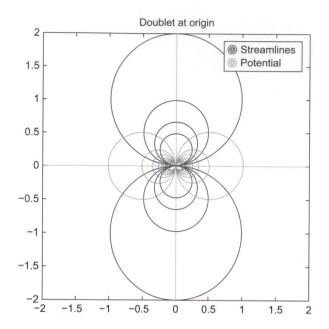

FIGURE 3.10

The potentials of an $-x$ directed doublet: streamlines and velocity potential lines are circles; flow along the x-axis is in the $-x$ direction.

Table 3.3 MATLAB Script Applied to Plot Figure 3.10

```
mu = 1;   % Doublet strength
% GRID:
x = -2:.02:2;
y = -2:.02:2;
for m = 1:length(x)
    for n = 1:length(y)
        xx(m,n) = x(m); yy(m,n) = y(n);
    % Velocity potential function:
    phi_Doublet(m,n) = mu * x(m)/(x(m)^2+(y(n)+.01)^2);
      % Stream function:
    psi_Doublet(m,n) = - mu * y(n)/(x(m)^2+(y(n)+.01)^2);
      end
end
% Plots
% Doublet at origin of coordinate system:\
figure(4)
contour(xx,yy,psi_Doublet,[-2:.5:2],'k'),hold on
contour(xx,yy,phi_Doublet,[-10:1:10],'r')
legend('streamlines' ,'potential')
title(' Doublet at origin')
axis image,hold off
```

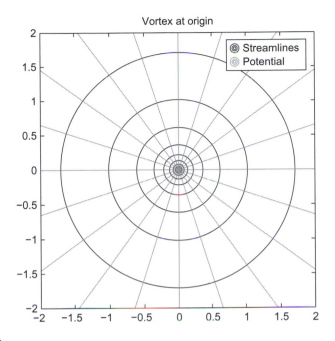

FIGURE 3.11

The potentials of a point vortex: streamlines are concentric circles; velocity potential lines are radial lines.

The potentials for this flow field are illustrated in Fig. 3.11. The MATLAB script used to generate this figure is given in Table 3.4.

Since the flow due to a line vortex gives streamlines that are concentric circles, $u_r = 0$ and u_θ is finite. In polar coordinates, the potentials are as follows:

$$\phi = \frac{\Gamma}{2\pi}\theta, \quad \psi = -\frac{\Gamma}{2\pi}\ln r \tag{3.29}$$

The corresponding components of the velocity field are

$$u_r = \frac{\partial \phi}{\partial r} = \frac{1}{r}\frac{\partial \psi}{\partial \theta} = 0, \quad u_\theta = \frac{1}{r}\frac{\partial \phi}{\partial \theta} = -\frac{\partial \psi}{\partial r} = \frac{\Gamma}{2\pi r} \tag{3.30}$$

3.2.5 Solid Boundaries and Image Systems

The fact that the flow is always along a streamline and not through it has an important fundamental consequence. A streamline of an *inviscid* flow can be replaced by a solid boundary of the same shape without affecting the remainder of the flow pattern. If, as is often the case, a streamline forms a closed curve that separates the flow pattern into two separate streams, one inside and one outside, then a solid body can replace

Table 3.4 MATLAB Script Applied to Plot Figure 3.11

```
gam = 1; % Vortex strength
% GRID:
x = -2:.02:2;
y = -2:.02:2;
for m = 1:length(x)
    for n = 1:length(y)
        xx(m,n) = x(m); yy(m,n) = y(n);
    % Velocity potential function:
    phi_Vortex(m,n) = (gam/2/pi)*atan2(y(n)+.01,x(m));
    % Stream function:
    psi_Vortex(m,n) = -(gam/4/pi)*log(x(m)^2+(y(n)+.01)^2);
    end
end
% Plots
% Vortex at origin of coordinate system:
contour(xx,yy,psi_Vortex,10,'k'),hold on
contour(xx,yy,phi_Vortex,[-.5:.05:.5],'r')
legend('streamlines' ,'potential')
title(' Vortex at origin')
axis image
```

the closed curve and the flow made outside without altering its shape. An example of this is illustrated in Fig. 3.12(a); this is the Rankine oval described in detail in Section 3.2.7. To represent the flow in the region of a contour or body, it is only necessary to replace the contour by a similarly shaped streamline; in this case, the closed streamline is oval shaped because it divides the flow between a source, S_o, of strength m followed by a sink, S_i, of strength $-m$ along the x-axis in a uniform stream flowing in the direction of the x-axis. All of the source flow is consumed by the sink; hence, a closed dividing streamline is formed that can be used to model a body of the same shape in a uniform inviscid flow. The following sections contain examples of simple flows that provide continuous streamlines in the shapes of circles and airfoils, and these emerge as consequences of the combinations of elementary solutions selected.

When arbitrary contours and their adjacent flows have to be replaced by identical flows containing similarly shaped streamlines, image systems must be placed within the contour that reflect the external flow system in the solid streamline.

Figure 3.12(b) shows the simple case of a source m placed a short distance A from an infinite plane wall. The effect of the solid boundary on the flow from the source is exactly represented by considering the effect of the image source m placed a distance A' reflected in the wall; the magnitude of the distance is the same as A. The source pair has a long straight streamline (i.e., the vertical axis of symmetry) that separates the flows from the two sources and that may be replaced by a solid boundary without

coordinates

There is

Substituting

This is the d
edge begins
as follows:

Thus

With two po
source flow
nose and so
$(r,\theta) = (r_t,$
the Rankine

Rearranging

This is an e
(x,y) coord
sian coordi
Fig. 3.13.
We can

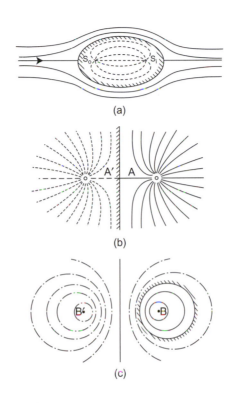

(a)

(b)

(c)

FIGURE 3.12

Solid boundaries constructed by image systems of singular solutions.

affecting the flow. Assume that the two sources are located on the x-axis and that the y-axis ($x = 0$) is the wall. The potential for this flow field is thus

$$\phi = \frac{m}{4\pi} \ln\left((x-A)^2 + y^2\right) + \frac{m}{4\pi} \ln\left((x+A)^2 + y^2\right)$$

where the second term on the right-hand side is the image source required to model the flow field of a source adjacent to the wall on the right side of the wall. The source is located at $(x,y) = (A,0)$. Its image is located at $(x,y) = (-A,0)$. Thus by symmetry the wall is located at $x = 0$. We can demonstrate that a wall is located at this position by solving for the velocity vector normal to the y-axis for all y as follows:

$$u = \frac{\partial \phi}{\partial x}\Big|_{x=0} = \left[-\frac{m}{2\pi}\frac{x-A}{(x-A)^2+y^2} - \frac{m}{2\pi}\frac{x+A}{(x+A)^2+y^2}\right]_{x=0} = 0$$

Thus, since the normal component of the velocity is zero at $y = 0$, the y-axis is a streamline and hence a model of a wall in inviscid flow. Examination of further

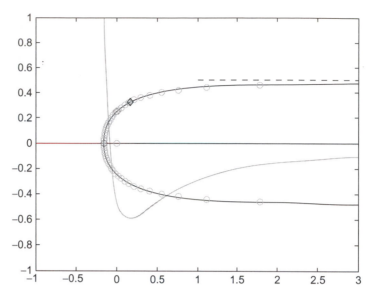

FIGURE 3.13

Rankine nose complete solution graphically displayed.

Substituting the velocity components, we get

$$C_p = -\left(\frac{m}{\pi U_\infty} \frac{\cos\theta}{r} + \frac{m^2}{4\pi^2 U_\infty^2} \frac{1}{r^2} \right)$$

From Eq. (3.33) on the Rankine nose, we find that

$$\frac{m}{U_\infty} = \frac{2\pi r \sin\theta}{\pi - \theta}$$

Substituting this into the equation for C_p, we get the pressure distribution on the Rankine nose:

$$-C_{pd} = \frac{2\sin\theta\cos\theta}{\pi - \theta} + \frac{\sin^2\theta}{(\pi - \theta)^2}$$

With this equation, we can determine the location of the minimum pressure and hence the minimum pressure exactly. To do this we take the derivative of this formula with respect to θ and set it equal to zero. The result of this operation (which is left as an exercise) leads to

$$\theta_m = \tan^{-1}\left(\frac{\pi - \theta_m}{\pi - \theta_m - 1} \right) \tag{3.37}$$

which may be solved by successive approximation with the initial guess of $\theta_m = 0$. The result is 62.9569 degrees and the minimum pressure is $C_{pmin} = -0.5866$. The numerical results compare with these results to within the significant figures shown. Table 3.5 presents the MATLAB solution to the entire problem.

The Rankine nose is a geometrically similar object because increasing or decreasing m and/or U_∞ does not change the location and the value of the minimum pressure coefficient. This is because changing the ratio m/U_∞ does not change the variation in curvature of the surface at any location on the body. The length of the nose and the thickness change proportionally when m and/or U_∞ are changed. Thus, we can say that all Rankine noses are geometrically similar.

3.2.7 Rankine Oval

A source of strength m upstream of a sink of strength $-m$ in a uniform stream U is used to construct the Rankine oval. The stream function due to this combination is

$$\psi = \frac{m}{2\pi} \tan^{-1} \frac{2cy}{x^2 + y^2 - c^2} - Uy \tag{3.38}$$

The uniform stream in this case is coming from the $-x$ direction. Also, the first term represents a source and sink combination set with the source to the right of the sink. For the source to be upstream of the sink, the uniform stream must be from right to left (i.e., flowing in the negative x direction). If the source is placed downstream of the sink, an entirely different stream pattern is obtained.

The velocity potential at any point in the flow due to this combination is given by

$$\phi = \frac{m}{2\pi} \ln \frac{r_1}{r_2} - Ur\sin\theta \tag{3.39}$$

or

$$\phi = \frac{m}{4\pi} \ln \frac{x^2 + y^2 + c^2 - 2xc}{x^2 + y^2 + c^2 + 2xc} - Ux \tag{3.40}$$

The streamline $\psi = 0$ gives a closed oval curve (not an ellipse) that is symmetrical about the x- and y-axes. Flow of stream function ψ greater than $\psi = 0$ shows the flow around such an oval set at zero incidence in a uniform stream. Streamlines can be obtained by plotting or by superposition of the separate standard flows (Fig. 3.14). The streamline $\psi = 0$ again separates the flow into two distinct regions. The first is wholly contained within the closed oval and consists of the flow out of the source and into the sink. The second is the approaching uniform stream that flows around the oval curve and returns to its uniformity. Again replacing $\psi = 0$ by a solid boundary, or indeed a solid body whose shape is given by $\psi = 0$, does not influence the flow pattern in any way.

Thus the stream function ψ of Eq. (3.35) can be used to represent the flow around a long cylinder of oval section set with its major axis parallel to a steady stream. To find the stream function representing a flow around such an oval cylinder, it must be

Table 3.5 MATLAB Script of Complete Solution of Rankine Nose Problem

```
%
% The Rankine nose (or leading edge) Feb. 2, 2010
% Source in a uniform stream: A 2D potential flow
%
    clear;clc
    disp(' Example: Rankine nose')
    m = 1; % Source strength for source at (x,y) = (0,0).
    V = 1; % Free stream velocity in the x-direction
    disp('    V      m ')
    disp([V m])
    disp(' Velocity potential:')
    disp('    phi = V*x + (m/4/pi)*log(x^2+y^2)')
    disp('Stream function:')
    disp(' psi = V*y + (m/2/pi)*atan2(y,x)')
    disp('The (x,y) components of velocity (u,v):')
    disp('    u = V + m/2/pi * xc/(x^2+y^2)')
    disp('    v = m/2/pi * y/(x^2+y^2)')
    %
    xstg = - m/2/pi/V; ystg = 0; % Location of stagnation point.
    %
    N = 1000;
    xinf = 3;
    xd = xstg:xinf/N:xinf;
    for n = 1:length(xd)
        if n==1
            yd(1) = 0;
        else
        yd(n) = m/2/V;
        for it = 1:2000
            yd(n) = (m/2/V)*( 1 - 1/pi*atan2(yd(n),xd(n)) );
        end
        end
    end
    xL(1) = xd(end); yL = -yd(end);
    for nn = 2:length(xd)-1
    xL(nn) = xd(end-nn); yL(nn) = -yd(end-nn);end
    plot([xd xL],[yd yL],'k',[-1 3],[0 0],'k'),axis([-1 3 -1 1])
     u = V + m/2/pi * xd./(xd.^2+yd.^2);
     v = m/2/pi * yd./(xd.^2+yd.^2);
     Cp = 1 - (u.^2+v.^2)/V^2;
     hold on
     plot(xd,Cp),axis([-1 3 -1 1])
    plot(0,m/V/4,'o')
    plot(xstg,ystg,'o')
    plot([1 3],[m/2/V m/2/V],'--k')
```

Table 3.5 (*Continued*)

```
    [Cpmin ixd] = min(Cp);
    xmin = xd(ixd);
    ymin = yd(ixd);
    plot(xmin,ymin,'+r')
    Cpmin
% Computation of normal and tangential velocity on (xd,yd):
    phi = V*xd + (m/4/pi).*log(xd.^2+yd.^2);
    dx = diff(xd); dy = diff(yd); ds = sqrt(dx.^2 + dy.^2);
    dph = diff(phi); ut = dph./ds; xm = xd(1:end-1) + dx/2;
    psi = V*yd + (m/2/pi).*atan2(yd,xd);
    plot(xm,1-ut.^2/V^2,'r')
%
% Check on shape equation
%
    th = 0:pi/25:2*pi;
    r = (m/2/pi/V)*(pi - th)./sin(th);
    xb = r.*cos(th);
    yb = r.*sin(th);
    plot(xb,yb,'om')
%
% Exact location of minimum pressure
    thm = 0;
    for nit = 1:1000
        thm = atan2(pi-thm,pi-thm-1);
    end
    thdegrees = thm*180/pi
        rm = (m/2/pi/V)*(pi - thm)/sin(thm);
            xm = rm*cos(thm);
                ym = rm*sin(thm);
    plot(xm,ym,'dk')
    um = V + m/2/pi * xmin/(xmin^2+ymin^2);
        vm = m/2/pi * ymin/(xmin^2+ymin^2);
            Cpm = 1 - (um^2+vm^2)/V^2
```

possible to obtain m and c (the strengths of the source and sink and distance apart) in terms of the size of the body and the speed of the incident stream.

Suppose there is an oval of breadth $2b_0$ and thickness $2t_0$ set in a uniform flow of U. The problem is to find m and c in the stream function, Eq. (3.35), which will then represent the flow around the oval.

1. The oval must conform to Eq. (3.35):

$$\psi = 0 = \frac{m}{2\pi} \tan^{-1} \frac{2cy}{x^2 + y^2 - c^2} - Uy$$

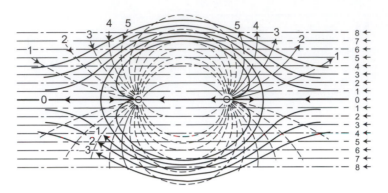

FIGURE 3.14

Rankine oval.

2. On streamline $\psi = 0$, maximum thickness t_0 occurs at $x = 0$, $y = t_0$. Therefore, substituting in the above equation, the oval must conform to Eq. (3.35):

$$0 = \frac{m}{2\pi} \tan^{-1} \frac{2ct_0}{t_0^2 - c^2} - Ut_0$$

and rearranging

$$\tan \frac{2\pi Ut + 0}{m} = \frac{2ct_0}{t_0^2 - c^2} \qquad (3.41)$$

3. A stagnation point (a point where the local velocity is zero) is situated at the "nose" of the oval (i.e., at the point $y = 0$, $x = b_0$):

$$u = 0 = \frac{\partial \psi}{\partial y} = \frac{\partial}{\partial y} \left(\frac{m}{2\pi} \tan^{-1} \frac{2cy}{x^2 + y^2 - c^2} - Uy \right)$$

$$\frac{\partial \psi}{\partial y} = \frac{m}{2\pi} \frac{1}{1 + \left(\frac{2cy}{x^2 + y^2 - c^2} \right)^2} \frac{\left(x^2 + y^2 - c^2 \right) 2c - 2y 2cy}{\left(x^2 + y^2 - c^2 \right)^2} - U$$

and putting $y = 0$ and $x = b_0$ with $\partial \psi / \partial y = 0$:

$$0 = \frac{m}{2\pi} \frac{\left(b_0^2 - c^2 \right) 2c}{\left(b_0^2 - c^2 \right)^2} - U = \frac{m}{2\pi} \frac{2c}{b_0^2 - c^2} - U$$

Therefore,

$$m = \pi U \frac{b_0^2 - c^2}{c} \qquad (3.42)$$

The simultaneous solution of Eqs. (3.38) and (3.39) will furnish values of m and c to satisfy any given set of conditions. Alternatively (a), (b), and (c) can be used to find the thickness and length of the oval formed by the streamline $\psi = 0$.

3.2.8 Circular Cylinder with Circulation in a Cross Flow

This section is on the detailed analysis of the potential flow around a circular cylinder in a cross flow. We construct the potential, solve for the velocity and pressure distributions around the cylinder, and integrate the latter to determine the lift coefficient.

The potential for the flow around a cylinder in a cross flow with circulation is the superposition of a uniform stream in the x direction, a doublet located at the origin pointing in the $-x$ direction, and a vortex located at the origin. Thus, the potentials for this system are

$$\phi' = Vx + \mu\frac{x}{x^2+y^2} + \frac{\Gamma}{2\pi}\tan^{-1}\frac{y}{x}$$

and

$$\psi' = Vy - \mu\frac{y}{x^2+y^2} - \frac{\Gamma}{4\pi}\ln(x^2+y^2)$$

where $U_\infty = V$ in this example. For convenience, we scale the problem such that all velocities are divided by the free-stream speed V; hence, the scaled free-stream speed is $V = 1$. We scale all distances by the radius of the cylinder, $R = \sqrt{\mu/V}$; hence, $R = 1$ implies $\mu = 1$. The potential may then be rewritten as

$$\phi = x + \frac{x}{x^2+y^2} + \frac{\Gamma}{2\pi}\tan^{-1}\frac{y}{x} \tag{3.43}$$

and the stream function as

$$\psi = y - \frac{y}{x^2+y^2} - \frac{\Gamma}{4\pi}\ln\left(x^2+y^2\right) \tag{3.44}$$

Taking the gradient of the velocity potential, ϕ, we get the components of the velocity:

$$u = \frac{\partial\phi}{\partial x} = \frac{\partial\psi}{\partial y} = 1 + \frac{y^2-x^2}{\left(x^2+y^2\right)^2} - \frac{\Gamma}{2\pi}\frac{y}{x^2+y^2} \tag{3.45}$$

and

$$v = \frac{\partial\phi}{\partial y} = -\frac{\partial\psi}{\partial x} = -\frac{2yx}{\left(x^2+y^2\right)^2} + \frac{\Gamma}{2\pi}\frac{x}{x^2+y^2} \tag{3.46}$$

Finally, from the Bernoulli theorem, we write the pressure coefficient as follows:

$$C_p = \frac{p - p_\infty}{\frac{1}{2}\rho V^2} = 1 - \left(u^2 + v^2\right) \qquad (3.47)$$

A program to solve these equations around the unit circle is given in Table 3.6. Fig. 3.15 shows the pressure distribution around the cylinder with zero circulation. The minimum pressure coefficient is $C_p = -3$ located dead center at the top and bottom. The maximum pressure coefficient is $C_p = 1$ located at the leading- and trailing-edge stagnation points. The pressure distribution along the top of the cylinder is the same as the pressure distribution along the bottom, so only one curve is shown in the figure.

The following are some important points to note about the pressure distribution for the potential flow around a cylinder illustrated in Fig. 3.15.

Table 3.6 MATLAB File Used to Produce Figures 3.15 and 3.17

```
V = 1; mu = 1; Gamma = -2; R = sqrt(mu/V);
N = 360;
    dthe = 2*pi/N;
        the = 0:dthe:2*pi;
for n = 1:length(the)-1
  them(n) = the(n) + dthe/2;
  x(n)  = R*cos(them(n));
  y(n)  = R*sin(them(n));
  u(n)  = V + V*R^2*(y(n)^2 - x(n)^2)/(x(n)^2 + y(n)^2)^2 ...
          - Gamma/2/pi*y(n)/(x(n)^2+y(n)^2);
  v(n)  = -2*V*R^2*y(n)*x(n)/(x(n)^2 + y(n)^2)^2 ...
          + Gamma/2/pi*x(n)/(x(n)^2+y(n)^2);
    ut(n) = -u(n)*sin(them(n)) + v(n)*cos(them(n));
    ur(n) = u(n)*cos(them(n)) + v(n)*sin(them(n));
    Cp(n) = 1 - (u(n)^2+v(n)^2)/V^2;
end
plot(x,y,'k',[-1 1],[0 0],'k'),axis image
        hold on
plot(x,Cp),axis([-2 2 -5 1])
CL = -sum(Cp.*sin(them))*dthe
CLexact = -2*Gamma/R/V
for mm=1:length(the)-1
    them(mm) = the(mm) + dthe/2;
    uthe(mm) = -2*sin(them(mm)) + Gamma/2/pi/R/V;
    Cp2(mm)  = 1 - uthe(mm)^2;
end
CL2 = -sum(Cp2.*sin(them))*dthe
```

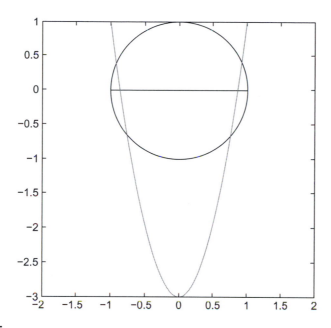

FIGURE 3.15

Pressure distribution around a circle; this is a plot of C_p versus x; the definition of C_p is given in Figure 3.16.

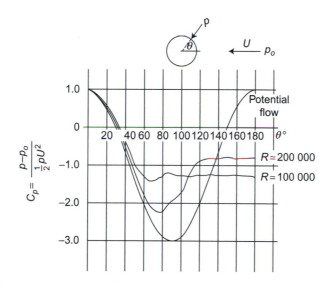

FIGURE 3.16

Pressure distribution around a circle; $C_L = 0$.

1. At the stagnation points ($0°$ and $180°$), the pressure difference ($p - p_0$) is positive and equal to $\rho U^2/2$.
2. At 30 and 150 degrees, where $\sin\theta = 1/2$, ($p - p_0$) is zero, and at these points the local velocity is the same as that of the free stream.
3. Between 30 and 150 degrees, C_p is negative, showing that p is less than p_0.
4. The pressure distribution is symmetrical about the vertical axis and there is therefore no drag force.

Fig. 3.16 compares this ideal pressure distribution with that obtained by experiment; it shows that the actual pressure distribution is similar to the theoretical value up to about 70 degrees, but departs radically from it thereafter. Furthermore, it can be seen that the pressure coefficient over the rear portion of the cylinder remains negative. This destroys the symmetry about the vertical axis and produces a force in the direction of the flow. This does *not* mean that the potential flow around a circular cylinder is useless. It is useful because this flow field can be mapped to the flow field around an airfoil as illustrated in Section 3.2.9.

Now we examine the same problem in polar coordinates. In addition, we explicitly keep V and R in the problem. This will help us interpret the meaning of the scaling applied previously. In polar coordinates, the velocity potential for this problem can be written as follows:

$$\phi = V\left(r + \frac{\mu}{rV}\right)\cos\theta + \frac{\Gamma}{2\pi}\theta$$

If we define the parameter $R = \sqrt{\mu/V}$, then

$$\phi = V\left(r + \frac{R^2}{r}\right)\cos\theta + \frac{\Gamma}{2\pi}\theta \tag{3.48}$$

The radial and tangential components of the velocity field are, respectively,

$$u_r = \frac{\partial\phi}{\partial r} = \left(1 - \frac{R^2}{r^2}\right)V\cos\theta \tag{3.49}$$

and

$$u_\theta = \frac{1}{r}\frac{\partial\phi}{\partial\theta} = -\left(1 + \frac{R^2}{r^2}\right)V\sin\theta + \frac{\Gamma}{2\pi r} \tag{3.50}$$

At $r = R$ the radial component of the velocity is zero. Hence, $r = R$ is the radius of a closed circular streamline and thus the surface of the cylinder in a cross flow V. On this surface $u_\theta = -2V\sin\theta + \Gamma/2\pi R$. Substituting this into the Bernoulli theorem, we get the exact solution of the pressure distribution on the surface of the cylinder, which is

$$p = p_\infty + \frac{1}{2}\rho\left[V^2 - \left(-2V\sin\theta + \frac{\Gamma}{2\pi R}\right)^2\right] \tag{3.51}$$

The lift and drag acting on the cylinder are defined as follows:

$$L = -\int_0^{2\pi} p\sin\theta\, R d\theta$$

$$D = -\int_0^{2\pi} p\cos\theta\, R d\theta$$

Substituting Eq. (3.48) into the integrals and doing the integrations, we get

$$L = -\rho V \Gamma \qquad\qquad (3.52)$$

$$D = 0 \qquad\qquad (3.53)$$

Equation (3.52) is the Kutta-Joukowski theorem. Equation (3.53) is what Euler called D'Alembert's paradox.

Let us rearrange Eq. (3.51) to get a formula for the dimensionless pressure coefficient:

$$C_p = \frac{p - p_\infty}{\frac{1}{2}\rho V^2} = 1 - \left(-2\sin\theta + \frac{\Gamma}{2\pi R V}\right)^2 \qquad\qquad (3.54)$$

If $\Gamma = 0$, then

$$C_p = \frac{p - p_\infty}{\frac{1}{2}\rho V^2} = 1 - 4\sin^2\theta$$

At $\theta = 0$ and π (the trailing and leading edges, respectively), this is unity. At $\theta = \pi/2$ and $3\pi/2$ (the top and bottom dead centers), this is -3. These results check with the results already given in Fig. 3.15.

Table 3.6 lists the MATLAB script for the case where R and V are explicitly allowed to be changed by the users. Figure 3.17 is the pressure distribution on the cylinder with $\Gamma = -1$. Note that the lift coefficient is equal to twice the magnitude of Γ because $V = 1$ and $R = 1$ were used. Further discussion on the lift coefficient and, in particular, the derivation of the exact result is described soon.

Circulation is not a geometric parameter. However, it can be related to a geometric parameter, α, illustrated in Fig. 3.18, the streamline plot of the flow around a cylinder. The script used to plot the streamlines is given in Table 3.7. The angle α is the angular location of the rear stagnation point as measured from the horizontal axis. It is related to the circulation in the following way. On the cylindrical surface the velocity is

$$\mathbf{u} = (u_r, u_\theta) = (0, -2V\sin\theta + \Gamma/2\pi R)$$

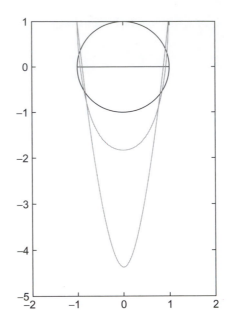

FIGURE 3.17

Pressure distribution around a circle with circulation; $C_L = 4$.

Table 3.7 MATLAB File Used to Produce Figure 3.18

```
mu = 1; gam = -2; V=1;
x = -3:.02:3;
y = -2:.02:2;
for m = 1:length(x)
  for n = 1:length(y)
   xx(m,n) = x(m); yy(m,n) = y(n);
   psis(m,n) = V * y(n) - mu * y(n)/(x(m)^2+(y(n)+.01)^2) ...
         - (gam/4/pi)*log(x(m)^2+(y(n)+.01)^2);
  end
end
contour(xx,yy,psis,[-3:.3:3],'k'), axis image
```

The stagnation points, if they occur on the cylinder, are located where $u_\theta = 0$. Thus, they are located at

$$\sin\theta \equiv \sin\alpha = \frac{\Gamma}{4\pi RV}$$

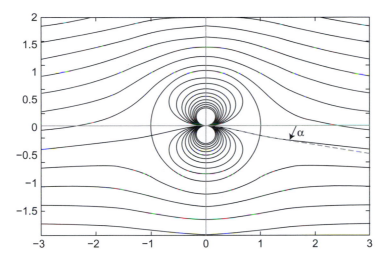

FIGURE 3.18

Streamlines and the definition of α.

Therefore,

$$\alpha = \sin^{-1}\left(\frac{\Gamma}{4\pi RV}\right) \tag{3.55}$$

or

$$\Gamma = 4\pi RV \sin\alpha \tag{3.56}$$

Substituting Eq. (3.56) into the Kutta-Joukowski theorem, we get

$$L = \rho V \Gamma = 8\pi \frac{1}{2}\rho V^2 R \sin\alpha$$

Thus, the lift coefficient in terms of α is

$$C_L = \frac{L}{\frac{1}{2}\rho V^2 R} = 8\pi \sin\alpha \tag{3.57}$$

Substituting for $\sin\alpha$ from Eq. (3.55) or (3.56), we can also write the lift coefficient as

$$C_L = 2\frac{\Gamma}{RV} \tag{3.58}$$

This coefficient of lift is the conventional definition for a cylinder with R used as the length scale. This illustrates the fact that there is zero lift if there is zero circulation.

3.2.9 Joukowski Airfoil and the Circular Cylinder

The solution of the flow around a circular cylinder with circulation in a cross flow can be used to predict the flow around thin airfoils. We can transform the local geometry of the cylinder into an ellipse, an airfoil, or a flat plate without influencing the geometry of the far-field flow. This procedure is known as conformal mapping. If we interpret the Cartesian coordinates as the coordinates of the plane of complex numbers $z = x + iy$, where x and y are real numbers, $i = \sqrt{-1}$, x is the real part of z, and y is the imaginary part of z, then doing this allows us to use complex variable theory to solve two-dimensional potential-flow problems. We are *not* going to examine complex variable theory here. Instead, we will give an example of applying one of the ideas to illustrate that the circle can be transformed into an airfoil. Table 3.8 is a MATLAB script that does this. MATLAB is very useful here because of its complex arithmetic capabilities. The steps are outlined in the table. The result of executing this code is illustrated in Fig. 3.19. The figure shows the circle and the airfoil onto which it is mapped. Each point on the circle corresponds to a unique point on the airfoil. The potential at each point on the cylinder is the same as the corresponding point on the airfoil. What is different is the distance between points, and thus the velocity and pressure distributions on the airfoil must be determined from the mapped distribution of the potential. Since the far field is not affected by the transformation, the lift on the airfoil is $L = -\rho V \Gamma$ (i.e., the same as the lift on the cylinder). However, the lift coefficient is not the same because the length of the airfoil is nearly four times the cylinder's radius. Using this fact, we multiply Eq. (3.54) by $R/c = 1/4$, where $R = 1$ is the radius of the unit circle examined in the previous section and $c = 4R$ is the length of the airfoil chord. Thus, for an airfoil with zero camber, we predict

$$C_L = \frac{L}{\frac{1}{2}\rho V^2 c} = 2\pi\alpha \tag{3.59}$$

where $\alpha \approx \sin\alpha$ for small angles was used. This formula illustrates the influence of angle of attack on the performance of an airfoil. The circulation was chosen such that the flow left the trailing edge smoothly (the Kutta condition).

This brief discussion is, of course, an overview of a connection between the flow around a circle and the flow around an airfoil. In the next chapter, we will deal with the details of thin-airfoil theory and show that this formula for lift is theoretically well founded. However, it illustrates only part of the story. We will also investigate the other geometric feature that contributes to lift—camber. It can be shown that, for cambered thin airfoils at small angles of attack,

$$C_L = \frac{L}{\frac{1}{2}\rho V^2 c} = 2\pi\left(\alpha + 2\frac{f}{c}\right)$$

where f is the maximum camber at midchord of a camber line that is symmetric fore and aft of the midchord. Aeronautical engineers find this equation and Eq. (3.59) quite useful, particularly in preliminary design and performance estimation analyses.

Table 3.8 MATLAB File Used to Produce Figure 3.19

```
    % Joukowski transformation MATLAB code
    %
    % Example of conformal mapping of a circle to an airfoil
    %                    AE425-ME425 Aerodynamics
    % Daniel T. Valentine ................... January 2009
    % Circle in (xp,yp) plane: R = sqrt(xp^2 + yp^2), R > 1
    % Complex variables of three complex planes of interest:
    %   zp = xp + i*yp ==> Circle plane
    %   z = x + i*y    ==> Intermediate plane
    %   w = u + i*v    ==> Airfoil (or physical) plane
    clear;clc
% Step 1: Select the parameters that define the airfoil of interest.
% (1) Select the a == angle of attack alpha
        a = -2;        % in degrees
        a = a*pi/180; % Conversion to radians
% (2) Select the parameter related to thickness of the airfoil:
        e = .1;
% (3) Select the shift of y-axis related to camber of the airfoil:
        f = .1;
% (4) Select the trailing edge angle parameter:
        te = .05;    % 0 < te < 1 (0 ==> cusped trailing edge)
        n = 2 - te;     % Number related to trailing edge angle.
        tea = (n^2-1)/3;    % This is a Karman-Trefftz extension.
% Step 2: Compute the coordinates of points on circle in zp-plane:
        R = 1 + e;
        theta = 0:pi/200:2*pi;
        yp = R * sin(theta);
        xp = R * cos(theta);
% Step 3: Transform coordinates of circle from zp-plane to z-plane:
        z = (xp - e) + i.*(yp + f);
% Step 4: Transform circle from z-plane to airfoil in w-plane
% (the w-plane is the "physical" plane of the airfoil):
   rot = exp(i*a); % Application of angle of attack.
   w = rot .* (z + tea*1./z); % Joukowski transformation.
% Step 5: Plot of circle in z-plane on top of airfoil in w-plane
        plot(xp,yp), hold on
        plot(real(w),imag(w),'r'),axis image, hold off
```

3.3 AXISYMMETRIC FLOWS (INVISCID AND INCOMPRESSIBLE FLOWS)

Consider now axisymmetric potential flows—that is, the flows around bodies such as cones aligned to the flow and spheres. To analyze and, for that matter, to define,

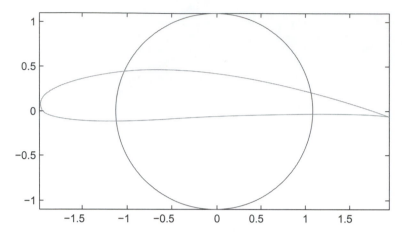

FIGURE 3.19

Streamlines and the definition of α.

axisymmetric flows, it is necessary to introduce cylindrical and spherical coordinate systems. Unlike the Cartesian system, these systems can exploit the underlying symmetry of the flows.

3.3.1 Cylindrical Coordinate System

The cylindrical coordinate system is illustrated in Fig. 3.20. The three coordinate surfaces are the planes $z =$ constant and $\theta =$ constant, with the surface of the cylinder having radius r. For the Cartesian system, in contrast, all three coordinate surfaces are planes. As a consequence for the Cartesian system, the directions (x, y, z) of the velocity components are fixed throughout the flow field. For the cylindrical coordinate system, though, only one of the directions (z) is fixed throughout the flow field; the other two $(r$ and $\theta)$ vary depending on the value of the angular coordinate θ. In this respect there is a certain similarity to the polar coordinates introduced earlier in the chapter. The velocity component q_r is always locally perpendicular to the cylindrical coordinate surface, and q_θ is always tangential to that surface. Once this elementary fact is properly understood, cylindrical and Cartesian coordinates are equally easy to use.

Like the relationships between velocity potential and velocity components derived for polar coordinates (see Section 3.1.3), the following relationships are obtained for cylindrical coordinates:

$$q_r = \frac{\partial \phi}{\partial r}, \ q_\theta = \frac{1}{r} \frac{\partial \phi}{\partial \theta}, \ q_z = \frac{\partial \phi}{\partial z} \tag{3.60}$$

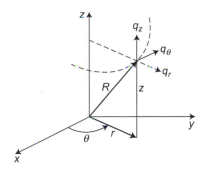

FIGURE 3.20

Cylindrical coordinates.

An axisymmetric flow is defined as one for which the flow variables (i.e., velocity and pressure) do not vary with the angular coordinate θ. This would be so, for example, for a body of revolution about the z-axis, along which the oncoming flow is directed. For such an axisymmetric flow a stream function can be defined. The continuity equation for axisymmetric flow in cylindrical coordinates can be derived in a similar manner as for two-dimensional flow in polar coordinates (see Section 2.4.3); it takes the form

$$\frac{1}{r}\frac{\partial rq_r}{\partial r} + \frac{\partial q_z}{\partial z} = 0 \tag{3.61}$$

The relationship between stream function and velocity component must be such as to satisfy Eq. (3.61); hence it can be seen that

$$q_r = -\frac{1}{r}\frac{\partial \psi}{\partial z}, \quad q_z = \frac{1}{r}\frac{\partial \psi}{\partial r} \tag{3.62}$$

3.3.2 Spherical Coordinates

For analyzing certain two-dimensional flows—for example, the flow over a circular cylinder with and without circulation—it is convenient to work with polar coordinates. The axisymmetric equivalents of polar coordinates are spherical coordinates, such as those used for analyzing the flow around spheres. Spherical coordinates are illustrated in Fig. 3.21. In this case none of the coordinate surfaces are plane and the directions of all three velocity components vary over the flow field, depending on the values of the angular coordinates θ and φ. The relationships between the velocity components and potential here are given by

$$q_R = \frac{\partial \phi}{\partial R}, \quad q_\theta = \frac{1}{R\sin\varphi}\frac{\partial \phi}{\partial \theta}, \quad q_\varphi = \frac{1}{R}\frac{\partial \phi}{\partial \varphi} \tag{3.63}$$

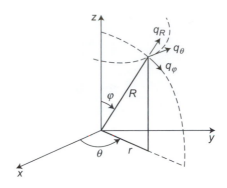

FIGURE 3.21

Spherical coordinates.

For axisymmetric flows, the variables are independent of θ, and in this case the continuity equation takes the form

$$\frac{1}{R^2}\frac{\partial\left(R^2 q_R\right)}{\partial R} + \frac{1}{R\sin\varphi}\frac{\left(\sin\varphi\, q_\varphi\right)}{\partial\varphi} = 0 \tag{3.64}$$

Again, the relationship between the stream function and the velocity components must be such as to satisfy the continuity equation (Eq. 3.64), so

$$q_R = \frac{1}{R^2\sin\varphi}\frac{\partial\psi}{\partial\varphi}, \quad q_\varphi = -\frac{1}{R\sin\varphi}\frac{\partial\psi}{\partial R} \tag{3.65}$$

3.3.3 Axisymmetric Flow from a Point Source (or towards a Point Sink)

The point source and sink are similar to the line source and sink discussed in Section 3.2. A close physical analogy can be found if one imagines the flow into or out of a very (strictly infinitely) thin round pipe, as depicted in Fig. 3.22. As suggested in this figure the streamlines would be purely radial in direction.

We will suppose that the flow rate out of the point source is given by Q. Q is usually referred to as the strength of the point source. Now, since the flow is purely radial away from the source, the total flow rate across the surface of any sphere having its center at the source is also Q. (Note that this sphere is purely notional and does not represent a solid body or in any way hinder the flow.) Thus the radial velocity component at any radius R is related to Q as follows:

$$4\pi R^2 q_R = Q$$

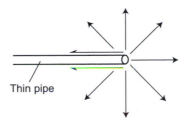

FIGURE 3.22

Illustration of a spherical source.

It therefore follows from Eq. (3.63) that

$$q_R = \frac{\partial \phi}{\partial R} = \frac{Q}{4\pi R^2}$$

Integration then gives the expression for the velocity potential of a point source as

$$\phi = -\frac{Q}{4\pi R} \qquad (3.66)$$

In a similar fashion, an expression for the stream function can be derived using Eq. (3.65), giving

$$\psi = -\frac{Q}{4\pi}\cos\varphi \qquad (3.67)$$

3.3.4 Point Source and Sink in a Uniform Axisymmetric Flow

Placing a point source and/or sink in a uniform horizontal stream of $-U$ leads to results very similar to those found in Section 3.2.5 for the two-dimensional case with line sources and sinks.

First the velocity potential and stream function for uniform flow $-U$ in the z direction must be expressed in spherical coordinates. The velocity components q_R and q_φ are related to $-U$ as follows:

$$q_R = -U\cos\varphi \quad \text{and} \quad q_\varphi = U\sin\varphi$$

Using Eq. (3.63) followed by integration then gives

$$\frac{\partial \phi}{\partial R} = -U\cos\varphi \rightarrow \phi = -UR\cos\varphi + f(\varphi)$$

$$\frac{\partial \phi}{\partial \varphi} = -UR\sin\varphi \rightarrow \phi = -UR\cos\varphi + g(R)$$

$f(\varphi)$ and $g(R)$ are arbitrary functions that take the place of constants of integration when partial integration is carried out. Plainly, in order for the two expressions for ϕ derived previously to be in agreement, $f(\varphi) = g(R) = 0$. The required expression for the velocity potential is thereby given as

$$\phi = -UR\cos\varphi \tag{3.68}$$

Similarly, using Eq. (3.65) followed by integration gives

$$\frac{\partial\psi}{\partial\varphi} = -UR^2\cos\varphi\sin\varphi \rightarrow \psi = -\frac{UR^2}{4}\cos 2\varphi + f(R)$$

$$\frac{\partial\psi}{\partial R} = -UR\sin^2\varphi \rightarrow \psi = -\frac{UR^2}{2}\sin^2\varphi + g(\varphi)$$

Recognizing that $\cos 2\varphi = 1 - 2\sin^2\varphi$, it can be seen that the two expressions given for ψ agree if the arbitrary functions of integration take the values $f(R) = -UR^2/4$ and $g(\varphi) = 0$. The required expression for the stream function is thus given as

$$\psi = -\frac{UR^2}{2}\sin^2\varphi \tag{3.69}$$

Using Eqs. (3.66) and (3.68) and Eqs. (3.67) and (3.69), it can be seen that for a point source at the origin placed in a uniform flow $-U$ along the z-axis,

$$\phi = -UR\cos\varphi - \frac{Q}{4\pi R} \qquad \psi = -\frac{1}{2}UR^2\sin^2\varphi - \frac{Q}{4\pi}\cos\varphi \tag{3.70}$$

The flow field represented by Eqs. (3.70) corresponds to the potential flow around a semi-finite body of revolution—very much like its two-dimensional counterpart described in Section 3.2.6. Similarly to the procedure described in Section 3.2.6, it can be shown that the stagnation point occurs at the point $(-a, 0)$, where

$$a = \sqrt{\frac{Q}{4\pi U}} \tag{3.71}$$

and that the streamlines passing through this stagnation point define a body of revolution given by

$$R^2 = 2a^2(1 + \cos\varphi)/\sin^2\varphi \tag{3.72}$$

The derivation of Eqs. (3.71) and (3.72) are left as an exercise (see Exercise 3.19 at the end of the chapter).

As in the two-dimensional case described in Section 3.2.7, a point source placed on the z-axis at $z = -a$, combined with an equal-strength point sink also placed on

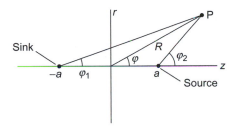

FIGURE 3.23

Source-sink pair.

the z-axis at $z = a$ (see Fig. 3.23) gives the following velocity potential and stream function at point P:

$$\phi = \frac{Q}{4\pi \left[(R\cos\varphi + a)^2 + R^2 \sin^2\varphi \right]^{1/2}}$$

$$- \frac{Q}{4\pi \left[(R\cos\varphi - a)^2 + R^2 \sin^2\varphi \right]^{1/2}} \tag{3.73}$$

$$\psi = \frac{Q}{4\pi}(\cos\varphi_1 - \cos\varphi_2) \tag{3.74}$$

where

$$\cos\varphi_1 = \frac{R\cos\varphi + a}{\left[(R\cos\varphi + a)^2 + R^2 \sin^2\varphi \right]^{1/2}}$$

$$\cos\varphi_2 = \frac{R\cos\varphi - a}{\left[(R\cos\varphi - a)^2 + R^2 \sin^2\varphi \right]^{1/2}}$$

If this source-sink pair is placed in a uniform stream $-U$ in the z direction, it generates the flow around a body of revolution known as a Rankine body. The shape is very similar to the two-dimensional Rankine oval shown in Fig. 3.14 and described in Section 3.2.7.

3.3.5 The Point Doublet and the Potential Flow around a Sphere

A point doublet is produced when the source-sink pair in Fig. 3.23 become infinitely close together. This is closely analogous to a line doublet, which was described in Section 3.2.8. Mathematically, the expressions for velocity potential and stream function for a point doublet can be derived, respectively, from Eqs. (3.73) and (3.74) by allowing $a \to 0$, keeping $\mu = 2Qa$ fixed. The latter quantity is known as the strength of the doublet.

If a is very small, a^2 may be neglected compared to $2Ra\cos\varphi$ in Eq. (3.73) and can then can be written as

$$\phi = \frac{Q}{4\pi R}\left[\frac{1}{[1+2\,(a/R)\cos\varphi]^{1/2}} - \frac{1}{[1-2\,(a/R)\cos\varphi]^{1/2}}\right] \qquad (3.75)$$

On expanding,

$$\frac{1}{\sqrt{1\pm x}} = 1 \mp \frac{1}{2}x + \dots$$

Therefore, as $a \to 0$, Eq. (3.75) reduces to

$$\phi = -\frac{\mu}{4\pi R^2}\cos\varphi \qquad (3.76)$$

Similarly, we can show that

$$\psi = \frac{\mu}{4\pi R}\sin^2\varphi \qquad (3.77)$$

The streamline patterns corresponding to the point doublet are similar to those depicted in Fig. 3.10. It is apparent from this streamline pattern and from the form of Eq. (3.77) that, unlike the point source, the flow field for the doublet is not omnidirectional. On the contrary, it is strongly directional. Moreover, the case analyzed previously is a somewhat special case in that the source-sink pair lie on the z-axis. In fact, the *axis* of the doublet can be in any direction in three-dimensional space.

For two-dimensional flow, it was shown in Section 3.2.8 that the line doublet placed in a uniform stream produces potential flow around a circular cylinder. It is similarly shown here that a point doublet placed in a uniform stream corresponds to the potential flow around a sphere.

From Eqs. (3.68) and (3.76), the velocity potential for a point doublet in a uniform stream, with both the stream and doublet axis aligned in the negative z direction, is given by

$$\phi = -UR\cos\varphi - \frac{\mu}{4\pi R^2}\cos\varphi \qquad (3.78)$$

From Eq. (3.63), the velocity components are given by

$$q_R = \frac{\partial\phi}{\partial R} = -\left(U - \frac{\mu}{2\pi R^3}\right)\cos\varphi \qquad (3.79)$$

$$q_\varphi = \frac{1}{R}\frac{\partial\phi}{\partial\varphi} = \left(U + \frac{\mu}{4\pi R^3}\right)\sin\varphi \qquad (3.80)$$

The stagnation points are defined by $q_R = q_\varphi = 0$. Let the coordinates of the stagnation points be denoted (R_s, φ_s). Then, from Eq. (3.80), it can be seen that

$$R_s^3 = -\frac{\mu}{4\pi U} \quad \text{or} \quad \sin\varphi_s = 0$$

The first of these two equations cannot be satisfied because it implies that R_s is not a positive number. Accordingly, the second of the two equations must hold, implying that

$$\phi_s = 0 \text{ and } \pi$$

It now follows from Eq. (3.76) that

$$R_s = \left(\frac{\mu}{2\pi U}\right)^{1/3} \tag{3.81}$$

Thus there are two stagnation points on the z-axis at equal distances from the origin.

From Eqs. (3.69) and (3.77), the stream function for a point doublet in a uniform flow is given by

$$\psi = -\frac{UR^2}{2}\sin^2\varphi + \frac{\mu}{4\pi R}\sin^2\varphi \tag{3.82}$$

It follows from substituting Eq. (3.81) into Eq. (3.82) that at the stagnation points $\psi = 0$. So the streamlines passing through the stagnation points are described by

$$\psi = -\left(\frac{UR^2}{2} - \frac{\mu}{4\pi R}\right)\sin^2\varphi = 0 \tag{3.83}$$

Equation (3.82) shows that, when $\varphi \neq 0$ or π, the radius R of the stream surface, containing the streamlines that pass through the stagnation points, remains fixed equal to R_s. R can take any value when $\phi = 0$ or π. Thus these streamlines define the surface of a sphere of radius R_s. This is very similar to the two-dimensional case of flow over a circular cylinder described in Section 3.2.8.

From Eqs. (3.80) and (3.81), it follows that the velocity on the surface of the sphere is given by

$$q = \frac{3}{2}U\sin\varphi$$

so that using the Bernoulli equation gives

$$p_0 + \frac{1}{2}\rho U^2 = p + \frac{1}{2}\rho q^2 = p + \frac{1}{2}\rho\left(\frac{3}{2}U\sin\varphi\right)^2$$

Therefore, the pressure variation over the sphere's surface is given by

$$p - p_0 = \frac{1}{2}\rho U^2 \left(1 - \frac{9}{4}\sin^2\varphi\right) \tag{3.84}$$

Again, this result is quite similar to that for the circular cylinder described in Section 3.2.8.

3.3.6 Flow around Slender Bodies

It was shown that the flow around a class of bodies of revolution can be modeled by the use of a source and sink of equal strength. Accordingly, it is natural to speculate whether the flow around more general body shapes can be obtained using several sources and sinks or a distribution of them along the z-axis. It is indeed possible to do this as Fuhrmann first demonstrated [5]. Two examples similar to those Fuhrmann presented are shown in Fig. 3.24. Although Fuhrmann's method could model the flow around realistic-looking bodies, it suffered an important defect from a design perspective. One could calculate the body of revolution corresponding to a specified distribution of sources and sinks, but a designer would want to solve the inverse problem of selecting the variation of source strength in order to obtain the flow around a given shape. This more practical approach became possible after Munk [6] introduced his slender-body theory for calculating the forces on airship hulls. We briefly describe this approach next.

For Munk's slender-body theory, it is assumed that the radius of the body is very much smaller than its total length. The flow is modeled by a distribution of sources and sinks placed on the z-axis as depicted in Fig. 3.25. In many respects, this theory is analogous to the theory for calculating the two-dimensional flow around symmetric wing sections, known as the thickness problem.

For an element of source distribution located at $z = z_1$, the velocity induced at point $P(r, z)$ is

$$q_R = \frac{\sigma(z_1)}{4\pi R^2}\,dz_1 \tag{3.85}$$

(a)   (b)

FIGURE 3.24

Two examples of flows around bodies of revolution generated by (a) point source plus linear distribution of sink strength, and (b) two linear distributions of source/sink strength. The distributions are denoted by broken lines.

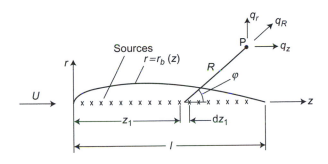

FIGURE 3.25

Flow over a slender body of revolution modeled by source distribution.

where $\sigma(z_1)$ is the source strength per unit length and $\sigma(z_1)dz_1$ takes the place of Q in Eq. (3.63). Thus, to obtain the velocity components in the r and z directions at P due to all sources, we resolve the velocity given by Eq. (3.82) in the two coordinate directions and integrate along the length of the body. Thus

$$q_r = \int_0^l q_R \sin\varphi \, dz_1 = \frac{1}{4\pi}\int_0^l \sigma(z_1)\frac{r}{\left[(z-z_1)^2+r^2\right]^{3/2}}dz_1 \qquad (3.86)$$

$$q_z = \int_0^l q_R \cos\varphi \, dz_1 = \frac{1}{4\pi}\int_0^l \sigma(z_1)\frac{z-z_1}{\left[(z-z_1)^2+r^2\right]^{3/2}}dz_1 \qquad (3.87)$$

Source strength can be related to body geometry by the following physical argument. Consider the elemental length of the body as shown in Fig. 3.26. If the body radius r_b is very small compared to the length, l, then the limit $r \to 0$ can be considered. For this limit, the flow from the sources may be considered purely radial so that the flow across the body surface of the element is entirely due to the sources within the element itself. Accordingly,

$$2\pi r q_r dz_1 = \sigma(z_1)dz_1 \text{ at } r = r_b \quad \text{provided} \quad r_b \to 0$$

But the effects of the oncoming flow as well as the sources must be considered. The net perpendicular velocity on the body surface due to both the oncoming flow and the sources must be zero. Provided that the slope of the body contour is very small (i.e., $dr_b/dz \ll 1$), the perpendicular and radial velocity components may be considered the same. Thus the requirement that the net normal velocity be zero becomes (see Fig. 3.26)

$$q_r = U\sin\beta = U\frac{dr_b}{dz_1}$$

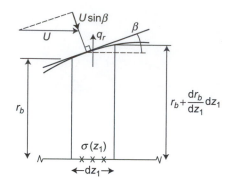

FIGURE 3.26

Element of slender body of revolution.

where q_r is associated with the sources and the two parts on the right-hand side are associated with the oncoming flow. The above relationship is such that the source strength per unit length and body shape are related as follows:

$$\sigma(z) = U\frac{dS}{dz} \tag{3.88}$$

where S is the frontal area of a cross-section and is given by $S = \pi r_b^2$.

In the limit, as $r \to 0$, Eq. (3.87) simplifies to

$$q_z = \frac{1}{4\pi} \int_0^l \frac{\sigma(z+1)}{(z-z_1)^2} dz_1 \tag{3.89}$$

Thus, once the variation of source strength per unit length has been determined according to Eq. (3.88), the axial velocity can be obtained by evaluating Eq. (3.89), and hence the pressure can be evaluated from the Bernoulli equation.

We see from the derivation of Eq. (3.89) that both r_b and dr_b/dz must be very small. Plainly, the latter requirement would be violated in the vicinity of $z = 0$ if the body had a rounded nose. This is a major drawback of the method.

The slender-body theory was extended by Munk [9] to the case of a body at an angle of incidence or yaw. This case is treated as a superposition of two distinct flows, as shown in Fig. 3.27(a). One of these is the slender body at zero angle of incidence as discussed previously. The other is the slender body in a cross flow. For such a slender body, the flow around a particular cross-section is closely analogous to that around a circular cylinder (see Section 3.2.8). Accordingly, this flow can be modeled by a distribution of point doublets with axes aligned in the direction of the cross flow,

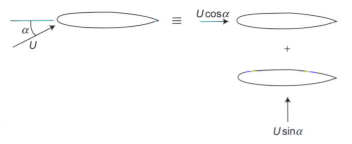

Flow at angle of yaw around a body of revolution as the superposition of two flows

(a)

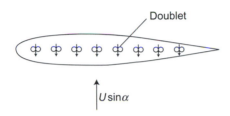

Cross flow over slender body of revolution modeled as distribution of doublets

(b)

FIGURE 3.27

Slender body in an arbitrary onset flow: at top (a) an angle of attack; at bottom (b) in a cross-flow.

as depicted in Fig. 3.27(b). Slender-body theory will not be taken further here. The student is referred to Thwaites and Karamcheti for further details [8, 9].

3.4 COMPUTATIONAL (PANEL) METHODS

In Section 3.2.7, it was shown how the two-dimensional potential flow around an oval-shaped contour, the Rankine oval, can be generated by the superposition of a source and sink on the x-axis and a uniform flow. An analogous three-dimensional flow can also be generated around a Rankine body (see Section 3.3.4) by using a point source and sink. Thus it can be demonstrated that the potential flow around certain bodies can be modeled by placing sources and sinks in the body's interior. However, it is only possible to deal with particular cases in this way. We can model the potential flow around slender bodies of any shape by distributing sources lying along the x-axis in the interior of the body. This slender-body theory is discussed in

Section 3.3.6. However, calculations based on this theory are only approximate unless the body is infinitely thin and the slope of the body contour is very small. Even in this case, the theory breaks down if the nose or leading edge is rounded because there the slope of the contour is infinite. The panel methods described here model the potential flow around a body by distributing sources over the body surface. In this way, the potential flow around a body of any shape can be calculated to a very high degree of precision. The method was developed by Hess and Smith [10] at Douglas Aircraft.

If a body is placed in a uniform flow of speed U, in exactly the same way as for the Rankine oval of Section 3.2.7, the velocity potential for the flow may be superimposed on that for the disturbed flow around the body to obtain a total velocity potential of the form

$$\Phi = Ux + \phi \tag{3.90}$$

where ϕ denotes the so-called disturbance potential (i.e., the departure from free-stream conditions). It can be shown that the disturbance potential flow around a body of any given shape can be modeled by a distribution of sources over its surface (Fig. 3.28). Let the source strength per unit arc of contour be σ_Q. In the two-dimensional case, $\sigma_Q ds_Q$ would replace $m/2\pi$ in Eq. (3.22). Thus the velocity potential at P due to sources on an element ds_Q of arc of contour centered at point Q is given by

$$\phi_{PQ} = \sigma_Q \ln R_{PQ} \, ds_Q \tag{3.91}$$

where R_{PQ} is the distance from P to Q. The velocity potential due to all of the sources on the surface is obtained by integrating Eq. (3.91), over the surface. Thus, following Eq. (3.90), the total velocity potential at P can be written (for the two-dimensional case) as

$$\phi_P = Ux + \oint \sigma_Q \ln R_{PQ} ds_Q \tag{3.92}$$

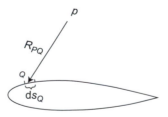

FIGURE 3.28

Surface distribution of singularities on a two-dimensional body.

where the integral is understood as being carried out over the contour of the body. With the power of desktop computers today, this is the basis of a class of computational techniques that is quite useful in aerodynamic design.

In order to use Eq. (3.92) for numerical modeling, it is first necessary to "discretize" the surface—that is, break it down into a finite but quite possibly large number of separate parts. This is achieved by representing the two-dimensional shape of the surface by a series of straight line segments (see Fig. 3.29).

The use of panel methods to calculate the potential flow around a body may be best illustrated with a concrete example. For this we select the two-dimensional flow around a symmetric airfoil (see Fig. 3.29).

The first step is to number all end points or *nodes* of the panels from 1 to N, as indicated in Fig. 3.29. Individual panels are assigned the same number as the node located to the left when facing in the outward direction from the panel. The midpoints of each panel are chosen as *collocation* points. We will see that the boundary condition of zero flow perpendicular to the surface is applied at these points. Then we define for each panel the unit normal and tangential vectors, $\hat{n}_i$ and $\hat{t}_i$, respectively. Consider panels i and j in the Fig. 3.29. The sources distributed over panel j induce a velocity, denoted by the vector $\mathbf{v}_{ij}$ at the collocation point of panel i. The components of $\mathbf{v}_{ij}$ perpendicular and tangential to the surface at the collocation point i are given by the scalar (or dot) products $\mathbf{v}_{ij} \cdot \hat{n}_i$ and $\mathbf{v}_{ij} \cdot \hat{t}_i$, respectively. Both of these quantities are proportional to the strength of the sources on panel j and therefore can be written in the forms

$$\mathbf{v}_{ij} \cdot \hat{n}_i = \sigma_j N_{ij} \text{ and } \mathbf{v}_{ij} \cdot \hat{t}_i = \sigma_j T_{ij} \qquad (3.93)$$

N_{ij} and T_{ij} are the perpendicular and tangential velocities induced at the collocation point of panel i by sources of unit strength distributed over panel j; they are known as the *normal* and *tangential influence coefficients*.

The actual velocity perpendicular to the surface at collocation point i is the sum of the perpendicular velocities induced by each of the N panels plus the contribution

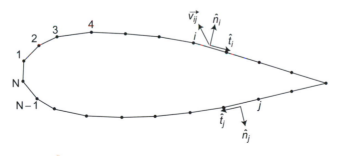

FIGURE 3.29

Discretization of a surface distribution of singularities.

due to the free stream. It is given by

$$v_{n_i} = \sum_{j=1}^{N} \sigma_j N_{ij} + \mathbf{U} \cdot \hat{n}_i \tag{3.94}$$

In a similar fashion, the tangential velocity at collocation point i is given by

$$v_{S_i} = \sum_{j=1}^{N} \sigma_j T_{ij} + \mathbf{U} \cdot \hat{t}_i \tag{3.95}$$

If the surface represented by the panels is to correspond to a solid surface, the actual perpendicular velocity at each collocation point must be zero. This condition may be expressed mathematically as $v_{n_i} = 0$. Equation (3.94) then becomes

$$\sum_{j=1}^{N} \sigma_j N_{ij} = -\mathbf{U} \cdot \hat{n}_i \quad (i = 1, 2, \dots, N) \tag{3.96}$$

Equation (3.96) is a system of linear algebraic equations for the N unknown source strengths, σ_i $(i = 1, 2, \dots, N)$. It takes the form of a matrix equation

$$\mathbf{N}\sigma = \mathbf{b} \tag{3.97}$$

where $\mathbf{N}$ is an $N \times N$ matrix composed of the elements N_{ij}, σ is a column matrix composed of the N elements σ_i, and $\mathbf{b}$ is a column matrix composed of the N elements $-\mathbf{U} \cdot \hat{n}_i$. Assuming for the moment that the perpendicular influence coefficients N_{ij} have been calculated and that the elements of the right-hand column matrix $\mathbf{b}$ have also been calculated, Eq. (3.97) may, in principle at least, be solved for the source strengths comprising the elements of the column matrix σ. Systems of linear equations such as (3.97) can be readily solved numerically using standard methods. For the results presented here, one of the matrix inversion procedures available in MATLAB was used to solve for the source strengths. This method is described by Press et al. [9], who also give listings for the necessary computational routines.

Once the influence coefficients N_{ij} have been calculated, the source strengths can be determined by solving the system of Eq. (3.96) by some standard numerical technique. If the tangential influence coefficients T_{ij} have also been calculated, then, once the source strengths have been determined, the tangential velocities may be obtained from Eq. (3.95). The Bernoulli equation can now be used to calculate the pressure acting at collocation point i; in particular, the coefficient of pressure is given by Eq. (2.24) as

$$C_{p_i} = 1 - \left(\frac{v_{S_i}}{U}\right)^2 \tag{3.98}$$

The calculation of the influence coefficient is a central and essential part of the panel method. This is the question now addressed. As a first step, consider the calculation of the velocity induced at a point P by sources of unit strength distributed over a panel centered at point Q.

In terms of a coordinate system (x_Q, y_Q) measured relative to the panel (Fig. 3.30), the disturbance potential is given by integrating Eq. (3.91) over the panel. Mathematically this is expressed as

$$\phi_{PQ} = \int_{-\Delta s/2}^{\Delta s/2} \ln \sqrt{(x_Q - \xi)^2 + y_Q^2} \, d\xi \tag{3.99}$$

The corresponding velocity components at P in the x_Q and y_Q directions can be readily obtained from Eq. (3.99) as

$$v_{x_Q} = \frac{\partial \phi_{PQ}}{\partial x_Q} = \int_{-\Delta s/2}^{\Delta s/2} \frac{x_Q - \xi}{(x_Q - \xi)^2 + y_Q^2} \, d\xi$$

$$= -\frac{1}{2} \ln \left[\frac{(x_Q + \Delta s/2)^2 + y_Q^2}{(x_Q - \Delta s/2)^2 + y_Q^2} \right] \tag{3.100}$$

$$v_{y_Q} = \frac{\partial \phi_{PQ}}{\partial y_Q} = \int_{-\Delta s/2}^{\Delta s/2} \frac{y_Q}{(x_Q - \xi)^2 + y_Q^2} \, d\xi$$

$$= -\left[\tan^{-1} \left(\frac{x_Q + \Delta s/2}{y_Q} \right) = \tan^{-1} \left(\frac{x_Q - \Delta s/2}{y_Q} \right) \right] \tag{3.101}$$

Armed with these results for the velocity components induced at point P due to the sources on a panel centered at point Q, we now return to the problem of calculating the influence coefficients. Suppose that points P and Q are chosen to be the collocation points i and j, respectively. Equations (3.100) and (3.101) give the velocity components in a coordinate system relative to panel j; whereas the velocity components perpendicular and tangential to panel i are required. In vector form, the velocity at collocation point i is given by

$$\mathbf{v}_{PQ} = v_{x_Q} \hat{t}_j + v_{y_Q} \hat{n}_j$$

Therefore, to obtain the components of this velocity vector perpendicular and tangential to panel i, we take the scalar product of the velocity vector with $\hat{n}_i$ and $\hat{t}_i$, respectively, to obtain

$$N_{ij} = \mathbf{v}_{PQ} \cdot \hat{n}_i = v_{x_Q} \hat{n}_i \cdot \hat{t}_j + v_{y_Q} \hat{n}_i \cdot \hat{n}_j \tag{3.102a}$$

$$T_{ij} = \mathbf{v}_{PQ} \cdot \hat{t}_i = v_{x_Q} \hat{t}_i \cdot \hat{t}_j + v_{y_Q} \hat{t}_i \cdot \hat{n}_j \tag{3.102b}$$

A COMPUTATIONAL ROUTINE IN MATLAB

To see how calculation of the influence coefficients works in practice, a function routine written in MATLAB is given in Table 3.9. This script file solves for the flow around an ellipse in a uniform stream. With the parameters selected, as given in the table, the ellipse is actually a circular cylinder in order to check the surface distribution of sources model of a cylinder without circulation. The results in Fig. 3.31 illustrate that the identical pressure distribution is obtained as compared with the model, previously examined, based on a doublet at the origin in a uniform stream.

The routine in this function, step by step, performs the following:

1. The surface is discretized by assigning numbers from 1 to N to points on the surface of the airfoil, as suggested in Fig. 3.29. The x and y coordinates of these points are function inputs XP and YP.
2. For each of the N panels, the collocation points are calculated by taking an average of the coordinates at either end of the panel in question.
3. For each of the panels, the length $S(J) = \Delta s_j$ is calculated.
4. The x and y components of the unit tangent vectors for each panel are calculated as follows:

$$t_{jx} = \frac{x'_j - x'_{j-1}}{\Delta s_j}, \quad t_{jy} = \frac{y'_j - y'_{j-1}}{\Delta s_j}$$

5. The unit normal vectors are then calculated from $n_{jx} = -t_{jy}$ and $n_{jy} = t_{jx}$. The main task of the function, that of calculating the influence coefficients, now begins.
6. For each possible combination of panels, I and $J = 1$ to N, the first step is dealing with the special case when $I = J$—that is, the velocity induced by the sources on the panel itself at its collocation point. From Eqs. (3.100), (3.101), (3.102a), and (3.102b) we see that, when $x_Q = y_Q = 0$,

$$v_{PQ_x} = \ln(1) = 0, v_{PQ_y} = \tan^{-1}(\infty) - \tan^{-1}(-\infty) = \pi$$

When $i \neq j$, the influence coefficients have to be calculated from Eqs. (3.100), (3.101), (3.102a), and (3.102b).

7. The components DX and DY of R_{PQ} are calculated in terms of the x and y coordinates.
8. The components of R_{PQ} in terms of the coordinate system based on panel j are then calculated as

$$X_Q = \vec{R}_{PQ} \cdot \hat{t}_j, \ Y_Q = \vec{R}_{PQ} \cdot \hat{n}_j$$

9. VX and VY (i.e., v_{x_Q} and v_{y_Q}) are evaluated using Eqs. (3.100) and (3.101).
10. $\hat{n}_i \cdot \hat{t}_j, \hat{n}_i \cdot \hat{n}_j, \hat{t}_i \cdot \hat{t}_j$, and $\hat{t}_i \cdot \hat{n}_j$ are evaluated.
11. Finally, the influence coefficients are evaluated from Eq. (3.102b).

The function just presented is primarily intended for educational purposes and has not been optimized to economize on computing time. Nevertheless, the MAT-LAB program in Table 3.10 produced the calculation for an NACA 0024 airfoil presented in Fig. 3.31. In this case, 160 panels were used for the complete airfoil consisting of upper and lower surfaces. The results compare exactly with results obtained from other potential-flow codes. For more details on other two-dimensional methods, on the lifting body problem (discussed at the end of the next chapter), and three-dimensional panel and other *boundary-integral* methods the book by Katz and Plotkin [12] is quite useful.

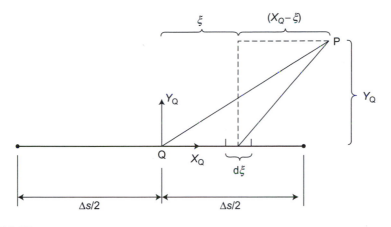

FIGURE 3.30

Local coordinate system of a surface panel.

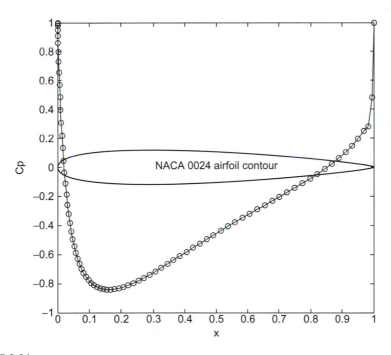

FIGURE 3.31

Comparison of pressure distributions.

Table 3.9 MATLAB Function to Compute Influence Coefficients of a Source Distribution

```
function [AN,AT,XC,YC,NHAT,THAT] = InfluSour(XP,YP,N)
% Influence coefficents for source distribution over a
% symmetric body.
% AN is the Nij matrix and AT is the Tij matix in Eqn (3.99).
%
for J = 1:N
if J==1
  XPL = XP(N);
  YPL = YP(N);
else
  XPL = XP(J-1);
  YPL = YP(J-1);
end
XC(J) = 0.5*(XP(J) + XPL);
YC(J) = 0.5*(YP(J) + YPL);
S(J) = sqrt( (XP(J) - XPL)^2 + (YP(J) - YPL)^2 );
THAT(J,1) = (XP(J) - XPL)/S(J);
THAT(J,2) = (YP(J) - YPL)/S(J);
NHAT(J,1) = - THAT(J,2);
NHAT(J,2) = THAT(J,1);
end
%Calculation of the influence coefficients.
for I = 1:N
for J = 1:N
if I==J
AN(I,J) = pi;
AT(I,J) = 0;
else
DX = XC(I) - XC(J);
DY = YC(I) - YC(J);
XQ = DX*THAT(J,1) + DY*THAT(J,2);
YQ = DX*NHAT(J,1) + DY*NHAT(J,2);
VX = -0.5*( log( (XQ + S(J)/2 )^2 + YQ*YQ )...
    -log( (XQ - S(J)/2 )^2 + YQ*YQ ) );
VY = -( atan2( (XQ + S(J)/2),YQ) - atan2((XQ - S(J)/2),YQ));
NTIJ = 0;
NNIJ = 0;
TTIJ = 0;
TNIJ = 0;
for K = 1:2
NTIJ = NHAT(I,K)*THAT(J,K) + NTIJ;
NNIJ = NHAT(I,K)*NHAT(J,K) + NNIJ;
TTIJ = THAT(I,K)*THAT(J,K) + TTIJ;
TNIJ = THAT(I,K)*NHAT(J,K) + TNIJ;
end
```

Table 3.9 (*Continued*)

```
AN(I,J) = VX*NTIJ + VY*NNIJ;
AT(I,J) = VX*TTIJ + VY*TNIJ;
end
end
end
```

Table 3.10 MATLAB Script to Compute Pressure Distribution on a
NACA 0024 Airfoil as illustrated in Figure 3.31

```
% This is the NACA 0024 offsets:
XY =[1.000000 0.000252
    0.990966 0.005041
    0.976102 0.009129
       ...      ...
    0.000160 0.004489
    0.000017 0.001486
    0.000017 -0.001486
    0.000160 -0.004489
       ...      ...
    0.941520 -0.018360
    0.959610 -0.013579
    0.976102 -0.009129
    0.990966 -0.005041
    1.000000 -0.000252];
 XP = XY(:,1); YP = XY(:,2); N = length(XP);
[AN,AT,XC,YC,NHAT,THAT] = InfluSour(XP,YP,N);
  uinfty = 1;
  for j=1:N
  b(j,1) = uinfty*NHAT(j,1) +0*NHAT(j,2);
  end
Sources = AN\b; ut = AT*Sources - THAT(:,1);
cp = 1 - ut.^2; plot(XC,cp,'-ok',XP,YP,'k')
xlabel(' x '),ylabel(' Cp ')
```

3.5 EXERCISES

3.1 Define vorticity in a fluid and obtain an expression for it at a point with polar
coordinates (r,θ), the motion being assumed two-dimensional. From the definition of a line vortex as irrotational flow in concentric circles determine the
variation of velocity with radius and thus obtain the stream function (ψ), and
the velocity potential (ϕ), for a line vortex. (U of L)

3.2   A sink of strength $120 \text{ m}^2\text{ s}^{-1}$ is situated 2 m downstream from a source of equal strength in an irrotational uniform stream of 30 m s^{-1}. Find the fineness ratio of the oval formed by the streamline $\psi = 0$.
(*Answer:* 1.51) (CU)

3.3   A sink of strength $20 \text{ m}^2\text{ s}^{-1}$ is situated 3 m upstream of a source of $40 \text{ m}^2\text{ s}^{-1}$ in a uniform irrotational stream. It is found that, at a point 2.5 m equidistant from both source and sink, the local velocity is normal to the line joining the source and sink. Find the velocity at this point and the velocity of the undisturbed stream.
(*Answer:* $1.02 \text{ m s}^{-1}, 2.29 \text{ m s}^{-1}$) (CU)

3.4   A line source of strength m and a sink of strength $2m$ are separated a distance c. Show that the field of flow consists in part of closed curves. Locate any stagnation points and sketch the field of flow. (U of L)

3.5   Derive the expression giving the stream function for irrotational flow of an incompressible fluid past a circular cylinder of infinite span. In this way, determine the position of generators on the cylinder at which the pressure is equal to that of the undisturbed stream.
(*Answer:* $\pm30°, \pm150°$) (U of L)

3.6   Determine the stream function for a two-dimensional source of strength m. Sketch the resultant field of flow due to three such sources, each of strength m, located at the vertices of an equilateral triangle. (U of L)

3.7   Derive the irrotational flow formula

$$p = p_0 = \frac{1}{2}\rho U^2(1 - 4\sin^2\theta)$$

giving the intensity of normal pressure p on the surface of a long circular cylinder, set at right angles to a stream of velocity U. The undisturbed static pressure in the fluid is p_0 and θ is the angular distance from the stagnation point. Describe briefly an experiment to test the accuracy of the formula and comment on the results obtained. (U of L)

3.8   A long right circular cylinder of diameter α m is set horizontally in a steady stream of velocity $U \text{ m s}^{-1}$ and caused to rotate at $\omega \text{ rad s}^{-1}$. Obtain an expression in terms of ω and U for the ratio of the pressure difference between the top and bottom of the cylinder to the dynamic pressure of the stream. Describe briefly the behavior of the stagnation lines of such a system as ω increases from zero, keeping U constant.
(*Answer:* $8a\omega/U$) (CU)

3.9   A line source is immersed in a uniform stream. Show that the resultant flow, if irrotational, may represent the flow past a two-dimensional fairing. If a maximum thickness of the fairing is 0.15 m and the undisturbed velocity of the

stream is 6.0 m/s, determine the strength and location of the source. Also obtain an expression for the pressure at any point on the surface of the fairing, taking the pressure at infinity as a datum.
(*Answer:* 0.9 m^2 s^{-1}, 0.0237 m) (U of L)

3.10 A long right circular cylinder of radius a m is held with its axis normal to an irrotational inviscid stream of U. Obtain an expression for the drag force acting on a unit length of the cylinder due to pressures exerted on the front half only.
(*Answer:* $-\rho U^2 a/3$) (CU)

3.11 Show that a velocity potential exists in a two-dimensional steady irrotational incompressible fluid motion. The stream function of a two-dimensional motion of an incompressible fluid is given by

$$\psi = \frac{a}{2}x^2 + bxy - \frac{c}{2}y^3$$

where a, b, and c are arbitrary constants. Show that, if the flow is irrotational, the lines of constant pressure never coincide with either the streamlines or the equipotential lines. Is this possible for rotational motion? (U of L)

3.12 State the stream function and velocity potential for each of the motions induced by a source, vortex, and doublet in a two-dimensional incompressible fluid. Show that a doublet may be regarded as either of the following:
(a) The limiting case of a source and sink.
(b) The limiting case of equal and opposite vortices, clearly indicating the direction of the resultant doublet. (U of L)

3.13 Define the stream function, the irrotational flow, and the velocity potential for two-dimensional motion of an incompressible fluid, indicating the conditions under which they exist. Determine the stream function for a point source of strength σ at the origin. Hence, or otherwise, show that, for the flow due to any number of sources at points on a circle, the circle is a streamline provided that the algebraic sum of the strengths of the sources is zero. (U of L)

3.14 A line vortex of strength Γ is mechanically fixed at the point $(l,0)$ in space described by a Cartesian coordinate system. It is in an inviscid incompressible fluid at rest at infinity bounded by a plane wall coincident with the y-axis. Find the velocity in the fluid at the point $(0, y)$, and determine the force that acts on the wall (per unit depth) if the pressure on the other side of the wall is the same as at infinity. Bearing in mind that this must be equal and opposite to the force acting on a unit length of the vortex, show that your result is consistent with the Kutta-Joukowski theorem. (U of L)

3.15 Write the velocity potential for the two-dimensional flow about a circular cylinder with a circulation Γ in an otherwise uniform stream of velocity U. Hence show that the lift on a unit span of the cylinder is $\rho U \Gamma$. Produce a brief but plausible argument that the same result should hold for the lift on a cylinder of arbitrary shape, basing your argument on consideration of the flow at large distances from the cylinder. (U of L)

3.16 Define the terms *velocity potential*, *circulation*, and *vorticity* as used in two-dimensional fluid mechanics, and show how they are related. The velocity distribution in the laminar boundary layer of a wide flat plate is given by

$$u = u_0 \left[\frac{3}{2} \frac{y}{\delta} - \frac{1}{2} \left(\frac{y}{\delta} \right)^3 \right]$$

where u_0 is the velocity at the edge of the boundary layer where y equals δ. Find the vorticity on the surface of the plate.
[*Answer:* $-(3/2)(u_0/\delta)$] (U of L)

3.17 A two-dimensional fluid motion is represented by a point vortex of strength Γ set at unit distance from an infinite straight boundary. Draw the streamlines and plot the velocity distribution on the boundary when $\Gamma = \pi$. (U of L)

3.18 The velocity components of a two-dimensional inviscid incompressible flow are given by

$$u = 2y - \frac{y}{\left(x^2 + y^2 \right)^{1/2}}, \quad v = -2x - \frac{x}{\left(x^2 + y^2 \right)^{1/2}}$$

Find the stream function and the vorticity, and sketch the streamlines.
[*Answer:* $\psi = x^2 + y^2 + (x^2 + y^2)^{1/2}$, $\zeta = -(4 + 1/\sqrt{(x^2 + y^2)})$] (U of L)

3.19 (a) Given that the velocity potential for a point source takes the form

$$\phi = -\frac{Q}{4\pi R}$$

where, in axisymmetric cylindrical coordinates (r, z), $R = \sqrt{z^2 + r^2}$, show that, when a uniform stream U is superimposed on a point source located at the origin, there is a stagnation point located on the z-axis upstream of the origin at distance

$$a = \sqrt{\frac{Q}{4\pi U}}$$

(b) Given that, in axisymmetric spherical coordinates (R, φ), the stream function for the point source takes the form

$$\psi = -\frac{Q}{4\pi R}$$

show that the streamlines passing through the stagnation point found in (a) define a body of revolution given by

$$R^2 = \frac{2a^2 \left(1 + \cos\varphi\right)}{\sin^2\varphi}$$

Make a rough sketch of this body.

Two-Dimensional Wing Theory

LEARNING OBJECTIVES

- Learn how to apply potential-flow theory to solve for the pressure distribution on the airfoil (or lifting wing section).

- Learn that potential-flow theories in and of themselves offer little further scope for this problem unless modified to simulate certain effects of real flows. The result is a powerful but elementary airfoil theory capable of wide exploitation.

- Learn how camber and angle of attack generate lift via circulation.

- Learn how flapped airfoils and jet flaps work.

- Learn how to deal with arbitrary shaped airfoils by applying surface distributions of singularities using relatively powerful computational panel methods.

4.1 INTRODUCTION

By the end of the nineteenth century, the theory of ideal, or potential, flow (see Chapter 3) was reasonably well developed. The motion of an inviscid fluid was a well-defined mathematical problem, satisfying a relatively simple linear partial differential equation, the Laplace equation (see Section 3.1), with well-defined boundary conditions. Owing to this state of affairs, many distinguished mathematicians were able to develop a wide variety of analytical methods for predicting such flows. Their work was and is very useful for many practical problems—for example, the flow around airships, ship hydrodynamics, and water waves. Early in the twentieth century, it was recognized that, for the important practical applications in aerodynamics (e.g., the flow around an airfoil), great care was required to successfully apply potential-flow theory.

Potential-flow theory predicted the flow field exactly for an inviscid fluid—that is, for an *infinite Reynolds number*. In two important respects, however, it did not correspond to the flow field of a real fluid, no matter how large the Reynolds number is. First, real flows have a tendency to separate from the surface of the body. This is especially pronounced when the bodies are bluff such as a circular cylinder, and in such cases real flow bears no resemblance to the corresponding potential flow.

Second, steady potential flow around a body can produce no force irrespective of the body's shape. This result is usually known as *d'Alembert's paradox* after the French mathematician who first discovered it in 1744. Thus there is no prospect of using potential-flow theory in its pure form to estimate the lift or drag of wings and thereby develop aerodynamic design methods.

Flow separation and d'Alembert's paradox both result from the subtle effects of viscosity on flows at high Reynolds numbers. The necessary understanding and knowledge of viscous effects came largely from work done during the first two decades of the twentieth century. It took several more decades, however, before this knowledge was fully exploited in aerodynamic design. The great German aeronautical engineer Prandtl [33] and his research team at the University of Göttingen deserve most of the credit both for explaining these paradoxes and for showing how potential-flow theory can be modified to yield useful predictions of flow around wings and thus predict their aerodynamic characteristics. His boundary-layer theory explained why flow separation occurs and how skin-friction drag can be calculated.

The boundary-layer theory was introduced in Chapter 2 to motivate the study of potential-flow theory in Chapter 3, this chapter, and Chapter 5. Its development is further elaborated in Chapter 8, which also introduces more advanced computational techniques available today that go under the name computational fluid dynamics (CFD). The potential-flow and boundary-layer theories help us gain insight when examining the data and the synthesized results of physical and computational experiments.

Prandtl also showed how a theoretical model based on vortices could be developed for the flow field of a wing having a large aspect ratio. This theory is described in Chapter 5, where it is shown how knowledge of the aerodynamic characteristics, principally the lift coefficient, of a wing of infinite span—an airfoil—can be adapted to give estimates of the aerodynamic characteristics of a wing of finite span. This work firmly established the relevance of two-dimensional flow around airfoils, which is the subject of the present chapter.

Both CFD and physical experiments require the geometry of a design to be already established and, hence, are analysis tools for design evaluation. The method of singularities to examine flows around airfoils and wings can help in the design of lifting surfaces. Therefore, it is still important in developing and applying analytical design approaches. Lifting surfaces (i.e., two-dimensional airfoils and finite-span wings) are useful because the force (the "lift") perpendicular to the direction of flow (or motion) is large in comparison with drag. Hence, these shapes are important in aircraft wings, turbine blades, and control surfaces among other "foils" that are part of successful aircraft design (airplanes, helicopters, etc.). Lifting surfaces are designed so that the flow adjacent the boundary of the lifting surface does not separate at design conditions. Prandtl's theories help up understand why relatively thin airfoils and thus wing sections can be properly designed (or selected from a set of available airfoils of known performance) and developed to satisfy the design requirements of a variety of problems requiring lifting surfaces to attain prescribed performance goals. This chapter is devoted to an examination of the potential flow around airfoils, the first step in the design of lifting surfaces.

4.1.1 The Kutta Condition

How can potential flow be adapted to provide a reasonable theoretical model for the flow around an airfoil that generates lift? The answer lies in an analogy between flow around an airfoil and that around a cylinder in inviscid flow with circulation induced by a vortex (see Section 3.2.8). For the latter it can be shown that, when a point vortex is superimposed with a doublet on a uniform flow, a lifting flow is generated. It was explained in Section 3.2.8 that the doublet and uniform flow alone constitute a non-circulatory irrotational flow with zero vorticity everywhere. In contrast, the vorticity is zero everywhere except at the origin when the vortex is present. Thus, although the flow is still irrotational everywhere save at the origin, the net effect is that circulation is nonzero. The generation of lift is always associated with circulation. In fact, it can be shown (see Eq. 3.52) that, for the flow around a cylinder with a vortex of finite strength, lift is directly proportional to circulation. We will show that this important result can also be extended to airfoils (which was hinted in Section 3.2.9). The other point to note, from Fig. 3.18, is that as vortex strength, and therefore circulation, rises, both the fore and aft stagnation points move downward along the surface of the cylinder.

Now suppose that in some way it is possible to use vortices to generate circulation, and thereby lift, for the flow around an airfoil. The result is shown schematically in Fig. 4.1. In part (a) of the figure, the pure noncirculatory potential flow is around an airfoil at an angle of incidence. If a small amount of circulation is added, the fore and aft stagnation points, SF and SA, move as shown in part (b). In this case the rear stagnation point remains on the upper surface. On the other hand, if the circulation is relatively large, the rear stagnation point moves to the lower surface, as shown in Fig. 4.1(c). For all three cases, the flow must pass around the trailing edge. For an inviscid flow, this implies that the flow speed becomes infinite at the trailing edge, which is evidently impossible in a real viscous fluid because viscous effects ensure that such flows cannot be sustained in nature. In fact, the only position for the rear stagnation point sustainable in a real flow is at the trailing edge, as illustrated in Fig. 4.1(d). Only with the rear stagnation point at the trailing edge does the flow leave the upper and lower surfaces smoothly at the trailing edge. This is the essence of the *Kutta condition* first introduced by the German mathematician Kutta [14].

Imposing the Kutta condition provides a unique way of choosing the circulation for an airfoil and thereby determining lift. This is extremely important because otherwise there would be an infinite number of different lifting flows, each corresponding to a different value of circulation, which we see in the flow around a circular cylinder with circulation for which the lift generated depends on vortex strength. The Kutta condition can be expressed as follows:

- For a given airfoil at a given angle of attack, the circulation must take the unique value that ensures that the flow leaves the trailing edge smoothly.
- For practical airfoils with trailing edges that subtend a finite angle (Fig. 4.2(a)), this condition implies that the rear stagnation point is located at the trailing edge.

All real airfoils are like that in Fig. 4.2(a), of course, but (as in Section 4.2) for theoretical reasons it is frequently desirable to consider infinitely thin airfoils

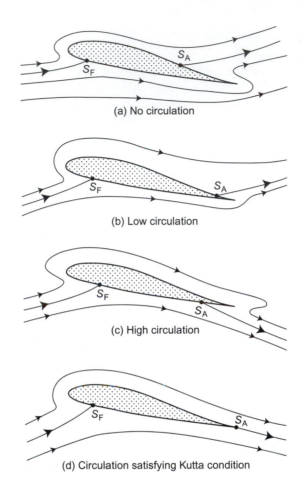

(a) No circulation

(b) Low circulation

(c) High circulation

(d) Circulation satisfying Kutta condition

FIGURE 4.1

Effect of circulation on flow around an airfoil at an angle of incidence.

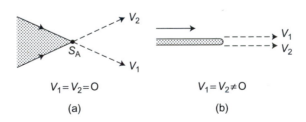

$V_1 = V_2 = 0$

(a)

$V_1 = V_2 \neq 0$

(b)

FIGURE 4.2

Flow near trailing edge details.

(Fig. 4.2(b)). In this case and for the more general case of a cusped trailing edge, the trailing edge need not be a stagnation point for the flow to leave the trailing edge smoothly.

- If the angle subtended by the trailing edge is zero, the velocities leaving the upper and lower surfaces at the trailing edge are finite and equal in magnitude and direction.

4.1.2 Circulation and Vorticity

From the previous discussion it is evident that circulation and vorticity, introduced in Section 2.7, are key concepts in the generation of lift. We now explore further and derive the precise relationship between lift force and circulation.

Consider an imaginary open curve AB drawn in a purely potential flow as in Fig. 4.3(a). The difference in the velocity potential evaluated at A and B is given by the line integral of the tangential velocity component of flow along the curve, In other words, if the flow velocity across AB at point P is q, inclined at angle α to the local tangent, then

$$\phi_A - \phi_B = \int_{AB} q \cos \alpha \, ds \tag{4.1}$$

which can also be written in the form

$$\phi_A - \phi_B = \int_{AB} (u dx + v dy)$$

Equation (4.1) may be regarded as an alternative definition of velocity potential.

Consider next a closed curve or circuit in a circulatory flow (Fig. 4.3(b)) (remember that the circuit is imaginary and does not influence the flow in any way—it is not a boundary). The *circulation* is defined in Eq. (2.83) as the line integral taken around the circuit and is denoted Γ:

$$\Gamma = \oint q \cos \alpha \, ds \quad \text{or} \quad \Gamma = \oint q \, (u dx + v dy) \tag{4.2}$$

It is evident from Eqs. (4.1) and (4.2) that, in a purely potential flow, for which ϕ_A must equal ϕ_B when the points coincide, the circulation must be zero.

Circulation implies a component of flow *rotation* in the system. This is not to say that there are circular streamlines, or that elements of fluid are actually moving around some closed loop, although this is possible. Rather, circulation in a flow means that the flow system can be resolved into a uniform irrotational portion and a circulating portion. (Figure 4.4 shows an idealized concept.) The implication is that, if circulation is present in a fluid motion, vorticity must be present, even though it may be confined to a restricted space (e.g., in the core of a point vortex). Alternatively, as with the circular cylinder with circulation, the vorticity at the center of the

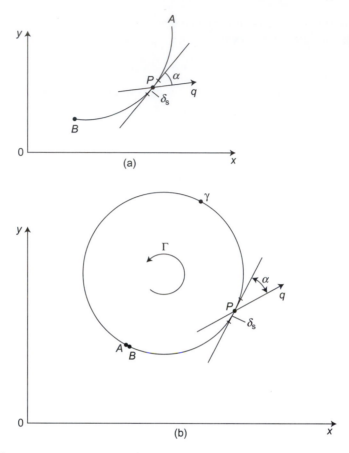

FIGURE 4.3

(a) Open curve in a potential flow. (b) Closed curve in a circularity flow; A and B coincide.

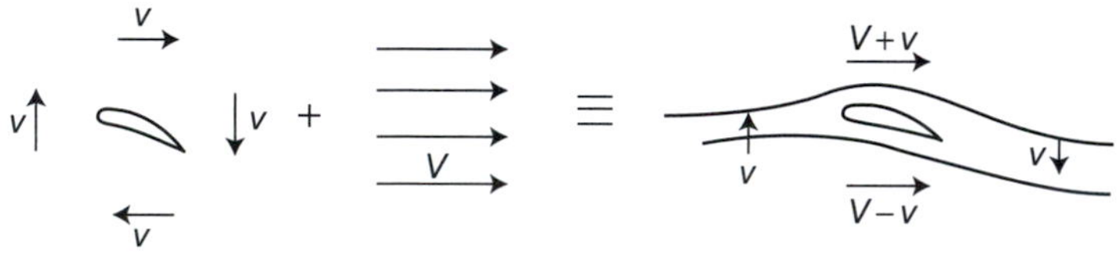

FIGURE 4.4

Uniform flow plus circulatory flow around an airfoil.

cylinder may actually be excluded from the region of flow considered—namely, that outside the cylinder.

Consider this by the reverse argument. Look again at Fig. 4.3(b). By definition the velocity potential of C relative to $A(\phi_{CA})$ must be equal to the velocity potential of C

relative to $B(\phi_{CB})$ in a potential flow. The integration continued around ACB gives

$$\Gamma = \phi_{CA} \pm \phi_{CB}$$

This is for a potential flow only. Thus, if Γ is finite, the definition of velocity potential breaks down and the curve ACB must contain a region of rotational flow.

An alternative equation for Γ is found by considering the circuit of integration to consist of a large number of rectangular elements of side $\delta x \delta y$ (e.g., see Section 2.7.7 and Example 2.2). Applying the integral $\Gamma = \int (udx + vdy)$ around abcd, say, which is the element at $P(x,y)$ where the velocity is u and v, gives (Fig. 4.5)

$$\Delta\Gamma = \left(v + \frac{\partial v}{\partial x}\frac{\delta x}{2}\right)\delta y - \left(u + \frac{\partial u}{\partial y}\frac{\delta y}{2}\right)\delta x - \left(v - \frac{\partial v}{\partial x}\frac{\delta x}{2}\right)\delta y + \left(u - \frac{\partial u}{\partial y}\frac{\delta y}{2}\right)\delta x$$

$$\Delta\Gamma = \left(\frac{\partial v}{\partial x} - \frac{\partial u}{\partial y}\right)\delta x \delta y$$

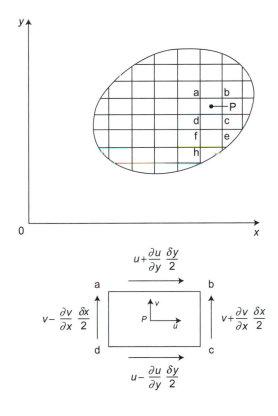

FIGURE 4.5

Illustration of procedure to integrate the vorticity over a surface to obtain the circulation around the surrounding contour.

The sum of the circulations of all of the areas is clearly the circulation of the circuit as a whole because, as the $\Delta\Gamma$ of each element is added to the $\Delta\Gamma$ of its neighbor, the contributions of the common sides disappear.

Applying this argument from element to neighboring element throughout the area, the only sides contributing to circulation when the $\Delta\Gamma$s of all areas are summed are those actually forming the circuit itself. This means that, for the circuit as a whole,

$$\Gamma = \int\int \left(\frac{\partial v}{\partial x} - \frac{\partial u}{\partial y}\right) dxdy = \oint (udx + vdy)$$

and

$$\frac{\partial v}{\partial x} - \frac{\partial u}{\partial y} = \zeta$$

which shows explicitly that the circulation is given by the integral of the vorticity in the region enclosed by the circuit.

If the strength of circulation Γ remains constant while the circuit shrinks to encompass an ever smaller area (i.e., until it shrinks to an area the size of a rectangular element) then

$$\Gamma = \zeta \times \delta x\delta y$$

where $\delta x\delta y$ is an element of area in the (x,y) plane. Recall that ζ is the z component of the vorticity vector, which is the only possible finite component of the vorticity vector in two-dimensional flow. Therefore,

$$\zeta = \lim_{\delta x\delta y \to 0} \frac{\Gamma}{\delta x\delta y} \tag{4.3}$$

Here the (potential) line vortex introduced in Section 3.2.4 is revisited and the definition (Fig. 4.2) of circulation is applied to two particular circuits around a point (Fig. 4.6). One of these is a circle, of radius r_1, at the center of the vortex. The second circuit is *ABCD*, composed of two circular arcs of radii r_1 and r_2 and two radial lines subtending the angle β at the vortex center. For the concentric circuit, the velocity is constant at the value

$$q = \frac{C}{r_1}$$

where C is the constant value of qr.

Since the flow is, by the definition of a vortex, along the circle, α is everywhere zero and therefore $\cos\alpha = 1$. Then, from Eq. (4.2),

$$\Gamma = \oint \frac{C}{r_1} ds$$

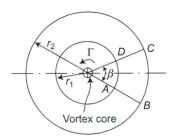

FIGURE 4.6

Two circuits in the flow around a point vortex.

Now suppose an angle θ is to be measured counterclockwise from some arbitrary axis, such as OAB. Then

$$\mathrm{d}s = r_1 \mathrm{d}\theta$$

whence

$$\Gamma = \int_0^{2\pi} \frac{C}{r_1} r_1 \mathrm{d}\theta \qquad (4.4)$$

Since C is a constant, it follows that Γ is also a constant, independent of the radius. We can show that, provided that the circuit encloses the center of the vortex, the circulation around it is equal to Γ, whatever the circuit's shape. The circulation Γ around a circuit enclosing the center of a vortex is called the strength of the vortex. The dimensions of circulation and vortex strength are, from Eq. (4.2), velocity times length (i.e., $L^2\,T^{-1}$), the units being m^2s^{-1}. Now $\Gamma = 2\pi C$, and, because C was defined as equal to qr,

$$\Gamma = 2\pi qr$$

and

$$q = \frac{\Gamma}{2\pi r} \qquad (4.5)$$

Taking now the second circuit $ABCD$, the contribution to the circulation from each part of the circuit is calculated as follows:

- *Radial line AB*: Since the flow around a vortex is in concentric circles, the velocity vector is everywhere perpendicular to the radial line (i.e., $\alpha = 90°$, $\cos \alpha = 0$). Thus the tangential velocity component is zero along AB, and there is therefore no contribution to circulation.
- *Circular arc BC*: Here $\alpha = 0$, $\cos \alpha = 1$. Therefore,

$$\delta\Gamma = \int_{BC} q \cos \alpha \, \mathrm{d}s = \int_0^\beta q r_2 \mathrm{d}\theta$$

But, by Eq. (4.5),

$$q = \frac{\Gamma}{2\pi r_2}$$

$$\delta\Gamma = \int_0^\beta \frac{\Gamma}{2\pi r_2} r_2 d\theta = \frac{\beta\Gamma}{2\pi}$$

- *Radial line CD*: As for *AB*, there is no contribution to circulation from this part of the circuit.
- *Circular arc DA*: Here the path of integration is from *D* to *A*, while the direction of velocity is from *A* to *D*. Therefore, $\alpha = 180°$, $\cos\alpha = -1$. Then

$$\delta\Gamma = \int_0^\beta \frac{\Gamma}{2\pi r_1}(-1)r_1 d\theta = -\frac{\beta\Gamma}{2\pi}$$

so the total circulation around the complete circuit *ABCD* is

$$\Gamma = 0 + \frac{\beta\Gamma}{2\pi} + 0 - \frac{\beta\Gamma}{2\pi} = 0 \tag{4.6}$$

The total circulation around this circuit, which does not enclose the core of the vortex, is zero. Now any circuit can be split into infinitely short circular arcs joined by infinitely short radial lines. Applying this process to such a circuit leads to the result that the circulation around a circuit of any shape that does not enclose the core of a vortex is zero.

4.1.3 Circulation and Lift (The Kutta–Joukowski Theorem)

In Eq. (3.52) it was shown that lift L, or l per unit span, and circulation Γ of a circular cylinder in a cross flow with circulation are simply related by

$$l = \rho V\Gamma$$

where ρ is the fluid density and V is the speed of the flow approaching the cylinder. In fact, as demonstrated independently by Kutta [14] and the Russian physicist Joukowski [15], at the beginning of the twentieth century, this result applies equally well to a cylinder of any shape and to an airfoil, in particular. This powerful and useful result is accordingly known as the *Kutta–Joukowski Theorem*. Its validity is demonstrated next.

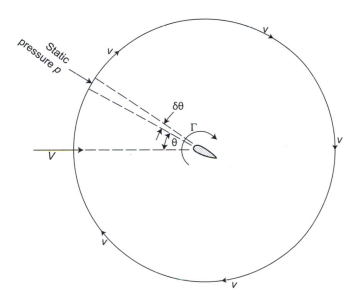

FIGURE 4.7

Circular control volume around an airfoil.

The lift on any airfoil moving relative to a bulk of fluid can be derived by direct analysis. Consider the airfoil in Fig. 4.7, which generates a circulation of Γ when in a stream of velocity V, density ρ, and static pressure p_0. The lift produced must be sustained by any boundary (imaginary or real) surrounding the airfoil.

For a circuit of radius r that is very large compared to the airfoil, the airfoil lift upward must be equal to the sum of the pressure force on the whole periphery of the circuit and the reaction to the rate of change of downward momentum of the air through the periphery. At this distance, the effects of airfoil thickness distribution may be ignored and the airfoil represented only by the circulation it generates.

The vertical static pressure force or buoyancy l_b on the circular boundary is the sum of the vertical pressure components acting on elements of the periphery. At the element subtending $\delta\theta$ at the center of the airfoil, the static pressure is p and the local velocity is the result of V and the velocity v induced by the circulation. By Bernoulli's equation,

$$p_0 + \frac{1}{2}\rho V^2 = p + \frac{1}{2}\rho\left[V^2 + v^2 + 2Vv\sin\theta\right]$$

giving

$$p = p_0 - \rho Vv\sin\theta$$

if v^2 may be neglected compared with V^2, which is permissible because r is large.

The vertical component of pressure force on this element is

$$-pr\sin\theta\,\delta\theta$$

and, on substituting for p and integrating, the contribution to lift from the force acting on the boundary is

$$l_b = -\int_0^{2\pi} (p_0 - \rho V v \sin\theta)r\sin\theta\,d\theta = +\rho V v r\pi \tag{4.7}$$

with p_0 and r constant.

The mass flow through the elemental area of the boundary is given by $\rho V r \cos\theta\,\delta\theta$. This mass flow has a vertical velocity increase of $v\cos\theta$, and therefore the rate of change of downward momentum through the element is $-\rho V v r \cos^2\theta\,\delta\theta$. By integrating around the boundary, then, the inertial contribution to the lift, l_i, is

$$l_i = +\int_0^{2\pi} \rho V v r \cos^2\theta\,d\theta = \rho V v r\pi \tag{4.8}$$

Thus the total lift is

$$l = 2\rho V v r\pi \tag{4.9}$$

From Eq. (4.5),

$$v = \frac{\Gamma}{2\pi r}$$

finally giving, for the lift per unit span l

$$l = \rho V \Gamma \tag{4.10}$$

We obtain this expression without considering the behavior of air in a boundary circuit, by integrating pressures directly on the surface of the airfoil.

It can be shown that this lift force is theoretically independent of the shape of the airfoil section, the main effect of which is to produce a pitching moment in potential flow plus drag in the practical case of motion in a real viscous fluid.

4.2 THE DEVELOPMENT OF AIRFOIL THEORY

The first successful airfoil theory, developed by Joukowski, was based on a very elegant mathematical concept—the conformal transformation—that exploits the theory

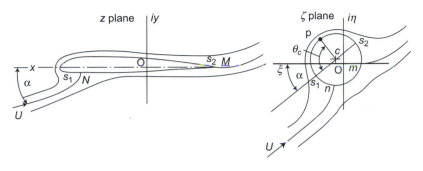

FIGURE 4.8

Joukovsky transformation of the flow around a circular cylinder with circulation to the flow around an airfoil generating lift.

of complex variables. That is, any two-dimensional potential flow can be represented by an analytical function of a complex variable. The basic idea behind Joukowski's theory is to take a circle in the complex $\zeta = (\xi + i\eta)$ plane (noting that here ζ does not denote vorticity) and map (or transform) it into an airfoil-shaped contour. This is illustrated in Fig. 4.8 (and was introduced in Section 3.2.9).

A potential flow can be represented by a complex potential defined by $\Phi = \phi + i\psi$, where, as previously, ϕ and ψ are the velocity potential and stream function, respectively. The same Joukowski mapping (or transformation), expressed mathematically as

$$z = \zeta + \frac{C^2}{\zeta}$$

(where C is a parameter), then maps the complex potential flow around the circle in the ζ plane to the corresponding flow around the airfoil in the z plane. This makes it possible to use the results for the cylinder with circulation (see Section 3.2.8) to calculate the flow around an airfoil. The magnitude of the circulation is chosen to satisfy the Kutta condition in the z plane.

From a practical point of view, Joukowski's theory suffers an important drawback. It applies only to a particular family of airfoil shapes. Moreover, all members of this family have a cusped trailing edge, whereas airfoils in practical aerodynamics have trailing edges with finite angles. Kármán and Trefftz [16] devised a more general conformal transformation, in 1918, that gave a family of airfoils with trailing edges of finite angle. Airfoil theory based on conformal transformation became a practical tool for aerodynamic design in 1931, when the American engineer Theodorsen [17] developed a method for airfoils of arbitrary shape, which continued to be developed well into the second half of the twentieth century. Advanced versions of the method exploited such modern computing techniques as fast Fourier transforms [18].

If aerodynamic design involved only two-dimensional flows at low speeds, a design method based on conformal transformation would be a good choice. However, the technique cannot be extended to three-dimensional or high-speed flows. For this reason, it is no longer widely used in aerodynamic design and is not discussed further here. Instead, two approaches, *thin-airfoil theory* and *computational boundary-element (or panel) methods*, that can be extended to three-dimensional flows are described.

The Joukowski theory introduced some features that are basic to practical airfoil theory. First, overall lift is proportional to the circulation generated; second, the magnitude of the circulation must be such as to keep the velocity finite at the trailing edge in accordance with the Kutta condition.

It is not necessary to suppose that the vorticity that gives rise to the circulation is due to a single vortex. The vorticity can instead be distributed throughout the region enclosed by the airfoil profile or even on the airfoil surface. But the magnitude of circulation generated by all of this vorticity must still be such as to satisfy the Kutta condition. A simple version of this concept is to concentrate the vortex distribution on the camber line, as suggested by Fig. 4.9. In this form, it becomes the basis of the classic thin-airfoil theory developed by Munk [19] and Glauert [20].

Glauert's version of this theory was based on a sort of reverse Joukowski transformation that exploited the not unreasonable assumption that practical airfoils are thin. He was thereby able to determine the airfoil shape required for specified characteristics. This made the theory a practical tool for aerodynamic design. However, as we remarked earlier, the use of conformal transformation is restricted to two dimensions. Fortunately, it is not necessary to use Glauert's approach to obtain his final results. In Section 4.3, we follow later developments using a method that does not depend on conformal transformation in any way and, accordingly (in principle at least), can be extended to three dimensions.

Thin-airfoil theory and its applications are described in Sections 4.3 through 4.9. As the name suggests, the method is restricted to thin airfoils with small camber at small angles of attack. This is not a major drawback since most practical wings are fairly thin. A modern computational method not restricted to thin airfoils, described in Section 4.10, is based on the extension of the panel method of Section 3.4 to lifting flows. It was developed in the late 1950s and early 1960s by Hess and Smith at Douglas Aircraft.

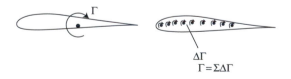

FIGURE 4.9

Examples of vortex models of lift on an airfoil.

4.3 GENERAL THIN-AIRFOIL THEORY

For the development of this theory, it is assumed that maximum airfoil thickness is small compared to chord length. It is also assumed that camber-line shape deviates only slightly from the chord line. A corollary of the second assumption is that the theory should be restricted to low angles of attack.

Consider a typical cambered airfoil as shown in Fig. 4.10. The upper and lower curves of its profile are denoted y_u and y_1, respectively. We denote the velocities in the x and y directions as u and v and write them in the form

$$u = U \cos \alpha + u', \quad v = U \sin \alpha + v'$$

where u' and v' represent the departure of the local velocity from the undisturbed free stream, and are commonly known as the *disturbance* or *perturbation* velocities. In fact, thin-airfoil theory is one example of small-perturbation theory.

The velocity component perpendicular to the airfoil profile is zero. This constitutes the boundary condition for the potential flow and can be expressed mathematically as

$$-u \sin \beta + v \cos \beta = 0 \quad \text{at} \quad y = y_u \quad \text{and} \quad y_l$$

Dividing both sides by $\cos \beta$, this boundary condition can be rewritten as

$$- \left(U \cos \alpha + u' \right) \frac{dy}{dx} + U \sin \alpha + v' = 0 \quad \text{at} \quad y = y_u \quad \text{and} \quad y_l \quad (4.11)$$

By the thin-airfoil assumptions mentioned earlier, Eq. (4.11) may be simplified. Mathematically, these assumptions can be written in the form

$$y_u \text{ and } y_l \ll c; \alpha, \quad \frac{dy_u}{dx}, \text{ and } \frac{dy_l}{dx} \ll 1$$

Note the additional assumption that the slope of the airfoil profile is small. The thin-airfoil assumptions imply that the disturbance velocities are small compared to the

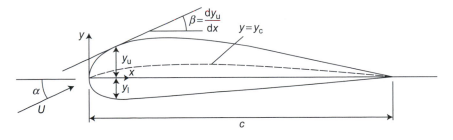

FIGURE 4.10

Illustration of the geometry and slope of the surface of an airfoil.

undisturbed free-stream speed:

$$u' \text{ and } v' \ll U$$

Given the earlier assumptions, Eq. (4.11) can be simplified by replacing $\cos\alpha$ and $\sin\alpha$ by 1 and α, respectively. Furthermore, products of small quantities can be neglected, allowing the term $u'dy/dx$ to be discarded so that Eq. (4.11) becomes

$$v' = U\frac{dy_u}{dx} - U\alpha \text{ and } v' = U\frac{dy_l}{dx} - U\alpha \tag{4.12}$$

We simplify further by recognizing that, if y_u and $y_l \ll c$, then to a sufficiently good approximation, the boundary conditions of Eq. (4.12) can be applied at $y = 0$ rather than at $y = y_u$ or y_1.

Since potential flow with Eq. (4.12) as a boundary condition is linear, the flow around a cambered airfoil at incidence can be regarded as the superposition of two separate flows, one circulatory and the other noncirculatory. This is illustrated in Fig. 4.11. The circulatory flow is that around an infinitely thin cambered plate, the noncirculatory flow is that around a symmetric airfoil at zero incidence. We demonstrate this formally as follows:

$$y_u = y_c + y_t \text{ and } y_l = y_c - y_t$$

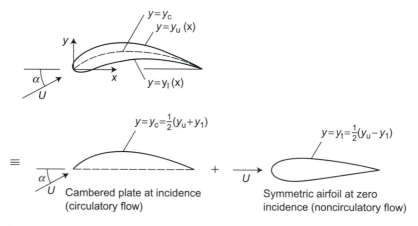

FIGURE 4.11

Cambered thin airfoil at incidence as the superposition of circulatory and noncirculatory flow.

where $y = y_c(x)$ is the function describing the camber line and $y = y_t = (y_u - y_1)/2$ is the thickness function. Now Eq. (4.12) can be rewritten in the form

$$v' = U\frac{dy_c}{x} - U\alpha \pm U\frac{dy_t}{dx}$$

where the plus sign applies to the upper surface and the minus sign to the lower surface. The first two terms on the right-hand side are the *circulatory* parts of v', the last term is the *noncirculatory* part of v'.

Thus the noncirculatory flow is given by the solution of potential flow subject to the boundary condition $v' = U dy_t/dx$, which is applied at $y = 0$ for $0 \leq x \leq c$. The solution to this problem is discussed in Section 4.9. The lifting characteristics of the airfoil are determined solely by the circulatory flow. Consequently, the solution to this problem is of primary importance. We turn now to the formulation and solution of the mathematical problem for circulatory flow.

It may be seen from Sections 4.1 and 4.2 that vortices can represent lifting flow. In the present case, the lifting flow generated by an infinitely thin cambered plate at incidence is represented by a string of line vortices, each of infinitesimal strength, along the camber line, as shown in Fig. 4.12. The camber line is thus replaced by a line of variable vorticity so that total circulation about the chord is the sum of the vortex elements. This can be written as

$$\Gamma = \int_0^c k\,ds \qquad (4.13)$$

where k is the distribution of vorticity over the element of camber line δs, and circulation is positive in the clockwise direction. The problem now becomes determining the function $k(x)$ such that the boundary condition as well as the Kutta condition is satisfied (see Section 4.1.1):

$$v' = U\frac{dy_c}{dx} - U\alpha \quad \text{at } y = 0,\ 0 \leq x \leq 1 \qquad (4.14)$$

The leading edge is the origin of a pair of coordinate axes x and y: x along the chord and y normal to it. The basic assumptions of the theory permit variation in vorticity along the camber line to be assumed the same as the variation along the x-axis, that is, δs differs negligibly from δx, so that Eq. (4.13) becomes

$$\Gamma = \int_0^c k\,dx \qquad (4.15)$$

Hence, from Eq. (4.10) for a unit span of this section, the lift is given by

$$l = \rho U \Gamma = \rho U \int_0^c k \, dx \qquad (4.16)$$

Alternatively, Eq. (4.16) can be written with $\rho U k = p$:

$$l = \int_0^c \rho U k \, dx = \int_0^c p \, dx \qquad (4.17)$$

Now, considering a unit spanwise length, p has the dimensions of force per unit area or pressure; the moment of these chordwise pressure forces about the leading edge or origin of the system is simply

$$M_{LE} = -\int_0^c p x \, dx = -\rho U \int_0^c k x \, dx \qquad (4.18)$$

Note that pitching "nose up" is positive.

The thin wing section has thus been replaced for analysis by a line discontinuity in flow in the form of a vorticity distribution. This gives rise to overall circulation, as does the airfoil, and produces a chordwise pressure variation.

For the airfoil in a flow of undisturbed velocity U and pressure p_0, the insert in Fig. 4.12 shows the static pressures p_1 and p_2 above and below the element δs, where the local velocities are $U + u_1$ and $U + u_2$, respectively. The overall pressure difference p is $p_2 - p_1$. By Bernoulli,

$$p_1 + \frac{1}{2}\rho(U + u_1)^2 = p_0 + \frac{1}{2}\rho U^2$$
$$p_2 + \frac{1}{2}\rho(U + u_2)^2 = p_0 + \frac{1}{2}\rho U^2$$

and subtracting

$$p_2 - p_1 = \frac{1}{2}\rho U^2\left[2\left(\frac{u_1}{U} - \frac{u_2}{U}\right) + \left(\frac{u_1}{U}\right)^2 - \left(\frac{u_2}{U}\right)^2\right]$$

With the airfoil thin and at small incidence, the perturbation velocity ratios u_1/U and u_2/U are so small compared with unity that $(u_1/U)^2$ and $(u_2/U)^2$ are neglected compared with u_1/U and u_2/U, respectively. Then

$$p = p_2 - p_1 = \rho U(u_1 - u_2) \qquad (4.19)$$

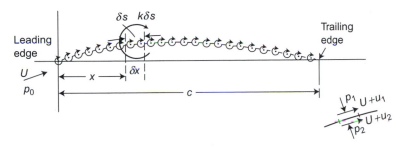

FIGURE 4.12

Vortex sheet model of an airfoil; insert shows the velocity and pressure above and below δs.

The equivalent vorticity distribution indicates that the circulation due to element δs is $k\,\delta x$ (δx because the camber line deviates only slightly from the x-axis). Evaluating the circulation around δs and taking clockwise as positive in this case, we take the algebraic sum of the flow of fluid along the top and bottom of δs:

$$k\delta x = +(U+u_1)\,\delta x - (U-u_2)\,\delta x = (u_1 - u_2)\,\delta x \qquad (4.20)$$

Comparing Eqs. (4.19) and (4.20) shows that $p = \rho U k$, as introduced in Eq. (4.17).

For a trailing-edge angle of zero, the Kutta condition (see Section 4.1.1) requires $u_1 = u_2$ at the trailing edge. It follows from Eq. (4.20) that this condition is satisfied if

$$k = 0 \quad \text{at} \quad x = c \qquad (4.21)$$

The induced velocity v in Eq. (4.14) can be expressed in terms of k by considering the effect of the elementary circulation $k\delta x$ at x, a distance $x - x_1$ from the point considered (Fig. 4.13). Circulation $k\delta x$ induces a velocity at point x_1 equal to (from Eq. (4.5)).

$$\frac{1}{2\pi}\frac{k\delta x}{x - x_1}$$

The effect of all such elements of circulation along the chord is the induced velocity v', where

$$v' = \frac{1}{2\pi}\int_0^c \frac{k\,dx}{x - x_1}$$

Introducing this in Eq. (4.14) gives

$$U\left[\frac{dy_c}{dx} - \alpha\right] = \frac{1}{2\pi}\int_0^c \frac{k\,dx}{x - x_1} \qquad (4.22)$$

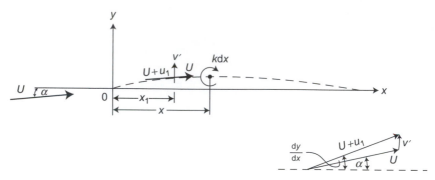

FIGURE 4.13

Velocities at the point x_1 on the camber line. Detail in lower rights shows $U + u_1$; V'. induced by chordwise variation in circulation U. Free-stream velocity inclined at angle α to $0x$.

The solution for $k\mathrm{d}x$ that satisfies Eq. (4.22) for a given shape of camber line (defining $\mathrm{d}y_c/\mathrm{d}x$) and incidence can be introduced in Eqs. (4.17) and (4.18) to obtain the lift and moment for the airfoil shape. The characteristics C_L and $C_{M_{LE}}$ follow directly, as do k_{CP}, the center-of-pressure coefficient, and the angle for zero lift.

4.4 SOLUTION TO THE GENERAL EQUATION

In the general case Eq. (4.22) must be solved directly to determine the function $k(x)$ that corresponds to a specified camber-line shape. Alternatively, the inverse design problem may be solved whereby the pressure distribution or, equivalently, the tangential velocity variation along the upper and lower surfaces of the airfoil is given. The corresponding $k(x)$ may then be simply found from Eqs. (4.19) and (4.20). The problem then becomes finding the requisite camber-line shape from Eq. (4.22). Our present approach is to work up to the general case through the simple case of the flat plate at incidence, and then consider some practical applications of the general case. To this end, we consider the integral in Eq. (4.22) and give expressions for some useful definite integrals.

To use certain trigonometric relationships, it is convenient to change variables from x to θ through $x = (c/2)(l - \cos\theta)$ and x_1 to θ_1; then the limits change as follows:

$$\theta \sim 0 \to \pi \quad \text{as} \quad x \sim 0 \to c, \quad \text{and} \quad \mathrm{d}x = \frac{c}{2}\sin\theta\,\mathrm{d}\theta$$

So

$$\int_0^2 \frac{k\,\mathrm{d}x}{x - x_1} = -\int_0^\pi \frac{k\sin\theta\,\mathrm{d}\theta}{(\cos\theta - \cos\theta_1)} \tag{4.23}$$

Also, the Kutta condition (4.21) becomes

$$k = 0 \quad \text{at} \quad \theta = \pi \tag{4.24}$$

The expressions found by evaluating two useful definite integrals are given here:

$$\int_0^\pi \frac{\cos n\theta}{(\cos\theta - \cos\theta_1)}\,d\theta = \pi\frac{\sin n\theta_1}{\sin\theta_1}; \quad n = 0,1,2,\ldots \tag{4.25}$$

$$\int_0^\pi \frac{\sin n\theta \sin\theta}{(\cos\theta - \cos\theta_1)}\,d\theta = -\pi\cos n\theta_1; \quad n = 0,1,2,\ldots \tag{4.26}$$

These are reasonably well-known results (see, e.g., Appendix C). We use Eqs. (4.25) and (4.26) in subsequent sections on applications of thin-airfoil theory.

4.4.1 Thin Symmetrical Flat-Plate Airfoil

In this simple case, the camber line is straight along x, and $dy_c/dx = 0$. Using Eq. (4.23), the general Eq. (4.22) becomes

$$U_\alpha = \frac{1}{2\pi}\int_0^\pi \frac{k\sin\theta}{(\cos\theta - \cos\theta_1)}\,d\theta \tag{4.27}$$

What value should k take on the right-hand side of Eq. (4.27) to give a left-hand side that does not vary with x or, equivalently, θ? To answer this question, consider the result of Eq. (4.25) with $n = 1$. From this we see that

$$\int_0^\pi \frac{\cos\theta\,d\theta}{(\cos\theta - \cos\theta_1)} = \pi$$

Comparing this result with Eq. (4.27), we see that, if $k = k_1 = 2U\alpha\cos\theta/\sin\theta$, Eq. (4.27) is satisfied. The only problem is that, far from satisfying the Kutta condition (4.24), this solution goes to infinity at the trailing edge. To overcome this problem, it is necessary to recognize that, if there exists a function k_2 such that

$$\int_0^\pi \frac{k_2\sin\theta\,d\theta}{(\cos\theta - \cos\theta_1)} = 0 \tag{4.28}$$

$k = k_1 + k_2$ also satisfies Eq. (4.27).

Consider Eq. (4.25) with $n = 0$ so that

$$\int_0^\pi \frac{1}{(\cos\theta - \cos\theta_1)} d\theta = 0$$

Comparing this to Eq. (4.28) shows that the solution is

$$k_2 = \frac{C}{\sin\theta}$$

where C is an arbitrary constant.

Thus the complete (or general) solution for the flat plate is given by

$$k = k_1 + k_2 = \frac{2U\alpha\cos\theta + C}{\sin\theta}$$

The Kutta condition Eq. (4.24) is satisfied if $C = 2U\alpha$, giving a final solution:

$$k = 2U\alpha\frac{(1 + \cos\theta)}{\sin\theta} \tag{4.29}$$

Aerodynamic Coefficients for a Flat Plate

The expression for k can now be put in the appropriate equations for lift and moment using the pressure

$$p = \rho U k = 2\rho U^2 \alpha \frac{1 + \cos\theta}{\sin\theta} \tag{4.30}$$

The lift per unit span is

$$l = \rho U \int_0^\pi 2U\alpha\left(\frac{1 + \cos\theta}{\sin\theta}\right)\frac{c}{2}\sin\theta\, d\theta = \alpha\rho U^2 c \int_0^\pi (1 + \cos\theta)\, d\theta = \pi\alpha\rho U^2 c$$

It therefore follows that, for a unit span,

$$C_L = \frac{l}{\frac{1}{2}\rho U^2 c} = 2\pi\alpha \tag{4.31}$$

The moment about the leading edge per unit span is

$$M_{LE} = -\int_0^c px\,dx$$

Changing the sign gives

$$-M_{LE} = 2\rho U^2 \alpha \int_0^\pi \frac{(1+\cos\theta)}{\sin\theta}\frac{c}{2}(1-\cos\theta)\frac{c}{2}\sin\theta\, d\theta = \frac{1}{2}\rho U^2 \alpha c^2$$

$$\times \int_0^\pi \left(1-\cos^2\theta\right) d\theta$$

Therefore, for a unit span,

$$-C_{M_{LE}} = -\frac{M_{LE}}{\frac{1}{2}\rho U^2 c^2} = \alpha \int_0^\pi \left(\frac{1}{2}-\frac{\cos 2\theta}{2}\right) d\theta = \alpha\frac{\pi}{2} \tag{4.32}$$

Comparing Eqs. (4.31) and (4.32) shows that

$$C_{M_{LE}} = -\frac{C_L}{4} \tag{4.33}$$

The center-of-pressure coefficient k_{CP} is given for small angles of incidence approximately by

$$k_{CP} = \frac{-C_{M_{LE}}}{C_L} = \frac{1}{4} \tag{4.34}$$

and this shows a fixed center of pressure coincident with the aerodynamic center, as is necessarily true for any symmetrical section.

4.4.2 General Thin-Airfoil Section

In general, the camber line can be any function of x (or θ) provided that $y_c = 0$ at $x = 0$ and c (i.e., at $\theta = 0$ and π). When trigonometric functions are involved, a convenient way to express an arbitrary function is with a Fourier series. Accordingly, the slope of the camber line appearing in Eq. (4.22) can be expressed in terms of a Fourier cosine series:

$$\frac{dy_c}{dx} = A_0 + \sum_{n=1}^\infty A_n \cos n\theta \tag{4.35}$$

Sine terms are not used here because practical camber lines must go to zero at the leading and trailing edges. Thus y_c is an odd function, which implies that its derivative is an even function.

Equation (4.22) now becomes

$$U(\alpha - A_0) - U\sum_{n=1}^{\infty} A_n \cos n\theta = \frac{1}{2\pi}\int_0^\pi k\frac{\sin\theta}{(\cos\theta - \cos\theta_1)}d\theta \qquad (4.36)$$

The solution for k as a function of θ has three parts so $k = k_1 + k_2 + k_3$, where

$$\frac{1}{2\pi}\int_0^\pi k_1(\theta)\frac{\sin\theta}{(\cos\theta - \cos\theta_1)}d\theta = U(\alpha - A_0) \qquad (4.37)$$

$$\frac{1}{2\pi}\int_0^\pi k_2(\theta)\frac{\sin\theta}{(\cos\theta - \cos\theta_1)}d\theta = 0 \qquad (4.38)$$

$$\frac{1}{2\pi}\int_0^\pi k_3(\theta)\frac{\sin\theta}{(\cos\theta - \cos\theta_1)}d\theta = -U\sum_{n=1}^{\infty} A_n \cos n\theta \qquad (4.39)$$

The solutions for k_1 and k_2 are identical to those given in Section 4.4.1 except that $U(\alpha - A_0)$ replaces $U\alpha$ in the case of k_1. Thus it is necessary to solve Eq. (4.39) only for k_3. By comparing Eq. (4.26) with Eq. (4.39), we see that the solution to Eq. (4.39) is given by

$$k_3(\theta) = 2U\sum_{n=1}^{\infty} A_n \sin n\theta$$

The complete solution is given by

$$k(\theta) = k_1 + k_2 + k_3 = 2U(\alpha - A_0)\frac{\cos\theta}{\sin\theta} + \frac{C}{\sin\theta} + 2U\sum_{n=1}^{\infty} A_n \sin n\theta$$

The constant C must be chosen so as to satisfy the Kutta condition Eq. (4.24), which gives $C = 2U(\alpha - A_0)$, So the final solution is

$$k(\theta) = 2U\left[(\alpha - A_0)\frac{\cos\theta + 1}{\sin\theta} + \sum_{n=1}^{\infty} A_n \sin n\theta\right] \qquad (4.40)$$

To obtain the coefficients A_0 and A_n in terms of the camber-line slope, the usual procedures for Fourier series are followed. On integrating both sides of Eq. (4.35) with respect to θ, the second term on the right-hand side vanishes, leaving

$$\int_0^\pi \frac{dy_c}{dx}d\theta = \int_0^\pi A_0 d\theta = A_0\pi$$

Therefore,

$$A_0 = \frac{1}{\pi} \int_0^\pi \frac{dy_c}{dx} \, d\theta \tag{4.41}$$

Multiplying both sides of Eq. (4.35) by $\cos m\theta$, where m is an integer, and integrating with respect to θ,

$$\int_0^\pi \frac{dy_c}{dx} \cos m\theta \, d\theta = \int_0^\pi \cos m\theta \, d\theta + \int_0^\pi \sum_{n=1}^\infty A_n \cos n\theta \cos m\theta \, d\theta$$

Note that

$$\int_0^\pi A_n \cos n\theta \cos m\theta \, d\theta = 0 \text{ except when } n = m$$

Then the first term on the right-hand side vanishes, as does the second term, except for $n = m$:

$$\int_0^\pi \frac{dy_c}{dx} \cos n\theta \, d\theta = \int_0^\pi A_n \cos^2 n\theta \, d\theta = \frac{\pi}{2} A_n$$

whence

$$A_n = \frac{2}{\pi} \int_0^\pi \frac{dy_c}{dx} \cos n\theta \, d\theta \tag{4.42}$$

Lift and Moment Coefficients for a General Thin Airfoil
From Eq. (4.7),

$$l = \int_0^c \rho U k \, dx = \int_0^\pi \rho U \frac{c}{2} k \sin\theta \, d\theta$$

$$= 2\rho U^2 \frac{c}{2} \int_0^\pi \left[(\alpha - A_0)(1 + \cos\theta) + \sum_1^\infty A_n \sin n\theta \sin\theta \right] d\theta$$

$$= 2\rho U^2 \frac{c}{2} \left[\pi (\alpha - A_0) + \frac{\pi}{2} A_1 \right] = C_L \frac{1}{2} \rho U^2 c$$

since

$$\int_0^\pi \sin n\theta \, d\theta = 0 \text{ when } n \neq 1$$

giving

$$C_L = \left(C_{L_0}\right) + \frac{dC_L}{d\alpha}\alpha = \pi\left(A_1 - 2A_0\right) + 2\pi\alpha \tag{4.43}$$

The first term on the right-hand side of Eq. (4.43) is the coefficient of lift at zero incidence. It contains the effects of camber and is zero for a symmetrical airfoil. It is worth noting that, according to general thin-airfoil theory, the lift-curve slope takes the same value 2π for all airfoils:

$$-M_{LE} = \rho U \int_0^\pi kx\,dx = -C_{M_{LE}}\frac{1}{2}\rho U^2 c^2$$

with the usual substitution

$$-C_{M_{LE}} = \frac{\frac{1}{2}\rho U^2 c^2}{\frac{1}{2}\rho U^2 c^2} \times \int_0^\pi \left[(\alpha - A_0)\frac{(1 + \cos\theta)}{\sin\theta} + \sum_1^\infty A_n \sin n\theta \right] \sin\theta\,(1 - \cos\theta)\,d\theta$$

$$= \int_0^\pi (\alpha - A_0)\left(1 - \cos^2\theta\right)d\theta + \int_0^\pi \sum_1^\infty A_n \sin n\theta \sin\theta\,d\theta$$

$$- \int_0^\pi \sum_1^\infty A_n \sin n\theta \cos\theta \sin\theta\,d\theta$$

$$= \frac{\pi}{2}(\alpha - A_0) + \frac{\pi}{2}A_1 - \frac{\pi}{4}A_2$$

since

$$\int_0^\pi \sin n\theta \sin m\theta\,d\theta = 0 \quad \text{when} \ \ n \neq m$$

Thus

$$C_{M_{LE}} = -\frac{\pi}{2}\left[(\alpha - A_0) + A_1 - \frac{A_2}{2} \right] \tag{4.44}$$

In terms of the lift coefficient, $C_{M_{LE}}$ becomes

$$C_{M_{LE}} = -\frac{C_L}{4}\left[1 + \frac{A_1 - A_2}{C_L/\pi} \right]$$

Then the center-of-pressure coefficient is

$$k_{CP} = -\frac{C_{M_{LE}}}{C_L} = \frac{1}{4} + \frac{\pi}{4C_L}(A_1 - A_2) \tag{4.45}$$

and, again, the center of pressure moves as the lift or incidence changes. Now, from Section 1.6.4,

$$k_{CP} = -\frac{C_{M_{1/4}}}{C_L} + \frac{1}{4} \qquad (4.46)$$

and comparing Eqs. (4.44) and (4.45) gives

$$-C_{M_{1/4}} = \frac{\pi}{4}(A_1 - A_2) \qquad (4.47)$$

This shows that the pitching moment about the quarter-chord point for a thin airfoil is theoretically constant, depending on the camber parameters only, and that the quarter-chord point is therefore the aerodynamic center.

It is apparent from this analysis that, no matter what the camber-line shape, only the first three terms of the cosine series describing the camber-line slope have any influence on the usual aerodynamic characteristics. This is indeed the case, but the terms corresponding to $n > 2$ contribute to the pressure distribution over the chord.

Owing to the quality of the basic approximations used in the theory, we find that the theoretical chordwise pressure distribution p does not closely agree with experimental data, especially near the leading edge and near stagnation points where the small-perturbation theory, for example, breaks down. Any local inaccuracies tend to vanish in the overall integration processes, however, and the airfoil coefficients are found to be reliable theoretical predictions.

4.5 THE FLAPPED AIRFOIL

Thin-airfoil theory lends itself readily to airfoils with variable camber, such as flapped airfoils. The distribution of circulation along the camber line for the general airfoil consists of the sum of a component due to a flat plate at incidence and a component due to the camber-line shape. It is sufficient for the theory's assumptions to consider the influence of a flap deflection as adding to the previous two components. Figure 4.14 shows how the three contributions can be combined. In fact, the deflection of the flap about a hinge in the camber line effectively alters the camber so that the contribution due to flap deflection is the effect of an additional camber-line shape.

The problem is thus reduced to the general case of finding a distribution to fit a camber line made up of the airfoil chord and the flap chord deflected through η (see Fig. 4.15). Thin-airfoil theory requires not that the leading and/or trailing edges be on the x-axis, only that the surface slope and the displacement from the x-axis be small.

With the camber defined as hc, the slope of part AB of the airfoil is zero; that of the flap, $-h/F$. To find the coefficients of k for the flap camber, substitute these slope values in Eqs. (4.41) and (4.42) but with the limits of integration confined to

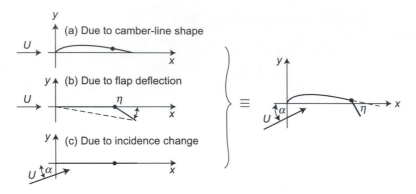

FIGURE 4.14

Subdivision of lift contributions to total lift of a cambered flapped airfoil.

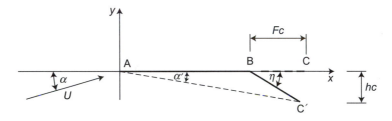

FIGURE 4.15

Geometric features of a flapped airfoil.

the airfoil parts over which the slopes occur. Thus

$$A_0 = \frac{1}{\pi} \int_0^\phi 0 \, d\theta + \frac{1}{\pi} \int_\phi^\pi -\frac{h}{F} \, d\theta \tag{4.48}$$

where ϕ is the value of θ at the hinge:

$$(1 - F)c = \frac{c}{2}(1 - \cos\phi)$$

from which $\cos\phi = 2F - 1$. Evaluating the integral,

$$A_0 = -\left(1 - \frac{\phi}{\pi}\right)\frac{h}{F}$$

That is, since all angles are small, $h/F = \tan\eta \approx \eta$, so

$$A_0 = -\left(1 - \frac{\phi}{\pi}\right)\eta \tag{4.49}$$

Similarly, from Eq. (4.42),

$$A_n = \frac{2}{\pi}\left[\int_0^\phi 0 \cos n\theta\, d\theta + \int_\phi^\pi -\frac{h}{F}\cos n\theta\, d\theta\right] = \frac{2\sin n\phi}{n\pi}\eta \tag{4.50}$$

Thus

$$A_1 = \frac{2\sin\phi}{\pi}\eta \quad \text{and} \quad A_2 = \frac{\sin 2\phi}{\pi}\eta$$

The distribution of chordwise circulation due to flap deflection becomes

$$k = 2U\alpha\frac{1 + \cos\theta}{\sin\theta} + 2U\left[\left(1 - \frac{\phi}{\pi}\right)\frac{1 + \cos\theta}{\sin\theta} + \sum_1^\infty \frac{2\sin n\phi}{n\pi}\sin n\theta\right]\eta \tag{4.51}$$

and this, for a constant incidence α, is a linear function of η, as is the lift coefficient (e.g., from Eq. 4.43):

$$C_L = 2\pi\alpha + 2\pi\eta\left(1 - \frac{\phi}{\pi}\right) + 2\eta\sin\phi$$

giving

$$C_L = 2\pi\alpha + 2(\pi - \phi + \sin\phi)\eta \tag{4.52}$$

Likewise, the moment coefficient from Eq. (4.44) is

$$-C_{M_{LE}} = \frac{\pi}{2}\alpha + \frac{\pi}{2}\left[\eta\left(1 - \frac{\phi}{\pi}\right) + \frac{2\sin\phi}{\pi}\eta - \frac{\sin 2\phi}{2\pi}\eta\right]$$

$$C_{M_{LE}} = -\frac{\pi}{2}\alpha - \frac{1}{2}[\pi - \phi + \sin\phi(2 - \cos\phi)]\eta \tag{4.53}$$

Note that a positive flap deflection (i.e., a downward deflection), decreases the moment coefficient, tending to pitch the main airfoil nose down and vice versa.

4.5.1 Hinge Moment Coefficient

A flapped-airfoil characteristic of great importance in stability and control calculations is the aerodynamic moment about the hinge line, shown as H in Fig. 4.16.

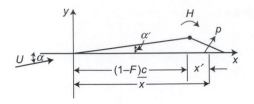

FIGURE 4.16

Illustration of a two-element airfoil; alternative description of a flapped airfoil.

Taking moments of elementary pressures p, acting on the flap about the hinge,

$$H = - \int_{hinge}^{trailing\ edge} px' dx$$

where $p = \rho U k$ and $x' = x - (1 - F)c$. Putting

$$x' = \frac{c}{2}(1 - \cos\theta) - \frac{c}{2}(1 - \cos\phi) = \frac{c}{2}(\cos\phi - \cos\theta)$$

and k from Eq. (4.51),

$$H = - \int_{\phi}^{\pi} 2\rho U^2 \left[\left(\alpha + \eta \left(1 - \frac{\phi}{\pi} \right) \right) \frac{1 + \cos\theta}{\sin\theta} + \eta \sum_{1}^{\infty} \frac{2\sin n\phi}{n\pi} \sin n\theta \right]$$

$$\times \frac{c}{2}(\cos\phi - \cos\theta)\frac{c}{2}\sin\theta\, d\theta$$

Substituting $H = C_H \frac{1}{2}\rho U^2 (Fc)^2$ and cancelling,

$$-C_H F^2 = \alpha \int_{\phi}^{\pi} (1 + \cos\theta)(\cos\phi - \cos\theta)d\theta$$

$$+ \eta \left[\left(1 - \frac{\phi}{\pi} \right) \cos\phi I_1 - \left(1 - \frac{\phi}{\pi} \right) I_2 + \sum_{1}^{\infty} \frac{2\sin n\phi}{n\pi} \cos\phi I_3 \right.$$

$$\left. + \sum_{1}^{\infty} \frac{2\sin n\phi}{n\pi} I_4 \right] \tag{4.54}$$

where

$$I_1 = \int_\phi^\pi (1 + \cos\theta)\mathrm{d}\theta = \pi - \phi - \sin\phi$$

$$I_2 = \int_\phi^\pi (1 + \cos\theta)\cos\theta\,\mathrm{d}\theta = \left[\frac{\pi - \phi}{2}\sin\phi - \frac{\sin 2\phi}{4}\right]$$

$$I_3 = \int_\phi^\pi \sin n\theta \sin\theta\,\mathrm{d}\theta = \frac{1}{2}\left[\frac{\sin(n+1)\phi}{n+1} - \frac{\sin(n-1)\phi}{n-1}\right]$$

$$I_4 = \int_\phi^\pi \sin n\theta \sin\theta \cos\theta\,\mathrm{d}\theta = \frac{1}{2}\left[\frac{\sin(n+2)\phi}{n+2} - \frac{\sin(n-2)\phi}{n-2}\right]$$

In the usual notation, $C_H = b_1\alpha + b_2\eta$, where

$$b_1 = \frac{\partial C_H}{\partial\alpha} \quad\text{and}\quad b_2 = \frac{\partial C_H}{\partial\eta}$$

From Eq. (4.54),

$$b_1 = -\frac{1}{F^2}\int_\phi^\pi (1 + \cos\theta)(\cos\phi - \cos\theta)\mathrm{d}\theta$$

giving

$$b_1 = -\frac{1}{4F^2}[2(\pi - \phi)(2\cos\phi - 1) + 4\sin\phi - \sin 2\phi] \tag{4.55}$$

Similarly, from Eq. (4.54), b_2 reduces to [21]

$$b_2 = -\frac{1}{4\pi F^2}\left[(1 - \cos 2\phi) - 2(\pi - \phi)^2(1 - 2\cos\phi) + 4(\pi - \phi)\sin\phi\right] \tag{4.56}$$

The parameter $a_1 = \partial C_L/\partial\alpha$ is 2π, and $a_2 = \partial C_L/\partial\eta$ from Eq. (4.52) becomes

$$a_2 = 2(\pi - \phi + \sin\phi) \tag{4.57}$$

Thus thin-airfoil theory provides an estimate of all parameters of a flapped airfoil.

Note that aspect-ratio corrections have not been included in this analysis, which is essentially two-dimensional. Following the conclusions of finite-wing theory in Chapter 5, the parameters a_1, a_2, b_1, and b_2 may be suitably corrected for end effects.

In practice, however, they are always determined from computational studies and wind-tunnel tests, and confirmed by flight tests.

4.6 THE JET FLAP

Consider the jet flap as a high-velocity sheet of air issuing from the trailing edge of an airfoil at some downward angle τ to the airfoil chord line. An analysis can be carried out by replacing the jet stream as well as the airfoil by a vortex distribution [22].

The flap contributes to lift in two ways. First, the downward deflection of the jet efflux produces a lifting component of reaction; second, the jet affects the pressure distribution on the airfoil in a similar manner to that obtained by adding to the circulation around the airfoil.

The jet is shown to be equivalent to a band of spanwise vortex filament, which for small deflection angles τ can be assumed to lie along the x-axis (Fig. 4.17). In the analysis, which is not carried out here, both components of lift are considered to arrive at the expression for C_L:

$$C_L = 4\pi A_0 \tau + 2\pi (1 + 2B_0)\alpha \tag{4.58}$$

where A_0 and B_0 are the initial coefficients in the Fourier series that are associated with the deflection of the jet and the incidence of the airfoil, respectively, and can be obtained in terms of the momentum (coefficient) of the jet.

It is interesting that in the experimental work on jet flaps at National Gas Turbine Establishment, in Pyestock, United Kingdom, good agreement was obtained with theoretical C_L even at large values of τ.

4.7 NORMAL FORCE AND PITCHING MOMENT DERIVATIVES DUE TO PITCHING

It is suggested that this section be omitted from general study until the student is familiar with these derivatives and their use.

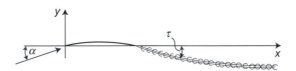

FIGURE 4.17

Illustration of vortex sheet wake behind an airfoil.

4.7.1 $(Z_q)(M_q)$ **Wing Contributions**

Thin-airfoil theory is a convenient basis for estimation of the important derivatives (Z_q) and (M_q). Although the use of these derivatives is beyond the scope of this volume, no text on thin-airfoil theory is complete without some reference to them.

When an airplane is rotating with pitch velocity q about an axis through the center of gravity (CG) normal to the plane of symmetry on the chord line produced (see Fig. 4.18), the airfoil's effective incidence changes with time; as a consequence, the aerodynamic forces and moments change as well.

The rates of change in these forces and moments with respect to pitching velocity q are two of the aerodynamic quasi-static derivatives that are commonly abbreviated to derivatives. Here the rate of change in normal force on the aircraft (i.e., resultant force in the normal or Z direction), with respect to pitching velocity is, in the conventional notation, $\partial Z/\partial q$. This is symbolized by Z_q. Similarly, the rate of change in M with respect to q is $\partial M/\partial q = M_q$.

In common with other aerodynamic forces and moments, these are reduced to nondimensional or coefficient form by dividing through, in this case, by $\rho V l_t$ and $\rho V l_t^2$, respectively, where l_t is the tail-plane moment arm, to give the nondimensional normal force derivative due to pitching z_q and the nondimensional pitching moment derivative due to pitching m_q.

The contributions to these two due to the mainplanes can be considered by replacing the wing by the equivalent thin airfoil. In Fig. 4.19, the center of rotation, which is the center of gravity (CG), is a distance hc behind the leading edge where c is the chord. At some point x from the leading edge of the airfoil, the velocity induced by airfoil rotation about CG is $v' = -q(hc - x)$. Because the vorticity replaces the camber line, a velocity v is induced. The incident flow velocity is V inclined at α to the chord line, from the condition that the local velocity at x must be tangential to the airfoil (camber line) (see Section 4.3). Eq. (4.14) becomes, for this case,

$$V\left(\frac{dy}{dx} - \alpha\right) = v - v'$$

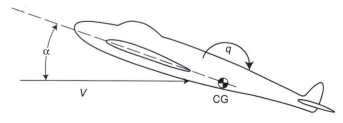

FIGURE 4.18

Illustration of location of the center of gravity (CG) of an airplane in flight.

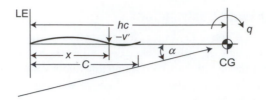

FIGURE 4.19

Illustration of cross-section of the main wing with respect to the center of gravity of the plane.

or

$$\frac{dy}{dx} - \alpha = \frac{v}{V} - \frac{q}{V}(hc - x) \tag{4.59}$$

and, with the substitution $x = \frac{c}{2}(1 - \cos\theta)$,

$$\frac{dy}{dx} - \alpha = \frac{v}{V} - \frac{qc}{V}\left(h - \frac{1}{2} + \frac{\cos\theta}{2}\right)$$

From the general case in steady straight flight, Eq. (4.35) gives

$$\frac{dy}{dx} - \alpha = A_0 - \alpha + \sum A_n \cos n\theta \tag{4.60}$$

but in the pitching case, the load distribution is altered to some general form given by, say,

$$\frac{v}{V} = B_0 + \sum B_n \cos n\theta \tag{4.61}$$

where the coefficients change because of the relative-flow changes, while the camber-line shape remains constant. That is, the form of the function remains the same but the coefficients change. Thus in the pitching case,

$$\frac{dy}{dx} - \alpha = B_0 + \sum B_n \cos n\theta - \frac{qc}{V}\left(h - \frac{1}{2} + \frac{\cos\theta}{2}\right) \tag{4.62}$$

Equations (4.60) and (4.62) give

$$B_0 = A_0 - \alpha - \frac{qc}{V}\left(\frac{1}{2} - h\right), \quad B_1 = A_1 + \frac{qc}{2V} \text{ and } B_n = A_n$$

Analogous to the derivation of Eq. (4.40), the vorticity distribution here can be written

$$k = 2V\left[-B_0\left(\frac{1 + \cos\theta}{\sin\theta}\right) + \sum B_n \sin n\theta\right]$$

and, following similar steps for those of the derivation of Eq. (4.43), this leads to

$$C_L = 2\pi(\alpha - B_0) + \pi B_1 = 2\pi \left[\alpha - A_0 + \frac{A_1}{2} + \left(\frac{3}{4} - h\right)\frac{qc}{V} \right] \qquad (4.63)$$

Remember that this is for a two-dimensional wing. However, the effect of the curvature of the trailing vortex sheet is negligible in three dimensions, so it remains to replace the ideal $\partial C_L/\partial \alpha = 2\pi$ by a reasonable value a that accounts for the aspect-ratio change (see Chapter 5). The lift coefficient of a pitching rectangular wing then becomes

$$C_L = a\left[\alpha - A_0 + \frac{A_1}{2} + \left(\frac{3}{4} - h\right)\frac{qc}{V} \right] \qquad (4.64)$$

Similarly the pitching-moment coefficient about the leading edge is found from Eq. (4.44):

$$C_{M_{LE}} = \frac{\pi}{4}(B_2 - B_1) - \frac{C_L}{4} = \frac{\pi}{4}(A_2 - A_1) - \frac{\pi qc}{8V} - \frac{1}{4}C_L \qquad (4.65)$$

which, for a rectangular wing, on substituting for C_L, becomes

$$C_{M_{LE}} = \frac{\pi}{4}(A_2 - A_1) - \frac{\pi}{8}\frac{c}{V}q - \frac{a}{4}\left[\alpha - A_0 + \frac{A_1}{2} + \left(\frac{3}{4} - h\right)\frac{qc}{V} \right] \qquad (4.66)$$

The moment coefficient of importance in the derivative is that about CG, and this is found from

$$C_{M_{CG}} = C_{M_{LE}} + hC_L \qquad (4.67)$$

Substituting appropriate values,

$$C_{M_{CG}} = \frac{\pi}{4}(A_2 - A_1) - \frac{2\pi}{16}\frac{qc}{V} + \left(h - \frac{1}{4}\right)a\left[\alpha - A_0 + \frac{A_1}{2} + \left(\frac{3}{4} - h\right)\frac{qc}{V} \right]$$

which can be rearranged in terms of a function of coefficients A_n plus a term involving q; thus

$$C_{M_{CG}} = f(A_n) - \left[\frac{a}{4}(1 - 2h)^2 + \frac{2\pi - a}{16} \right]\frac{qc}{V} \qquad (4.68)$$

In this way, the contribution of the wings to Z_q or z_q becomes

$$Z_q = \frac{\partial Z}{\partial q} = -\frac{\partial L}{\partial q} = -\frac{\partial C_L}{\partial q}\frac{1}{2}\rho V^2 S = \frac{1}{2}\rho V^2 Sa\left(\frac{3}{4} - h\right)\frac{c}{V}$$

by differentiating Eq. (4.64) with respect to q.

Therefore, for a rectangular wing, defining z_q by $Z_q/(\rho V S l_t)$,

$$z_q = -\frac{a}{2}\left(\frac{3}{4} - h\right)\frac{c}{l_t} \tag{4.69}$$

For other than rectangular wings, an approximate expression can be obtained using strip theory, for example,

$$Z_q = -\rho V \int_{-s}^{s} \frac{a}{2}\left(\frac{3}{4} - h\right)c^2 dy$$

giving

$$z_q = -\frac{1}{Sl_t}\int_{-s}^{s} \frac{a}{2}\left(\frac{3}{4} - h\right)c^2 dy \tag{4.70}$$

In a similar fashion, the contribution to M_q and m_q can be found by differentiating the expression for M_{CG} with respect to q:

$$M_q = \frac{\partial M_{CG}}{\partial q} = \frac{\partial C_{MCG}}{\partial q}\frac{1}{2}\rho V^2 Sc = -\frac{1}{2}\rho V^2 Sc\left[\frac{a}{4}(1-2h)^2 + \frac{2\pi - a}{16}\right]\frac{c}{V}$$

where Eq. (4.68) was used; hence

$$M_q = -\left[\frac{a}{8}(1-2h)^2 + \frac{2\pi - a}{32}\right]VSc^2 \tag{4.71}$$

giving for a rectangular wing

$$m_q = \frac{M_q}{\rho V S l_t^2} = -\left[\frac{a}{8}(1-2h)^2 + \frac{2\pi - a}{32}\right]\frac{c^2}{l_t^2} \tag{4.72}$$

For other than rectangular wings, the contribution becomes, by strip theory,

$$M_q = -\rho V \int_{-s}^{s}\left(\frac{a}{8}(1-2h)^2 + \frac{2\pi - a}{32}\right)c^3 dy \tag{4.73}$$

and

$$m_q = -\frac{1}{Sl_t^2}\int_{-s}^{s}\left(\frac{a}{8}(1-2h)^2 + \frac{2\pi - a}{32}\right)c^3 dy \tag{4.74}$$

For the theoretical estimation of z_q and m_q of the complete aircraft, the contributions of the tailplane must be added. These are given here for completeness:

$$z_{q_{tail}} = -\frac{1}{2}\frac{S'}{S}\left(\frac{\partial C'_L}{\partial \alpha'} + C'_0\right), \quad m_{q_{tail}} = -\frac{1}{2}\frac{S'}{S}\left(\frac{\partial C'_L}{\partial \alpha'} + C'_0\right) \qquad (4.75)$$

where the terms with primes refer to tailplane data.

4.8 PARTICULAR CAMBER LINES

We saw that quite general camber lines may be used in the theory satisfactorily and that reasonable predictions of airfoil characteristics can be obtained. The reverse problem may be of more interest to the airfoil designer who wishes to obtain the camber-line shape to produce certain desirable characteristics. The general design problem is more comprehensive than this simple statement suggests, and the theory as dealt with so far is capable of considerable extension involving the introduction of thickness functions to give shape to the camber line. This is outlined in Section 4.9.

4.8.1 Cubic Camber Lines

Starting with a desirable aerodynamic characteristic, the simpler problem will be considered here. Numerous authorities [24] take a cubic equation as the general shape and evaluate the coefficients required to give the airfoil a fixed center of pressure. The resulting camber line has the reflex trailing edge, which is well-known.

Example 4.1

Find the cubic camber line that provides zero pitching moment about the quarter-chord point for a given camber.

The general equation for a cubic can be written $y = a'x(x+b')(x+d')$, with the origin at the leading edge. For convenience, the new variables $x_1 = x/c$ and $y_1 = y/\delta$ can be introduced, with δ as the camber. The conditions to be satisfied are that

1. $y = 0$ when $x = 0$; i.e., $y_1 = x_1 = 0$ at the leading edge.
2. $y = 0$ when $x = c$; i.e., $y_1 = 0$ when $x_1 = 1$.
3. $dy/dx = 0$ and $y = \delta$; i.e., $dy_1/dx_1 = 0$ when $y_1 = 1$ (when $x_1 = x_0$).
4. $C_{M_{1/4}} = 0$; i.e., $A_1 - A_2 = 0$.

Rewriting the cubic in the dimensionless variables x_1 and y_1,

$$y_1 = ax_1(x_1 + b)(x_1 + d) \qquad (a)$$

Formula (a) satisfies condition 1.

To satisfy condition 2, $(x_1 + d) = 0$ when $x_1 = 1$; therefore, $d = -1$, giving

$$y_1 = ax_1(x_1 + b)(x_1 - 1) \tag{b}$$

or multiplying out,

$$y_1 = ax_1^3 + a(b-1)x_1^2 - abx_1 \tag{c}$$

Differentiating Eq. (c) to satisfy condition 3,

$$\frac{dy_1}{dx_1} = 3ax_1^2 + 2a(b-1)x_1 - ab = 0 \quad \text{when} \quad y_1 = 1 \tag{d}$$

and if x_0 corresponds to the value of x_1 when $y_1 = 1$ (i.e., at the point of maximum displacement from the chord), the two simultaneous equations are

$$1 = ax_0^3 + a(b-1)x_0^2 - abx_0 \quad \text{and} \quad 0 = 3ax_0^2 + 2a(b-1)x_0 - ab \tag{e}$$

To satisfy condition 4, A_1 and A_2 must be found; dy_1/dx_1 can be converted to expressions suitable for comparison with Eq. (4.35) by writing

$$x = \frac{c}{2}(1 - \cos\theta) \quad \text{or} \quad x_1 = \frac{1}{2}(1 - \cos\theta)$$

$$\frac{dy_1}{dx_1} = \frac{3}{4}a\left(1 - 2\cos\theta + \cos^2\theta\right) + a(b-1) - a(b-1)\cos\theta - ab$$

$$= \left(\frac{3}{4}a + ab - a - ab\right) - \left(\frac{3}{2} + ab - a\right)\cos\theta + \frac{3}{4}a\cos^2\theta$$

Thus

$$\frac{dy_c}{dx} = \frac{\delta}{c}\frac{dy_1}{dx_1} = \frac{\delta}{c}\left[\frac{3}{4}a\cos^2\theta - \left(\frac{a}{2} + ab\right)\cos\theta - \frac{a}{4}\right] \tag{f}$$

Comparing Eqs. (f) and (4.35) gives

$$A_0 = -\frac{a}{4}\frac{\delta}{c} + \frac{3}{8}a\frac{\delta}{c} = \frac{a}{8}\frac{\delta}{c}$$

$$A_1 = -\left(\frac{a}{2} + ab\right)\frac{\delta}{c}$$

$$A_2 = a\frac{3}{8}\frac{\delta}{c}$$

Condition 4 is satisfied if $A_1 = A_2$, that is,

$$-\left(\frac{a}{2} + ab\right)\frac{\delta}{c} = a\frac{3}{8}\frac{\delta}{c}, \quad \text{giving} \quad b = -\frac{7}{8} \tag{g}$$

The quadratic in Eq. (e) can be solved for x_0 after cancelling a. With $b = -7/8$ from Eq. (g), we get $x_0 = 22.55/24$ or $x_0 = 7.45/24$—that is, taking the smaller value since the larger gives only the point of reflexure near the trailing edge: $y = \delta$ when $x = 0.31\times$ chord. Substituting $x_0 = 0.31$ in the cubic of Eq. (e) gives $a = 1/0.121 = 8.28$. The camber-line equation is then

$$y = 8.28\delta x_1 \left(x_1 - \frac{7}{8} \right)(x_1 - 1) \qquad (h)$$

This cubic camber-line shape is plotted in Fig. E4.1, and the ordinates are given in the inset table.

Lift Coefficient
The lift coefficient is given from Eq. (4.43) by

$$C_L = 2\pi \left(\alpha - A_0 + \frac{A_1}{2} \right)$$

So, with the values of A_0 and A_1 given previously,

$$C_L = 2\pi \left[\alpha - \frac{a}{8}\frac{\delta}{c} - \frac{1}{2}\left(\frac{a}{2} + ab \right)\frac{\delta}{c} \right]$$

Substituting for $a - 8.28$ and $b - -7/8$, we get

$$C_L = 2\pi \left(\alpha + 0.518\frac{\delta}{c} \right)$$

which gives a no-lift angle $\alpha_0 = -0.518\delta/c$ radians, or with $\beta = 100\delta/c$, which is the percentage camber, $\alpha_0 = -0.3\beta$ degrees.

Load Distribution
From Eq. (4.40),

$$k = 2U \left[\left(\alpha - \frac{1.04\delta}{c} \right)\frac{1 + \cos\theta}{\sin\theta} + \frac{3.12\delta}{c}\sin\theta + \frac{3.12\delta}{c}\sin 2\theta \right]$$

for the first three terms. This was evaluated for the incidence $\alpha_o = 29.6(\delta/c)$ degrees, and the result is plotted and tabulated in Fig. E4.1.

Note that the leading-edge value is omitted since it is infinite according to this theory. This is due to the term

$$\left(\alpha - \frac{1.04\delta}{c} \right)\frac{1 + \cos\theta}{\sin\theta}$$

becoming infinite at $\theta = 0$. When

$$\alpha = \frac{a}{8}\left(\frac{\delta}{c} \right) = 1.04\frac{\delta}{c}$$

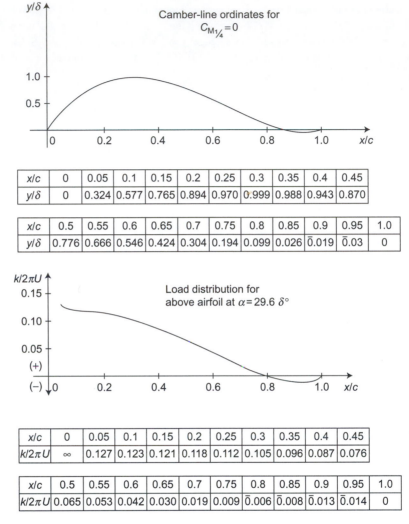

FIGURE E4.1

Illustration of predictions of performance of the cubic-camber airfoil.

x/c	0	0.05	0.1	0.15	0.2	0.25	0.3	0.35	0.4	0.45
y/δ	0	0.324	0.577	0.765	0.894	0.970	0.999	0.988	0.943	0.870

x/c	0.5	0.55	0.6	0.65	0.7	0.75	0.8	0.85	0.9	0.95	1.0
y/δ	0.776	0.666	0.546	0.424	0.304	0.194	0.099	0.026	$\bar{0}$.019	$\bar{0}$.03	0

x/c	0	0.05	0.1	0.15	0.2	0.25	0.3	0.35	0.4	0.45
$k/2\pi U$	∞	0.127	0.123	0.121	0.118	0.112	0.105	0.096	0.087	0.076

x/c	0.5	0.55	0.6	0.65	0.7	0.75	0.8	0.85	0.9	0.95	1.0
$k/2\pi U$	0.065	0.053	0.042	0.030	0.019	0.009	$\bar{0}$.006	$\bar{0}$.008	$\bar{0}$.013	$\bar{0}$.014	0

the coefficient in the previous equation becomes zero, then the intensity of circulation at the leading edge is zero and the stream flows smoothly onto the camber line at the leading edge, which is a stagnation point. This is the so-called Theodorsen condition, and the appropriate C_L is the ideal, optimum, or design lift coefficient, $C_{L_{opt}}$.

4.8.2 NACA Four-Digit Wing Sections

According to Abbott and von Doenhoff [25], when the NACA four-digit wing sections were first derived in 1932, it was found that the thickness distributions of efficient wing sections, such as the Goettingen 398 and the Clark Y, were nearly the same when the maximum thicknesses were set equal to the same value. The thickness distribution for the NACA four-digit sections was selected to correspond closely to those for these earlier wing sections; it is given by the following equation:

$$y_t - \pm 5ct \left[0.2969\sqrt{\xi} - 0.1260\xi - 0.3516\xi^2 + 0.2843\xi^3 - 0.1015\xi^4 \right] \quad (4.76)$$

where t is the maximum thickness expressed as a fraction of the chord, and $\xi = x/c$. The leading-edge radius is

$$r_t = 1.1019ct^2 \quad (4.77)$$

Note from Eqs. (4.76) and (4.77) that the ordinate at any point is directly proportional to the thickness ratio and that the leading-edge radius varies as the square of the thickness ratio.

To systematically study the effect of variation in the amount of camber and the shape of the camber line, the camber line shapes are expressed analytically as two parabolic arcs tangent at the position of the maximum camber-line ordinate. The equations to define the camber line are

$$y_c = \frac{mc}{p^2}(2p\xi - \xi^2), \, \xi \leq p; \quad y_c = \frac{mc}{(1-p)^2}\left[(1-2p) + 2p\xi - \xi^2\right], \xi \geq p \quad (4.78)$$

where m is the maximum value of y_c expressed as a fraction of the chord c, and p is the value of x/c corresponding to this maximum.

The numbering system for the NACA four-digit wing sections is based on section geometry. The first integer equals $100m$, the second equals $10p$, and the final two taken together equal $100t$. Thus the NACA 4412 wing section has 4% camber at $x = 0.4c$ from the leading edge and is 12% thick.

To determine the lifting characteristics using thin-airfoil theory, the camber-line slope must be expressed as a Fourier series. Differentiating Eq. (4.78) with respect to x gives

$$\frac{dy_c}{dx} = \frac{d(y_c/c)}{d\xi} = \frac{2m}{p^2}(p - \xi), \quad \xi \leq p$$

$$\frac{dy_c}{dx} = \frac{d(y_c/c)}{d\xi} = \frac{2m}{(1-p)^2}(p - \xi), \quad \xi \geq p$$

Changing variables from ξ to θ where $\xi = (1 - \cos\theta)/2$ gives

$$\frac{dy_c}{dx} = \frac{m}{p^2}(2p - 1 + \cos\theta), \ \theta \le \theta_p; \quad \frac{dy_c}{dx} = \frac{m}{(1-p)^2}(2p - 1 + \cos\theta),$$

$$\theta \ge \theta_p \qquad (4.79)$$

where θ_p is the value of θ corresponding to $x = pc$.

Substituting Eq. (4.79) into Eq. (4.41) gives

$$A_0 = \frac{1}{\pi}\int_0^\pi \frac{dy_c}{dx}\, d\theta = \frac{1}{\pi}\int_0^{\theta_p} \frac{m}{p^2}(2p - 1 + \cos\theta)\, d\theta$$

$$+ \frac{1}{\pi}\int_{\theta_p}^\pi \frac{m}{(1-p)^2}(2p - 1 + \cos\theta)\, d\theta$$

Integrating, we have

$$A_0 = \frac{m}{\pi p^2}\left[(2p-1)\theta_p + \sin\theta_p\right] + \frac{m}{\pi(1-p)^2}\left[(2p-1)(\pi - \theta_p) - \sin\theta_p\right] \quad (4.80)$$

Similarly from Eq. (4.42),

$$A_1 = \frac{1}{\pi}\int_0^\pi \frac{dy_c}{dx}\cos\theta\, d\theta$$

hence

$$A_1 = \frac{2m}{\pi p^2}\left[(2p-1)\sin\theta_p + \frac{1}{4}\sin 2\theta_p + \frac{\theta_p}{2}\right]$$

$$- \frac{2m}{\pi(1-p)^2}\left[(2p-1)\sin\theta_p + \frac{1}{4}\sin 2\theta_p - \frac{1}{2}(\pi - \theta_p)\right] \qquad (4.81)$$

$$A_2 = \frac{1}{\pi}\int_0^\pi \frac{dy_c}{dx}\cos 2\theta\, d\theta$$

and so

$$A_2 = \frac{2m}{\pi p^2}\left[(2p-1)\left(\frac{1}{4}\sin 2\theta_p + \frac{\theta_p}{2}\right) + \sin\theta_p - \frac{1}{3}\sin^3\theta_p\right]$$

$$- \frac{2m}{\pi(1-p)^2}\left[(2p-1)\left(\frac{1}{4}\sin 2\theta_p - \frac{\pi - \theta_p}{2}\right) + \sin\theta_p - \frac{1}{3}\sin^3\theta_p\right] \quad (4.82)$$

Example 4.2

NACA 4412

For a NACA 4412 wing section, $m = 0.04$ and $p = 0.4$ so that

$$\theta_p = \cos^{-1}(1 - 2 \times 0.4) = 1.3694 \text{ radians}$$

Making these substitutions into Eqs. (4.80) through (4.82) gives

$$A_0 = 0.0090, \quad A_1 = 0.1630, \quad \text{and} \quad A_2 = 0.0228$$

Thus Eqs. (4.43) and (4.47) give

$$C_L = \pi(A_1 - 2A_0) + 2\pi\alpha = 0.456 + 6.2832\alpha \tag{a}$$

$$C_{M_{1/4}} = -\frac{\pi}{4}(A_1 - A_2) = -0.110 \tag{b}$$

4.9 THE THICKNESS PROBLEM FOR THIN-AIRFOIL THEORY

Before extending the theory to take into account the thickness of airfoil sections, it is useful to review parts of the method. Briefly, in thin-airfoil theory the two-dimensional thin wing is replaced by the vortex sheet, which occupies the camber surface or, to the first approximation, the chordal plane. Vortex filaments comprising the sheet extend to infinity in both directions normal to the plane, and all velocities are confined to the xy plane. In such a situation, as shown in Fig. 4.12, the sheet supports a pressure difference producing a normal (upward) increment of force $(p_1 - p_2)\delta s$ per unit spanwise length. Subscripts 1 and 2 refer to the lower and upper sides of the sheet, respectively. However, from Bernoulli's equation,

$$p_1 - p_2 = \frac{1}{2}\rho(u_2^2 - u_1^2) = \rho(u_2 - u_1)\frac{u_2 + u_1}{2} \tag{4.83}$$

Writing $(u_2 + u_1)/2 \approx U$, the free-stream velocity, and $(u_2 - u_1) = k$, the local loading on the wing, we have

$$(p_1 - p_2)\delta s = \rho U k \delta s \tag{4.84}$$

The lift may then be obtained by integrating the normal component and, similarly, the pitching moment. It remains now to relate the local vorticity to the thin shape of the airfoil, by introducing the solid boundary condition of zero velocity normal to the surface. For the vortex sheet to completely simulate the airfoil, the velocity component induced locally by the distributed vorticity must be sufficient to make the resultant velocity tangential to the surface. In other words, the component of the free-stream velocity normal to the surface at a point on the airfoil must be completely nullified by the normal-velocity component induced by the distributed vorticity. This condition, which is satisfied completely by replacing the surface line by a streamline, results in an integral equation that relates vortex distribution strength to airfoil shape.

So far in this review, no assumptions or approximations have been made. However, in addition to the thin assumption of zero thickness and small camber, thin-airfoil theory employs the following:

- That the magnitude of total velocity at any point on the airfoil is that of the local chordwise velocity $\equiv U + u'$.
- That chordwise perturbation velocities u' are small in relation to the chordwise component of the free-stream U.
- That the vertical perturbation velocity v anywhere on the airfoil may be taken as that (locally) at the chord.

Making use of these restrictions gives

$$v = \int_0^c \frac{k}{2\pi} \frac{dx}{x - x_1}$$

and thus we get Eq. (4.42):

$$U\left[\frac{dy_c}{dx} - \alpha\right] = \int_0^c \frac{k}{2\pi} \frac{dx}{x - x_1}$$

This last integral equation relates the chordwise loading (i.e., vorticity) to the shape and incidence of the thin airfoil and, by the insertion of a suitable series expression for k in the integral, is capable of solution for both the direct and indirect airfoil problems. The airfoil is reduced to, in essence, a thin lifting sheet, infinitely long in span, and is replaced by a distribution of singularities that satisfies the same conditions at the boundaries of the airfoil system (i.e., at the surface and at infinity). Further, the theory is linearized, permitting, for example, the velocity at a point in the vicinity of the airfoil to be taken to be the sum of the velocity components due to the various characteristics of the system, each treated separately. As shown in Section 4.3, these linearization assumptions extend the theory by allowing a perturbation velocity contribution due to thickness to be added to the other effects.

4.9.1 Thickness Problem for Thin Airfoils

A symmetrical closed contour of small thickness-chord ratio may be obtained from a distribution of sources, and sinks, confined to the chord and immersed in a uniform undisturbed stream parallel to the chord. The typical model is shown in Fig. 4.20,

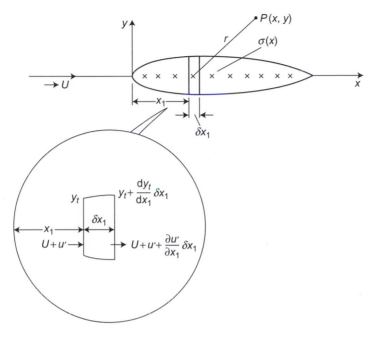

FIGURE 4.20

Illustration of chordwise source distribution.

where $\sigma(x)$ is the chordwise source distribution. Recall that a system of discrete sources and sinks in a stream may result in a closed streamline.

Consider the influence of the sources in the element δx_1 of chord x_1 from the origin. The strength of these sources is

$$\delta m = \sigma(x_1)\delta x_1$$

Since the elements of the upper and lower surfaces are impermeable, the strength of the sources between x_1 and $x_1 + \delta x_1$ are found from continuity:

$$\delta m = 2\left[\left(U + u' - \frac{\partial u'}{\partial x_1}\delta x_1\right)\left(y_t + \frac{dy_t}{dx_1}\delta x_1\right) - (U + u')y_t\right] \qquad (4.85)$$

which is the outflow across the boundary at $y_t \pm (dy_t/dx_1)\delta x_1$ minus the inflow across the boundary $\pm y_t$. Neglecting second-order quantities,

$$\delta m = 2U\frac{dy_t}{dx_1}\delta x_1 \qquad (4.86)$$

The velocity potential at a general point P for a source of this strength is given by (see Eq. 3.22)

$$\delta\phi = \frac{\delta m}{2\pi}\ln r = \frac{U}{\pi}\frac{dy_t}{dx_1}\delta x_1 \ln r \tag{4.87}$$

where $r = \sqrt{(x-x_1)^2 + y^2}$. The velocity potential for the complete distribution of sources lying between 0 and c on the x-axis becomes

$$\phi = \frac{U}{\pi}\int_0^c \frac{dy_t}{dx_1}\ln r\,dx_1 \tag{4.88}$$

and adding the free stream gives

$$\phi = Ux + \frac{U}{\pi}\int_0^c \frac{dy_t}{dx_1}\ln r\,dx_1 \tag{4.89}$$

Differentiating to find the velocity components,

$$u = \frac{\partial\phi}{\partial x} = U + \frac{U}{\pi}\int_0^c \frac{dy_t}{dx_1}\frac{(x-x_1)}{(x-x_1)^2 + y^2}\,dx_1 \tag{4.90}$$

$$v = \frac{\partial\phi}{\partial y} = \frac{U}{\pi}\int_0^c \frac{dy_t}{dx_1}\frac{y}{(x-x_1)^2 + y^2}\,dx_1 \tag{4.91}$$

To obtain the tangential velocity at the surface of the airfoil, the limit as $y \to 0$ is taken for Eq. (4.90) so that

$$u = U + u' = U + \frac{U}{\pi}\int_0^c \frac{dy_t}{dx_1}\frac{1}{x-x_1}\,dx_1 \tag{4.92}$$

The coefficient of pressure is then given by

$$C_p = -2\frac{u'}{U} = -\frac{2}{\pi}\int_0^c \frac{dy_t}{dx_1}\frac{1}{x-x_1}\,dx_1 \tag{4.93}$$

The theory in this form is of limited usefulness for practical airfoil sections because most of these have rounded leading edges. At a rounded leading edge, dy_t/dx_1 becomes infinite and so violates the assumptions made to develop the thin-airfoil theory. In fact, from Example 4.3 to come, we see that the theory breaks down even when dy_t/dx_1 is finite at the leading and trailing edges. Various refinements to the theory partially overcome this problem [26] and permit its extension to moderately thick airfoils [27].

Example 4.3

Find the pressure distribution on the biconvex airfoil

$$\frac{y_t}{c} = \frac{t}{2c}\left[1 - \left(\frac{2x}{c}\right)^2\right]$$

(with the origin at midchord) set at zero incidence in an otherwise undisturbed stream. For the given airfoil,

$$\frac{y_t}{c} = \frac{t}{2c}\left[1 - \left(\frac{2x_1}{c}\right)^2\right]$$

and

$$\frac{dy_t}{dx_1} = -4t\frac{x_1}{c^2}$$

From before,

$$u' = \frac{u}{\pi}\int_{-c/2}^{c/2} -4\frac{t}{c^2}\frac{x_1}{x-x_1}\,dx_1$$

hence

$$C_p = \frac{2}{\pi}\frac{4t}{c^2}\int_{-c/2}^{c/2}\frac{x_1}{x-x_1}\,dx_1 = \frac{-8}{\pi}\frac{t}{c^2}[x\ln(x-x_1)+x_1]_{-c/2}^{c/2}$$

Thus

$$C_p = -\frac{8}{\pi}\frac{t}{c}\left[1 + \frac{x}{c}\ln\frac{2x-c}{2x+c}\right]$$

At the midchord point,

$$x = 0, \quad C_p = -\frac{8t}{\pi c}$$

At the leading and trailing edges, $x = \pm c$, $C_p \to -\infty$. The latter result shows that the approximations in the linearization do not permit the method to apply to local effects in the region of stagnation points, even when the slope of the thickness shape is finite.

4.10 COMPUTATIONAL (PANEL) METHODS FOR TWO-DIMENSIONAL LIFTING FLOWS

We now describe the extension of the computational method described in Section 3.4 to two-dimensional lifting flows. The basic panel method was developed by a group led by Hess and Smith [28] at Douglas Aircraft in the late 1950s and early 1960s. Their method applied surface distributions of sources and doublets (the latter required to solve the lifting problem). A panel method that applies vortex panels appears to have been first developed by Rubbert [29] at Boeing to solve lifting flows. Its two-dimensional version can be applied to airfoil sections of any thickness and camber. In essence, to generate the circulation necessary for the production of lift, vorticity in some form must be introduced into the modeling of the flow.

In the panel method for nonlifting bodies, as described in Section 3.4, a surface distribution of sources is applied. For the lifting case, the airfoil section can be modeled by vortex panels in the form of straight-line segments with the strength, which can be distributed over panels that model the airfoil contour itself, as shown in Fig. 4.21. This procedure allows us to model the airfoil shape completely (thickness and camber) using vortex panels only. This is the approach we use in this section to examine two-dimensional airfoils.

The central problem of extending the panel method to lifting flows is satisfying the Kutta condition (see Section 4.1.1). It is not possible with a computational scheme to do this directly; instead, the aim is to satisfy some of the implied conditions:

1. The streamline leaves the trailing edge with a direction along the bisector of the trailing-edge angle.

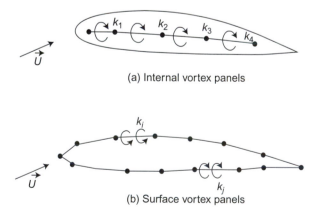

(a) Internal vortex panels

(b) Surface vortex panels

FIGURE 4.21

Vortex panels: (a) internal; (b) surface.

2. As the trailing edge is approached, the velocity magnitudes on the upper and lower surfaces approach the same limiting value.
3. In the practical case of an airfoil with a finite trailing-edge angle, the trailing edge must be a stagnation point; the common limiting value of (b), then, must be zero.
4. The source strength per unit length must be zero at the trailing edge.

Computational schemes use either conditions (a) or (b). It is not generally possible to satisfy (c) and (d) as well because, as will be shown, an overspecification of the problem results. The methods for satisfying (a) and (b) are illustrated in Fig. 4.22. For condition (a), an additional panel must be introduced oriented along the bisector of the trailing-edge angle. The value of the circulation is then fixed by requiring the normal velocity to be zero at the collocation point of the additional $(N+1)$ panel. For condition (b), the magnitudes of the tangential velocity vectors at the collocation points of the two panels, which define the trailing edge, are required to be equal. Hess [30] showed that condition (b) gives more accurate results than condition (a), other things being equal. The use of surface, rather than interior, vorticity panels is also preferable for computational accuracy.

To use vortex panels alone, each of the N panels carries a vortex distribution of uniform strength per unit length, γ_i $(i = 1, 2, \ldots, N)$. In general, the vortex strength varies from panel to panel. Let $i = t$ for the panel on the upper surface at the trailing edge so that $i = t + 1$ for the panel on the lower surface at the trailing edge. Condition (b) is equivalent to requiring that

$$\gamma_t = -\gamma_{t+1} \tag{4.94}$$

The normal velocity component at the collocation point of each panel must be zero, as it is for the nonlifting case. This gives N conditions to be satisfied for each of the N panels. Thus, when we take into account condition Eq. (4.94), there are $N+1$ conditions to be satisfied in total. Unfortunately, there are only N unknown vortex strengths, so it is not possible to satisfy all $N+1$ conditions. Thus, in order to proceed further, it is necessary to ignore the requirement that the normal velocity be zero for

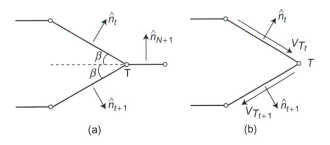

(a)   (b)

FIGURE 4.22

Two methods of implementing the Kutta condition at the trailing edge T.

one panel. This is unsatisfactory since it is not at all clear which panel is the best choice.

An alternative and more satisfactory method is to distribute both sources and vortices of uniform strength per unit length over each panel. In this case, though, the vortex strength is the same for all panels:

$$\gamma_i = \gamma \, (i = 1, 2, \ldots, N) \tag{4.95}$$

Thus there are now $N + 1$ unknown quantities—namely, the N source strengths and the uniform vortex strength per unit length γ to match the $N + 1$ conditions. With this approach it is perfectly feasible to use internal vortex instead of surface panels. However, the internal panels must carry vortices that are either of uniform strength or of predetermined variable strength, providing that the variation is characterized by a single unknown parameter. Generally, however, surface vortex panels lead to better results. Also condition (a) can be used in place of condition (b). Again, however, condition (b) generally gives more accurate results.

A practical panel method for lifting flows around airfoils is described in some detail next. It uses condition (b) and is based on a combination of surface vortex panels, of uniform strength, and source panels. First, however, it is necessary to show how the normal and tangential influence coefficients for vortex panels may be evaluated. It turns out that the procedure is very similar to that for source panels.

The velocity at point P due to vortices on an element of length $\delta\xi$ in Fig. 4.23 is given by

$$\delta V_\theta = \frac{\gamma}{R} d\xi \tag{4.96}$$

where $\gamma d\xi$ replaces $\Gamma/(2\pi)$ used in Section 3.2.4. δV_θ is oriented at angle θ as shown. Therefore, the velocity components in the panel-based coordinate directions

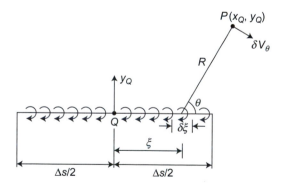

FIGURE 4.23

Thin airfoil modeled as a vortex sheet.

(i.e., in the x_Q and y_Q directions) are given by

$$\delta V_{x_Q} = \delta V_\theta \sin\theta = \frac{\gamma y_Q}{(x_Q - \xi)^2 + y_Q^2}\delta\xi \qquad (4.97)$$

$$\delta V_{y_Q} = -\delta V_\theta \cos\theta = -\frac{\gamma(x_Q - \xi)}{(x_Q - \xi)^2 + y_Q^2}\delta\xi \qquad (4.98)$$

To obtain the corresponding velocity components at P due to all vortices on the panel, we integrate along the length of the panel to get

$$V_{x_Q} = \gamma \int_{-\Delta s/2}^{\Delta s/2} \frac{y_Q}{(x_Q - \xi)^2 + y_Q^2}d\xi = \gamma\left[\tan^{-1}\left(\frac{x_Q + \Delta s/2}{y_Q}\right)\right.$$

$$\left. - \tan^{-1}\left(\frac{x_Q - \Delta s/2}{y_Q}\right)\right] \qquad (4.99)$$

$$V_{y_Q} = -\gamma \int_{-\Delta s/2}^{\Delta s/2} \frac{x_Q - \xi}{(x_Q - \xi)^2 + y_Q^2}d\xi = -\frac{\gamma}{2}\ln\left[\frac{(x_Q + \Delta s/2)^2 + y_Q^2}{(x_Q - \Delta s/2)^2 + y_Q^2}\right] \qquad (4.100)$$

Following the basic method described in Section 3.4, we introduce normal and tangential influence coefficients, N'_{ij} and T'_{ij}; the primes distinguish these coefficients from those introduced in Section 3.4 for the source panels. N'_{ij} and T'_{ij} represent the normal and tangential velocity components at collocation point i due to vortices of unit strength per unit length distributed on panel j.

Let $\hat{t}_i$ and $\hat{n}_i (i - 1, 2, \ldots, N)$ denote the unit tangent and normal vectors for each panel, and let point P correspond to collocation point i. Then in vector form we get the velocity at collocation point i:

$$\vec{V}_{PQ} = V_{x_Q}\hat{t}_j + V_{y_Q}\hat{n}_j$$

To obtain the components of this velocity vector perpendicular and tangential to panel i, we take the scalar product of the velocity vector with $\hat{n}_i$ and $\hat{t}_i$, respectively. If furthermore γ is set equal to 1 in Eqs. (4.99) and (4.100), we obtain the following expressions for the influence coefficients:

$$N'_{ij} = \vec{V}_{PQ}\cdot\hat{n}_i = V_{x_Q}\hat{n}_i\cdot\hat{t}_j + V_{y_Q}\hat{n}_i\cdot\hat{n}_j; \quad T'_{ij} = \vec{V}_{PQ}\cdot\hat{t}_i = V_{x_Q}\hat{t}_i\cdot\hat{t}_j + V_{y_Q}\hat{t}_i\cdot\hat{n}_j \quad (4.101)$$

Comparing Eqs. (4.99) and (4.101) for the vortices and the corresponding expressions (3.97) and (3.99) for the source panels shows that

$$[V_{x_q}]_{\text{vortices}} = [V_{y_q}]_{\text{sources}} \quad \text{and} \quad [V_{y_q}]_{\text{vortices}} = -[V_{x_q}]_{\text{sources}} \qquad (4.102)$$

With these results, it is now possible to describe how the basic panel method of Section 3.4 may be extended to lifting airfoils.

Each of the N panels now carries a source distribution of strength σ_i per unit length and a vortex distribution of strength γ per unit length. Thus there are now $N+1$ unknown quantities, so the $N \times N$ influence coefficient matrices N_{ij} and T_{ij} corresponding to the sources must be expanded to $N \times (N+1)$. The $(N+1)$ column at this point contains the velocities induced at the collocation points by vortices of unit strength per unit length on all of the panels. Thus $N_{i,N+1}$ represents the normal velocity at the ith collocation point induced by the vortices over all panels and similarly for $T_{i,N+1}$. Thus, using Eqs. (4.101),

$$N_{i,N+1} = \sum_{j=1}^{N} N'_{i,j}, \quad \text{and} \quad T_{i,N+1} = \sum_{j=1}^{N} T'_{i,j} \qquad (4.103)$$

In a similar fashion as for the nonlifting case described in Section 3.4, the total normal velocity at each collocation point—due to the net effect of all sources, vortices, and the oncoming flow—must be zero. This can be written in the form

$$\sum_{j=1}^{N} \sigma_j N_{ij} + \gamma N_{i,N+1} + \vec{U} \cdot \hat{n}_i = 0, \quad (i = 1, 2, \ldots, N) \qquad (4.104)$$

where the first term represents the sources, the second term represents the vortices, and the third term is the onset (or oncoming) flow. These N equations are supplemented by imposing condition (b). The simplest way to do this is to equate the magnitudes of the tangential velocities at the collocation point of the two panels defining the trailing edge (see Fig. 4.22(b)). Remember that the unit tangent vectors $\hat{t}_t$ and $\hat{t}_{t+1}$ are in opposite directions, so condition (b) can be expressed mathematically as

$$\sum_{j=1}^{N} \sigma_j T_{t,j} + \gamma T_{t,N+1} + \vec{U} \cdot \hat{t}_t = - \left(\sum_{j=1}^{N} \sigma_j T_{t+1,j} + \gamma T_{t+1,N+1} + \vec{U} \cdot \hat{t}_{t+1} \right) \qquad (4.105)$$

Equations (4.104) and (4.105) combine to form a matrix equation written as

$$\mathbf{Ma} = \mathbf{b} \qquad (4.106)$$

where $\mathbf{M}$ is an $(N+1) \times (N+1)$ matrix and $\mathbf{a}$ and $\mathbf{b}$ are $(N+1)$ column vectors. The elements of the matrix and vectors are as follows:

$$M_{i,j} = N_{i,j}, \; i = 1, 2, \ldots, N, \; j = 1, 2, \ldots, N+1$$
$$M_{N+1,j} = T_{t,j} + T_{t+1,j}, \; j = 1, 2, \ldots, N+1$$

$$a_i = \sigma_i, \quad i = 1, 2, \ldots, N, \quad \text{and} \quad a_{N+1} = \gamma$$

$$b_i = -\vec{U} \cdot \hat{n}_i, \quad i = 1, 2, \ldots, N, \quad \text{and} \quad b_{N+1} = -\vec{U} \cdot \left(\hat{t}_t + \hat{t}_{t+1}\right)$$

Systems of linear equations such as Eq. (4.106) can be readily solved numerically for unknowns a_i using standard methods (see Section 3.4). Also, it is now possible to see condition (c), requiring that the tangential velocities on the upper and lower surfaces tend to zero at the trailing edge, cannot be satisfied in this sort of numerical scheme. Condition (c) could be imposed approximately by requiring, say, that the tangential velocities on panels t and $t+1$ both be zero. Referring to Eq. (4.105), this approximate condition can be expressed mathematically as

$$\sum_{j=1}^{N} \sigma_j T_{t,j} + \gamma T_{t,N+1} + \vec{U} \cdot \hat{t}_t = 0$$

and

$$\sum_{j=1}^{N} \sigma_j T_{t+1,j} + \gamma T_{t+1,N+1} + \vec{U} \cdot \hat{t}_{t+1} = 0$$

Equation (4.105) is replaced by these two equations so that M in Eq. (4.106) is now a $(N+2) \times (N+1)$ matrix. At this point, the problem is overdetermined (i.e., there is one more equation than the number of unknowns), and Eq. (4.106) can no longer be solved for the vector **a**—that is, for the source and vortex strengths.

Calculating the influence coefficients is at the heart of any panel method. In Section 3.4, a MATLAB function was given for computing the influence coefficients for the nonlifting case and was used to solve the problem of flow around a NACA 0024 symmetric airfoil. In this section, we provide a MATLAB function to compute the influence coefficients due to additional vortices required for a lifting flow. It is applied to solve the flow around a NACA 4412 airfoil to demonstrate that the numerics agree with experiments and with the theoretical results already described.

Two modifications to `function InfluSour` from Section 3.4 are required to extend it to the lifting case.

1. The following loop is added at the begininig of the function:

```
NP1 = N + 1
for I = 1,N
  AN(I,NP1) = 0;
  AT(I,NP1) = pi;
end
```

The additional lines initialize the values of the influence coefficients, $N_{i,N+1}$ and $T_{i,N+1}$, preparing them for calculation later in the program. Note that the initial value of $T_{i,N+1}$ is set at π because, in Eq. (4.103),

$$T'_{N+1,N+1} = N_{i,j} = \pi$$

that is, the tangential velocity induced on a panel by vortices of unit strength per unit length distributed over the same panel is, from Eq. (4.102), the same as the normal velocity induced by sources of unit strength per unit length. This takes the value π in Eq. (3.100).

2. The two lines of instruction that calculate the additional influence coefficients according to Eq. (4.103) are inserted below the last two command lines in the function:

```
AN(I,J) = VX * NTIJ + VY * NNIJ; % existing line
AT(I,J) = VX * TTIJ + VY * TNIJ; % existing line
AN(I,NP1) = AN (I,NP1) + VY * NTIJ - VX * NNIJ; % added line
AT(I,NP1) = AT (I,NP1) + VY * TTIJ - VX * TNIJ; % added line
```

The modified function is given here:

```
function [AN,AT,XC,YC,NHAT,THAT] = InfluSourV(XP,YP,N,NT,NTP1)
% Influence coefficients for source distribution over a
% symmetric body.
%
NP1 = N + 1;
for J = 1:N
    AN(J,NP1) = 0;
    AT(J,NP1) = pi;
if J==1
  XPL = XP(N);
  YPL = YP(N);
else
  XPL = XP(J-1);
  YPL = YP(J-1);
end
XC(J) = 0.5*(XP(J) + XPL);
YC(J) = 0.5*(YP(J) + YPL);
S(J) = sqrt( (XP(J) - XPL)^2 + (YP(J) - YPL)^2 );
THAT(J,1) = (XP(J) - XPL)/S(J);
THAT(J,2) = (YP(J) - YPL)/S(J);
NHAT(J,1) = - THAT(J,2);
NHAT(J,2) =   THAT(J,1);
end
%Calculation of the influence coefficients.
for I = 1:N
for  J = 1:N
if I==J
AN(I,J) = pi;
```

```
AT(I,J) = 0;
else
DX = XC(I) - XC(J);
DY = YC(I) - YC(J);
XQ = DX*THAT(J,1) + DY*THAT(J,2);
YQ = DX*NHAT(J,1) + DY*NHAT(J,2);
VX = -0.5*(  log( (XQ + S(J)/2 )^2 + YQ*YQ )...
      -log( (XQ - S(J)/2 )^2 + YQ*YQ )  );
VY = -( atan2((XQ + S(J)/2 ),YQ) - atan2((XQ - S(J)/2),YQ ));
NTIJ = 0;
NNIJ = 0;
TTIJ = 0;
TNIJ = 0;
for K = 1:2
NTIJ = NHAT(I,K)*THAT(J,K) + NTIJ;
NNIJ = NHAT(I,K)*NHAT(J,K) + NNIJ;
TTIJ = THAT(I,K)*THAT(J,K) + TTIJ;
TNIJ = THAT(I,K)*NHAT(J,K) + TNIJ;
end
AN(I,J) = VX*NTIJ + VY*NNIJ;
AT(I,J) = VX*TTIJ + VY*TNIJ;
AN(I,NP1) = AN(I,NP1) + VY*NTIJ - VX*NNIJ;
AT(I,NP1) = AT(I,NP1) + VY*TTIJ - VX*TNIJ;
end
end
end
for n = 1:NP1
    AN(NP1,n) = -(AT(NT,n) + AT(NTP1,n));
    AT(NP1,n) = 0;
end
AT(NP1,NP1) = pi;
```

Like the original routine presented in Section 3.4, this modified routine is primarily intended for educational purposes. Nevertheless, as shown by the example computation for the NACA 4412 airfoil presented next, a computer program based on this function and MATLAB matrix-inversion methods gives accurate results for pressure distribution and hence for coefficients of lift and pitching moment. The predictions of pressure distribution were compared with other computations and experiments at the same lift coefficient since the effective angle of attack was not known in the experiment.

The essential parts of the script used to create Fig 4.24 are as follows:

```
%
% NACA 4-digit airfoil: 4412
%
% clear;clc
c = 1;
Nthe = 201;
```

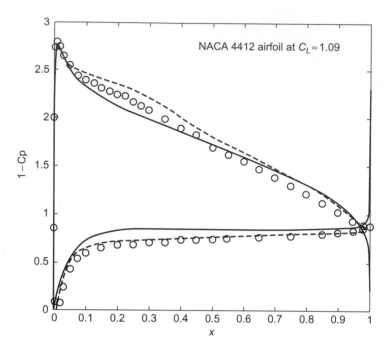

FIGURE 4.24

Solid line is the present prediction; dotted line is a prediction based on a code with viscous effects included; circles are experimental data.

```
the = 0:-2*pi/Nthe:-pi;
x = (1 + cos(the) )/2;
m=4/100; p=4/10; t = 0.12;
yt = 5*t*c .* (.2969*x.^0.5 - .126*x ...
    - .3516*x.^2 + .2843*x.^3 - .1015*x.^4); %yt(length(x))=0;
for n=1:length(x)
  if x <= p
    yc(n) = (m*c/p)*(2*p*x(n)-x(n)^2);
  else
    yc(n) = (m*c/(1-p)^2)*((1-2*p) + 2*p*x(n) - x(n)^2);
  end
end
XPU = x; YPU = yt + yc;
for n = 1:length(x)-1
    XPL(n) = x(length(x)-n);
    YPL(n) = yc(length(x)-n) - yt(length(x)-n);
end
XP = [XPU(1:end) XPL(1:end-1)];
YP = [YPU(1:end) YPL(1:end-1)];
% Inherent camber angle of attack = atan(0.022)*180/pi = 1.3
```

```
alpha = 4.2; alp = pi*alpha/180; N = length(XP);
NT = 1; NTp1 = N;
% for m = 1:N
%       YP(m) = YP(m) - sin(alpha*pi/180)*XP(m);
% end
figure(1);plot(XP,YP,'-o')
%
[AN,AT,XC,YC,NHAT,THAT] = InfluSourV(XP,YP,N,NTp1,NT);
% uinfty = 1;
 for j=1:N
 b(j,1) =  cos(alp)*NHAT(j,1) + sin(alp)*NHAT(j,2);
 end
  b(N+1,1) = - ( THAT(NT,1) + THAT(NT+1,1));
 Sources = AN\b;
 THAT(N+1,1) = 0; THAT(N+1,2) = 0;
 ut = AT*Sources - THAT(:,1);
 cp = 1 - ut.^2;
figure(2);
plot(XC,1-cp(1:end-1),'k')
xlabel(' x '),ylabel('  1-Cp '),axis([0 1 0 3])
title(' NACA 4412 airfoil at C_L \approx 1.09')
```

The NACA 4412 wing section was chosen to illustrate the panel method. It is moderately thick with moderate camber. The variation in pressure coefficient around this wing section at an angle of attack of 8 degrees is presented in the figure. Experimental data are compared with the computed results for 200 panels. Agreement between the computed data and the experimental data is satisfactory. Note that the predictions of pressure distribution are for the same lift coefficient, as determined by integrating the measured pressure. Agreement between the data and the predictions is remarkably good for this airfoil.

4.11 EXERCISES

4.1   A thin two-dimensional airfoil of chord c is operating at its ideal lift coefficient C_{Li}. Assume that the loading (i.e., the pressure difference across the airfoil) varies linearly with its maximum value at the leading edge. Show that

$$\frac{dy_c}{dx} - \alpha = \frac{C_{Li}}{2\pi}\left[(1-\xi)\ln\left(\frac{1-\xi}{\xi}\right) - 1\right]$$

where y_c defines the camber line, α is the angle of incidence, and $\xi = x/c$.

Hint: Do not attempt to make the transformation $x = (c/2)(1 - \cos\theta)$; instead, write the singular integral as follows:

$$\int_0^1 \frac{1}{\xi - \xi_1} d\xi = \lim_{\epsilon \to 0} \left[\int_0^{\xi_1 - \epsilon} \frac{1}{\xi - \xi_1} d\xi + \int_{\xi_1 + \epsilon}^1 \frac{1}{\xi - \xi_1} d\xi \right]$$

Then, using this result, show that the angle of incidence and the camber-line shape are given by

$$\alpha = \frac{C_L}{4\pi}; \quad \frac{y_c}{c} = \frac{C_L}{4\pi} \left[-(1 - \xi)^2 \ln(1 - \xi) + \xi(\xi - 2) \ln\xi \right]$$

Hint: Write $-1 = C - 1 - C$, where $1 + C = 2\pi\alpha/C_L$ and the constant C is determined by requiring that $y_c = 0$ at $\xi = 0$ and $\xi = 1$.

4.2 A thin airfoil has a camber line defined by the relation $y_c = kc\xi(\xi - 1)$ $(\xi - 2)$. Show that, if the maximum camber is 2% of the chord, $k = 0.052$. Determine the coefficients of lift and pitching moment (C_L and $C_{M_{1/4}}$) at 3 degrees of incidence.
(Answer: 0.535, −0.046)

4.3 Use thin-airfoil theory to estimate the coefficient of lift at zero incidence and the pitching-moment coefficient $C_{M_{1/4}}$ for a NACA 8210 wing section.
(Answer: 0.789, −0.172)

4.4 Use thin-airfoil theory to select a NACA four-digit wing section with a coefficient of lift at zero incidence approximately equal to unity. The maximum camber must be located at 40% chord and the thickness ratio is to be 0.10. Estimate the required maximum camber as a percentage of chord to the nearest whole number.
Hint: Use a computer program to solve by trial and error.
(Answer: NACA 9410)

4.5 Use thin-airfoil theory to select a NACA four-digit wing section with a coefficient of lift at zero incidence approximately equal to unity and a pitching-moment coefficient $C_{M_{1/4}} = -0.025$. The thickness ratio is to be 0.10. Estimate the required maximum camber as a percentage of chord to the nearest whole number and its position to the nearest tenth of a chord. The C_L value must be within 1% of the required value and $C_{M_{1/4}}$ within 3%.
Hint: Use a computer program to solve by trial and error.
(Answer: NACA 7610, but NACA 9410 and NACA 8510 are also close.)

4.6 A thin two-dimensional flat-plate airfoil is fitted with a trailing-edge flap of chord $100e\%$ of the airfoil chord. Show that the flap effectiveness,

$$\frac{a_2}{a_1} \equiv \frac{\partial C_L / \partial \eta}{\partial C_L / \partial \alpha}$$

where α is the angle of incidence and η is the flap angle, is approximately $4\sqrt{e}/\pi$ for flaps of small chord.

4.7 A thin airfoil has a circular-arc camber line with a maximum camber of 0.025 chord. Determine the theoretical pitching-moment coefficient $C_{M_{1/4}}$ and indicate methods by which this could be reduced without changing maximum camber. The camber line may be approximated by the expression

$$y_c = kc \left[\frac{1}{4} - \left(\frac{x'}{c} \right)^2 \right]$$

where $x' = x - 0.5c$. (*Answer:* -0.025π)

4.8 The camber line of a circular-arc airfoil is given by

$$\frac{y_c}{c} = 4h \frac{x}{c} \left(1 - \frac{x}{c} \right)$$

Derive an expression for the load distribution (pressure difference across the airfoil) at incidence α. Show that the zero-lift angle $\alpha_0 = -2h$, and sketch the load distribution at this incidence. Compare the lift curve of this airfoil with that of a flat plate.

4.9 A flat-plate airfoil is aligned along the x-axis with the origin at the leading edge and trailing edge at $x = c$. The plate is at an angle of incidence α to a free stream of air speed U. A vortex of strength Γ_v is located at (x_v, y_v). Show that the distribution $k(x)$ of vorticity along the airfoil from $x = 0$ to $x = c$ satisfies the integral equation

$$\frac{1}{2\pi} \int_0^c \frac{k(x)}{x - x_1} dx = -U\alpha - \frac{\Gamma_v(x_v - x_1)}{(x_v - x_1)^2 + y_v^2}$$

where $x = x_1$ is a particular location on the chord of the airfoil. If $x_v = c/2$ and $y_v = h \gg x_v$, show that the additional increment of lift produced by the vortex (which can represent a nearby airfoil) is given approximately by

$$\frac{4\Gamma}{3\pi h^2}$$

Wing Theory

LEARNING OBJECTIVES

- Learn the concept of a vortex line and explore features of the velocity field induced by the vortex line.

- Understand the logical extension of a vortex line to a horseshoe vortex, and apply the horseshoe vortex model to develop Prandtl's lifting-line theory for modeling wing performance.

- Understand the phenomenon of induced drag and how Prandtl's lifting-line theory predicts induced drag quantitatively based on fundamental wing properties.

Whatever the operating requirements of an airplane may be in terms of speed, endurance, payload, and so on, a critical stage in its eventual operation is the low-speed flight regime, and this must be accommodated in the overall design. The fact that low-speed flight was the classic flight regime means that over the years a vast array of empirical data has been accumulated from flight and other tests, and a range of theories and hypotheses put forth to explain and extend these observations. Some theories have survived to provide successful working processes for wing design that are capable of further exploitation by computational methods.

In this chapter such a classic theory is developed to the stage of preliminary low-speed aerodynamic design of straight, swept, and delta wings. Theoretical fluid mechanics of vortex systems are employed to model the loading properties of lifting wings in terms of their geometric and attitudinal characteristics and the properties of associated flow processes.

The basis on which historical solutions to the wing problem were arrived at is explained in detail, and the work is refined and extended to take advantage of more modern computing techniques.

A great step forward in aeronautics came with the vortex theory of a lifting airfoil due to Lanchester [34, 35] and subsequent development of this work by Prandtl [36]. Previously, all airfoil data had to be obtained from experimental work and fitted to other aspect ratios, planforms, and so forth, by empirical formulae based on experience with other airfoils.

Among other things, the Lanchester-Prandtl theory showed how knowledge of two-dimensional airfoil data could be used to predict the aerodynamic characteristics of (three-dimensional) wings. This derivation of the aerodynamic characteristics of wings is the topic of this chapter. Airfoil data can be obtained empirically from wind-tunnel tests, by means of the theories such as described in Chapter 4, or by computational modeling. Provided that the aspect ratio is fairly large (typically 6 or greater) and that the assumptions of thin-airfoil theory are met (see Section 4.3), the theory can be applied to wing planforms and sections of any shape.

5.1 THE VORTEX SYSTEM

Lanchester's contribution was essentially the replacement of the lifting wing by a theoretical model consisting of a system of vortices that imparted to the surrounding air a motion similar to the actual flow, and that sustained a force equivalent to the lift known to be created. The vortex system can be divided into three main parts: the starting vortex, the trailing vortex system, and the bound vortex system. Each of these may be treated separately, but it should be remembered that they are all component parts of one whole.

5.1.1 Starting Vortex

When a wing is accelerated from rest, the circulation around it, and therefore the lift, is not produced instantaneously. Instead, at the instant of starting, the streamlines over the rear part of the wing section are as shown in Fig. 5.1, with a stagnation point occurring on the rear upper surface. At the sharp trailing edge, the air is required to change direction suddenly while still moving at high speed. This high speed calls for extremely high local accelerations that produce very large viscous forces, and the air is unable to turn around the trailing edge to the stagnation point. Instead, the flow of air leaves the surface and produces a vortex just above the trailing edge. The (upper surface) stagnation point moves toward the trailing edge, as the circulation

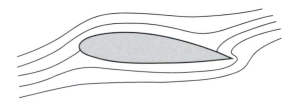

FIGURE 5.1

Streamlines of the flow around an airfoil with zero circulation, resulting in a stagnation point located on the rear upper surface.

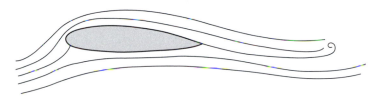

FIGURE 5.2

Streamlines of the flow around an airfoil with full circulation, resulting in a stagnation point at the trailing edge. The initial eddy is left far behind and rapidly becomes negligible to flight.

around the wing and therefore its lift increases. When the stagnation point reaches the trailing edge, the air is no longer required to flow around the trailing edge. Instead it decelerates gradually along the airfoil surface, comes to rest at the trailing edge, and then accelerates from rest in a different direction (Fig. 5.2). The vortex is left behind at the point reached by the wing when the stagnation point reaches the trailing edge. Its reaction, the circulation around the wing, is stabilized at the value necessary to place the stagnation point at the trailing edge (see Section 4.1.1).[1] The vortex left behind is equal in strength and opposite in sense to the circulation around the wing and is called the starting vortex or initial eddy.

5.1.2 Trailing Vortex System

The pressure on the upper surface of a lifting wing is lower than that of the surrounding atmosphere, while the pressure on the lower surface is greater than that on the upper surface, and may be greater than that of the surrounding atmosphere. Thus, over the upper surface air tends to flow inward toward the root from the tips, being replaced by air that was originally outboard of the tips. Similarly, on the undersurface air either tends to flow inward to a lesser extent or tends to flow outward. Where these two streams combine at the trailing edge, the difference in spanwise velocity causes the air to roll up into a number of small streamwise vortices distributed along

---

[1]There appears to be no fully convincing physical explanation for the production of the starting vortex and the generation of the circulation around the airfoil. Various incomplete explanations will be found in the references quoted in the bibliography. A common one is that just given, based on the large viscous forces associated with the high velocities around the trailing edge. It may be, however, that local flow acceleration is equally important, without invoking viscosity, in accounting for the failure of the flow to turn around the sharp trailing edge. It is known from the work of T. Weis-Fogh [37] and M. J. Lighthill [38] on the hovering flight of the small wasp *Encarsia formosa*, that it is apparently possible to generate circulation and lift in the complete absence of viscosity.

In practical aeronautics, air has nonzero viscosity, and the complete explanation of this phenomenon must take into account viscosity and the consequent growth of the boundary layer as well as high local velocities as motion is generated. For example, Reynolds numbers much greater than unity indicate that vorticity in the boundary layer is transported downstream far quicker than it can diffuse, such as around the trailing edge to interact with the flow on the other surface.

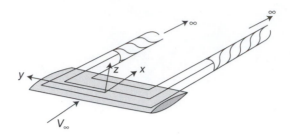

FIGURE 5.3

Horseshoe vortex. Because air is largely transparent, such flow structures are generally not visible; however, the mathematics applied in this chapter describes flow structures such as horseshoe vortices, for which very real results (e.g., lift force), can be measuered.

the entire span. These small vortices roll up into two large vortices just inboard of the wingtips as illustrated in Fig. 5.3. The strength of each of these two vortices will equal the strength of the vortex replacing the wing itself.

The existence of the trailing and starting vortices may be visually verified. When a fast airplane pulls out of a dive in humid air, the reduction in pressure and temperature at the centers of the trailing vortices is often sufficient to cause some of the water vapor to condense into droplets, which are seen as a thin streamer for a short distance behind each wingtip (see frontispiece). To see the starting vortex, all that is needed is a tub of water and a small piece of board or even a hand. If the board is placed upright in the water cutting the surface and then suddenly moved through the water at a moderate incidence, an eddy will be seen to leave the rear and move forward and away from the "wing." This is the starting vortex, and its movement is induced by the circulation around the plate.

5.1.3 Bound Vortex System

The starting vortex and the trailing system of vortices are physical entities that can be explored and seen if conditions are right. The bound vortex system, on the other hand, is a hypothetical arrangement of vortices that we use to replace the physical wing, neglecting thickness as in thin-airfoil theory, in the theoretical treatments to come. This is the essence of wing theory: developing the equivalent bound vortex system that simulates accurately, at least a little distance away, the properties, effects, disturbances, forces, and so forth, from the real wing as much as possible.

Consider a wing in steady flight. What effect does it have on the surrounding air, and how will changes in basic wing parameters such as span, planform, aerodynamic or geometric twist, and the like, alter these disturbances? The replacement bound vortex system must create the same disturbances, and this mathematical model must be sufficiently versatile to allow for the effects of parameter changes.

That is, if the wing span is changed or the aircraft flies faster, the model we use should accurately reflect the effects of these changes on the flow field and forces.

A real wing produces a trailing vortex system and thus the hypothetical bound vortex must do the same. A consequence of the tendency to equalize the pressures acting on the top and bottom surfaces of an airfoil is that the lift force per unit span is less near both wingtips. The bound vortex system must produce the same variation of lift along the span.

For complete equivalence, the bound vortex system should consist of a large number of spanwise vortex elements of differing spanwise lengths, all turned backward at each end to form a pair of the vortex elements in the trailing system. The varying spanwise lengths accommodate the grading of lift toward the wingtips, the ends turned back produce the trailing system, and the two physical attributes of a real wing are thus simulated.

For partial equivalence, the wing can be considered replaced by a single bound vortex of a strength equal to the mid-span circulation, which, bent back at each end, forms the trailing vortex pair. This concept is adequate for good estimations of wing effects at distances greater than about two chord lengths from the center of pressure.

5.1.4 Horseshoe Vortex

The vortex system associated with a wing, including its replacement bound vortex, forms a complete vortex ring that satisfies all physical laws (see Section 5.2.1). The starting vortex, however, is soon left behind, and the trailing pair stretches effectively to infinity as steady flight proceeds. Thus the velocities induced on the wing by the starting vortex rapidly become negligible as the aircraft flies away from the starting vortex. For practical purposes, then, the vortex system model of the wing consists of the bound vortices in the wing and the trailing vortex on either side close to it. This three-sided vortex has been called the *horseshoe vortex* (Fig 5.3).

Study of the completely equivalent vortex system is largely confined to wing effects in close proximity to the wing. For estimation of distant phenomena, the system is simplified to a single bound vortex and trailing pair, known as the *simplified horseshoe vortex* (Fig 5.4). This is dealt with in Section 5.3, before the more involved and complete theoretical treatments of wing aerodynamics.

5.2 LAWS OF VORTEX MOTION

The theoretical modeling of the flow around wings was discussed in the previous section. There an equivalent vortex system to model the lifting effects of a wing was described. To use this theoretical model in quantitative predictions of the aerodynamic characteristics of a wing, it is necessary first to study the laws of vortex

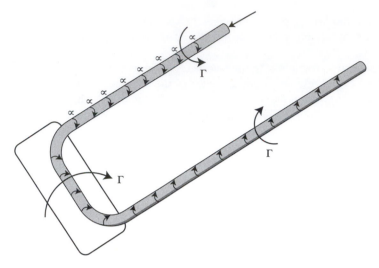

FIGURE 5.4

Simplified horseshoe vortex as a crude model for a lifting wing.

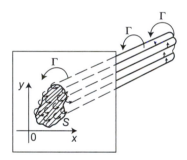

FIGURE 5.5

Vorticity of a section of vortex tube in the (x, y) plane.

motion. These laws act as a guide for developing modern computationally based wing theories.

The analysis of the point vortex (Chapter 3) considered it to be a string of rotating particles surrounded by fluid moving irrotationally under their influence. Further, the flow investigation was confined to a plane section normal to the length or axis of the vortex. A more general definition is that a vortex is a flow system in which a finite area in a normal section plane contains vorticity. Figure 5.5 shows the section area S of a vortex so-called because it possesses vorticity. The axis of the vortex (or of the vorticity or spin) is clearly always normal to the two-dimensional flow plane considered previously, and the influence of the so-called line vortex is that, in a section plane, of an infinitely long, straight-line vortex of a vanishingly small area.

In general, the vortex axis is a curve in space and area S has finite size. It is convenient to assume that S is made up of several elemental areas or, alternatively, that the vortex consists of a bundle of elemental vortex lines or filaments. Such a bundle is often called a *vortex tube* (compare to a *stream tube*, which is a bundle of streamlines) because it is a tube bounded by vortex filaments.

Since the vortex axis is a curve winding about within the fluid, capable of flexure and motion as a whole, the estimation of its influence on the fluid at large is somewhat complex and beyond the present study. All of the vortices of significance to the present theory are fixed relative to some axes in the system or free to move in a very controlled fashion, and they can be assumed to be linear. Nonetheless, they are not all of infinite length, and therefore some three-dimensional or end influence must be accounted for.

Vortices conform to certain laws of motion. A rigorous treatment of these is precluded from a text of this level, but may be acquired with additional study of the basic references [39].

5.2.1 Helmholtz's Theorems

The four fundamental theorems of vortex motion in an inviscid flow are named after their author, Helmholtz. The first theorem was discussed in part in Sections 2.7 and 4.1 and refers to a fluid particle in general motion possessing all or some of the following: linear velocity, vorticity, and distortion. The second theorem demonstrates the constancy of a vortex's strength along its length. This is sometimes referred to as the equation of vortex continuity. It is not difficult to prove that the strength of a vortex cannot grow or diminish along its axis or length. This strength is the magnitude of the circulation around it, which is equal to the product of the vorticity ζ and area S. Thus

$$\Gamma = \zeta S$$

It follows from the second theorem that ζS is constant along the vortex tube (or filament), so that, if the section area diminishes, the vorticity increases and vice versa. Since infinite vorticity is unacceptable, the cross-sectional area S cannot diminish to zero.

In other words, a vortex line cannot end in the fluid but rather must form a closed loop or originate (or terminate) in a discontinuity in the fluid such as a solid body or a surface of separation. A refinement of this is that a vortex tube cannot change in strength between two sections unless vortex filaments of equivalent strength join or leave the tube (Fig. 5.6). This is of great importance in the vortex theory of lift.

The third and fourth theorems, not elaborated on here, demonstrate respectively that a vortex tube consists of the same particles of—fluid that is, there is no fluid interchange between the tube and the surrounding fluid—the strength of a vortex remains constant as the vortex moves through the fluid.

The theorem of most consequence to the present chapter is the second one, although the third and fourth are tacitly accepted as the development proceeds.

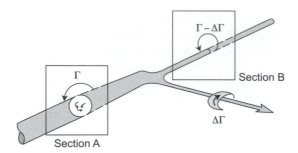

FIGURE 5.6

Implication of Helmholtz's law—vortex lines of equal but opposite strength combine (or split from) the vortex line of strength Γ.

5.2.2 The Biot-Savart Law

The original application of the Biot-Savart law was in electromagnetism, where it related the intensity of the magnetic field in the vicinity of an electric current to the magnitude of the current. In the present application, velocity and vortex strength (circulation) are analogous to the magnetic field strength and electric current, respectively, and a vortex filament replaces the electrical conductor. Thus the Biot-Savart law can also be interpreted as the relationship between the velocity *induced* by a vortex tube and the tube's strength (circulation). Only the fluid motion aspects will be pursued here, except to remark that the term *induced velocity*, which describes the velocity generated at a distance by the vortex tube, was borrowed from electromagnetism. Derivation of the application of the Biot-Savart law to fluids is lengthy and can be found in few texts. We proceed with example applications of the law:

$$\delta v = \frac{\Gamma}{4\pi R^2} \sin\theta \delta s \tag{5.1}$$

where the directions of δv, δs, and q are shown in Fig. 5.7.

Special Cases of the Biot-Savart Law

Usually, integration of Eq. (5.1) in a specific geometry is needed to yield applicable results. This integration, of course, varies with the length and shape of the finite vortex being studied. The vortices of immediate elementary interest are all straight lines that vary only in their overall length.

Example 5.1: A Linear Vortex of Finite Length AB

Figure 5.7 shows a length AB of a vortex with an adjacent point P located by the angular displacements α and β from A and B, respectively. Further, point P has coordinates r and θ with respect to any elemental length δs of the length AB that may be defined as a distance s

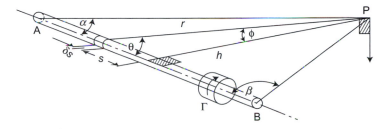

FIGURE 5.7

Finite-length segment of a straight vortex line. Ends A and B can be viewed as limits of integration within a longer vortex line or as physical limits of this vortex segment in this direction, such as for the head of a horseshoe vortex.

from the foot of the perpendicular h. From Eq. (5.1), the velocity at P induced by the elemental length δs is

$$\delta v = \frac{\Gamma}{4\pi r^2}\sin\theta\,\delta s \tag{5.2}$$

in the direction shown (i.e., normal to the plane APB).

To find the velocity at P due to the length AB, we sum the velocities induced by all such elements within the vortex line. This sum is computed as an integral; however, before integrating, all the variables must be written in terms of a single one. A convenient variable is the angle φ, as shown in Fig. 5.7, and the limits of the integration are

$$\phi_A = -\left(\frac{\pi}{2}-\alpha\right) \quad \text{to} \quad \phi_B = +\left(\frac{\pi}{2}-\beta\right)$$

since φ passes through zero when integrating from A to B. Additionally,

$$\sin\theta = \cos\phi, \quad r^2 = h^2\sec^2\phi$$

$$ds = d(h\tan\phi) = h\sec^2\phi\,d\phi$$

Following substitutions, the integration of Eq. (5.2) becomes

$$v = \int_{-(\pi/2-\alpha)}^{+(\pi/2-\beta)} \frac{\Gamma}{4\pi h}\cos\phi\,d\phi = \frac{\Gamma}{4\pi h}\left[\sin\left(\frac{\pi}{2}-\beta\right)+\sin\left(\frac{\pi}{2}-\alpha\right)\right]$$

$$= \frac{\Gamma}{4\pi h}(\cos\alpha+\cos\beta) \tag{5.3}$$

This result is of the utmost importance in aerodynamics, and in what follows in this chapter. Moreover, as it is a simple equation, it should be committed to memory. All of the values for induced velocity used in this chapter can be derived from Eq. (5.3), which describes a limited length of a straight-line vortex from A to B.

Example 5.2: The Influence of a Semi-Infinite Vortex (Fig. E5.2(a))

If one end of the vortex stretches to infinity (e.g., end B), then $\beta = 0$ and $\cos \beta = 1$. Thus, Eq. (5.3) describes the velocity induced at any point A:

$$v = \frac{\Gamma}{4\pi h}(\cos\alpha + 1) \tag{5.4}$$

In wing theory you will find that when point A is in a plane normal to the vortex line and containing the end of the vortex (Fig. E5.2(b)), $\alpha = \pi/2$ and $\cos\alpha = 0$. Equation (5.3) then becomes

$$v = \frac{\Gamma}{4\pi h} \tag{5.5}$$

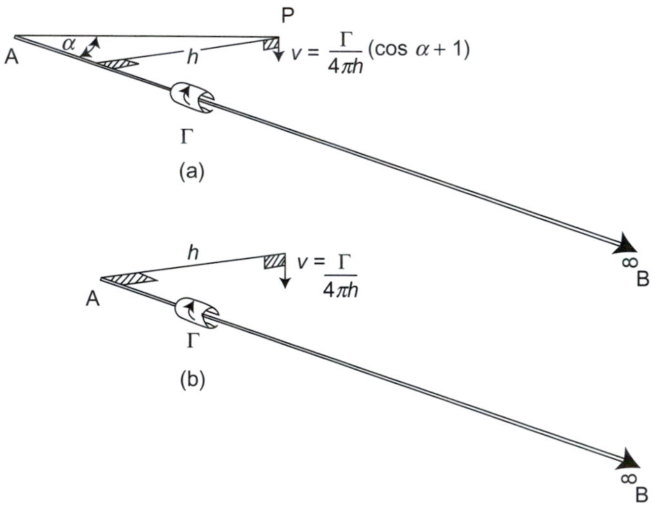

FIGURE E5.2

Two interesting cases of infinite vortex lines. The "semi-infinite" line in (a) with a special case in (b) for which the velocity is computed in a plane normal to the end points A.

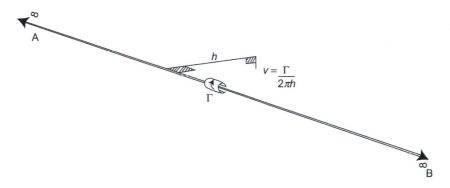

FIGURE E5.3

A "fully" infinite vortex line.

Example 5.3: The Influence of an Infinite Vortex (Fig. E5.3)

When point A in Example 5.2 is sent to infinity in the direction opposite from point B, $\alpha = \beta = 0$, $\cos \alpha = 1$, and $\cos \beta = 0$. Thus Eq. (5.3) becomes

$$v = \frac{\Gamma}{2\pi h} \tag{5.6}$$

You should recognize this as the familiar expression for velocity due to the line vortex of Section 3.3.2. Note that it is twice the velocity induced by a semi-infinite vortex, which can be seen intuitively.

The vortex line just studied is, in reality, a very useful and productive approximation of a slightly more complex real vortex flow. In nature, a vortex is a core of fluid rotating as though it were solid and around which air flows in concentric circles. The vorticity associated with the vortex is confined to its core and so, although an element of outside air is flowing in circles, the element itself does not rotate. This is not easy for all students to visualize, but a good analogy is a car on a ferris wheel. Although the car circulates around the axis of the wheel, it does not rotate about its own axis—that is, the passengers are never upside down. The elements of air in the flow outside a vortex core behave in a very similar way.

5.2.3 Variation of Velocity in Vortex Flow

To confirm how the velocity outside a vortex core varies with distance from the center, consider an element in a thin shell of air (Fig. 5.8). Here flow conditions

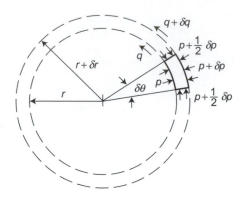

FIGURE 5.8

Definitions for circular motion of an element outside a vortex core.

depend only on the distance from the center and are constant around the vortex at any given radius. The small element, which subtends the angle $\delta\theta$ at the center, circulates around the center in steady motion under the influence of the force due to the radial pressure gradient.

The inward force per unit axial length due to the pressures is

$$(p + \delta p)(r + \delta r)\delta\theta - pr\delta\theta - 2\left(p + \frac{1}{2}\delta p\right)\delta r \frac{1}{2}\delta\theta$$

which reduces to $\delta p(r - \frac{1}{2}\delta r)\delta\theta$. Recognizing that $\frac{1}{2}\delta r$ is negligible in comparison with r, this becomes $r\delta p\,\delta\theta$. The volume per unit length of the element is $r\delta r\delta\theta$, and therefore its mass per unit length is $\rho r\delta r\delta\theta$. The centripetal acceleration is (velocity)^2/radius, and the force per unit axial length required to produce this acceleration is the product of mass and acceleration. Denoting the tangential component of the velocity field as V_θ,

$$\text{mass}\left[\frac{\text{velocity}^2}{\text{radius}}\right] = \rho r\delta r\delta\theta\frac{V_\theta^2}{r} = \rho V_\theta^2\delta r\delta\theta$$

Equating this to the force per unit axial length produced by the pressure gradient, we recognize that $\delta\theta \neq 0$ leads to

$$r\delta p = \rho V_\theta^2\delta r \tag{5.7}$$

Now, since the flow outside the vortex core is assumed to be inviscid, Bernoulli's equation for incompressible flow can be used to give, in this case,

$$p + \frac{1}{2}\rho V_\theta^2 = (p + \delta p) + \frac{1}{2}\rho (V_\theta + \delta V_\theta)^2$$

Expanding the term $(V_\theta + \delta V_\theta)^2$, neglecting smaller terms such as $(\delta V_\theta)^2$, and cancelling leads to

$$\delta p + \rho V_\theta \delta V_\theta = 0$$

which is also

$$\delta p = -\rho V_\theta \delta V_\theta \qquad (5.8)$$

Substituting this value for δp in Eq. (5.7) gives

$$\rho V_\theta^2 \delta r + \rho V_\theta r \delta V_0 = 0$$

which, when divided by ρV_θ, becomes

$$V_\theta \delta r + r \delta V_\theta = 0$$

But the left-hand side of this equation is $\delta(V_\theta r)$. Thus

$$\delta(V_\theta r) = 0 \text{ or } V_\theta r = \text{constant} \qquad (5.9)$$

The velocity is therefore inversely proportional to the radius in the inviscid flow around a vortex core (see also Section 3.3.2).

When the radius of the core is small, or assumed concentrated on a line axis, it is apparent from Eq. (5.15) that when r is small V_θ can be very large. However, within the core the air behaves as though it were a solid cylinder and rotates at a uniform angular velocity; thus V_θ tends toward zero at the axis. Figure 5.9 shows the variation of velocity with radius for a typical vortex. The solid line in the figure represents the idealized case; in reality, however, the boundary is not so distinct and the velocity peak is rounded off, after the style of the dotted lines.

5.3 THE WING AS A SIMPLIFIED HORSESHOE VORTEX

A simplified system may replace the complete vortex system of a wing when considering the influence of the lifting system on distant points in the flow. Many such problems do exist, and simple solutions, although not all exact, can be readily obtained using the suggested simplification. This necessitates replacing the wing

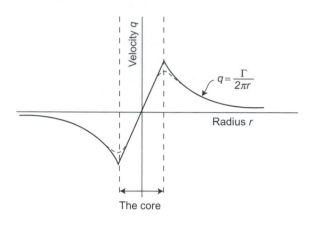

FIGURE 5.9

Velocity distribution in a real vortex with a core. Viscosity is important in the core of the vortex, leading to a solid-body rotation as shown by the straight line.

with a spanwise bound vortex of constant strength that is turned through a right angle at each end to form the trailing vortices that extend effectively to infinity behind the wing. The general vortex system and this simplified equivalent vortex must have two things in common:

- Each must provide the same total lift.
- Each must have the same value of circulation about the trailing vortices and hence the same total lift.

These equalities provide for the complete definition of the simplified system.

Two (or more) conventions are useful for this purpose, as shown in Fig. 5.10. We use a full-span vortex with strength less than the mid-span vortex strength of the actual wing, and we use a sub-span vortex with strength equal to the circulation at mid-span of the actual wing. At this date, both conventions are oversimplifications that have aphysical symptoms, but they remain of great use in beginning to understand the aerodynamic phenomena at play in a problem. Both are simple to improve on by vortex-lattice or panel methods, and both will be used to construct a more physically relevant general vortex system to represent the wing. They are also useful when estimating velocities far from the vortex.

Full-span vortex: The simpler and most common of the two approximations is the full-span horseshoe vortex, shown in Fig. 5.10. The total lift of the wing is divided by span to give a mean lift per span that is used as the vortex strength. The problem here is that the induced velocity, or "downwash," is infinite downward at the tips.

Sub-span vortex: The spanwise distributions created for the general vortex system and its sub-span simplified equivalent are also shown in Fig. 5.10. Both have the same mid-span circulation Γ_0 that is now constant along part of the span of the simplified equivalent case. For equivalence in area under the curve, which is

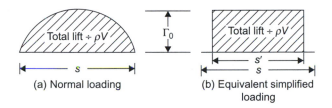

(a) Normal loading

(b) Equivalent simplified loading

FIGURE 5.10

Two conventions for defining the spanwise extent of a vorticity distribution.

proportional to total lift, the span length of the single vortex must be less than that of the wing.

Thus

$$\Gamma_0 s' = \text{area under general distribution} = \frac{lift}{\rho V}$$

Hence

$$\frac{s'}{s} = \frac{total\ lift}{s\rho V\Gamma_0} \tag{5.10}$$

where s' is the distance between the trailing vortex core centers. Solving for areas under the curves, one can show that the two spans of interest are simply related:

$$\Gamma_0 = 4sV \sum A_n \sin n\frac{\pi}{2}$$

$$= \frac{\pi}{4} \frac{A_1}{[A_1 - A_3 + A_5 - A_7 \dots]}$$

Thus,

$$s' = \left(\frac{\pi}{4}\right) s \tag{5.11}$$

In the absence of other information, it is usual to assume that the separation of the trailing vortices is given by the elliptic case. This model also has the infinite downwash problem, but now within the span at $y = \pm\frac{s'}{2}$ rather than at the tips, which are at $y = \pm\frac{s}{2}$.

Example 5.4: Formation Flying Effects

Aircraft flying in close proximity experience mutual interference effects, and good estimates of these effects are obtained by replacing each aircraft in the formation with its equivalent simplified horseshoe vortex.

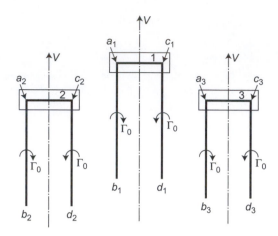

FIGURE E5.4

Simple formation flight geometry. Note that the Canadian Goose formation is not the only option.

Consider the problem shown in Fig. E5.4, where three identical aircraft are flying in a vee formation at a forward speed V in the same horizontal plane. The total mutual interference is the sum of (1) that of the followers on the leader (plane 1), (2) that of the leader and follower (plane 2) on (3), and (3) that of leader and follower (plane 3) on (plane 2). Interferences (1) and (2) are identical.

1. The leader is flying in a flow regime that has additional vertical flow components induced by the following vortices. Upward components appear from the bound vortices a_2c_2 and a_3c_3, trailing vortices c_2d_2 and a_3b_3, and downward components from the trailing vortices a_2b_2 and c_3d_3. The net result is an upwash on the leader.
2. These wings have additional influences to their own trails from the leader and the other follower. Bound vortex a_1c_1 and trailing vortices a_1b_1 and a_2b_2 produce downwashes. Again, the net influence is an upwash.

From these simple considerations it appears that each aircraft is flying in a regime in which upward components are induced by the presence of the other aircraft. The upwash components reduce the downward velocities induced by the aircraft's own trail and hence its trailing vortex drag. Because of the drag reduction, less power is required to maintain forward velocity, and the well-known operational fact emerges that each aircraft in a formation performs better than when flying singly. In most problems it is usual to assume that the wings have an elliptic distribution and that the influence calculated for mid-span position is typical of the whole wingspan. Also, any curvature of the trails is neglected and the special forms of the Biot-Savart law (Section 5.2.2) are used unreservedly. A recent NASA research task actually measured formation flight performance with F-18 aircraft [40]. Results showed that mutual benefit is possible with carefully controlled spacing.

5.3.1 Influence of Downwash on the Tailplane

On most aircraft the tailplane is between the trailing vortices springing from the wing ahead, and the flow around it is considerably influenced by them. Forces on airfoils are proportional to the square of the velocity and the angle of attack. Small velocity changes, therefore, have negligible effect unless they alter the incidence airfoil. Small velocities that alter the angle of attack of the airfoil include vertical velocities such as those induced by the trailing vortices. These velocities act to tilt the relative wind and thus the lift. The solution to a particular problem shows the method.

Example 5.5

Let the tailplane of an aircraft be at distance x behind the wing center of pressure and in the plane of the vortex trail (Fig. E5.5). Assuming elliptic distribution and the sub-span horseshoe vortex method, the span of the bound vortex is given by Eq. (5.11) as

$$s' = \left(\frac{\pi}{4}\right)s$$

The downwash at the mid-span point P of the tailplane caused by the wing is the sum of that caused by the bound vortex ac and that of each of the trailing vortices ab and cd. Using a special form of the Biot-Savart equations (Section 5.2.2) the downwash at P is

$$w_p \downarrow = \frac{\Gamma_0}{4\pi x} 2\sin\beta + 2\left[\frac{\Gamma_0}{4\pi (s'/2)}(1+\cos\beta)\right]$$

$$w_p \downarrow = \frac{\Gamma_0}{\pi}\left(\frac{\sin\beta}{2x} + \frac{1+\cos\beta}{s'}\right)$$

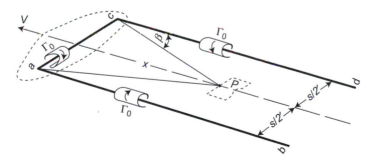

FIGURE E5.5

Geometry and definitions for computing downwash on a tailplane (or any other object) in the same plane as the wing.

From Fig. E5.5, $x = (s'/2)\cot\beta$ and $s\prime = (\pi/4)s$:

$$w_p \downarrow = \frac{\Gamma_0}{\pi}\left(\frac{\sin\beta}{s'\cot\beta} + \frac{1+\cos\beta}{s'}\right) = \frac{\Gamma_0}{\pi s'}(1+\sec\beta) = \frac{4\Gamma_0}{\pi^2 s}(1+\sec\beta)$$

Now, using the Kutta–Zhukovsky theorem, Eq. (4.10), and downwash angle,

$$\varepsilon = \frac{w_P}{V}$$

$$\varepsilon = \frac{8C_L VS}{\pi^3 s^2 V}(1+\sec\beta)$$

or

$$\varepsilon = \frac{8C_L}{\pi^3(AR)}(1+\sec\beta)$$

The derivative of the downwash angle with respect to angle of attack is

$$\frac{\partial\varepsilon}{\partial\alpha} = \frac{\partial\varepsilon}{\partial C_L}\frac{\partial C_L}{\partial\alpha} = \frac{\partial\varepsilon}{\partial C_L}a_1$$

Here $a_1 \leq 2\pi$ is the slope of the lift curve for this wing. Thus

$$\frac{\partial\varepsilon}{\partial\alpha} = \frac{8a_1}{\pi^3(AR)}(1+\sec\beta) \tag{5.12}$$

For cases where the distribution is nonelliptic or the tailplane is above or below the wing center of pressure, the arithmetic of the problem is altered from that just given, which applies only to this restricted problem. Again, the mid-span point is taken as representative of the whole tailplane.

5.3.2 Ground Effects

In this example, the influence of solid boundaries on airplane (or model) performance is estimated, and once again the wing is replaced by the equivalent simplified horseshoe vortex. Since this is a linear problem, the method of superposition may be used in the following way. If a point vortex is placed at height h above a horizontal plane (Fig. 5.11(b)), and an equal but opposite vortex is placed at depth h below it, the vertical velocity component induced at any point on the plane by one of the vortices is equal and opposite to that due to the other. Thus the net vertical velocity, induced at any point on the plane, is zero. This shows that the superposition of the image vortex is equivalent in effect to the presence of a solid boundary. In exactly the same way, the effect of a solid boundary on the horseshoe vortex can be modeled with an image horseshoe vortex (Fig. 5.11(a)). In this case, the boundary is the level ground and its influence on an aircraft h above is the same as that of the "inverted" aircraft flying "in formation" h below ground level (Figs. 5.11(a) and 5.12).

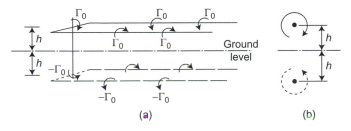

FIGURE 5.11

Horseshoe vortex for the wing (solid line) and image vortex (dashed lines) for ground effects. Part (b) is a view looking upwind for the left wingtip vortex.

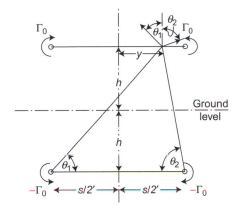

FIGURE 5.12

Upwind view showing geometry and definitions for ground effects.

Before working out a particular problem, it is clear from the figure that the image system reduces the downwash on the wing and hence the drag and power required; it also materially changes the downwash angle at the tail and hence the overall pitching equilibrium of the airplane.

An airplane of weight W and span s is flying horizontally near the ground at altitude h and speed V. Estimate the reduction in drag due to ground effect. If $W = 22 \times 10^4$ N, $h = 15.2$ m, $s = 27.4$ m, and $V = 45\,\mathrm{m\,s^{-1}}$, calculate the reduction in Newtons.

With the notation of Fig. 5.12, the change in downwash at y along the span is $\Delta w \uparrow$, where

$$\Delta w \uparrow = \frac{\Gamma_0}{4\pi r_1}\cos\theta_1 + \frac{\Gamma_0}{4\pi r_2}\cos\theta_2$$

$$\Delta w = \frac{\Gamma_0}{4\pi}\left(\frac{s'/2+y}{r_1^2} + \frac{s'/2-y}{r_2^2}\right)$$

On a strip of span δy at y from the centerline, assuming the sub-span horseshoe vortex method,

$$\text{lift } l = \rho V \Gamma_0 \delta y$$

and change in vortex drag,

$$\Delta d_v = \frac{l \Delta w}{V}$$

$$= \frac{\rho V \Gamma_0 \delta y \Delta w}{V} \qquad (5.13)$$

Total change in drag ΔD_v across the span is the integral of Eq. (5.13) from $-s'/2$ to $s'/2$ (or twice that from 0 to $s'/2$). Therefore,

$$-\Delta D_v = 2 \int_0^{s'/2} \frac{\rho \Gamma_0^2}{4\pi} \left(\frac{s'/2 + y}{r_1^2} + \frac{s'/2 - y}{r_2^2} \right) dy$$

From the geometry, $r_1^2 = 4h^2 + \left(s'/2 + y \right)^2$ and $r_2^2 = 4h^2 + \left(s'/2 - y \right)^2$. Making these substitutions and evaluating the integral,

$$-\Delta D_v = \frac{\rho \Gamma_0^2}{4\pi} \left[\ln \frac{4h^2 + \left(s'/2 + y \right)^2}{4h^2 + (s'/2 - y)^2} \right] \Bigg|_0^{s'/2}$$

$$-\Delta D_v = \frac{\rho \Gamma_0^2}{4\pi} \ln \left[1 + \left(\frac{s'}{2h} \right)^2 \right]$$

With $W = \rho V \Gamma_0 \pi s / 2$ and $s' = (\pi/4)s$ (assuming elliptic distribution):

$$\Delta D_v = \frac{8 W^2}{\rho V^2 s^2 \pi^3} \ln \left(1 + \frac{\pi^2}{64} \frac{s^2}{h^2} \right)$$

and substituting the values given

$$\Delta D_v = 1390 \, \text{N}$$

A simpler approach is to assume that mid-span conditions are typical of the entire wing. With this the case,

$$\theta_1 = \theta_2 = \theta = \arccos \frac{s'/2}{\sqrt{(s'/2)^2 + 4h^2}} = \arccos \frac{s'}{\sqrt{s'^2 + 16h^2}}$$

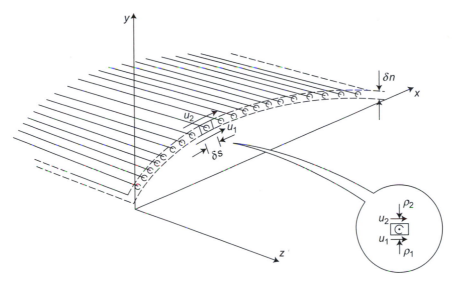

FIGURE 5.13

Many vortex lines, which create a vortex sheet. The vortex sheet has a continuous distribution of vorticity, but only a finite number can be shown.

and the change in drag is 1524 N (a difference of about 10% from the first answer).

5.4 VORTEX SHEETS

To estimate the influence of the near wake on the aerodynamic characteristics of a lifting wing, it is useful to investigate the "hypothetical" bound vortex in greater detail. For this analysis, the wing is replaced by a sheet of vortex filaments. In order to satisfy Helmholtz's second theorem (Section 5.2.1), each filament must either be part of a closed loop or form a horseshoe vortex with trailing vortex filaments running to infinity. Even with this restriction, there are still infinitely many ways of arranging such vortex elements for modeling the flow field associated with a lifting wing. For illustrative purposes, consider the simple arrangement where a sheet of vortex filaments passes in the spanwise direction through a given wing section (Fig. 5.13). It should be noted, however, that at two, here unspecified, spanwise locations, each of these filaments must be turned back to form trailing vortex filaments.

Consider the flow in the vicinity of a sheet of fluid moving irrotationally in the xy plane, as in Fig. 5.13. In this stylized figure, the "sheet" is seen to have a section curved in the xy plane and to be of thickness δn; the vorticity is represented by a number of vortex filaments normal to the xy plane. The circulation around the element of fluid having sides δs and δn is, by definition, $\Delta\Gamma = \zeta\delta s.\delta n$, where ζ is the vorticity of the fluid within the area $\delta s\delta n$.

Now, for a sheet $\delta n \to 0$ and if ζ is so large that the product $\zeta\delta n$ remains finite, the sheet is termed a vortex sheet of strength $k = \zeta\delta n$. The circulation around the element can thus be written

$$\Delta\Gamma = k\delta s \tag{5.14}$$

An alternative way to find the circulation around the element is to integrate the tangential flow components. Thus

$$\Delta\Gamma = (u_2 - u_1)\delta s \tag{5.15}$$

Comparison of Eqs. (5.14) and (5.15) shows that the local strength k of the vortex sheet is the tangential-velocity jump through the sheet. Alternatively, a flow situation in which the tangential velocity changes discontinuously in the normal direction may be mathematically represented by a vortex sheet of a strength proportional to the velocity change.

The vortex sheet concept has important applications in wing theory.

5.4.1 Use of Vortex Sheets to Model the Lifting Effects of a Wing

In Section 4.3, it was shown that the flow around a thin wing can be regarded as a superposition of rotational and irrotational flow. In a similar fashion, the same can be established for the flow around a thin wing. For a wing to be classified as thin, the following must hold:

- The maximum thickness-to-chord ratio, usually located at mid-span, must be much less than unity.
- The camber lines of all wing sections must deviate only slightly from the corresponding chord line.
- The wing may be twisted, but the angles of attack of all wing sections must remain small and the rate of change in twist must be gradual.
- The rate of change in wing taper must be gradual.

These conditions are met for most practical wings. If they are satisfied, the velocities at any point over the wing differ only by a small amount from those of the oncoming flow.

For the thin airfoil, noncirculatory flow corresponds to that around a symmetrical airfoil at zero incidence. Similarly for the thin wing, it corresponds to that around an untwisted wing, having the same planform shape as the actual wing but with symmetrical sections at zero angle of attack. Like its two-dimensional counterpart in airfoil theory, this so-called *displacement* (or *thickness*) *effect* makes no contribution to the wing's lifting characteristics. The circulatory flow—the so-called *lifting effect*—corresponds to that around an infinitely thin, cambered, and possibly twisted plate at an angle of attack. The plate takes the same planform shape as the mid-plane of the actual wing. This circulatory part of the flow is modeled by a vortex sheet. The lifting characteristics of the wing are determined solely by it, so, the lifting effect is of much greater practical interest than the displacement effect. Accordingly, much of the rest of this chapter will be devoted to the former, with the displacement effect briefly considered at the end.

AERODYNAMICS AROUND US

Visible Streamwise Vortices

Have you ever seen a horseshoe vortex? We haven't either. Yet you have likely seen evidence of the trailing-vortex portion of a horseshoe vortex. The accompanying photograph shows such a visible vortex, created from a vane on the outer side of the engine nacelle on an Airbus 380 on landing approach. Consider that to keep air flowing in a circle, the pressure at the center of the circle must be lower than the pressure outside. For a trailing, or tip, vortex like this one (the vane is essentially a very short wing), the pressure outside the vortex flow is atmospheric. Thus the pressure in the core is below atmospheric pressure. As air from upstream flows into the vortex core, the pressure and the temperature drop (see the discussion of isentropic relations in Chapter 6). When the temperature drops enough on a humid day, the water vapor in the air condenses into minute droplets and is visible as the "cloud" in the vortex core. As such, the cloud marks the low-pressure center of the vortex. You will see this most often on landing when the air speed is low and the lift coefficient is high. High lift coefficient leads to greater vortex strength, which is faster spin and thus lower pressure and temperature, leading more often to visible vortex cores.

Photo courtesy of James Kallimani.

Lifting Effect

To understand the fundamental concepts in modeling the lifting effect of a vortex sheet, consider first the simple rectangular wing depicted in Fig. 5.14. Here the vortex sheet is constructed from a collection of horseshoe vortices located in the $y = 0$ plane. From Helmholtz's second theorem (Section 5.2.1), the strength of the circulation around any section of the vortex sheet (or wing) is the sum of the strengths of the vortex filaments cut by the section plane. As the section plane is progressively moved outward from the center section to the tips, fewer and fewer bound vortex filaments are left for successive sections to cut, so the circulation around the sections diminishes. In this way, the spanwise change in circulation around the wing is related to the spanwise lengths of the bound vortices. Now, as the section plane is moved outward along the bound bundle of filaments, and as the strength of the bundle decreases, the strength of the vortex filaments so far shed must increase because the overall strength of the system cannot diminish. Thus the change in circulation from section to section is equal to the strength of the vorticity shed between sections.

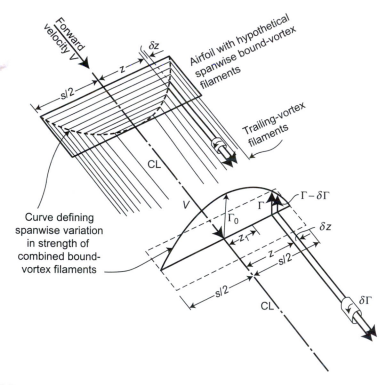

FIGURE 5.14

Relation between spanwise load variation and trailing vortex strength for a planar wing in steady level flight.

Figure 5.14 shows a simple rectangular wing shedding a vortex trail with each pair of trailing vortex filaments completed by a spanwise bound vortex. Notice that a line joining the ends of all the spanwise vortices forms a curve that, assuming each vortex is of equal strength and given a suitable scale, represents the total strengths of the bound vortices at any section plotted against the span. This curve has been plotted for clarity on a spanwise line through the center of pressure of the wing. It is a plot of (chordwise) circulation (Γ) measured on a vertical ordinate, against spanwise distance from the centerline (CL) measured on the horizontal ordinate. Thus at a section z from the centerline, sufficient hypothetical bound vortices are cut to produce a chordwise circulation around that section equal to Γ. At a further section $z + \delta z$ from the centerline, the circulation falls to $\Gamma - \delta\Gamma$, indicating that between sections z and $z + \delta z$, trailing vorticity of the strength of $\delta\Gamma$ has been shed.

If the circulation curve can be described as some function of z—$f(z)$, say—then the strength of circulation shed

$$\delta\Gamma = -\frac{df(z)}{dz}\delta z \qquad (5.16)$$

Now, at any section the lift per span is given by the Kutta–Zhukovsky theorem Eq. (4.10):

$$l = \rho V \Gamma$$

and for a given flight speed and air density, Γ is thus proportional to l. But l is the local intensity of lift or lift grading, which is known or is the required quantity in the analysis.

The substitution of the wing by a system of bound vortices is not rigorously justified at this stage. The idea allows a relation to be built between the physical load distribution on the wing, which depends, as will be shown, on the wing geometric and aerodynamic parameters, and the trailing vortex system.

Figure 5.14 illustrates two further points:

- The leading sketch shows that the trailing filaments are closer together when they are shed from a rapidly diminishing or changing distribution curve. Where the filaments are closer, the strength of the vorticity is greater. Near the tips, therefore, the shed vorticity is the strongest; at the center, where the distribution curve is flattened out, the shed vorticity is weak to infinitesimal.
- A wing infinitely long in the spanwise direction, or in two-dimensional flow, has constant spanwise loading. The bundle has filaments all of equal length, and none is turned back to form trailing vortices. Thus there is no trailing vorticity associated with two-dimensional wings. This can be deducted by a more direct process. That is, as the wing is infinitely long in the spanwise direction, the lower-surface (high) and upper-surface (low) pressures cannot tend to equalize by spanwise components of velocity, so the streams of air meeting at the trailing edge after sweeping under and over the wing have no opposite spanwise motions, but join

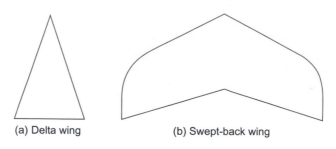

(a) Delta wing   (b) Swept-back wing

FIGURE 5.15

Two examples of planforms for which the vortex sheet modeled in Fig. 5.14 may not be suitable.

up in symmetrical flow in the direction of motion. Again, no trailing vorticity is formed.

A more rigorous treatment of vortex-sheet modeling is now considered. In Section 4.3, it was shown that for thin airfoils, without loss of accuracy, the vortices can be considered as distributed along the chord line (i.e., the x-axis rather than the camber line). Similarly, in the present case the vortex sheet can be located on the (x, z) plane rather than on the cambered and possibly twisted mid-surface of the wing. This greatly simplifies the details of theoretical modeling.

One of the infinite ways of constructing a suitable vortex-sheet model is suggested by Fig. 5.14. It is certainly suitable for wings with a simple planform shape (e.g., a rectangular wing). Some wing shapes for which it is not at all suitable are presented in Fig. 5.15, which shows that for the general case an alternative model is required. It is usually preferable to assign an individual horseshoe vortex of strength $k(x, z)$ per unit chord to each element of wing surface (Fig. 5.16). However, this method of constructing the vortex sheet leads to certain mathematical difficulties when calculating induced velocity. These can be overcome by recombining the elements in the way depicted in Fig. 5.17, where it is recognized that partial cancellation occurs for two elemental horseshoe vortices occupying adjacent spanwise positions, z and $z + \delta z$. Accordingly, the horseshoe-vortex element can be replaced by the L-shaped vortex element shown in the figure. Note that although this arrangement appears to violate Helmholtz's second theorem, it is merely a mathematically convenient way of expressing the model depicted in Fig. 5.16, which fully satisfies this theorem.

5.5 RELATIONSHIP BETWEEN SPANWISE LOADING AND TRAILING VORTICITY

In Section 5.5.1 use show how to calculate the velocity induced by the elements of the vortex sheet that notionally replace the wing. This is an essential step in the development of a general wing theory. Initially, the general case is considered. Then

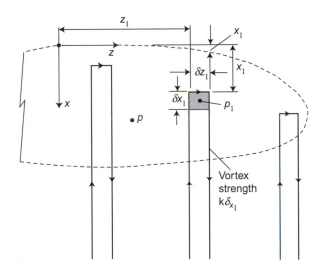

FIGURE 5.16

Modeling the lifting effect by a distribution of horseshoe vortex elements. The shape of each element does not depend on the planform.

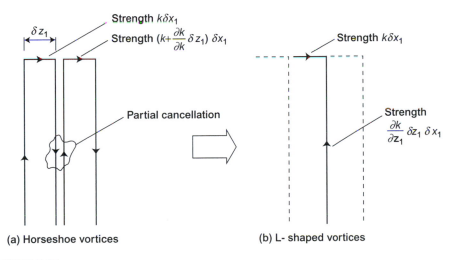

(a) Horseshoe vortices   (b) L- shaped vortices

FIGURE 5.17

Equivalence between distributions of (a) horseshoe vortices and (b) L-shaped vortices.

we show in Sections 5.5.2 through 5.6 how the general case can be considerably simplified in the special case of wings of high aspect ratio. We discuss important design options of swept and of smaller aspect ratio wings in Section 5.7 and in Section 5.8 examine wings, as found on some aircraft and rocket tail fins, using a vortex-panel method.

5.5.1 Induced Velocity (Downwash)

Suppose that it is required to calculate the velocity induced at the point $P_1(x_1, z_1)$ in the $y = 0$ plane by the L-shaped vortex element associated with the element of wing surface located at point $P(x, z)$, now relabeled A (Fig. 5.18). Making use of Eq. (5.3), it can be seen that this induced velocity is perpendicular to the $y = 0$ plane and can be written as

$$\delta v_i(x_1, z_1) = (\delta v_i)_{AB} + (\delta v_i)_{BC}$$

$$= -\frac{k\delta x}{4\pi(x - x_1)}\left[\cos\theta_1 - \cos\left(\theta_2 + \frac{\pi}{2}\right)\right] + \frac{1}{4\pi}\frac{\partial k}{\partial z}\delta z\delta x\frac{(1 + \cos\theta_2)}{(z + \delta z - z_1)}$$

(5.17)

From the geometry of Fig. 5.25 the various trigonometric expressions in Eq. (5.17) can be written as

$$\cos\theta_1 = \frac{z - z_1}{\sqrt{(x - x_1)^2 + (z - z_1)^2}}$$

$$\cos\theta_2 = -\frac{x - x_1}{\sqrt{(x - x_1)^2 + (z + \delta z - z_1)^2}}$$

$$\cos\left(\theta_2 + \frac{\pi}{2}\right) = -\sin\theta_2 = \frac{z + \delta z - z_1}{\sqrt{(x - x_1)^2 + (z + \delta z - z_1)^2}}$$

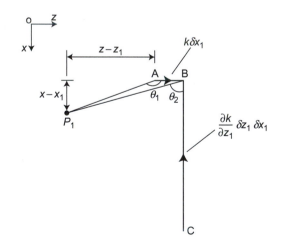

FIGURE 5.18

Geometric notation for an L-shaped vortex element used to compute the downwash at point P_1.

The binomial expansion

$$(a+b)^n = a^n + na^{n-1}b + \cdots$$

can be used to expand some of the terms:

$$[(x-x_1)^2 + (z+\delta z - z_1)^2]^{-1/2} = \frac{1}{r} - \frac{(z-z_1)}{r^3}\delta z + \cdots$$

where $r = \sqrt{(x-x_1)^2 + (z-z_1)^2}$. In this way, the trigonometric expressions given can be rewritten as

$$\cos\theta_1 = \frac{z-z_1}{r} \tag{5.18}$$

$$\cos\theta_2 = -\frac{x-x_1}{r} + \frac{(x-x_1)(z-z_1)}{r^3}\delta z + \cdots \tag{5.19}$$

$$\cos\left(\theta_2 + \frac{\pi}{2}\right) = \frac{z-z_1}{r} + \left[\frac{1}{r} - \frac{(z-z_1)^2}{r^3}\right]\delta z + \cdots \tag{5.20}$$

Equations (5.18) through (5.20) are now substituted into Eq. (5.17), and terms involving $(\delta z)^2$ and higher powers are ignored to give

$$\delta v_i = \frac{k}{4\pi}\delta x \delta z \frac{(x-x_1)}{r^3} + \frac{1}{4\pi}\frac{\partial k}{\partial z}\delta x \delta z\left[\frac{1}{z-z_1} - \frac{x-x_1}{r(z-z_1)}\right] \tag{5.21}$$

To obtain the velocity induced at P_1 due to all of the horseshoe vortex elements, δv_i is integrated over the entire wing surface projected onto the (x, z) plane. Eq. (5.21) thus leads to

$$v_1(x_1, z_1) = \frac{1}{4\pi}\int\limits_{-s/2}^{s/2}\int\limits_{x_l}^{x_l+c}\left\{\frac{\partial k}{\partial z}\underbrace{\left[\underbrace{\frac{1}{z-z_1}}_{(a)} - \underbrace{\frac{x-x_1}{r(z-z_1)}}_{(b)}\right]} + k\underbrace{\frac{(x-x_1)}{r^3}}_{(c)}\right\}dx\,dz \tag{5.22}$$

The induced velocity at the wing itself and in its wake is usually in a downward direction and accordingly is often called the *downwash w* with the convention that $w = -v_i$.

It would be a difficult and involved process to develop wing theory based on Eq. (5.22) in its present general form. Nowadays, similar vortex-sheet models are in panel methods (described in Section 5.8) to provide computationally based models of the flow around a wing or an entire aircraft. For this reason, a discussion of the theoretical difficulties involved in using vortex sheets to model wing flows will be postponed until Section 5.8. The remainder of the present section and Section 5.6 are devoted solely to the special case of unswept wings having high aspect ratio. This is by no means unrealistically restrictive, since aerodynamic considerations tend to dictate the use of wings with moderate to high aspect ratio for low-speed applications such as gliders, light airplanes, and commuter passenger aircraft. In this special case, Eq. (5.22) can be considerably simplified.

This simplification is achieved as follows. To determine the aerodynamic characteristics of the wing, it is only necessary to evaluate the induced velocity at the wing itself. Accordingly, ranges for the variables of integration are given by $-s/2 \leq z \leq s/2$ and $0 \leq x \leq (c)_{max}$. For high aspect ratios, $s/c \gg 1$ so that $|x - x_1| \ll r$ over most of the range of integration. Consequently, the contributions of terms (b) and (c) to the integral in Eq. (5.22) are very small compared to that of term (a) and can therefore be neglected. This allows Eq. (5.22) to be simplified to

$$v_i(z_1) = -w(z_1) = \frac{1}{4\pi} \int_{-s/2}^{s/2} \frac{d\Gamma}{dz} \frac{1}{z - z_1} dz \tag{5.23}$$

where, as explained in Section 5.4.1, from Helmholtz's second theorem,

$$\Gamma(z) = \int_{x_l}^{c(z)+x_l} k(x, z) dx \tag{5.24}$$

is the total circulation due to all the vortex filaments passing through the wing section at z.

The approximate theoretical model implicit in Eq. (5.23) and (5.24) physically corresponds to replacing the wing by a single bound vortex having variable strength Γ, the so-called *lifting line* (Fig. 5.19). This model, together with Eqs. (5.23) and (5.24), is the basis of Prandtl's general wing theory, which is described in Section 5.6. The more involved theories based on the full version of Eq. (5.22) are usually referred to as *lifting-surface* theories.

Equation (5.23) can also be deduced directly from the simple, less general theoretical model illustrated in Fig. 5.14. Looking at this figure, consider now the influence of the trailing vortex filaments of strength $\delta\Gamma$ shed from the wing section at z. At some other point z_l along the span, according to Eq. (5.5), an induced velocity equal to

$$\delta v_i(z_1) = \frac{1}{4\pi(z - z_1)} \frac{df}{dz} \delta z$$

will be felt in the downward direction in the usual case of positive vortex strength. All elements of shed vorticity along the span add their contribution to the induced velocity at z_1, so the total influence of the trailing system there is given by Eq. (5.23).

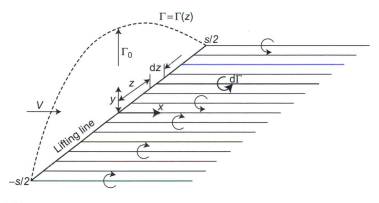

FIGURE 5.19

Prandtl's lifting-line model—a simple and powerful definition of vortex lines or sheets.

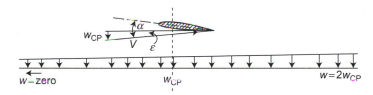

FIGURE 5.20

Variation in magnitude of downwash in front of and behind the wing.

5.5.2 The Consequences of Downwash—Trailing Vortex Drag

The induced velocity at z_1 is generally in a downward direction and is sometimes called downwash. It has two very important consequences that modify the flow about the wing and alter its aerodynamic characteristics.

First, the downwash obtained for the particular point z_1 is felt to a lesser extent ahead of z_1 and to a greater extent behind (see Fig. 5.20). It has the effect of tilting the resultant oncoming flow at the wing (or anywhere else within its influence) through an angle

$$\varepsilon = \tan^{-1}\frac{w}{V} \cong \frac{w}{V}$$

where w is the local downwash. This reduces the effective incidence so that, for the same lift as the equivalent infinite wing or two-dimensional wing at incidence α_∞, an incidence $\alpha = \alpha_\infty + \varepsilon$ is required at that section on the finite wing. This is illustrated in Fig. 5.21, which in addition shows how the two-dimensional lift L_∞ is normal to the resultant velocity V_R and is therefore tilted back against the actual

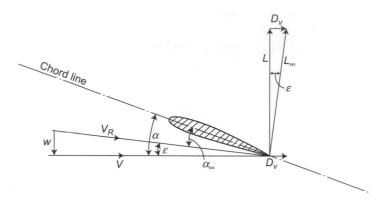

FIGURE 5.21

Influence of downwash on wing velocities and forces: W—downwash; V—forward speed of the wing; V_R—resultant oncoming flow at the wing; α—incidence; ε—downwash angle = W/V; α_∞—$(\alpha - \varepsilon)$ = equivalent two-dimensional incidence; L_∞—two-dimensional lift; L—wing lift; D_v—trailing vortex drag.

direction of motion of the wing V. The two-dimensional lift L_∞ is resolved into the aerodynamic forces L and D_v, respectively, normal to and against the direction of the wings forward velocity. Thus the second important consequence of downwash emerges, the generation of a drag force D_v. This is so important that the sequence just given will be explained in an alternative way.

A section of wing generates a circulation of strength Γ. Superimposed on an apparent oncoming flow velocity V, this circulation produces a lift force $L_\infty = \rho V \Gamma$ according to the Kutta–Zhukovsky theorem Eq. (4.10), which is normal to the apparent oncoming flow direction. The apparent oncoming flow felt by the wing section is the result of the forward velocity and the downward induced velocity arising from the trailing vortices. Thus the aerodynamic force L_∞ produced by the combination of Γ and V appears as a lift force L normal to the forward motion and a drag force D_v against the normal motion. This drag force is the *trailing vortex drag*, abbreviated to *vortex drag* or more commonly *induced drag* (see Section 1.6.7). A very descriptive name that is used more frequently by pilots than by aerodynamicists is *drag due to lift*.

Considering for a moment the wing as a whole moving through air at rest at infinity, two-dimensional wing theory suggests that, taking air as being of small to negligible viscosity, the static pressure of the free stream ahead is recovered behind the wing. This means roughly that the kinetic energy induced in the flow is converted back to pressure energy, and zero drag results. The existence of a thin boundary layer and narrow wake is ignored, but this does not really modify the argument.

In addition to this motion of the airstream, a finite wing spins the airflow near the tips into what eventually becomes two trailing vortices of considerable core

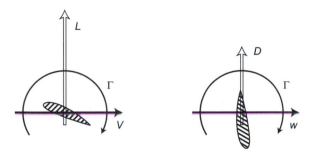

FIGURE 5.22

Circulation superimposed on forward wind velocity and downwash to give lift and vortex drag (induced drag), respectively.

size. The generation of these vortices requires a quantity of kinetic energy that is not recovered by the wing system and that in fact is lost to the wing by being left behind. The continuing expenditure of energy appears to the wing as induced drag. In what follows, a third explanation of this important consequence of downwash will be of use. Figure 5.22 shows the two velocity components of the apparent oncoming flow superimposed on the circulation produced by the wing. The forward-flow velocity produces the lift, and the downwash produces the vortex drag per unit span.

Thus the lift per unit span of a finite wing l is, by the Kutta–Zhukovsky theorem,

$$l = \rho V \Gamma$$

with the total lift being

$$L = \int_{-s/2}^{s/2} \rho V \Gamma \, dz \tag{5.25}$$

The induced drag per unit span (d_v), or the induced drag grading, again by the Kutta–Zhukovsky theorem, is

$$d_v = \rho w \Gamma \tag{5.26}$$

and by similar integration over the span

$$D_v = \int_{-s/2}^{s/2} \rho w \Gamma \, dz \tag{5.27}$$

This expression for induced drag force D_v shows conclusively that if w is zero all along the span, then D_v is zero also. Clearly, if there is no trailing vorticity, there will be no induced drag. This condition arises when a wing is working under two-dimensional conditions or if all sections are producing zero lift.

As a consequence of the trailing vortex system, which is produced by the basic lifting action of a (finite-span) wing, the wing characteristics are considerably modified, almost always adversely, from those of the equivalent two-dimensional wing of the same section. Equally, a wing with flow systems that more nearly approach the two-dimensional case has better aerodynamic characteristics than one where the end effects are more prominent. It seems therefore that a wing that is large in the spanwise dimension (i.e., large aspect ratio) is a better wing—nearer the ideal—than a short-span wing of the same airfoil section. It thus appears that a wing with a large aspect ratio has better aerodynamic characteristics than one of the same section with a lower aspect ratio. This is the case with aircraft for which aerodynamic efficiency is paramount. Good examples are sailplanes and the Global Hawk UAV. In nature this is also true; consider the albatross, which has wings with high aspect ratios.

In general, induced velocity also varies in the chordwise direction, as is evident from Eq. (5.21). In effect, the assumption of high aspect ratio, leading to Eq. (5.23), permits the chordwise variation to be neglected. Accordingly, the lifting characteristics of a section from a wing with a high aspect ratio at a local angle of attack $\alpha(z)$ are identical to those for a two-dimensional wing at an effective angle of attack $\alpha(z) - \varepsilon$. Thus Prandtl's theory shows how two-dimensional airfoil characteristics can be used to determine the lifting characteristics of wings of finite span. The calculation of the *induced angle of attack* ε now becomes the central problem, posing certain difficulties because ε depends on circulation, which in turn is closely related to lift per unit span. The problem is therefore to some degree circular, which makes a simple direct approach to its solution impossible. The required solution procedure is described in Section 5.6.

Before developing the general theory, we consider in some detail the much simpler inverse problem of a specified spanwise circulation distribution in the next subsection. Although this is a special case, it nevertheless leads to many results of practical interest. In particular, a simple quantitative result emerges that reinforces the qualitative arguments given earlier concerning the greater aerodynamic efficiency of wings with high aspect ratios.

5.5.3 Characteristics of Simple Symmetric Loading—Elliptic Distribution

To demonstrate the general method of obtaining the aerodynamic characteristics of a wing from its loading distribution, the simplest load expression for symmetric flight is taken—that is, a semi-ellipse. In addition, this will be found to be a good approximation of many more mathematically complicated distributions and is thus suitable for first predictions in performance estimates.

The spanwise variation in circulation is taken to be represented by a semi-ellipse having the span s as the major axis and the circulation at mid-span Γ_0 as the semi-minor axis (Fig. 5.23). From the general expression for an ellipse

$$\frac{\Gamma^2}{\Gamma_0^2} + \frac{4z^2}{s^2} = 1$$

or

$$\Gamma = \Gamma_0 \sqrt{1 - 4\left(\frac{z}{s}\right)^2} \tag{5.28}$$

This expression can now be substituted in Eqs. (5.23), (5.25), and (5.27) to find the lift, downwash, and vortex drag on the wing.

Lift for Elliptic Distribution

From Eq. (5.25),

$$L = \int_{-s/2}^{s/2} \rho V \Gamma \, dz = \int_{-s/2}^{s/2} \rho V \Gamma_0 \sqrt{1 - 4\left(\frac{z}{s}\right)^2} \, dz$$

or

$$L = \frac{\rho V \Gamma_0 \pi s}{4} \tag{5.29}$$

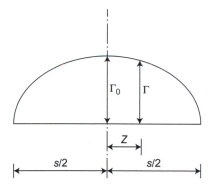

FIGURE 5.23

Elliptic loading, or a spanwise elliptic distribution of vortex strength.

And thus

$$\Gamma_0 = \frac{4L}{\rho V \Gamma_0 \pi s}$$

or, introducing the lift coefficient,

$$L = C_L \frac{1}{2} \rho V^2 S$$

$$\Gamma_0 = \frac{2 C_L V S}{\pi s} \tag{5.30}$$

giving the mid-span circulation in terms of the overall airfoil lift coefficient and geometry.

Downwash for Elliptic Distribution
Here

$$\frac{d\Gamma}{dz} = -\Gamma_0 \frac{z}{(s/2)^2} \left[1 - \left(\frac{z}{s/2} \right)^2 \right]^{-1/2} = -\frac{\Gamma_0}{(s/2)} \frac{z}{\sqrt{(s/2)^2 - z^2}} = -\frac{4\Gamma_0}{s} \frac{z}{\sqrt{s^2 - z^2}}$$

Substituting this in Eq. (5.23),

$$w_{z_1} = \frac{\Gamma_0}{4\pi \, (s/2)} \int_{-s/2}^{s/2} \frac{z \, dz}{\sqrt{(s/2)^2 - z^2} \, (z - z_1)} = \frac{\Gamma_0}{2\pi S} \int_{-s/2}^{s/2} \frac{z \, dz}{\sqrt{(s/2)^2 - z^2} \, (z - z_1)}$$

Writing the numerator as $(z - z_1) + z_1$,

$$w_{z_1} = \frac{\Gamma_0}{4\pi \, (s/2)} \int_{-s/2}^{s/2} \frac{(z - z_1) + z_1}{\sqrt{(s/2)^2 - z^2} \, (z - z_1)} \, dz$$

$$= \frac{\Gamma_0}{2\pi s} \left[\int_{-s/2}^{s/2} \frac{dz}{\sqrt{(s/2)^2 - z^2}} + z_1 \int_{-s/2}^{s/2} \frac{dz}{\sqrt{(s/2)^2 - z^2} \, (z - z_1)} \right]$$

Evaluating the first integral, which is standard, and writing I for the second,

$$w_{z_1} = \frac{\Gamma_0}{2\pi s} [\pi + z_1 I] \tag{5.31}$$

Now, as this is a symmetric flight case, the shed vorticity is the same from each side of the wing and the value of the downwash at some point z_1 is identical to that at the corresponding point—z_1 on the other wing. So, substituting for $\pm z_1$ in Eq. (5.31) and equating,

$$w_{\pm z_1} = \frac{\Gamma_0}{2\pi s}[\pi + z_1 I] = \frac{\Gamma_0}{2\pi s}[\pi - z_1 I]$$

This identity is satisfied over all z_1 only if $I = 0$, so that for any point $z - z_1$ along the span

$$w = \frac{\Gamma_0}{2s} \qquad (5.32)$$

This important result shows that downwash is constant along the span. Uniform downwash exists only for elliptic lift distribution.

Induced Drag (Vortex Drag) for Elliptic Distribution
From Eq. (5.27)

$$D_v = \int_{-s/2}^{s/2} \rho w \Gamma dz = \int_{-s/2}^{s/2} \rho \frac{\Gamma_0}{2s} \Gamma_0 \sqrt{1 - \left(\frac{2z}{s}\right)^2} \, dz$$

whence

$$D_V = \frac{\pi}{8} \rho \Gamma_0^2 \qquad (5.33)$$

Introducing

$$D_V = C_{Dv} \frac{1}{2} \rho V^2 S$$

and from Eq. (5.30)

$$\Gamma_0 = \frac{C_L V S}{\pi s}$$

Eq. (5.33) gives

$$C_{Dv} \frac{1}{2} \rho V^2 S = \frac{\pi}{8} \rho \left(\frac{C_L V S}{\pi s}\right)^2$$

or

$$C_{D_v} = \frac{C_L^2}{\pi (AR)} \qquad (5.34)$$

because

$$\frac{s^2}{S} = \frac{\text{span}^2}{\text{area}} = \text{aspect ratio } (AR)$$

Equation (5.34) establishes quantitatively how C_{D_v} falls with a rise in (AR) and confirms the previous conjecture given in Eq. (5.27), that at zero lift in symmetric flight C_{D_v} is zero and the other condition that as (AR) increases (to infinity for two-dimensional flow) C_{D_v} decreases (to zero).

The elliptic lift distribution can be shown to be the optimal distribution on a planar wing. That is, it produces the minimum induced drag of all possible lift distributions on a wing of that span. Nonplanar features of a wing may, however, produce even better results. Winglets are now a common site at airports. Winglet design is a difficult task that is beyond this text, but a general description of winglet function is possible. A well-designed winglet reduces induced drag by altering lift distribution, typically by increasing lift near the tips. This nonzero lift at the tips is why a winglet is often described as "making the wing act like it has a longer span." A longer span has a greater aspect ratio, which, as you see from above, leads to lower induced drag. In addition to winglets, substantial nonplanar effects are possible for reducing induced drag. The work of research groups such as Kroo at Stanford shows this clearly [41], yet of course the structural requirements of a real wing present the engineer with additional opportunities for creative solutions.

5.5.4 General (Series) Distribution of Lift

In the previous section attention was directed to distributions of circulation (or lift) along the span in which the load is assumed to fall symmetrically about the centerline according to a particular family of load distributions. For steady symmetric maneuvers this is quite satisfactory, and the previous distribution formula may be arranged to suit certain cases. However, its use is strictly limited, and it is necessary to seek an expression that satisfies every possible combination of wing design parameter and flight maneuver. For example, it has so far been assumed that the wing is an isolated lifting surface that in straight steady flight has a load distribution rising steadily from zero at the tips to a maximum at mid-span (Fig. 5.24(a)). The general wing, however, has a fuselage located in the center sections that modifies the loading in that region (Fig. 5.24(b)), and engine nacelles or other protuberances or appendages may locally deform the remainder of the curve.

The load distributions on both the isolated wing and the general airplane wing are considerably changed in antisymmetric flight. In rolling, for instance, the upgoing wing suffers a large reduction in lift, which may become negative at some incidences (Fig. 5.24(c)). With ailerons in operation, the curve of spanwise loading is no longer smooth and symmetrical, but can be rugged and distorted in shape (Fig. 5.24(d)).

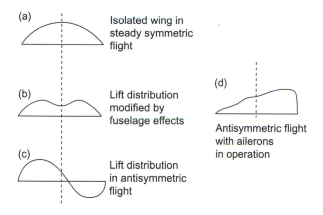

(a) Isolated wing in steady symmetric flight

(b) Lift distribution modified by fuselage effects

(c) Lift distribution in antisymmetric flight

(d) Antisymmetric flight with ailerons in operation

FIGURE 5.24

Examples of several possible spanwise distributions of lift on a wing in steady and maneuvering flight.

It is clearly necessary to find an expression that accommodates these various possibilities. From previous work, the formula $l = \rho V \Gamma$ for any section of span is familiar. Writing l in the form of the nondimensional lift coefficient and equating to $\rho V \Gamma$,

$$\Gamma = \frac{C_L}{2} V c \tag{5.35}$$

is easily obtained. This shows that, for a given steady flight state, the circulation at any section can be represented by the product of the forward velocity and the local chord. Now, in addition, the local chord can be expressed as a fraction of the semi-span s, and with this fraction absorbed in a new number and the numeral 4 introduced for later convenience, Γ becomes

$$\Gamma = 2C_\Gamma s$$

where C_Γ is dimensionless circulation that varies similarly to Γ across the span. In other words, C_Γ is the shape parameter or variation of the Γ curve and, being dimensionless, can be expressed as the Fourier sine series $\sum_1^\infty A_n \sin n\theta$, in which the coefficients A_n represent the amplitudes and the sum of the successive harmonics describes the shape. The sine series was chosen to satisfy the end conditions of the curve reducing to zero at the tips, where $y = \pm s$. These correspond to the values of $\theta = 0$ and π. It is well understood that such a series is unlimited in angular measure, but the portions beyond 0 and π can be disregarded here. Further, the series can fit any shape of curve but, in general, for rapidly changing distributions as shown by a

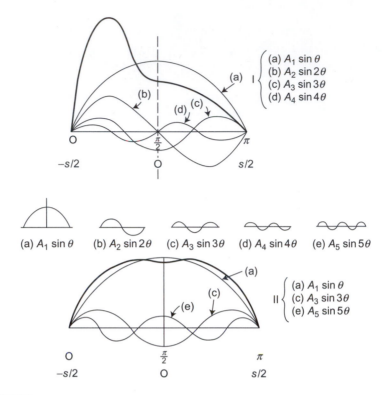

$$\begin{cases} \text{(a) } A_1 \sin\theta \\ \text{(b) } A_2 \sin 2\theta \\ \text{(c) } A_3 \sin 3\theta \\ \text{(d) } A_4 \sin 4\theta \end{cases}$$

(a) $A_1 \sin\theta$ (b) $A_2 \sin 2\theta$ (c) $A_3 \sin 3\theta$ (d) $A_4 \sin 4\theta$ (e) $A_5 \sin 5\theta$

$$\text{II} \begin{cases} \text{(a) } A_1 \sin\theta \\ \text{(c) } A_3 \sin 3\theta \\ \text{(e) } A_5 \sin 5\theta \end{cases}$$

FIGURE 5.25

Creation of a spanwise vorticity distribution $\Gamma(\theta)$ by superposition of selected sine waves.

rugged curve, for example, many harmonics are required to produce a sum that is a good representation.

In particular, the series is simplified for the symmetrical loading case when the even terms disappear (Fig. 5.25(II)). For the symmetrical case, a maximum or minimum must appear at the midsection. This is possible only for sines of odd values of $\pi/2$. That is, the symmetrical loading must be the sum of symmetrical harmonics. Odd harmonics are symmetrical. Even harmonics, on the other hand, return to zero again at $\pi/2$, where in addition there is always a change in sign. For any asymmetry in the loading, one or more even harmonics are necessary.

With the number and magnitude of harmonics effectively giving all possibilities, the general spanwise loading can be expressed as

$$\Gamma = 2sV \sum_{1}^{\infty} A_n \sin n\theta \tag{5.36}$$

It should be noted that since $l = \rho V \Gamma$, the spanwise lift distribution can be expressed as

$$l = 2\rho V^2 s \sum_{1}^{\infty} A_n \sin n\theta \tag{5.37}$$

The aerodynamic characteristics for symmetrical general loading are derived in the next subsection. The case of asymmetrical loading is not included, but it may be dealt with in a very similar manner, and in this way expressions can be derived for such quantities as rolling and yawing moment.

5.5.5 Aerodynamic Characteristics for Symmetrical General Loading

The operations to obtain lift, downwash, and drag vary only in detail from the previous cases.

Lift on the Wing

Lift on the wig is the sum, or total, of all lift-per-span forces along the span of the wing. This is computed as an integral of lift per span from one wing tip to the other wing tip:

$$L = \int_{-s/2}^{s/2} \rho V \Gamma dz$$

Setting the variable $z = -(s/2)\cos\theta$,

$$L = \frac{1}{2} \int_{0}^{\pi} \rho V \Gamma s \sin\theta d\theta$$

and substituting for the general series expression,

$$L = \int_{0}^{\pi} \rho V^2 s^2 \sum A_n \sin n\theta \sin\theta d\theta$$

$$= s^2 \rho V^2 \int_{0}^{\pi} \sum A_n [\cos(n-1)\theta + \cos(n+1)\theta] d\theta$$

$$= s^2 \rho V^2 \frac{1}{2} \sum A_n \left[\frac{\sin(n-1)\theta}{n-1} + \frac{\sin(n+1)\theta}{n+1} \right]_{0}^{\pi}$$

The sum in the squared bracket equals zero for all values of n other than unity when it becomes

$$\left[\lim_{(n-1)\to 0} A_1 \frac{\sin(n-1)\theta}{n-1}\right]_0^\pi = A_1 \pi$$

Thus

$$L = A_1 \pi \frac{1}{2} \rho V^2 s^2 = C_L \frac{1}{2} \rho V^2 S$$

and writing aspect ratio $(AR) = s^2/S$ gives

$$C_L = \pi A_1 (AR) \tag{5.38}$$

This indicates the rather surprising result that lift depends on the magnitude of the coefficient of the first term only, no matter how many more may be present in the series describing the distribution. The reason is that the terms $A_3 \sin 3\theta$, $A_5 \sin 5\theta$, and so forth, provide positive lift on some sections and negative lift on others. The overall effect thus is zero. These terms provide the characteristic variations in the spanwise distribution, but do not affect the total lift of the whole, which is determined solely from the amplitude of the first harmonic. Thus

$$C_L = \pi (AR) A_1 \quad \text{and} \quad C_L = \frac{\pi}{2} \rho V^2 s^2 A_1 \tag{5.39a}$$

Downwash

Changing the variable and limits of Eq. (5.23), the equation for the downwash is

$$w_{\theta_1} = \frac{1}{2\pi s} \int_0^\pi \frac{\frac{d\Gamma}{d\theta} d\theta}{\cos\theta - \cos\theta_1}$$

In this case $\Gamma = 2sV \sum A_n \sin n\theta$, and thus on differentiating

$$\frac{d\Gamma}{d\theta} = 2sV \sum nA_n \cos n\theta$$

Introducing this into the integral expression gives

$$w_{\theta_1} = \frac{2sV}{2\pi s} \int_0^\pi \frac{\sum nA_n \cos n\theta \, d\theta}{\cos\theta - \cos\theta_1}$$

$$= \frac{V}{\pi} \sum nA_n G_n$$

and writing in $G_n = \pi \sin n\theta_1 / \sin\theta_1$ from Appendix 3, and reverting to the general point θ,

$$w = V \frac{\sum nA_n \sin n\theta}{\sin\theta} \qquad (5.39b)$$

This involves all the coefficients of the series and is symmetrically distributed about the centerline for odd harmonics.

Induced Drag (Vortex Drag)

The induced drag increment is given by $d_v = \rho w \Gamma$. Integrating along the span from tip to tip gives the total induced drag

$$D_v = \int_{-s/2}^{s/2} \rho w \Gamma \, dz$$

or, in the polar variable,

$$D_v = \int_0^\pi \rho \underbrace{\frac{V \sum nA_n \sin n\theta}{\sin\theta}}_{w} \underbrace{sV \sum A_n \sin n\theta}_{\Gamma} \underbrace{s \sin\theta \, d\theta}_{dz}$$

$$= \rho V^2 s^2 \int_0^\pi \sum nA_n \sin n\theta \sum A_n \sin n\theta \, d\theta$$

The integral becomes

$$I = \int_0^\pi (A_1^2 \sin^2 \theta + 3A_2^2 \sin^2 3\theta + 5A_5^2 \sin^2 5\theta + \cdots) d\theta$$

$$= \frac{\pi}{2}[A_1^2 + 3A_3^2 + 5A_5^2 + \cdots] = \frac{\pi}{2} \sum nA_n^2$$

which can be demonstrated by multiplying out the first three (say) odd harmonics, thus

$$I = \int_0^\pi (A_1 \sin\theta + 3A_3 \sin 3\theta + 5A_5 \sin 5\theta)(A_1 \sin\theta + A_3 \sin 3\theta + A_5 \sin\theta) d\theta$$

$$= \int_0^\pi \{A_1^2 \sin^2 \theta + 3A_3^2 \sin^2 \theta + 5A_5^2 \sin^2 \theta + [A_1 A_3 \sin\theta \sin 3\theta \text{ and}$$

other like terms that are products of different multiples of θ]$\}d\theta$

On carrying out the integration from 0 to π, all terms other than the squared terms vanish, leaving

$$I = \int_0^\pi \left(A_1^2 \sin^2 \theta + 3A_2^2 \sin^2 3\theta + 5A_5^2 \sin^2 5\theta + \cdots \right) d\theta$$

$$= \frac{\pi}{2} \left[A_1^2 + 3A_3^2 + 5A_5^2 + \cdots \right] = \frac{\pi}{2} \sum nA_n^2$$

This gives

$$D_V = \rho V^2 s^2 \frac{\pi}{2} \sum nA_n^2 = C_{Dv} \frac{1}{2} \rho V^2 S$$

whence

$$C_{Dv} = \pi(AR) \sum nA_n^2 \tag{5.40}$$

From Eq. (5.38),

$$A_1^2 = \frac{C_L^2}{\pi^2 (AR)^2}$$

and introducing this into Eq. (5.40),

$$C_{D_V} = \frac{C_L^2}{\pi (AR)} \sum n \left(\frac{A_n}{A_1}\right)^2$$

$$= \frac{C_L^2}{\pi (AR)} \left[1 + \left(\frac{3A_3^2}{A_1^2} + \frac{5A_5^2}{A_1^2} + \frac{7A_7^2}{A_1^2} + \cdots\right)\right]$$

Writing the symbol δ for the term $\left(\frac{3A_3^2}{A_1^2} + \frac{5A_5^2}{A_1^2} + \frac{7A_7^2}{A_1^2} + \cdots\right)$,

$$C_{D_V} = \frac{C_L^2}{\pi (AR)}[1 + \delta] \qquad (5.41a)$$

Plainly, δ is always a positive quantity because it consists of squared terms that must always be positive. C_{D_v} can be a minimum only when $\delta = 0$—that is, when $A_3 = A_5 = A_7 = \ldots = 0$ and the only term remaining in the series is $A_1 \sin \theta$.

It has become common to use a span efficiency factor, $\varepsilon = \frac{1}{1+\delta}$. This results in $0 < \varepsilon \leq 1$, where the upper limit is for the elliptic lift distribution. Equation (5.41a) then becomes

$$C_{D_V} = \frac{C_L^2}{\pi \varepsilon (AR)} \qquad (5.41b)$$

You may find ε called the Oswald, or span, efficiency factor.

Minimum Induced Drag Condition

Comparing Eq. (5.41a) with the induced-drag coefficient for the elliptic case Eq. (5.34), it can be seen that modifying the spanwise distribution away from the elliptic increases the drag coefficient by the fraction δ that is always positive. It follows that for the induced drag to be a minimum, δ must be zero so that the distribution for minimum induced drag is the semi-ellipse. It will also be noted that the minimum drag distribution produces a constant downwash along the span whereas all other distributions produce a spanwise variation in induced velocity. This is no coincidence, but is part of the physical explanation of the elliptic distribution having minimum induced drag.

To see this, consider two wings (Fig. 5.26(a) and 5.26(b)) of equal span with spanwise distributions in downwash velocity $w = w_0 = $ constant along (a) and $w = f(z)$ along (b). Without altering the latter downwash variation, it can be expressed as the sum of two distributions w_0 and $w_1 = f_1(z)$, as shown in Fig. 5.26(c).

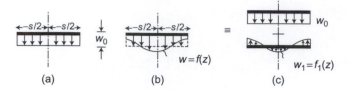

FIGURE 5.26

(a) Elliptic distribution gives constant downwash and minimum drag; (b) nonelliptic distribution gives varying downwash; (c) equivalent variation for comparison.

If the lift due to both wings is the same under given conditions, the rate of change in vertical momentum in the flow is the same for both. Thus for (a)

$$L \propto \int_{-s/2}^{s/2} \dot{m} w_0 dx \qquad (5.42)$$

and for (b)

$$L \propto \int_{-s/2}^{s/2} \dot{m} \left(w_0 + f_1(z) \right) dz \qquad (5.43)$$

where $\dot{m}$ is a representative mass flow meeting unit span. Since L is the same on each wing

$$\int_{-s/2}^{s/2} \dot{m} f_1(z) \, dz = 0 \qquad (5.44)$$

Now the energy transfer or rate of change in the kinetic energy of the representative mass flows is the induced drag (or vortex drag). For (a),

$$D_{v(a)} \propto \frac{1}{2} \dot{m} \int_{-s/2}^{s/2} w_0^2 dz \qquad (5.45)$$

For (b),

$$D_{V(b)} \propto \frac{1}{2} \dot{m} \int_{-s/2}^{s/2} \left(w_0 + f_1(z) \right)^2 dz$$

$$\propto \frac{1}{2} \dot{m} \int_{-s/2}^{s/2} w_0^2 + 2 w_0 f_1(z) + \left(f_1(z) \right)^2 dz$$

and since $\int_{-s/2}^{s/2} \dot{m} f_1(z) \, dz = 0$ in Eq. (5.44),

$$D_{V(b)} \propto \frac{1}{2}\dot{m} \int_{-s/2}^{s/2} w_0^2 \, dz + \frac{1}{2}\dot{m} \int_{-s/2}^{s/2} (f_1(z))^2 \, dz \qquad (5.46)$$

Comparing Eqs. (5.45) and (5.46),

$$D_{V(b)} \propto D_{V(a)} + \frac{1}{2}\dot{m} \int_{-s/2}^{s/2} (f_1(z))^2 \, dz$$

and since $f_1(z)$ is an explicit function of z,

$$\int_{-s/2}^{s/2} (f_1(z))^2 \, dz > 0$$

since $(f_1(z))^2$ is always positive whatever the sign of $f_1(z)$. Hence $D_{V(b)}$ is always greater than $D_{V(a)}$.

5.6 DETERMINATION OF LOAD DISTRIBUTION ON A GIVEN WING

This is the direct problem broadly facing designers who wish to predict the performance of a projected wing before the long and costly process of model testing begins. This does not imply that such tests need not be carried out. On the contrary, they may be important steps in the design process toward a production aircraft.

The problem can be rephrased to suggest that designers want some indication of how the wing characteristics vary as, for example, the geometric parameters of the project wing are changed. In this way, they can balance the aerodynamic effects of their changing ideas against the basic specification—provided there is a fairly simple process relating changes in design parameters to aerodynamic characteristics. Of course, this is stating one of the design problems in its baldest and simplest terms, but as in any design work, plausible theoretical processes yielding reliable predictions are very comforting.

The loading on the wing was described in the most general terms available, and the overall characteristics are immediately on hand in terms of the coefficients of the loading distribution (Section 5.5). It remains to relate the coefficients (or the series as a whole) to the basic airfoil parameters of planform and airfoil section characteristics.

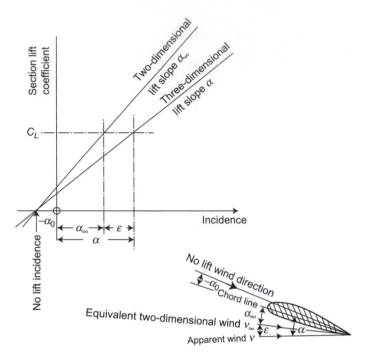

FIGURE 5.27

Lift-versus-incidence curve for an airfoil section of a certain profile, working two-dimensionally and in a flow regime influenced by end effects (i.e., working at some point along the span of a finite lifting wing).

5.6.1 General Theory for Wings of High Aspect Ratio

As a start, consider the influence of the end effect, or downwash, on the lifting properties of an airfoil section at some distance z from the centerline of the wing. Figure 5.27 shows the lift-versus-incidence curve for an airfoil section of a certain profile working two-dimensionally and in a flow regime influenced by end effects—that is, working at some point along the span of a finite lifting wing.

We assume that both curves are linear over the range considered (i.e., the working range) and that under both flow regimes the zero-lift incidence is the same. Then

$$C_L = a_\infty[\alpha_\infty - \alpha_0] = a[\alpha - \alpha_0] \tag{5.47}$$

Taking the first equation with $\alpha_\infty = \alpha - \varepsilon$,

$$C_L = a_\infty[(\alpha - \alpha_0) - \varepsilon] \tag{5.48}$$

But equally from Eq. (4.10),

$$C_L = \frac{\text{lift per unit span}}{\frac{1}{2}\rho V^2 c} = \frac{l}{\frac{1}{2}\rho V^2 c}$$

$$= \frac{\rho V \Gamma}{\frac{1}{2}\rho V^2 c} \qquad (5.49)$$

$$C_L = \frac{2\Gamma}{Vc}$$

Equating Eqs. (5.48) and (5.49) and rearranging,

$$\frac{2\Gamma}{ca_\infty} = V[(\alpha - \alpha_0) - \varepsilon]$$

and since

$$V\varepsilon = w = -\frac{1}{4\pi}\int_{-s/2}^{s/2} \frac{(d\Gamma/dz)}{z - z_1}\,dz \quad \text{from Eq. (5.23)}$$

$$\frac{2\Gamma}{ca_\infty} = V(\alpha - \alpha_0) + \frac{1}{4\pi}\int_{-s/2}^{s/2} \frac{(d\Gamma/dz)}{z - z_1}\,dz \qquad (5.50)$$

This is Prandtl's integral equation for the circulation Γ at any section along the span in terms of all airfoil parameters. These will be discussed when Eq. (5.50) is reduced to a form more amenable to numerical solution. To do this, we take the general series expression (5.36) for Γ:

$$\Gamma = 2sV \sum A_n \sin n\theta$$

The previous section gave Eq. (5.39b):

$$w = \frac{V \sum n A_n \sin n\theta}{\sin \theta}$$

which, substituted in Eq. (5.50), gives together

$$2\frac{2sV \sum A_n \sin n\theta}{ca_\infty} = V(\alpha - \alpha_0) - \frac{V \sum n A_n \sin n\theta}{\sin \theta}$$

Cancelling V and collecting $ca_\infty/4s$ into the single parameter μ, this equation becomes

$$\mu(\alpha - \alpha_0) = \sum_{n=1}^{\infty} A_n \sin n\theta \left(1 + \frac{\mu n}{\sin \theta}\right) \tag{5.51}$$

the solution to which cannot in general be found analytically for all points along the span, but only numerically at selected spanwise stations and at each end.

5.6.2 General Solution to Prandtl's Integral Equation

This is best understood if a particular value of θ, or position along the span, is taken in Eq. (5.51). For example, the position $z = -0.25s$ is midway between the mid-span sections and the tip. Beginning with

$$z = -\frac{s \cos \theta}{2}$$

and, substituting for z, the result is

$$\theta = \cos^{-1}\left(\frac{1}{2}\right) = 60°$$

Then, if the value of the parameter μ is μ_1 and the incidence from no lift is $(\alpha_1 - \alpha_{01})$, Eq. (5.51) becomes

$$\mu_1(\alpha_1 - \alpha_{01}) = A_1 \sin 60° \left[1 + \frac{\mu_1}{\sin 60°}\right] + A_2 \sin 120° \left[1 + \frac{2\mu_1}{\sin 60°}\right]$$

$$+ A_3 \sin 180° \left[1 + \frac{3\mu_1}{\sin 60°}\right] + \text{etc.}$$

This is obviously an equation with A_1, A_2, A_3, A_4, and so on, as the only unknowns.

Other equations in which A_1, A_2, A_3, A_4, and so on, are the unknowns can be found by considering other points z along the span, bearing in mind that the value of μ and of $(\alpha - \alpha_0)$ may also change from point to point. If, say, four terms in the series are desired, an equation of the previous form must be obtained at each of four values of θ, noting that normally the values $\theta = 0$ and π (i.e., the wingtips) lead to the trivial equation $0 = 0$ and are therefore useless for the present purpose. Generally, four coefficients are sufficient in the symmetrical case to produce a spanwise distribution that is insignificantly altered by the addition of further terms. In the case of symmetric flight the coefficients would be A_1, A_3, A_5, and A_7, since the even harmonics do not appear. Also, the arithmetic need only be concerned with values of θ between 0 and $\pi/2$ since the curve is symmetrical about the mid-span section.

If the spanwise distribution is irregular, more harmonics are necessary in the series to describe it adequately, and more coefficients must be found from the integral equation. This becomes quite a tedious and lengthy operation by "hand," but being a simple mathematical procedure, the simultaneous equations can be easily programmed for a computer.

The airfoil parameters are contained in the expression

$$\mu = \frac{\text{chord} \times \text{ two-dimensional lift slope}}{8 \times \text{semi-span}}$$

and the absolute incidence $(\alpha - \alpha_0)$. μ clearly allows for any spanwise variation in the chord—that is, a change in plan shape—or in the two-dimensional slope of the airfoil profile—that is, a change in airfoil section. α is the local geometric incidence and will vary if there is any geometric twist present on the wing. α_0, the zero-lift incidence, may vary if there is any aerodynamic twist present (if the airfoil section is changing along the span).

Example 5.6

Consider a tapered airfoil. For completeness in the example, every parameter is allowed to vary in a linear fashion from mid-span to wingtips.

Mid-Span Data		Wingtip Data
3.048	Chord m	1.524
5.5	$\left(\frac{\partial C_L}{\partial \alpha}\right)_\infty$ per radian	5.8
5.5	Absolute incidence α°	3.5
	Total span of wing: 12.192 m	

Obtain the airfoil characteristics of the wing, the spanwise distribution of circulation, comparing it with the equivalent elliptic distribution for the wing flying straight and level at 89.4 $\mathrm{m\,s^{-1}}$ at low altitude.

From the data,

$$\text{Wing area } S = \frac{3.048 + 1.524}{2} \times 12.192 = 27.85 \mathrm{m^2}$$

$$\text{Aspect ratio } (AR) = \frac{\text{span}^2}{\text{area}} = \frac{12.192^2}{27.85} = 5.333$$

At any section z from the centerline (θ from the wingtip),

$$\text{chord } c = 3.048\left[1 - \frac{3.048 - 1.524}{3.048}\left(\frac{z}{s}\right)\right] = 3.048[1 + 0.5\cos\theta]$$

$$\left(\frac{\partial C_L}{\partial \alpha}\right)_\infty = a = 5.5\left[1 + \frac{5.5 - 5.8}{5.5}\left(\frac{z}{s}\right)\right] = 5.5[1 - 0.05455\cos\theta]$$

$$\alpha^\circ = 5.5\left[1 - \frac{5.5 - 3.5}{5.5}\left(\frac{z}{s}\right)\right] = 5.5[1 + 0.36364\cos\theta]$$

This gives at any section

$$\mu = \frac{ca_\infty}{4s} = 0.34375\,(1+0.5\cos\theta)\,(1-0.05455\cos\theta)$$

and

$$\mu\alpha = 0.032995\,(1+0.5\cos\theta)\,(1-0.05455\cos\theta)\,(1+0.36364\cos\theta)$$

where α is now in radians. For convenience Eq. (5.51) is rearranged to

$$\mu\alpha\sin\theta = A_1\sin\theta\,(\sin\theta+\mu)+A_3\sin3\theta\,(\sin\theta+3\mu)+A_5\sin5\theta\,(\sin\theta+5\mu)$$
$$+A_7\sin7\theta\,(\sin\theta+7\mu)$$

and since the distribution is symmetrical, only the odd coefficients appear. Four coefficients are evaluated and, because of symmetry, it is only necessary to take values of θ between 0 and $\pi/2$ (i.e., $\pi/8$, $\pi/4$, $3\pi/8$, $\pi/2$).

Table E5.6 gives values of $\sin\theta$, $\sin n\theta$, and $\cos\theta$ for the angles just given, and these substituted in the rearranged Eq. (5.51) lead to the following four simultaneous equations in the unknown coefficients:

$$0.004739 = 0.22079A_1 + 0.89202A_3 + 1.25100A_5 + 0.66688A_7$$

$$0.011637 = 0.66319A_1 + 0.98957A_3 - 1.31595A_5 - 1.64234A_7$$

$$0.021665 = 1.11573A_1 - 0.67935A_3 - 0.89654A_5 + 2.68878A_7$$

$$0.032998 = 1.34375A_1 - 2.03125A_3 - 2.71875A_5 - 3.40625A_7$$

When solved these equations give

$$A_1 = 0.020329,\ A_3 = -0.000955,\ A_5 = 0.001029,\ A_7 = -0.0002766$$

Thus

$$\Gamma = 4sV\{0.020329\sin\theta - 0.000955\sin3\theta + 0.001029\sin5\theta - 0.0002766\sin7\theta\}$$

Table E5.6 Numerical Values for Example 5.6

θ	$\sin\theta$	$\sin 3\theta$	$\sin 5\theta$	$\sin 7\theta$	$\cos\theta$
$\pi/8$	0.38268	0.92388	0.92388	0.38268	0.92388
$\pi/4$	0.70711	0.70711	−0.70711	−0.70711	0.70711
$3\pi/8$	0.92388	−0.38268	−0.38268	0.92388	0.38268
$\pi/2$	1.00000	−1.00000	1.00000	−1.00000	0.00000

and substituting the values of θ taken previously, the circulation takes the following values:

θ	0	$\pi/8$	$\pi/4$	$3\pi/8$	$\pi/2$
z/s	1/2	0.462	0.354	0.192	0
$\Gamma\text{m}^2\text{s}^{-1}$	0	16.85	28.7	40.2	49.2
Γ/Γ_0	0	0.343	0.383	0.82	1.0

As a comparison, the equivalent elliptic distribution with the same coefficient of lift gives a series of values:

$\Gamma\text{ m}^2\text{ s}^{-1}$	0	14.9	27.6	36.0	38.8

The aerodynamic characteristics follow from the equations given in Section 5.5.4. Thus

$$C_L = \pi(AR)A_1 = 0.3406$$

$$C_D = \frac{C_L^2}{\pi(AR)}[1+\delta] = 0.007068$$

since

$$\delta - 3\left(\frac{A_3}{A_1}\right)^2 + 5\left(\frac{A_5}{A_1}\right)^2 + 7\left(\frac{A_7}{A_1}\right)^2 = 0.02073$$

That is, the induced drag is 2% greater than the minimum.

For completeness the total lift and drag may be given:

$$\text{Lift} = C_L \frac{1}{2}\rho V^2 S = 0.3406 \times 139910 = 47.72\text{kN}$$

$$\text{Drag (induced)} = C_{Dv}\frac{1}{2}\rho V^2 S = 0.007068 \times 139910 = 988.82\text{N}$$

Example 5.7

A wing is untwisted and of elliptic planform with a symmetrical airfoil section. It is rigged symmetrically in a wind tunnel at incidence α_1 to a wind stream having an axial velocity V. In addition, the wind has a small uniform angular velocity ω about the tunnel axis. Show that the distribution of circulation along the wing is given by

$$\Gamma = 4sV[A_1 \sin\theta + A_2 \sin 2\theta]$$

and determine A_1 and A_2 in terms of the wing parameters. Neglect wind-tunnel constraints.

From Eq. (5.51),

$$\mu(\alpha - \alpha_0) = \sum A_n \sin n\theta \left(1 + \frac{\mu n}{\sin \theta}\right)$$

In this case $\alpha_0 = 0$, and the effective incidence at any section z from the centerline is

$$\alpha = \alpha_1 + z\frac{\omega}{V} = \alpha_1 - \frac{\omega}{V} s \cos \theta$$

Also, since the planform is elliptic and untwisted, $\mu = \mu_0 \sin \theta$ (Section 5.5.3) and the equation becomes for this problem

$$\mu_0 \sin \theta \left[\alpha_1 - \frac{\omega}{2V} s \cos \theta\right] = \sum A_n \sin n\theta \left(1 + \frac{\mu_0 n \sin \theta}{\sin \theta}\right)$$

Expanding both sides,

$$\mu_0 \alpha_1 \sin \theta - \frac{\mu_0 \omega s}{2V} \frac{\sin 2\theta}{2} = A_1 \sin \theta \left(1 + \mu_0\right) + A_2 \sin 2\theta \left(1 + 2\mu_0\right) + A_3 \sin 3\theta \left(1 + 3\mu_0\right) + \text{etc.}$$

Equating like terms,

$$\mu_0 \alpha_1 \sin \theta = A_1 \left(1 + \mu_0\right) \sin \theta$$

$$-\frac{\mu_0 \omega s}{2V} \frac{\sin 2\theta}{2} = A_2 \sin 2\theta \left(1 + 2\mu_0\right)$$

$$0 = A_3 \sin 3\theta \left(1 + 3\mu_0\right) \text{ and similar for } n > 3$$

Thus the spanwise distribution for this case is

$$\Gamma = 2sV \left[A_1 \sin \theta + A_2 \sin 2\theta\right]$$

and the coefficients are

$$A_1 = \left(\frac{\mu_0}{1 + \mu_0}\right) \alpha_1$$

and

$$A_2 = \frac{\mu_0}{(1 + 2\mu_0)} \frac{ws}{4V}$$

5.6.3 Load Distribution for Minimum Drag

Minimum induced drag for a given lift will occur if C_D is a minimum, and this is so only if δ is zero since δ is always a positive quantity. Because δ involves squares of all coefficients other than the first, it follows that the minimum drag condition coincides with the distribution that provides $A_3 = A_5 = A_7 = A_n = 0$. Such a distribution is $\Gamma = 2sVA_1 \sin\theta$, and substituting $z = -(s/2)\cos\theta$ gives

$$\Gamma = 2sVA_1 \sqrt{1 - \left(\frac{2z}{s}\right)^2}$$

which is an elliptic spanwise distribution and in accordance with the findings in Section 5.5.3. This elliptic distribution can be pursued in an analysis involving the general Eq. (5.51) to give a far-reaching expression. Putting $A_n = 0$, $n \neq 1$ in Eq. (5.60) gives

$$\mu(\alpha - \alpha_0) = A_1 \sin\theta \left(1 + \frac{\mu}{\sin\theta}\right)$$

and rearranging gives

$$A_1 = \frac{\mu}{\sin\theta + \mu}(\alpha - \alpha_0) \qquad (5.52)$$

Now consider an untwisted wing producing an elliptic load distribution and hence minimum induced drag. By Section 5.5.3, the downwash is constant along the span and so the equivalent incidence $(\alpha - \alpha_0 - w/V)$ anywhere along the span is constant. This means that the lift coefficient is constant. Therefore, in the equation as l and Γ vary elliptically so must c:

$$\text{lift per unit span } l = \rho V\Gamma = C_L \frac{1}{2}\rho V^2 c \qquad (5.53)$$

since, on the right-hand side, $C_L \frac{1}{2}\rho V^2$ is a constant along the span. Thus

$$c = c_o \sqrt{1 - \left(\frac{2z}{s}\right)^2}$$

and the general inference emerges that for a spanwise elliptic distribution an untwisted wing will have an elliptic chord distribution, although the planform may not be a true ellipse—for example, the one-third chord line may be straight whereas, for a true ellipse, the midchord line will be straight (see Fig. 5.28).

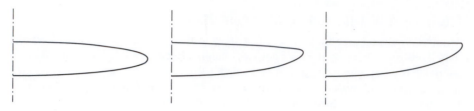

FIGURE 5.28

Three wing planforms with the same elliptic chord distribution showing that spanwise lift distribution is not the same as the shape of the wing. The rightmost shape can be flipped front to back and remain an elliptic distribution.

It should be noted that an elliptic spanwise variation can vary the other parameters in Eq. (5.53). For example, Eq. (5.53) can be rearranged as

$$\Gamma = C_L \frac{V}{2} c$$

and putting

$$C_L = a_\infty [(\alpha - \alpha_0) - \varepsilon] \text{ from Eq. (5.48)}$$

$$\Gamma \propto c a_\infty [(\alpha - \alpha_0) - \varepsilon]$$

Thus to make Γ vary elliptically, geometric twist (varying $(\alpha - \alpha_0)$) or change in airfoil section (varying α_∞ and/or α_0) may be employed in addition to, or instead of, changing the planform.

Returning to an untwisted elliptic planform, the important expression can be obtained by including $c = c_0 \sin \theta$ in μ to give

$$\mu = \mu_0 \sin \theta \text{ where } \mu_0 = \frac{c_0 a_\infty}{4s}$$

Then Eq. (5.52) gives

$$A_1 = \frac{\mu_0}{1 + \mu_0} (\alpha - \alpha_0) \tag{5.54}$$

But

$$A_1 = \frac{C_L}{\pi (AR)} \text{ from Eq. (5.38)}$$

Now

$$\frac{C_L}{(\alpha - \alpha_0)} = a = \text{three-dimensional lift slope}$$

and

$$\mu_0 = \frac{c_0 a_\infty}{4s} = \frac{a_\infty}{\pi AR}$$

for an elliptic chord distribution. Thus on substituting in Eq. (5.54) and rearranging,

$$a = \frac{a_\infty}{1 + [a_\infty/\pi(AR)]} \tag{5.55}$$

This equation gives the lift-curve slope a for a given aspect ratio (AR) in terms of the two-dimensional slope of the airfoil section used. It was derived with regard to the particular case of an elliptic planform producing minimum drag conditions and is strictly true only for this case. However, most practical airfoils diverge so little from the elliptic in this respect that Eq. (5.55) and its inverse,

$$a_\infty = \frac{a}{1 - [a/\pi(AR)]}$$

can be used with confidence in performance predictions, forecasting of wind-tunnel results, and like problems.

Probably the most famous elliptically shaped wing belongs to the Supermarine Spitfire, the British World War II fighter. It would be pleasing to report that the wing shape was chosen with due regard to aerodynamic theory. Unfortunately, it is extremely doubtful whether the Spitfire's chief designer, R. D. Mitchell, was even aware of Prandtl's theory. In fact, the elliptic wing was a logical way to meet the structural demands arising from the requirement that the Spitfire's wings carry four large machine guns. The elliptic shape allowed the wings to be as thin as possible, so the true aerodynamic benefits were rather more indirect than wing theory would suggest. Also, the elliptic shape gave rise to considerable manufacturing problems, greatly reducing the rate at which the aircraft could be made. For this reason, the Spitfire's elliptic wing was probably not a good engineering solution when all the relevant factors were taken into account [42].

5.7 SWEPT AND DELTA WINGS

Owing to the dictates of modern flight, many modern aircraft have swept-back or slender delta wings. Such wings are used for the benefits they confer in high-speed flight (see Section 6.8.2). Nevertheless, because aircraft have to land and take off, a text on aerodynamics should contain at least a brief discussion of the low-speed aerodynamics of such wings.

5.7.1 Yawed Wings of Infinite Span

For a swept-back wing of fairly high aspect ratio it is reasonable to expect that away from the wingtips the flow is similar to that over a yawed (or sheared) wing of infinite span (Fig. 5.29). To understand the fundamentals of such flow, it is helpful to use the coordinate system (x', y, z') (see Fig. 5.29). In this coordinate system the free stream has two components—$U_\infty \cos \Lambda$ and $U_\infty \sin \Lambda$—respectively perpendicular and parallel to the wing's leading edge. As the flow approaches the wing, it departs from the free-stream conditions. The total velocity field can be thought of as the superposition of the free stream and a perturbation field $(u', v', 0)$ corresponding to the departure from free-stream conditions. Note that the velocity perturbation $w' \equiv 0$ because the shape of the wing remains constant in the z' direction.

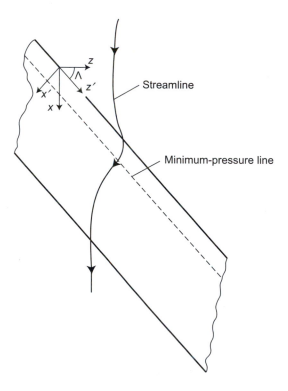

FIGURE 5.29

Streamline over a sheared wing of infinite span as viewed from above the wing.

An immediate consequence of this method of constructing the velocity field is that it can be readily shown that, unlike for infinite-span straight wings, the streamlines do not follow the free-stream direction in the x–z plane. This is an important characteristic of swept wings. The streamline direction is determined by

$$\left(\frac{dx'}{dz'}\right)_{SL} = \frac{U_\infty \cos \Lambda + u'}{U_\infty \sin \Lambda} \tag{5.56}$$

When $u' = 0$, downstream of the trailing edge and far upstream of the leading edge, the streamlines follow the free-stream direction. As the flow approaches the leading edge, the streamlines are increasingly deflected in the outboard direction, reaching a maximum deflection at the fore stagnation point (strictly a stagnation line), where $u' = U_\infty$. Thereafter the flow accelerates rapidly over the leading edge so that u' quickly becomes positive, and the streamlines are then deflected in the opposite direction—the maximum being reached on the line of minimum pressure.

Another advantage of the (x', y, z') coordinate system is that it allows the theory and data for two-dimensional airfoils to be applied to the infinite-span yawed wing. So, for example, the lift developed by the yawed wing is given by adapting Eq. (4.43) to read

$$L = \frac{1}{2}\rho(U_\infty \cos \Lambda)^2 S \left(\frac{dC_L}{d\alpha}\right)_{2D}(\alpha_n - \alpha_{0n}) \tag{5.57}$$

where α_n is the angle of attack defined with respect to the x' direction and α_{0n} is the corresponding angle of attack for zero lift. Thus

$$\alpha_n = \alpha/\cos \Lambda \tag{5.58}$$

so the lift-curve slope for the infinite yawed wing is given by

$$\frac{dC_L}{d\alpha} = \left(\frac{dC_L}{d\alpha}\right)_{2D}\cos \Lambda \cong 2\pi \cos \Lambda \tag{5.59}$$

and

$$L \propto \cos \Lambda \tag{5.60}$$

5.7.2 Swept Wings of Finite Span

The yawed wing of infinite span gives an indication of the flow over part of a swept wing, provided it has a reasonably high aspect ratio. But, as with unswept wings, three-dimensional effects dominate near the wingtips. In addition, unlike for straight wings, for swept wings three-dimensional effects predominate in the mid-span region, which has highly significant consequences for the aerodynamic

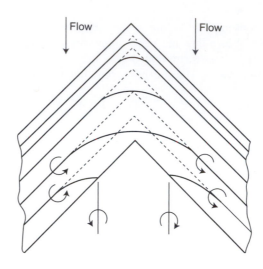

FIGURE 5.30

Vortex sheet model for a swept wing. This uses the same physics as in Prandtl's lifting-line theory, but the analytical effort required for this model is substantially greater.

characteristics of swept wings and can be demonstrated in the following way. Suppose that the simple lifting-line model that was shown in Fig. 5.19 is adapted for a swept wing merely by making a kink in the bound vortex at the mid-span position. This approach is illustrated by the broken lines in Fig. 5.30. There is, however, a crucial difference between straight and kinked bound vortex lines. For the former there is no self-induced velocity or downwash whereas for the latter there is, as is readily apparent from Eq. (5.1). Moreover, this self-induced downwash approaches infinity near the kink at mid-span. Large induced velocities imply a significant loss in lift.

Nature does not tolerate infinite velocities, and a more realistic vortex-sheet model is shown in Fig. 5.30 (full lines). It is evident from this figure that the assumptions leading to Eq. (5.23) cannot be made in the mid-span region even for high aspect ratios. Thus, for swept wings, simplified vortex-sheet models are inadmissible and the complete expression in Eq. (5.22) must be used to evaluate the induced velocity. The bound vortex lines must change direction and curve around smoothly in the mid-span region. Some may even turn back into trailing vortices before reaching mid-span. All this is likely to occur within about one chord from the mid-span. Further away, conditions approximate those for an infinite-span yawed wing. In effect, the flow in the mid-span region is more like that for a wing of low aspect ratio. Accordingly, the generation of lift will be considerably impaired in that region. This effect is evident in the comparison of pressure coefficient distributions over straight and swept wings shown in Fig. 5.31. The reduction in peak pressure over the mid-span region is shown to be very pronounced.

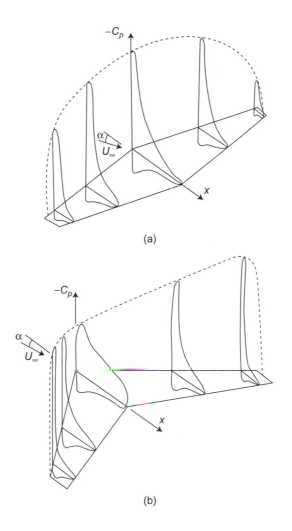

(a)

(b)

FIGURE 5.31

Comparison of pressure distributions over straight and swept-back wings.

The pressure variation depicted in Fig. 5.31(b) has important consequences. First, if it is borne in mind that suction pressure is plotted in the figure, it can be seen that there is a pronounced positive pressure gradient outward along the wing. This tends to promote flow in the direction of the wingtips, which is highly undesirable. Second, since the pressure distributions near the wingtips are much peakier than those further inboard, flow separation leading to wing stall tends to occur near the wingtips first. For straight wings, on the other hand, the opposite situation prevails and stall usually first occurs near the wing root—a much safer state of affairs because rolling moments are smaller and flow over the ailerons, needed to control rolling moments, remains largely attached. These difficulties make the design of swept wings considerably more challenging than the design of straight wings.

5.7.3 Wings of Small Aspect Ratio

For the wings of large aspect ratio considered in Sections 5.5 and 5.6, it was assumed that the flow around each wing section is approximately two-dimensional. Much the same assumption is made at the opposite extreme of small aspect ratio. The crucial difference is that now the wing sections are taken as being in the spanwise direction (see Fig. 5.32). Let the velocity components in the (x, y, z) directions be separated into free-stream and perturbation components:

$$(U_\infty \cos\alpha + u', U_\infty \sin\alpha + v', w') \tag{5.61}$$

Then let the velocity potential associated with the perturbation velocities be denoted φ'. For slender-wing theory φ' corresponds to the two-dimensional potential flow

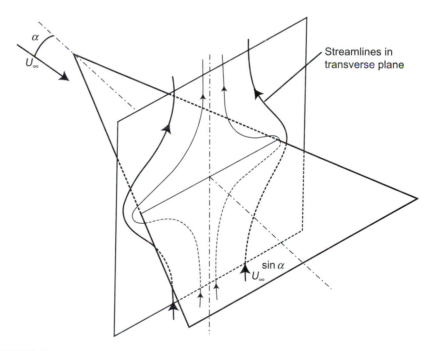

FIGURE 5.32

Approximate flow in the transverse plane of a slender delta wing within the scope of two-dimensional potential-flow theory.

around the spanwise wing section so that

$$\frac{\partial^2 \varphi'}{\partial y^2} + \frac{\partial^2 \varphi'}{\partial z^2} = 0 \tag{5.62}$$

Thus, for an infinitely thin uncambered wing, this is the flow around a two-dimensional flat plate that is perpendicular to the oncoming flow component $U_\infty \sin\alpha$. The solution to this problem can be readily obtained by means of the potential-flow theory described in Chapter 3. On the surface of the plate the velocity potential is given by

$$\varphi' = \pm U_\infty \sin\alpha \sqrt{(b/2)^2 - z^2} \tag{5.63}$$

where the plus and minus signs correspond to the upper and lower surfaces, respectively.

As previously with thin-wing theory (Eq. (4.103) for example) the coefficient of pressure depends only on $u' = \partial\varphi'/\partial x$. x does not appear in Eq. (5.62), but it does appear in parametric form in Eq. (5.63) through the variation of the wing-section width b.

Example 5.8

Consider the slender delta wing shown in Fig. 5.32. Obtain expressions for the coefficients of lift and drag using slender-wing theory.

From Eq. (5.63), assuming that b varies with x,

$$u' = \frac{\partial\varphi'}{\partial x} = \pm\frac{U_\infty \sin\alpha}{2} \frac{b}{\sqrt{b^2 - 4z^2}} \frac{db}{dx} \tag{5.64}$$

From Bernoulli's equation the surface pressure is given by

$$p = p_0 - \frac{1}{2}\rho\left(U_\infty + u' + v' + w'\right)^2 \cong p_\infty - \rho U_\infty u' + O\left(u'^2\right)$$

so the pressure difference acting on the wing is given by

$$\Delta p = \rho U_\infty^2 \frac{\sin\alpha}{2} \frac{b}{\sqrt{b^2 - 4z^2}} \frac{db}{dx}$$

The lift is obtained by integrating Δp over the wing surface and resolving perpendicularly to the free stream. Thus by changing variables to $\zeta = 2z/b$, the lift is given by

$$L = \frac{1}{2}\rho U_\infty^2 \sin\alpha \cos\alpha \int_0^c b\frac{db}{dx}\int_{-1}^1 \frac{d\zeta}{\sqrt{1-\zeta^2}} dx \tag{5.65}$$

Evaluating the inner integral first,

$$\int_{-1}^{1} \frac{d\zeta}{\sqrt{1-\zeta^2}} = \sin^{-1}(1) - \sin^{-1}(-1) = \pi$$

Therefore Eq. (5.65) becomes

$$L = \frac{\pi}{2} \sin\alpha \cos\alpha \rho U_\infty^2 \int_0^c b\frac{db}{dx} dx \qquad (5.66)$$

For the delta wing, $b = 2x \tan \Lambda$ so that

$$\int_0^c b\frac{db}{dx} dx = 4\tan^2\Lambda \int_0^c xdx = 2c^2\tan^2\Lambda$$

Eq. (5.66) then gives

$$C_L = \frac{L}{\frac{1}{2}\rho U_\infty^2 c^2 \tan\Lambda} = 2\pi \tan\Lambda \sin\alpha \cos\alpha \qquad (5.67)$$

The drag is found in a similar fashion except that now the pressure force has to be resolved in the direction of the free stream, so $C_D \propto \sin\alpha$ whereas $C_L \propto \cos\alpha$; therefore,

$$C_D = C_L \tan\alpha \qquad (5.68)$$

For small α, $\sin\alpha \cong \tan\alpha \cong \alpha$. Note also that the aspect ratio $(AR) = 4\tan\Lambda$ and that, for small α, Eq. (5.67) can be rearranged to give

$$\alpha \cong \frac{C_L}{2\pi \tan\Lambda}$$

Thus, for small α, Eq. (5.68) can also be written in the form

$$C_D = \alpha C_L = \frac{2C_L^2}{\pi (AR)^2} \qquad (5.69)$$

Note that this is exactly twice the corresponding drag coefficient given in Eq. (5.34) for an elliptic wing of high aspect ratio.

At first sight the procedure just outlined seems to violate d'Alembert's law (see Section 4.1), which states that no net force is generated by a purely potential flow around a body. For airfoils and wings it has been found necessary to introduce circulation in order to generate lift and induced drag. Circulation was not introduced in the previous procedure in any apparent way. However, it should be noted that although the flow around each spanwise wing section is assumed to be noncirculatory potential flow, the integrated effect of summing the contributions of each wing section do not, necessarily, approximate the noncirculatory potential flow around the wing as a whole. In fact, the purely noncirculatory potential flow around a chordwise wing section, at the centerline for example, looks something like what was shown in Fig. 4.1(a). By constructing the flow around the wing in the way described previously, it is assured that there is no flow reversal at the trailing edge and, in fact, a kind of Kutta condition is implicitly imposed, meaning that the flow as a whole does indeed possess circulation.

The so-called *slender-wing theory* is of limited usefulness because, for wings of small aspect ratio, the "wingtip" vortices tend to roll up and dominate the flow field for all but very small angles of attack. An example is the flow field around a slender delta wing as depicted in Fig. 5.33. In this case, the flow separates from the leading edges and rolls up to form a pair of stable vortices over the upper surface. The vortices first appear at the apex of the wing and increase in strength on moving downstream, becoming fully developed by the time the trailing edge is reached. The low pressures generated by these vortices contribute much of the lift.

Pohlhamus [43] offered a simple way to estimate the contribution of the vortices to lift on slender deltas (see Figs 5.34 and 5.35) He suggested that, at higher angles of

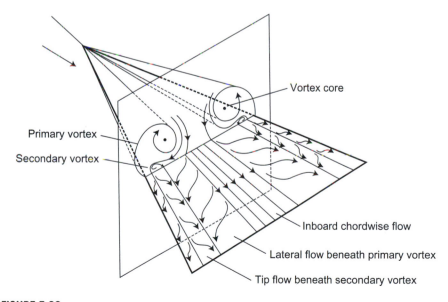

Vortex core

Primary vortex

Secondary vortex

Inboard chordwise flow

Lateral flow beneath primary vortex

Tip flow beneath secondary vortex

FIGURE 5.33

Real flow field around a slender delta wing, showing the vortex structure and surface-flow pattern.

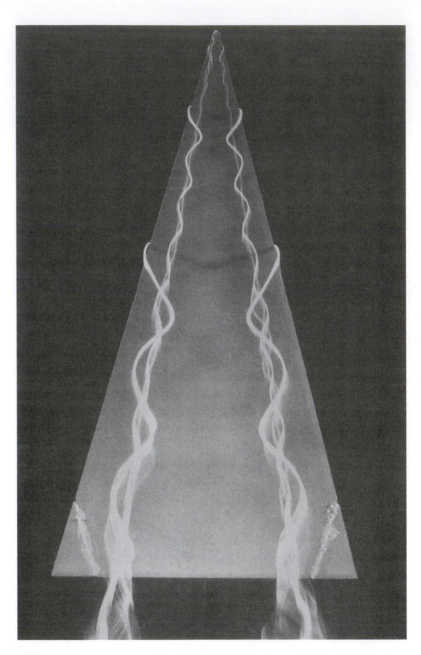

FIGURE 5.34

Vortices above a delta wing. The symmetrical pair of vortices over a delta wing are made visible by dye in water flow. The wing is made of thin plate and has a semi-vertex angle of 15 degrees. The angle of attack is 20 degrees, and the Reynolds number is 20,000 based on chord. The flow direction is from top to bottom. See also Fig. 5.33. (Source: *H. Werlé, ONERA, France.*)

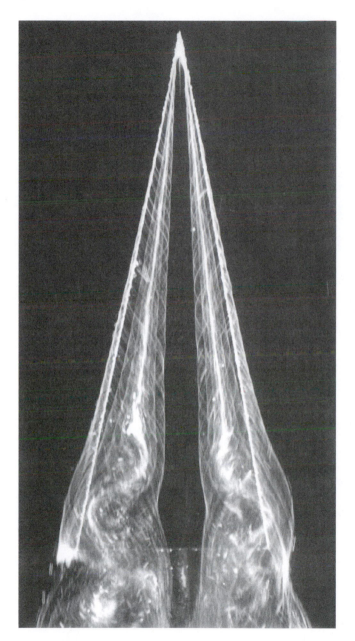

FIGURE 5.35

Vortices above a delta wing. The symmetrical pair of vortices over a delta wing are made visible by electrolysis in water flow. The wing is made of thin plate and has a semi-vertex angle of 10 degrees. The angle of attack is 35 degrees, and the Reynolds number is 3000 based on chord. The flow direction is from top to bottom. Vortex breakdown occurs at about 0.7 maximum chord. See also Fig. 5.33. (*Source: J.-L. Solignac, ONERA, France.*)

attack, the potential-flow pattern of Fig. 5.32 be replaced by a separated-flow pattern similar to that found for real flow around a flat plate oriented perpendicular to the oncoming flow. In effect, this transverse flow generates a "drag force" (per unit chord) of magnitude

$$\frac{1}{2}\rho U_\infty^2 \sin^2\alpha\, b C_{DP}$$

where C_{DP} has the value appropriate to real flow past a flat plate of infinite span placed perpendicular to the free stream (i.e., $C_{DP} \simeq 1.95$). Now this force acts perpendicularly to the wing, and the lift is the component perpendicular to the actual free stream. Thus

$$L = \frac{1}{2}\rho U_\infty^2 \sin^2\alpha\cos\alpha\, b C_{DP} \int_0^c b\,\mathrm{d}x, \text{ or } C_L = C_{DP}\sin^2\alpha\cos\alpha \qquad (5.70)$$

This component of the lift is called the *vortex* lift, and the component given in Eq. (5.67) is called the *potential-flow* lift.

The total lift acting on a slender delta wing is assumed to be the sum of the vortex and potential-flow lifts. Thus

$$C_L = \underbrace{K_P \sin\alpha\cos\alpha}_{\text{Potential-flow lift}} + \underbrace{K_V \sin^2\alpha\cos\alpha}_{\text{Vortex lift}} \qquad (5.71)$$

where K_P and K_V are coefficients that are given approximately by $2\pi\tan\Lambda$ and 1.95, respectively, or they can be determined from experimental data. The potential-flow term dominates at small angles of attack; the vortex lift, at greater angles of attack. The mechanism for generating the vortex lift is probably nonlinear to a significant extent, so there is really no theoretical justification for simply summing the two effects. Nevertheless, Eq. (5.71) fits the experimental data reasonably well, as shown in Fig. 5.36, where the separate contributions of potential-flow and vortex lift are plotted.

It can be seen from Fig. 5.36 that there is no conventional stalling phenomenon for a slender delta in the form of a sudden catastrophic loss of lift when a certain angle of attack is reached. Rather, there is a gradual loss of lift at around $\alpha = 35$ degrees. This phenomenon is not associated directly with boundary-layer separation, but is caused by the vortices bursting at locations that move progressively further upstream as the angle of attack is increased. The phenomenon of vortex breakdown is illustrated in Fig. 5.38 (see also Figs 5.35 and 5.37).

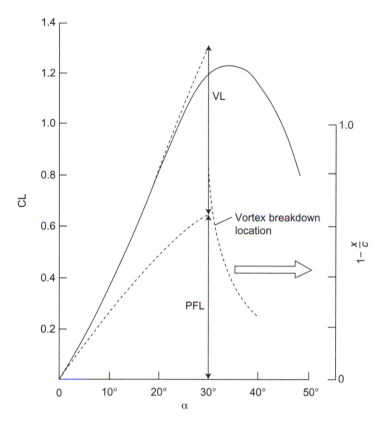

FIGURE 5.36

Typical variation of lift coefficient with angle of attack for a slender delta wing. "PFL" and "VL" denote, respectively, contributions from the first and second terms on the right-hand side of Eq. (5.71).

5.8 COMPUTATIONAL (PANEL) METHODS FOR WINGS

The application of the panel method, described in Sections 3.5 and 4.10, to entire aircraft leads to additional problems and complexities. For example, it can be difficult to define the trailing edge precisely at the wingtips and roots. In more unconventional lifting-body configurations, there may be more widespread difficulties in identifying a trailing edge for the purpose of applying the Kutta condition. In most conventional aircraft configurations, however, it is relatively straightforward to divide the aircraft into lifting and nonlifting portions (see Fig. 5.39), which allows most of the difficulties to be readily overcome. The computation of whole-aircraft aerodynamics is now routine in the aircraft industry.

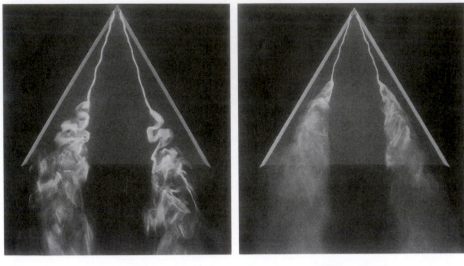

(a) *Re* = 5000   (b) *Re* = 10,000

FIGURE 5.37

Vortex breakdown above a delta wing. The wing is made of thin plate, and its planform is an equilateral triangle. The vortices are made visible by dye filaments in water flow. The angle of attack is 20 degrees. In (a), where the Reynolds number based on chord is 5000, the laminar vortices that form after separation from the leading edge abruptly thicken and initially describe a larger-scale spiral motion followed by turbulent flow. For (b), the Reynolds number based on chord is 10,000. At this higher Reynolds number the vortex breakdown moves upstream and appears to change form. The flow direction is from top to bottom. See also Fig. 5.35. (*Source: H. Werlé, ONERA, France.*)

In Section 4.10, the bound vorticity was modeled by either internal or surface vortex panels (see Fig. 4.21). Analogous methods have been used for the three-dimensional wings. There are, however, certain difficulties in using vortex panels. For example, it can often be difficult to avoid violating Helmholtz's theorem (see Section 5.2.1) when constructing vortex paneling. For this and other reasons, most modern methods are based on source and doublet distributions. They have a firm theoretical basis because Eq. (3.89b) can be generalized to lifting flows to read

$$\varphi = Ux + \underset{\text{Wing}}{\iint} \left[\frac{1}{r}\sigma + \frac{\partial}{\partial n}\left(\frac{1}{r}\right)\mu \right] dS + \underset{\text{Wake}}{\iint} \frac{\partial}{\partial n}\left(\frac{1}{r}\right)\mu \, dS \qquad (5.72)$$

where n denotes the local normal direction to the surface, and σ and μ are the source and doublet strengths, respectively.

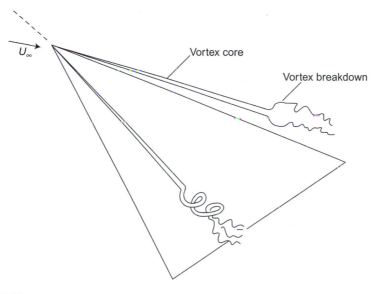

FIGURE 5.38

Schematic view of the vortex breakdown over a slender delta wing, showing both axisymmetric and spiral forms. Vortices from leading-edge strakes show similar behavior.

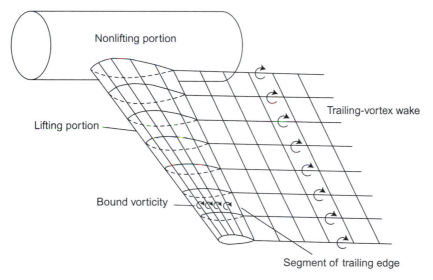

FIGURE 5.39

Panel method applied to a wing-body combination.

For a given application there is no unique mix of sources and doublets. For many methods [44] in common use, each panel of the lifting surface is assigned a distribution of constant-strength sources. The doublet distribution must now be such that it provides one additional independent parameter for each segment of the trailing edge. Once the doublet strength is known at the trailing edge, it is determined on the panels comprising the trailing vorticity. The initially unknown doublet strength at the trailing-edge segments represents the spanwise load distribution of the wing. With this arrangement each chordwise segment of wing comprises N panels and 1 trailing-edge segment. There are therefore N unknown source strengths and one unknown doublet parameter. For each chordwise segment, then, the $N + 1$ unknowns are determined by satisfying the N zero-normal-velocity conditions at the collocation points of the panels on the wing, plus the Kutta condition.

As in Section 4.10, the Kutta condition may be implemented either by adding an additional panel at the trailing edge or by requiring that the pressure be the same for the upper and lower panels defining the trailing edge (see Fig. 4.22). The former method is much less accurate since in the three-dimensional case the streamline leaving the trailing edge does not, in general, follow the trailing edge bisector. On the other hand, in the three-dimensional case, equating the pressures on the two trailing-edge panels leads to a nonlinear system of equations because the pressure is related by Bernoulli's equation to the square of the velocity. Nevertheless, this method is the preferred one if computational inaccuracy is to be avoided.

Displacement Effect

In Section 4.9, it was shown how the noncirculatory component of the flow around an airfoil could be modeled by a distribution of sources and sinks along the chord line. Similarly, in the case of the wing this flow component can be modeled by distributing sources and sinks over the entire mid-plane of the wing (Fig. 5.40). In much the same way as Eq. (4.103) was derived (referring to Fig. 5.40 for the geometric notation), it can be shown that the surface pressure coefficient at point (x_1, y_1) due to the thickness effect is given by

$$C_p = -2\frac{u'}{U} = \frac{1}{\pi} \int\limits_{-s/2}^{s/2} \int\limits_{x_i(z)}^{x_i(z)+c(z)} \frac{dy_t(x,z)}{dx} \frac{x - x_1}{\left[(x - x_1)^2 + (z - z_1)^2\right]^{3/2}} \, dx \, dz \quad (5.73)$$

where $x_1(z)$ denotes the leading edge of the wing.

In general, Eq. (5.73) is fairly cumbersome, and nowadays modern computational techniques like the panel method (see Section 5.8) are used. In the special case of wings having high aspect ratio, intuition suggests that the flow over most of the wing behaves as if it were two-dimensional. Plainly this is not a good approximation near the wingtips, where the formation of the trailing vortices leads to highly three-dimensional flow. However, away from the wingtip region, Eq. (5.73) reduces approximately to Eq. (4.103), and, to a good approximation, the C_p distributions

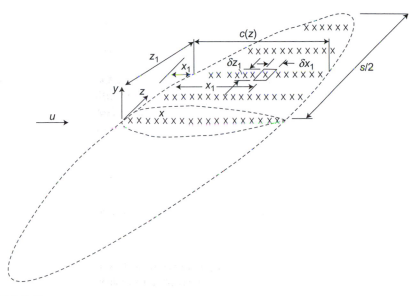

FIGURE 5.40

Modeling the displacement effect by a distribution of sources.

obtained for symmetrical airfoils can be used for the wing sections. For completeness this result is demonstrated formally here.

Change the variables in Eq. (5.73) to $\bar{x} = (x - x_1)/c$, $\bar{z}_1 = z_1/c$, and $\bar{z} = (z - z_1)/c$. Now, provided that the nondimensional shape of the wing section does not change along the span or, at any rate, changes very slowly, $S_t \equiv d(y_t/c)/d\bar{x}$ does not vary with $\bar{z}$ and the integral I_1 in Eq. (5.73) becomes

$$I_1 = \frac{1}{c} \int_0^1 S_t(\bar{x})\bar{x} \underbrace{\int_{-(s/2+z_1)/c}^{(s/2-z_1)/c} \frac{d\bar{z}}{(\bar{x}^2 + \bar{z}^2)^{3/2}} \, d\bar{x}}_{I_2}$$

To evaluate the integral I_2, change the variable to $\chi = 1/\bar{z}$ so that

$$I_2 = -\int_{-c/(s+z_1)}^{-\infty} \frac{\chi}{(\bar{z}^2\chi^2 + 1)^{3/2}} d\chi - \int_{\infty}^{c/(s-z_1)} \frac{\chi}{(\bar{z}^2\chi^2 + 1)^{3/2}} d\chi$$

$$= \frac{1}{\bar{z}^2}\left[-\frac{1}{\sqrt{\left(\frac{\bar{z}_1 c}{s/2+z}\right)^2 + 1}} - \frac{1}{\sqrt{\left(\frac{\bar{z}_1 c}{s/2-z}\right)^2 + 1}} \right]$$

For large aspect ratios $s \gg c$, so, provided z_1 is not close to $\pm s/2$ (i.e., near the wingtips),

$$\left(\frac{\bar{z}_1 c}{s/2 + z_1}\right)^2 \ll 1 \text{ and } \left(\frac{\bar{z}_1 c}{s/2 - z_1}\right)^2 \ll 1$$

giving

$$I_2 \cong -\frac{2}{z_1^2}$$

Thus Eq. (5.73) reduces to the two-dimensional result, Eq. (4.103):

$$C_p \cong -\frac{2}{\pi} \int_{x_l}^{c+x_l} \frac{dy_t}{dx} \frac{1}{x - x_1} dx \qquad (5.74)$$

5.9 EXERCISES

5.1 An airplane weighing 73.6 kN has elliptic wings 15.23 m in span. For a speed of 90 ms^{-1} in straight and level flight at low altitude, find (a) the induced drag and (b) the circulation around sections halfway along the wings. (*Answer*: 1.37 kN, 44 m^{2}s^{-1})

5.2 A glider has wings of elliptical planform with an aspect ratio of 6. The total drag is given by $C_D = 0.02 + 0.06C_L^2$. Find the change in minimum angle of glide if the aspect ratio is increased to 10.

5.3 Discuss the statement that the minimum induced drag of a wing is associated with elliptic loading, and plot a curve of induced drag coefficient against lift coefficient for a wing of aspect ratio 7.63.

5.4 Obtain an expression for the downward induced velocity behind a wing of span $2s$ at a point of distance y from the center of span, the circulation around the wing at any point y being denoted Γ. If the circulation is parabolic, that is,

$$\Gamma = \Gamma_0 \left(1 - \left(\frac{2y}{s}\right)^2\right)$$

calculate the value of the induced velocity w at mid-span, and compare it with that obtained when the same lift is distributed elliptically.

5.5 For a wing with modified elliptic loading such that, at distance y from the center of the span, the circulation is given by

$$\Gamma = \Gamma_0 \left(1 - \frac{1}{6}\left(\frac{2y}{s}\right)^2\right)\sqrt{1 - \left(\frac{2y}{s}\right)^2}$$

where s is the semi-span, show that the downward induced velocity at y is

$$\frac{\Gamma_0}{2s}\left(\frac{11}{12} + \frac{2y^2}{s^2}\right)$$

Also prove that for such a wing with aspect ratio AR, the induced drag coefficient at lift coefficient C_L is

$$C_{D_0} = \frac{628}{625}\frac{C_L^2}{\pi AR}$$

5.6 A rectangular, untwisted wing of aspect ratio 3 has an airfoil section for which the lift-curve slope is 6 in two-dimensional flow. Take the distribution of circulation across the span of a wing to be given by

$$\Gamma = 2sU\sum A_n \sin n\theta$$

and use the general theory for wings of high aspect ratio to determine the approximate circulation distribution in terms of angle of attack by retaining only two terms in the expression for circulation and satisfying the equation at $\theta = \pi/4$ and $\pi/2$. (*Answer:* $A_1 = 0.372\alpha$, $A_2 = 0.0231\alpha$)

5.7 A wing of symmetrical cross-section has an elliptical planform and is twisted so that, when the attack at the center of the span is $2°$, the circulation Γ at a distance y from the wing root is given by

$$\Gamma = \Gamma_0\left[1 - \left(\frac{2y}{s}\right)^2\right]^{3/2}$$

Find a general expression for the downwash velocity along the span, and determine the corresponding incidence at the wingtips. The aspect ratio is 7, and the lift-curve slope for the airfoil section in two-dimensional flow is 5.8. (*Answer:* $\alpha_{tip} = 0.566$ deg)

5.8 A straight wing is elliptic and untwisted, and is installed symmetrically in a wind tunnel with its centerline along the tunnel axis. If the air in the wind tunnel has an axial velocity V and also a small uniform angular velocity ω about its axis, show that the distribution of circulation along the wing is given by

$$\Gamma = 2sA_2 \sin 2\theta$$

and determine A_2 in terms of ω and the wing parameters. (The wind-tunnel wall corrections should be ignored.)

5.9 The spanwise distribution of circulation along an untwisted rectangular wing of aspect ratio 5 can be written in the form

$$\Gamma = 2sv\alpha\,[0.02340\sin\theta + 0.00268\sin 3\theta + 0.00072\sin 5\theta + 0.00010\sin 7\theta]$$

Calculate the lift and induced drag coefficients when the incidence α measured to no lift is $10°$. (*Answer*: $C_L = 0.691$, $C_{Di} = 0.0317$)

5.10 An airplane weighing 250 kN has a span of 34 m and is flying at 40 m/s with its tailplane level with its wings and at height 6.1 m above the ground. Estimate the change due to ground effect in the downwash angle at the tailplane, which is 18.3 m behind the center of pressure of the wing. (*Answer*: $3.83°$)

5.11 Three airplanes of the same type, having elliptical wings with an aspect ratio of 6, fly in vee formation at 67 m s^{-1} with $C_L = 1.2$. The followers keep a distance of one span length behind the leader and the same distance apart from each another. Estimate the percentage savings in induced drag due to flying in this formation. (*Answer*: 22%)

5.12 An airplane weighing 100 kN is 24.4 m in span. Its tailplane, which has a symmetrical section and is located 15.2 m behind the center of pressure of the wing, is required to exert zero pitching moment at a speed of 67 ms^{-1}. Estimate the required tail-setting angle assuming elliptic loading on the wings. (*Answer*: $1.97°$)

5.13 Show that the downwash angle at the center span of the tailplane is given to a good approximation by

$$\varepsilon = \text{constant} \times \frac{C_L}{AR}$$

where AR is the aspect ratio of the wing. Determine the numerical value of the constant for a tailplane located at $s/3$ behind the center of pressure, s being the wing span. (*Answer*: 0.723 for angle in radians)

5.14 An airplane weighing 100 kN has a span of 19.5 m and a wing loading of 1.925 k Nm^{-2}. The wings are rather sharply tapered, having around the center of span a circulation 10% greater than that for elliptic wings of the same span and lift. Determine the downwash angle one-quarter of the span behind the center of pressure, which is located at the quarter-chord point. The air speed is 67 m/s. Assume the trailing vorticity to be completely rolled up just behind the wings. (*Answer*: 4.67°)

5.15 Given a horseshoe vortex of strength Γ, find the induced velocity field at all points in (x, z) space where $y = 0$. The length of the head of the horseshoe vortex is b. There will be singularities (infinite velocities) on the vortex lines and on extensions of those lines, but they are absent from any real vortex, so we do not worry about them. This result will be applicable to every horseshoe vortex you may use in wing theory or elsewhere. *Hint*: Draw a top view looking straight down from the $+y$ direction, as this will be more useful for setting up your analysis than the isometric view from below is likely to be.

5.16 If we call your answer to Exercise 5.15 $w_1(x, z)$ and add in a uniform flow field of velocity $\bar{V}_\infty = U\hat{i}$, where $\hat{i}$ is the x-direction unit vector, what is the new velocity field in the $y = 0$ plane in terms of U and $w_1(x, z)$?

5.17 Simplest vortex-lattice method project: symmetric rectangular wing with no twist. Parts a and b are structured to lead you through the analysis necessary to begin computer coding in part c. This is typical of computation model development—that is, the more pencil pushing you do to begin with, the better the coding goes.

    a. For the horseshoe vortex of strength per length Γ_{ij} in the $z = 0$ plane, sketched in Fig. Ex5.17(a), determine the z-component of the velocity field induced by the vortex at a point (x, y) in the $z = 0$ plane. The coordinates of the two corners of the vortex are $(x, y) = (\xi_i, \eta_j)$ and $(x, y) = (\xi_i, \eta_{j+1})$. Remember, because of the principle of linear superposition, we can add up either the velocity potentials or the velocity fields of solutions and have a new solution.

    b. Repeat part a for a pair of horseshoe vortices located symmetrically about the x-axis, as sketched in Fig. Ex5.17(b).

    c. Create a functioning vortex-lattice code and use it for a grid-convergence study for span efficiency factor e.

        i. Find a good way to plot the results of your study.

        ii. Your code should use no toolboxes in MATLAB. Check this with the "Show Dependency Report" in the "Tools" pulldown menu in the MATLAB editor.

        iii. Running your code once should run one angle of attack and aspect ratio pair for one choice of number of spanwise panels and one choice of number of chordwise panels.

        iv. Demonstrate the code by computing the lift distribution on a rectangular wing of aspect ratio 10. Find a good way to plot the resulting lift distribution.

        v. For the same number of lattice cells (patches, points, etc.) in task iv, repeat for aspect ratios of 25 and 4. Present the analogous plots as

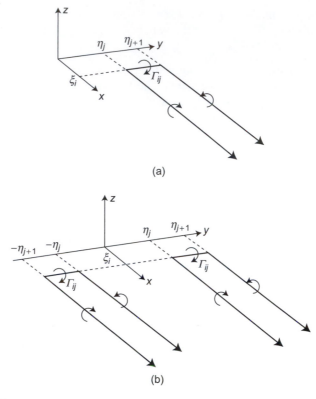

(a)

(b)

FIGURE Ex5.17

Horseshoe vortex geometry for Exercises 5.17(a) and 5.17(b).

in task iv, plus induced drag distribution plots for all three aspect ratios.

vi. Discuss what you observe in all three results.

vii. Discuss how to modify the b vector to include a linear wing twist of -2 degrees if twist is zero at the root and varies linearly with spanwise position. That is, each wingtip is down 2 degrees compared to the root, and halfway to the tip the twist is 1 degree down, for example.

Note: Keep in mind that the Kutta-Zhukovsky thereom, $L' = \rho V_\infty \Gamma$, gives us lift per span. Thus the lift on each patch of your vortex-lattice model needs to be found as the product of lift per span and span of the patch; call it Δb. That is, the lift on patch (i,j) is found, $L_{ij} = \Delta b L'_{ij} = \rho V_\infty^2 (\Delta b) \Gamma_{ij}$, but keep in mind, you are solving for Γ/V_∞ as the x vector and thus $L_{ij} = \Delta b L'_{ij} = \rho V_\infty^2 (\Delta b) \Gamma_{ij} = \rho V_\infty^2 (\Delta b) x_?$. Here the question-mark subscript indicates that the actual index number depends on how you have arranged your matrix. As there is more than one correct way to arrange the A-matrix, there is more than one correct answer for what the question-mark subscript is. However, for each choice of A-matrix organization, there is just one correct answer for what the question-mark subscript should be.

When you nondimensionalize to get the lift coefficient, you see that this works out as

$$L = 2 \sum_{j=1}^{nsp} \sum_{i=1}^{nch} \rho V_\infty^2 (\Delta b) (\Gamma_{ij}/V_\infty) = 2\rho V_\infty^2 (\Delta b) \sum_{j=1}^{nsp} \sum_{i=1}^{nch} (\Gamma_{ij}/V_\infty)$$

(remember why the 2 is in this formula?) and therefore

$$C_L = \frac{L}{qS} = \frac{2\rho V_\infty^2 (\Delta b) \sum_{j=1}^{nsp} \sum_{i=1}^{nch} (\Gamma_{ij}/V_\infty)}{(1/2)\rho V_\infty^2 bc} = \frac{4(\Delta b) \sum_{j=1}^{nsp} \sum_{i=1}^{nch} (\Gamma_{ij}/V_\infty)}{bc} \quad \text{or}$$

$$C_L = \frac{2}{nsp\, c} \sum_{j=1}^{nsp} \sum_{i=1}^{nch} (\Gamma_{ij}/V_\infty)$$

where $nsp = \frac{b/2}{\Delta b} = \frac{b}{2\Delta b}$ is the number of spanwise patches on our half-span and c is the chord. You can choose $c = 1$ when coding the problem and retain all relevant physics.

Induced drag is computed similarly, including the patch span Δb to get from induced drag per span to induced drag.

Use a similar approach for computing total induced drag as a summation of induced drag contributions on each vortex.

Compressible Flow

- Learn how elementary thermodynamics is incorporated into fluid dynamics to describe compressible flow.

- Develop one-dimensional analysis to solve for fluid and thermodynamics property changes through a steady normal shock wave.

- Apply results of the analysis of steady normal shock to understand flow through oblique shock waves.

- Learn about Prandtl-Meyer expansion flows and how they differ from oblique shocks.

Thus far in this text the study of aerodynamics has been almost exclusively restricted to incompressible flow. For incompressible flow the density and temperature of the fluid are assumed invariant throughout the flow field, and energy is exchanged between kinetic energy and pressure only. This is only suitable for the aerodynamics of low-speed flight and similar applications. As flow speeds rise, thermal energy per mass (or per volume) also begins to change, leading to a more complex fluid model and to even more interesting aerodynamic phenomena. One prominent contemporary example of fluid physics and flight research in compressible flows is known as "quiet supersonic" flight, in which the aircraft shape, typically the underside toward the nose, is designed to produce as weak a shock wave as possible in hopes of enabling regular overland supersonic flight [45]. Analyzing, computing, or measuring the impact of small changes in aircraft shape on the pressure wave miles away from the aircraft is a true challenge. The generation and transfer of heat due to viscous effects and heat conduction are also significant in the boundary layer, but these and other viscous effects are not considered in this chapter.

The chapter begins with what is known as quasi-one-dimensional (Q1D) flow. This is an approximate approach suitable for flows through ducts and nozzles when changes in the cross-sectional area are gradual. Under this circumstance, the flow variables are assumed to be uniform across a cross-section so that they vary only in the streamwise direction. Despite its apparently restrictive nature, quasi-one-dimensional-flow theory is applicable to a wide range of practical problems. Quasi-one-dimensional flow is also a good introduction to the concepts and phenomena of compressible flow, such as creation of supersonic nozzle flows and the development of shock waves. The chapter proceeds to a description of the formation of Mach and shock waves in two-dimensional flow.

6.1 INTRODUCTION

In previous chapters, the study of aerodynamics was almost exclusively restricted to incompressible flow. This theoretical model is suitable only for the aerodynamics of low-speed flight and similar applications. For incompressible flow, air density and temperature are assumed to be invariant throughout the flow field, but as flight speeds rise, greater pressure changes are generated, leading to the compression of fluid elements and causing in turn a rise in internal energy and thus temperature. The resulting variation in these flow variables throughout the flow field makes the results obtained from incompressible-flow theory less and less accurate as flow speeds rise. For example, in Section 2.3.4 we showed how use of the incompressibility assumption led to errors in estimating the stagnation pressure coefficient of 2% at $M = 0.3$, rising to 6% at $M = 0.5$ and to 28% at $M = 1$.

However, these errors in estimating pressures and other flow variables are not the most important disadvantage of the incompressible-flow model. Far more significant is the marked qualitative changes in the flow field that take place when local flow speed exceeds the speed of sound. The formation of shock waves is a particularly important phenomenon and is a consequence of the propagation of sound through the air. In incompressible flow the fluid elements are not permitted to change in volume as they pass through the flow field. Moreover, since sound waves propagate by alternately compressing and expanding the medium (see Section 1.3.7), this is tantamount to assuming an infinite speed of sound, which has important consequences when a body such as a wing moves through air otherwise at rest (or, equivalently, when a uniform flow of air approaches the body). The presence of the body is signaled by sound waves propagating in all directions. If the speed of sound is infinite, the presence of the body instantly propagates to the farthest extent of the flow field and the flow instantly begins to adjust to it.

The consequences of a finite speed of sound for the flow field will be illustrated in Fig. 6.12. Part (b) of the figure depicts the wave pattern generated when a source of disturbances (e.g., part of a wing) moves at subsonic speed into still air. It can be seen that the wave fronts are closer together in the direction of flight. Otherwise, however, the flow field is qualitatively little different from the flow field (analogous to incompressible flow) corresponding to the stationary source shown in part (a). In both cases the sound waves eventually reach all parts of the flow field (instantly in the case of incompressible flow). Contrast this with the case depicted in part (c) where a source is moving at supersonic speed. Now the waves propagating in the forward direction line up to make planar wave fronts. The flow field remains undisturbed outside the regions reached by these fronts, and waves no longer propagate to all parts of the flow field. Planar wave fronts are formed from a superposition of many sound waves and are therefore much stronger than an individual wave. In many cases they correspond to shock waves, across which the flow variables change almost discontinuously. At supersonic speeds the flow field is fundamentally wavelike, meaning that information propagates from one part of it another along wave fronts. In subsonic flow fields, in contrast, which are not wavelike, information propagates to all parts of the flow field.

This wavelike character of supersonic flow fields makes them qualitatively different from the low-speed flow fields studied in previous chapters. Furthermore, shock waves cause additional drag and other undesirable changes from the viewpoint of wing aerodynamics. As a consequence, flow compressibility has a strong influence on wing design for high-speed flight even at subsonic speeds. It might at first be assumed that shock waves only affect wing aerodynamics at supersonic flight speeds, but this is not so. Recall that local flow speeds near the point of minimum pressure over a wing are substantially greater than free-stream flow speeds. The local flow speed first reaches the speed of sound, what is known as critical flow speed. Thus, at flight speeds above critical, regions of supersonic flow appear over the wing and shock waves are generated. These lead to wave drag and other undesirable effects. For this reason, swept-back wings are used for high-speed subsonic aircraft. It is also worth pointing out that, for such aircraft, wave drag typically contributes 20% to 30% of the total drag.

In recent decades great advances were made in computing solutions to equations of motion for compressible flow. These give the design engineer freedom to explore a wide range of possible configurations. It might be thought that the ready availability of computational solutions makes approximate analytical solutions unnecessary, and, up to a point, there is some truth in this. There is certainly no longer any need to learn the more complex and involved of the traditional methods of approximation. Nevertheless, the fundamentals, as approximate analytical methods, are still of great value. First and foremost, relatively simple model flows, such as the one-dimensional flows described in Sections 6.2 and 6.3, enable the essential flow physics to be properly understood. In addition, these approaches offer approximate methods that can be used to give reasonable estimates within a few minutes. The flow physics of a normal shock in a one-dimensional flow, developed in Section 6.4, are, moreover, present in all two- and three-dimensional shocks that may be computed. These basic methods are valuable tools for diagnosing problems in computer model development and for checking the reliability of complex computer-generated solutions.

6.2 ISENTROPIC ONE-DIMENSIONAL FLOW

For many applications in aeronautics, viscous effects can be neglected to a good approximation and, moreover, no significant heat transfer occurs. Under these circumstances, thermodynamic processes are termed *adiabatic process*. Provided no other irreversible processes occur, we can also assume that the entropy remains unchanged. Such processes are termed *isentropic*. We can therefore refer to "isentropic flow." At this point it is convenient to recall the special relationships between the main thermodynamic and flow variables that hold when flow processes are isentropic.

In Section 1.3.8 we saw that, for isentropic processes $p = k\rho^\gamma$ (Eq. 1.24), where k is a constant. When this relationship is combined with the equation of state for a perfect gas (see Eq. 1.9), $p/(\rho T) = R$, where R is the specific gas constant, we can write the following relationships linking the variables at two different states (or stations) of an isentropic flow:

$$\frac{p_1}{\rho_1 T_1} = \frac{p_2}{\rho_2 T_2}, \qquad \frac{p_1}{\rho_1^\gamma} = \frac{p_2}{\rho_2^\gamma} \tag{6.1}$$

From these it follows that

$$\frac{T_2}{T_1} = \left(\frac{\rho_2}{\rho_1}\right)^{\gamma-1} = \left(\frac{p_2}{p_1}\right)^{(\gamma-1)/\gamma} \tag{6.2}$$

A useful simplified model flow is *one-dimensional* or, more precisely *quasi-one-dimensional*. This is an internal flow through ducts or passages having slowly varying cross-sections so that, to a good approximation, the flow is uniform at each cross-section and the flow variables vary only with x in the streamwise direction. Despite the seemingly restrictive nature of these assumptions, this is a very useful model flow with several important applications. Also, it provides a good way to learn about the fundamental features of compressible flow.

The equations of conservation and state for *steady* quasi-one-dimensional, adiabatic flow in differential form become

$$\frac{\mathrm{d}(\rho u A)}{\mathrm{d}x} = 0 \quad \text{(for conservation of mass)} \tag{6.3}$$

$$\frac{\mathrm{d}(\rho u^2 A)}{\mathrm{d}x} + A\frac{\mathrm{d}p}{\mathrm{d}x} = 0 \quad \text{(for momentum)} \tag{6.4}$$

$$\frac{d(c_p T + u^2/2)}{dx} = 0 \quad \text{(for energy)} \tag{6.5}$$

$$\frac{d(p/\rho T)}{dx} = 0 \quad \text{(for the equation of state)} \tag{6.6}$$

where u is the streamwise, and only non-negligible, velocity component. Expanding Eq. (6.3) and rearranging gives

$$\frac{d\rho}{\rho} + \frac{du}{u} + \frac{dA}{A} = 0 \tag{6.7}$$

Similarly, for Eq. (6.6),

$$\frac{dp}{p} - \frac{d\rho}{\rho} - \frac{dT}{T} = 0 \tag{6.8}$$

From Eq. (6.4), using Eq. (6.3),

$$\rho u A\, du + A\, dp = 0 \tag{6.9}$$

which, on dividing through by $u^2 A$ and using the identity $M^2 = u^2/a^2 = \rho u^2/(\gamma p)$,[1] Eq. (1.6d) for the speed of sound in isentropic flow becomes

$$\frac{dp}{p} = -\gamma M^2 \frac{du}{u} \tag{6.10}$$

Likewise, the energy Eq. (6.5), with $c_p T = a^2/(\gamma - 1)$, found by combining Eqs. (1.12) and (1.6d), becomes

$$\frac{dT}{T} = -(\gamma - 1)M^2 \frac{du}{u} \tag{6.11}$$

Then, combining Eqs. (6.7) and (6.8) to eliminate $d\rho/\rho$ and substituting for dp/p and dT/T gives

$$(M^2 - 1)\frac{du}{u} = \frac{dA}{A} \tag{6.12}$$

---

[1] M is the symbol for Mach number, named for the Austrian physicist Ernst Mach. It is defined as the ratio of flow speed to speed of sound at a point in a fluid flow. The Mach number of an airplane in flight is the ratio of flight speed to speed of sound in the surrounding atmosphere (see also Section 1.4.2).

Equation (6.12) indicates how the cross-sectional area of the stream tube must change to produce a change in velocity for a given mass flow. Note that a change in sign occurs at $M = 1$.

For subsonic flow, dA must be negative for an increase (i.e., positive change) in velocity. At $M = 1$, dA is zero and a throat appears in the tube. For acceleration to supersonic flow, a positive change in area is required—that is, the tube diverges from the point of minimum cross-sectional area.

Equation (6.12) indicates that a stream tube along which gas accelerates from subsonic to supersonic velocity must have a converging-diverging shape. For the reverse process, slowing down, a similar change in tube area is theoretically required, but a continuous deceleration from supersonic flow is generally not possible in practice. Note the flow structure specified in the first sentence in this paragraph—a stream tube. To think of the flow physics described by Eq. (6.12) only in the context of a supersonic nozzle is too limiting. A stream tube at the nozzle wall is indeed a useful description of the nozzle, but the stream tube could be anywhere. For example, the flow over a modern transport aircraft wing at high subsonic-flight Mach numbers usually has a region of slightly supersonic flow—hence the label "super-critical wing." The same flow physics in the stream tube in a converging-diverging nozzle exist if such a stream tube is present over a wing.

Factors other than a simple convergence control the flow in the tube, the driving pressure difference being perhaps the most important. To investigate the change in other parameters along the tube it is convenient to consider the model flow shown in Fig. 6.1. Here the air expands from a high-pressure reservoir (where the conditions may be identified by suffix 0) to a low-pressure reservoir through a constriction, or throat, in a convergent–divergent tube. With conditions at two separate points along the tube denoted by subscripts 1 and 2, respectively, the equations of state, continuity, motion, and energy become

$$\frac{p_1}{\rho_1 T_1} = \frac{p_2}{\rho_2 T_2} \tag{6.13}$$

$$\rho_1 u_1 A_1 = \rho_2 u_2 A_2 \tag{6.14}$$

$$\rho_1 u_1^2 A_1 - \rho_2 u_2^2 A_2 + p_1 A_1 - p_2 A_2 + \tfrac{1}{2}(p_1 + p_2)(A_2 - A_1) = 0 \tag{6.15}$$

$$c_p T_1 + \frac{u_1^2}{2} = c_p T_2 + \frac{u_2^2}{2} \tag{6.16}$$

The last of these equations, on taking into account the various ways in which acoustic speed can be expressed in isentropic flow (see Eqs. (1.6c, d)),

$$a = \sqrt{\frac{\gamma p}{\rho}} = \sqrt{\gamma R T} = \sqrt{(\gamma - 1) c_p T} = u / M$$

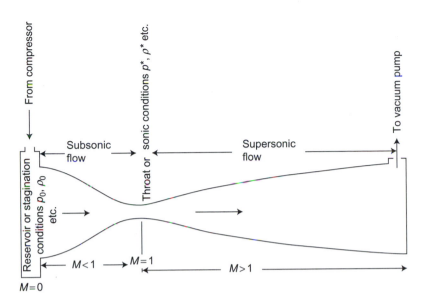

FIGURE 6.1

Typical geometry for quasi-one-dimensional isentropic expansive flow.

may be rewritten in several forms for quasi-one-dimensional isentropic flow:

$$
\left.
\begin{array}{c}
\dfrac{u_1^2}{2} + \dfrac{\gamma}{\gamma - 1}\dfrac{p_1}{\rho_1} = \dfrac{u_2^2}{2} + \dfrac{\gamma}{\gamma - 1}\dfrac{p_2}{\rho_2} \\[2ex]
\text{or}\quad \dfrac{u_1^2}{2} + \dfrac{a_1^2}{\gamma - 1} = \dfrac{u_2^2}{2} + \dfrac{a_2^2}{(\gamma - 1)} \\[2ex]
\text{or}\quad M_1^2 + \dfrac{2}{\gamma - 1} = \left(M_2^2 + \dfrac{2}{\gamma - 1}\right)\left(\dfrac{a_2}{a_1}\right)^2
\end{array}
\right\}
\qquad (6.17)
$$

6.2.1 Pressure, Density, and Temperature Ratios along a Streamline in Isentropic Flow

Occasionally, a further manipulation of Eq. (6.17) is of use. Rearrangement gives successively

$$
\frac{u_2^2 - u_1^2}{2} = \frac{\gamma}{\gamma - 1}\left(\frac{p_1}{\rho_1} - \frac{p_2}{\rho_2}\right) = \frac{\gamma}{\gamma - 1}\frac{p_2}{\rho_2}\left[\left(\frac{p_1}{p_2}\right)^{(\gamma - 1)/\gamma} - 1\right]
$$

since it follows from the relationship Eq. (6.1) for isentropic processes that $p_1/p_2 = (\rho_1/\rho_2)^\gamma$.

Finally, with $a_2^2 = (\gamma p_2/\rho_2)$, this equation can be rearranged to give

$$\frac{p_1}{p_2} = \left[1 + \frac{\gamma - 1}{2} \frac{u_2^2 - u_1^2}{a_2^2}\right]^{\gamma/(\gamma-1)} \tag{6.18}$$

If condition 1 refers to stagnation or reservoir conditions, $u_1 = 0$, $p_1 = p_0$, and then the pressure ratio is

$$\frac{p_0}{p} = \left[1 + \frac{\gamma - 1}{2}M^2\right]^{\gamma/(\gamma-1)} = \left[1 + \frac{M^2}{5}\right]^{7/2} \quad \text{for air} \tag{6.18a}$$

where the quantity without suffix refers to any point in the flow. This ratio is plotted in Fig. 6.2 over the Mach number range 0 to 4. More particularly, taking the ratio between the pressure in the reservoir and the throat, where $M = M^* = 1$ and the superscript $*$ denotes that state where $M = 1$,

$$\frac{p_0}{p^*} = \left[\frac{\gamma + 1}{2}\right]^{\gamma/(\gamma-1)} = 1.89 \quad \text{for airflow} \tag{6.18b}$$

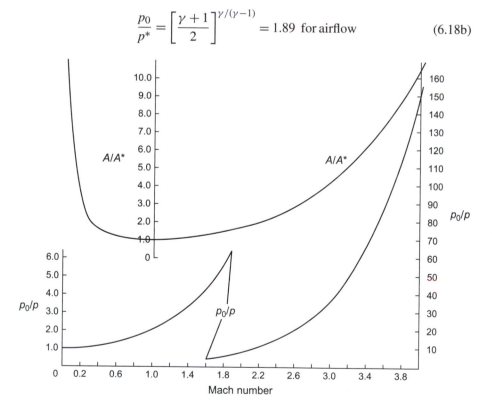

FIGURE 6.2

Plots of Eq. (6.18a) (left and right axes), and Eq. (6.25) (central axis).

Note that this is the minimum pressure ratio that will permit sonic flow. A greater value is required to produce supersonic flow. The ratios of the other parameters follow from Eqs. (6.18) and (6.2):

$$\frac{\rho_1}{\rho_2} = \left(\frac{p_1}{p_2}\right)^{1/\gamma} = \left[1 + \frac{\gamma-1}{2}\frac{u_2^2 - u_1^2}{a_2^2}\right]^{1/(\gamma-1)} \tag{6.19}$$

$$\frac{\rho_0}{\rho} = \left[1 + \frac{\gamma-1}{2}M^2\right]^{1/(\gamma-1)} \tag{6.19a}$$

$$\frac{\rho_0}{\rho^*} = \left[\frac{\gamma+1}{2}\right]^{1/(\gamma-1)} = 1.58 \text{ for airflow} \tag{6.19b}$$

and

$$\frac{T_1}{T_2} = \left(\frac{p_1}{p_2}\right)^{(\gamma-1)/\gamma} = 1 + \frac{\gamma-1}{2}\frac{u_2^2 - u_1^2}{a_2^2} \tag{6.20}$$

$$\frac{T_0}{T} = 1 + \frac{\gamma-1}{2}M^2 \tag{6.20a}$$

$$\frac{T_0}{T^*} = \frac{\gamma+1}{2} = 1.2 \text{ for airflow} \tag{6.20b}$$

Example 6.1

In streamline airflow near the upper surface of an airplane wing, the velocity just outside the boundary layer changes from 257 $km h^{-1}$ at point A near the leading edge to 466 $km h^{-1}$ at point B to the rear of A. If the temperature at A is 281 K, calculate the temperature at B. Take $\gamma = 1.4$. Also, find the value of the local Mach number at point B.

Assume that the flow outside the boundary layer closely approximates quasi-one-dimensional, isentropic flow. Then

$$\frac{T_B}{T_A} = 1 + \frac{\gamma-1}{2}\frac{u_A^2 - u_B^2}{a_A^2}$$

$$= 1 + \frac{1}{5}\frac{u_A^2 - u_B^2}{a_A^2}$$

$$a_A = 72.4\sqrt{T_A} \quad \text{and} \quad T_A = 8 + 273 = 281\,K$$

$$a_A = 1215 \text{ km h}^{-1}$$

$$M_A = \frac{u_A}{a_A} = \frac{257}{1215} = 0.212$$

$$\frac{u_B}{a_A} = \frac{466}{1215} = 0.385$$

$$\frac{T_B}{T_A} = 1 + \frac{1}{5}[0.212^2 - 0.385^2] = 0.979 = 1 - 0.021$$

Therefore,

$$T_B = 0.979 \times 281 = 275\,\text{K} = \text{temperature at B}$$

$$a_B = 72.4\sqrt{275} = 1200\,\text{km h}^{-1}$$

$$M_B = \frac{466}{1200} = 0.386$$

Example 6.2

An airfoil is tested in a high-speed wind tunnel at a Mach number of 0.7, and, at a point on the upper surface, the pressure drop is found to be numerically equal to twice the stagnation pressure of the undisturbed stream. Calculate from first principles the Mach number found just outside the boundary layer at the point in question. Take $\gamma = 1.4$.

Let suffix ∞ refer to the undisturbed stream; then, from previously,

$$\frac{\gamma}{\gamma - 1}\frac{p_\infty}{\rho_\infty} + \frac{u_\infty^2}{2} = \frac{\gamma}{\gamma - 1}\frac{p}{\rho} + \frac{u^2}{2}$$

With $\gamma = 1.4$, this becomes

$$5a_\infty^2 + u_\infty^2 = 5a^2 + u^2$$

Dividing by a^2,

$$M^2 + 5 = \frac{M_\infty^2 + 5}{(a/a_\infty)^2} \tag{a}$$

but

$$(a/a_\infty)^2 = \frac{p}{\rho}\bigg/\frac{p_\infty}{\rho_\infty}, \quad \left(\frac{p}{p_\infty}\right)^{(\gamma-1)/\gamma} = \left(\frac{p}{p_\infty}\right)^{2/7}$$

With the data given,

$$\frac{p}{p_\infty} = 1 - \gamma M_\infty^2 = 0.314$$

and Eq. (a) gives

$$M^2 + 5 = \frac{M_\infty^2 + 5}{(1 - \gamma M_\infty^2)^{2/7}} = \frac{5.49}{0.314^{2/7}} = 7.635$$

$$M^2 = 2.635 \text{ giving } M = 1.63$$

6.2.2 Ratio of Areas at Different Sections of the Stream Tube in Isentropic Flow

To this point we have not dealt with continuity (Eq. 6.41) and the mass flow $\dot{m}$. Recall that $\dot{m} = \rho u A$ for the general section at axial location x, and thus there are no subscripts here. Introducing again the reservoir or stagnation conditions and using Eq. (6.1),

$$\dot{m} = \frac{\rho}{\rho_0} \rho_0 A u = \rho_0 A u \left(\frac{p}{p_0} \right)^{1/\gamma} \qquad (6.21)$$

Now the energy equation (Eq. 6.17) gives the pressure ratio (Eq. 6.18), which, when referred to the appropriate sections of flow, is rearranged to

$$u = a_0 \sqrt{\left[1 - \left(\frac{p}{p_0} \right)^{(\gamma-1)/\gamma} \right] \frac{2}{\gamma - 1}}$$

Substituting $\sqrt{\gamma p_0 / \rho_0}$ for a_0 and introducing both into Eq. (6.21), the equation of continuity gives

$$\frac{\dot{m}}{A} = \left(\frac{p}{p_0} \right)^{1/\gamma} \sqrt{\frac{2\gamma}{\gamma - 1} p_0 \rho_0 \left[1 - \left(\frac{p}{p_0} \right)^{(\gamma-1)/\gamma} \right]} \qquad (6.22)$$

Now consider the flow at the throat, which we will identify by an asterisk as in the cross-sectional area $A*$. Flow speed at this point is $u* = a*$, and the continuity Eq. (6.22) becomes

$$\frac{\dot{m}}{A*} = \left(\frac{p*}{p_0} \right)^{1/\gamma} \sqrt{\frac{2\gamma}{\gamma - 1} p_0 \rho_0 \left[1 - \left(\frac{p*}{p_0} \right)^{(\gamma-1)/\gamma} \right]} \qquad (6.23)$$

But from Eq. (6.19b) the ratio $p*/p_0$ has the explicit value

$$\frac{p*}{p_0} = \left[\frac{\gamma + 1}{2} \right]^{-\gamma/(\gamma-1)}$$

and so

$$\frac{\dot{m}}{A*} = \sqrt{\gamma p_0 \rho_0 \left(\frac{2}{\gamma + 1} \right)^{(\gamma+1)/(\gamma-1)}} = \frac{p_0}{\sqrt{T_0}} \sqrt{\frac{\gamma}{R} \left(\frac{2}{\gamma + 1} \right)^{(\gamma+1)/(\gamma-1)}} \qquad (6.24)$$

If the constant quantities $\dot{m}, p_0$, and ρ_0 are eliminated from Eqs. (6.22) and (6.24), the area ratio becomes

$$\frac{A}{A^*} = \left(\frac{p_0}{p}\right)^{1/\gamma} \sqrt{\frac{\gamma-1}{2}\left(\frac{2}{\gamma+1}\right)^{(\gamma+1)/(\gamma-1)}\left[1-\left(\frac{p}{p_0}\right)^{(\gamma-1)/\gamma}\right]^{-1}}$$

and, substituting from Eqs. (6.20) and (6.20a),

$$\left(\frac{p_0}{p}\right)^{-\gamma/(\gamma-1)} = 1 + \frac{\gamma-1}{2}M^2$$

From Eq. (6.18a), the expression reduces to the area-Mach number relation,

$$\frac{A}{A^*} = \frac{1}{M}\left[\frac{1+\frac{\gamma-1}{2}M^2}{\frac{\gamma+1}{2}}\right]^{(\gamma+1)/[2(\gamma-1)]} = \frac{1}{M}\left[\frac{2+(\gamma-1)M^2}{\gamma+1}\right]^{(\gamma+1)/[2(\gamma-1)]}$$

$$(6.25)$$

Fig. 6.2 shows this ratio plotted against Mach number over the range $0 < M < 4$.

Example 6.3

Derive from first principles the following expression for the rate of change in stream-tube area with Mach number in the isentropic flow of air, with $\gamma = 1.4$:

$$\frac{dA}{dM} = \frac{5A}{M}\frac{M^2-1}{M^2+5}$$

Then, at the station where $M = 1.4$, the area of the stream tube is increased by 1% over the distance dx. Find the corresponding change in pressure.

From Eq. (6.25),

$$\left(\frac{A}{A^*}\right)^2 = \frac{1}{M^2}\left[\frac{2}{\gamma+1}\left(1+\frac{\gamma-1}{2}M^2\right)\right]^{(\gamma+1)/(\gamma-1)}$$

For $\gamma = 1.4$,

$$\left(\frac{A}{A^*}\right)^2 = \left(\frac{5+M^2}{6}\right)^6 \frac{1}{M^2}$$

or

$$\frac{A}{A^*} = \frac{\left(\frac{5+M^2}{6}\right)^3}{M} \tag{a}$$

Differentiating this expression with respect to M,

$$\frac{1}{A^*}\frac{dA}{dM} = \left(\frac{5+M^2}{6}\right)^3 \frac{5}{M^2}\frac{M^2-1}{M^2+5} = \frac{A}{A^*}\frac{5}{M}\frac{M^2-1}{M^2+5}$$

which, rearranged, gives

$$\frac{dA}{dM} = \frac{5A}{M}\frac{M^2-1}{M^2+5} \tag{b}$$

Similarly, from Eq. (6.18a), with $\gamma = 1.4$,

$$\frac{p}{p_0} = \left[\frac{6}{M^2+5}\right]^{7/2}$$

$$\frac{1}{p_0}\frac{dp}{dM} = -7\left(\frac{6}{M^2+5}\right)^{7/2}\frac{M}{M^2+5} = -\frac{p}{p_0}\frac{7M}{M^2+5} \tag{c}$$

Thus

$$\frac{dp}{p} = \frac{dM}{M}\frac{-7M^2}{M^2+5} \tag{d}$$

From Eq. (b) above,

$$\frac{dM}{M} = \frac{dA}{A}\frac{M^2+5}{5(M^2-1)} \tag{e}$$

Then substitute Eq. (e) into Eq. (d) to find the nondimensional pressure change in terms of the Mach number and area change:

$$\frac{dp}{p} = \frac{dA}{A}\frac{-7M^2}{5(M^2-1)} \tag{f}$$

In the previous question, $M = 1.4$, $M^2 = 1.96$, and $dA/A = 0.01$, so we can compute

$$\frac{dp}{p} = -0.0286$$

6.2.3 Velocity along an Isentropic Stream Tube

The velocity at any point may be best nondimensionalized by either the critical speed of sound a^* or the ultimate velocity c, both of which may be used as a flow parameter that is constant through isentropic flow processes.

The critical speed of sound a^* is the local acoustic speed at the throat—that is, where the local Mach number is unity. Thus the local velocity is equal to the local speed of sound, which can be expressed in terms of the reservoir conditions by applying the energy Eq. (6.17) between reservoir and throat. Thus

$$\frac{\gamma}{\gamma - 1}\frac{p_0}{\rho_0} = c_p T_0 = \frac{u^{*2}}{2} + \frac{a^{*2}}{\gamma - 1}$$

which, with $u^* = a^*$, yields

$$a^{*2} = \frac{2(\gamma - 1)}{\gamma + 1}c_p T_0 = \frac{2\gamma}{\gamma + 1}\frac{p_0}{\rho_0} \tag{6.26}$$

The ultimate velocity c is the maximum speed to which the flow can accelerate from the given reservoir conditions p_0 and T_0. It indicates a flow state in which all of the gas energy is converted to kinetic energy of linear motion. It follows from the definition that this state has zero pressure and zero temperature and thus is not practically attainable. Nevertheless, ultimate velocity is a useful reference condition for some analyses.

Again applying the energy Eq. (6.17) between reservoir and ultimate conditions,

$$\frac{\gamma}{\gamma - 1}\frac{p_0}{\rho_0} = c_p T_0 = \frac{c^2}{2},$$

so the ultimate, or maximum possible, velocity is

$$c^2 = \frac{2\gamma}{\gamma - 1}\frac{p_0}{\rho_0} = 2c_p T_0 \tag{6.27}$$

Expressing velocity as a ratio of the ultimate velocity and introducing the Mach number,

$$\frac{u^2}{c^2} = \left(\frac{Ma}{c}\right)^2 = M^2\frac{(\gamma - 1)c_p T}{2c_p T_0}$$

or

$$\frac{u}{c} = M\sqrt{\frac{\gamma - 1}{2}}\sqrt{\frac{T}{T_0}}$$

and substituting Eq. (6.20a) for T/T_0,

$$\frac{u}{c} = M\sqrt{\frac{\gamma - 1}{2 + (\gamma - 1)M^2}} \tag{6.28a}$$

Another useful reference condition is found by accelerating or decelerating the flow isentropically to the state where the Mach number is unity. This is the *sonic reference condition*, and analysis similar to the above produces

$$\frac{u}{a^*} = M\sqrt{\frac{(\gamma+1)}{2+(\gamma-1)M^2}} \tag{6.28b}$$

This ratio, often called the *characteristic Mach number*, is labeled M^* and is obviously related to $\frac{u}{c}$ by a constant factor $\sqrt{\frac{\gamma+1}{\gamma-1}}$. Both $\frac{u}{c}$ and M^* are often advantageous over M in analysis because the denominator of each is a constant throughout an isentropic process.

Example 6.4

How do the relationships in Eqs. (6.28a) and (6.28b) compare to the more familiar Mach number M? Colloquially, what do these equations "look like"? A simple MATLAB example will show how to plot these equations and provide the answer. To begin with, we want to plot the two quantities $\frac{u}{c}$ and M^* versus the Mach number M for the range $0 \le M \le 4$. We define a vector M = (0:0.05:4); and define the ratio of specific heats as gam = 7/5; Then we compute the ratios $\frac{u}{c}$ and M^*, call them uc and Mstar, respectively:

```
uc = M .* sqrt((gam-1) ./ (2+(gam-1)*M.^2));
Mstar = M .* sqrt((gam+1) ./ (2+(gam-1)*M.^2));
```

Next we plot them:

```
plot(M, uc, 'k-', M, Mstar, 'k--')
```

and label the axes and the lines:

```
xlabel('Mach number, M')
ylabel('uc or Mstar')
legend('uc', 'Mstar', 'Location', 'NW')
```

The resulting simple plot is shown in Fig. E6.4.

Note that the characteristic Mach number, M^*, is less than 1 when the Mach number is less than 1, and greater than 1 when the Mach number is greater than 1. Also, $M^* = 1$ when $M = 1$. The ratio $\frac{u}{c}$ is always less than 1 for finite Mach numbers.

6.2.4 Variation of Mass Flow with Pressure

Consider a converging tube (Fig. 6.3) exhausting a source of air at high stagnation pressure p_0 into a large receiver at a lower pressure. The mass flow in the nozzle is given by the equation of continuity (Eq. 6.22) in terms of pressure ratio p/p_0 and the tube's area of exit A, That is,

$$\frac{\dot{m}}{A} = \left(\frac{p}{p_0}\right)^{1/\gamma} \sqrt{\frac{2\gamma}{\gamma-1}p_0\rho_0\left[1-\left(\frac{p}{p_0}\right)^{(\gamma-1)/\gamma}\right]}$$

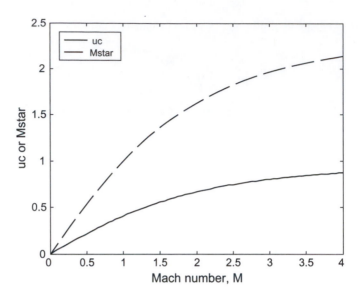

FIGURE E6.4

Comparison of $\frac{u}{c}$ and $\frac{u}{a^*}$ (Eqs. (6.28a) and (6.28b)) with the more familiar Mach number M.

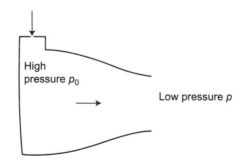

FIGURE 6.3

Compressible flow in a converging duct. Note that the walls of the duct, the inlet, and the outlet for a stream tube are used as a control volume.

A slight rearrangement allows the mass flow, in nondimensional form, to be expressed solely in terms of the pressure ratios,

$$\left(\frac{\dot{m}}{A\sqrt{p_0\rho_0}}\right)^2 = \frac{2\gamma}{\gamma-1}\left[\left(\frac{p}{p_0}\right)^{2/\gamma} - \left(\frac{p}{p_0}\right)^{(\gamma+1)/\gamma}\right] \qquad (6.29)$$

Inspection of Eq. (6.29) or Eq. (6.22) reveals the obvious fact that $\dot{m} = 0$ when $p/p_0 = 1$; that is, no flow takes place for zero pressure difference along the duct. Further inspection shows that $\dot{m}$ is also apparently zero when $p/p_0 = 0$ (i.e., under maximum pressure drop conditions). This apparent paradox is resolved by considering the behavior of the flow as p gradually decreases from p_0. As p is lowered, mass flow increases until maximum mass flow occurs.

Maximum flow may be found by the usual differentiation process (Eq. 6.29):

$$\frac{d}{d\left(\frac{p}{p_0}\right)}\left[\left(\frac{p}{p_0}\right)^{2/\gamma} - \left(\frac{p}{p_0}\right)^{(\gamma+1)/\gamma}\right] = 0 \quad \text{when} \quad \left(\frac{\dot{m}}{A\sqrt{p_0\rho_0}}\right)^2 \quad \text{is a maximum}$$

In other words,

$$\frac{2}{\gamma}\left(\frac{p}{p_0}\right)^{(2/\gamma)-1} - \frac{\gamma+1}{\gamma}\left(\frac{p}{p_0}\right)^{[(\gamma+1)/\gamma]-1} = 0$$

which gives

$$\frac{p}{p_0} = \left(\frac{2}{\gamma+1}\right)^{\gamma/(\gamma-1)} \tag{6.30}$$

Recall that this is the value of the pressure ratio for the condition $M = 1$, and thus the maximum mass flow occurs when the pressure drop is sufficient to produce sonic flow at the exit of this converging nozzle. The choking phenomenon and the associated mass flow limit remain when a diverging section is added to the nozzle to create supersonic flow.

A further decrease in pressure will not result in a further increase in mass flow, which retains its maximum value. When these conditions occur, the nozzle is said to be choked. The pressure at the exit section remains that given by Eq. (6.30), and as the pressure continues to lower, the gas expands from the exit in a supersonic jet.

The condition for sonic flow, which is the condition for maximum mass flow, implies a throat, or section of minimum area, in the stream. Further expansion to a lower pressure and acceleration to supersonic flow is accompanied by an increase in the jet's section area. It is impossible for the pressure ratio in the exit section to fall below that given by Eq. (6.30), and solutions of Eq. (6.29) have no physical meaning for values of

$$\frac{p}{p_0} < \left(\frac{2}{\gamma+1}\right)^{\gamma/(\gamma-1)}$$

Equally, it is necessary for the convergent-divergent tube of Fig. 6.1 to be choked so that the flow in the divergent portion remains supersonic. If this condition is not realized, the flow accelerates to a maximum value in the throat that is less than the local sonic speed, and then decelerates in the divergent portion, accompanied by a pressure recovery. This condition can be schematically shown by the curves A in Fig. 6.4, which plot p/p_0 against tube length for increasing mass-flow magnitudes. Curves B and C result when the tube is carrying its maximum flow. Branch B indicates the

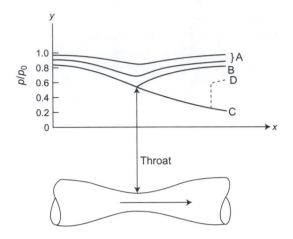

FIGURE 6.4

Possibilities for variation in pressure ratio with position in a converging-diverging stream tube.

pressure recovery resulting from the flow that has just reached sonic conditions in the throat and has then been retarded to subsonic flow in the divergent portion. This branch is the limiting curve for subsonic flow in the duct and for mass flows less than the maximum or choked value. Curve C represents the choked flow accelerated to supersonic velocities downstream of the throat.

Figure 6.4 suggests that pressure ratios of a value between those of curves B and C are unattainable at a given station downstream of the throat. This is, in fact, the case if isentropic flow conditions are to be maintained. To arrive at some intermediate value D between B and C implies that a recompression from some point on supersonic branch C is required. This is not compatible with isentropic flow, and the earlier equations no longer apply. The necessary mechanism is called shock recompression.

Example 6.5

A wind tunnel has a smallest section measuring 1.25 m × 1 m and a largest section of 4 m per side. The smallest section is vented so that it is at atmospheric pressure. A pressure tapping at the largest section is connected to an inclined tube manometer, sloped at 30 degrees to the horizontal. The manometer's reservoir is vented to the atmosphere, and its liquid has a relative density of 0.85. What is the manometer reading when the speed at the smallest section is 80 ms^{-1} and 240 ms^{-1}? In the latter case, assume that the static temperature in the smallest section is 0°C (273 K).

Denote conditions at the smallest section suffix 2, and conditions at the largest section suffix 1. Since both the smallest section and the reservoir are vented to the same pressure, the reservoir may be regarded as connected directly to the smallest section.

Area of smallest section, $A_2 = 1.25 \text{ m}^2$

Area of largest section, $A_1 = 16 \text{ m}^2$

Since the maximum speed is 80 m s^{-1}, the flow may be regarded as incompressible. Then

$$v_1 A_1 = v_2 A_2$$

That is,

$$v_1 \times 16 = 80 \times 1.25$$

giving

$$v_1 = 6.25 \text{ m s}^{-1}$$

By Bernoulli's equation, and assuming standard temperature and pressure,

$$p_1 + \frac{1}{2}\rho v_1^2 = p_2 + \frac{1}{2}\rho v_2^2$$

Then

$$p_1 - p_2 = \frac{1}{2}\rho(v_2^2 - v_1^2) = 0.613(80^2 - 6.25^2)$$

$$= 0.613 \times 86.25 \times 73.25$$

$$= 3900 \text{ N m}^{-2}$$

This is the pressure across the manometer, and therefore

$$\Delta p = \rho_m g \Delta h$$

where Δh is the head of liquid and ρ_m is the manometric fluid density:

$$3900 = (1000 \times 0.85) \times 9.807 \times \Delta h$$

This gives

$$\Delta h = 0.468 \text{ m}$$

but

$$\Delta h = r \sin \theta$$

where r is the manometer reading and θ is the manometer slope. Then

$$0.468 = r \sin 30° = \frac{1}{2}r$$

and therefore

$$r = 0.936 \text{ m}$$

The speed of 240 m s^{-1} is well into the range where compressibility becomes important, and we see how much more complicated the solution is. At the smallest section, $T_2 = 0°C = 273$ K,

$$a_2 = (1.4 \times 287.1 \times 273)^{\frac{1}{2}} = 334 \, \text{m s}^{-1}$$

From the equation for conservation of mass,

$$\rho_1 A_1 v_1 = \rho_2 A_2 v_2$$

that is,

$$\frac{\rho_1}{\rho_2} = \frac{A_2 v_2}{A_1 v_1}$$

Also, from the isentropic flow relation Eq. (6.19) for compressible flow,

$$\frac{\rho_1}{\rho_2} = \left(1 - \frac{1}{5} \frac{v_1^2 - v_2^2}{a_2^2} \right)^{2.5}$$

Equating these expressions for ρ_1/ρ_2, and putting in the known values for A_1, A_2, v_2, and a_2,

$$\frac{1.25 \times 240}{16 v_1} = \left[1 - \frac{1}{5} \frac{v_1^2 - (240)^2}{(334)^2} \right]^{2.5}$$

or

$$\frac{18.75}{v_1} = \left[1.1035 - \frac{v_1^2}{557780} \right]^{2.5}$$

A first approximation to v_1 can be obtained by assuming incompressible flow, for which

$$v_1 = 240 \times 1.25/16 = 18.75 \, \text{m s}^{-1}$$

With this value, $v_1^2/557,780 \approx 0.0006$. Therefore, the second term in the brackets on the right-hand side can be ignored, and

$$18.75/v_1 = (1.1035)^{2.5} = 1.278$$

So

$$v_1 = 14.7 \, \text{m s}^{-1}$$

which makes the ignored term even smaller.

Further,

$$\rho_1/\rho_2 = 18.75/v_1 = 1.278$$

and therefore

$$\frac{p_1}{p_2} = \left(\frac{\rho_1}{\rho_2} \right)^{\gamma} = (1.278)^{1.4} = 1.410$$

So

$$p_1 - p_2 = p_2 \left(\frac{p_1}{p_2} - 1 \right)$$

$$= 101,325 \times 0.410$$

$$= 41,500 \, \mathrm{N\,m^{-2}}$$

Then the reading of the manometer is given by

$$r = \frac{\Delta p}{\rho_m g \sin\theta} = \frac{41\,500 \times 2}{1000 \times 0.85 \times 9.807}$$

$$= 9.95 \, \mathrm{m}$$

This result shows that, for speeds of this order, a manometer using a low-density liquid is unsuitable. Historically, mercury was used, which reduced the reading to $9.95 \times 0.85/13.6 = 0.62$ m— a far more manageable figure. Often in modern labs, the toxicity of mercury rules out its use in manometers, but a number of companies manufacture electronic pressure transducers that work for this application, with a variety of resolutions and prices. A suitable transducer that converts the pressure into an electrical signal is even more probable in a modern laboratory.

Example 6.6

The manometer reading in Example 6.5 at a certain tunnel speed is 710 mm. Another tube is connected at the manometers's free end to a point on an airfoil model in the smallest section of the tunnel, while a third is connected to the total pressure tube of a Pitôt-static tube. If the liquid in the second tube is 76 mm above the zero level, calculate the pressure coefficient and the flow speed at that point on the model. Calculate also the reading, including sense, of the third tube.

(i) To find the speed of flow at the smallest section:

$$\text{Manometer reading} = 0.710\,\mathrm{m}$$

Therefore,

$$\text{Pressure difference} = 1000 \times 0.85 \times 9.807 \times 0.71 \times \frac{1}{2}$$

$$= 2960 \, \mathrm{N\,m^{-2}}$$

but

$$p_1 - p_2 = \frac{1}{2}\rho_0 \left(v_2^2 - v_1^2 \right)$$

and

$$v_1 = 1.25 v_2/16 = 5 v_2/64$$

So

$$2960 = 0.613v_2^2 \left[1 - \left(\frac{5}{64}\right)^2\right]$$

$$= 0.613v_2^2 \left[\frac{4071}{4096}\right]$$

and

$$v_2^2 = \frac{2960 \times 4096}{0.613 \times 4071} = 4860\,(\mathrm{m\,s^{-1}})^2$$

$$v_2 = 69.7\,\mathrm{m\,s^{-1}}$$

Hence, dynamic pressure at the smallest section is

$$= \frac{1}{2}\rho_0 v_2^2 = 0.613v_2^2$$

$$= 2980\,\mathrm{N\,m^{-2}}$$

(ii) To find the pressure coefficient:

Since static pressure at the smallest section = atmospheric pressure, the pressure difference between the airfoil and the tunnel stream = the pressure difference between the airfoil and the atmosphere. This pressure difference is 76 mm on the manometer, or

$$\Delta_p = 1000 \times 0.85 \times 9.807 \times 0.076 \times \tfrac{1}{2} = 317.5\,\mathrm{N\,m^{-2}}$$

Now the manometer liquid has been drawn upward from the zero level, showing that the pressure on the airfoil is less than that of the undisturbed tunnel stream; therefore, the pressure coefficient is negative:

$$C_p = \frac{p - p_0}{\frac{1}{2}\rho V^2} = \frac{-317.5}{2980} = -0.1068$$

Now

$$C_p = 1 - \left(\frac{q}{v}\right)^2$$

$$-0.1068 = 1 - \left(\frac{q}{v_2}\right)^2$$

Hence,

$$q = v_2(1 - C_p)^{1/2} = 69.7(1.1068)^{1/2}$$

$$= 73.2\,\mathrm{m\,s^{-1}}$$

(iii) The total pressure is equal to the stream static pressure plus the dynamic pressure; therefore, the pressure difference corresponding to the reading of the third tube is $\left(p_0 + \frac{1}{2}\rho v_2^2\right) - p_0$

(i.e., equal to $\frac{1}{2}\rho v_2^2$). If, then, the reading is r_3,

$$\frac{1}{2}\rho v_2^2 = \rho_m g r_3 \sin\theta$$

$$2980 = 1000 \times 0.85 \times 9.807 \times r_3 \times \frac{1}{2}$$

whence

$$r_3 = 0.712\,\text{m}$$

Since the total head is greater than the stream static pressure and therefore greater than the atmospheric pressure, the liquid in the third tube is depressed below the zero level—that is, the reading is –0.712 m.

Example 6.7

An aircraft is flying at 6100 m, where the pressure, temperature, and relative density are 46,500 N m^{-2}, –24.6°C, and 0.533, respectively. The wing is vented so that its internal pressure is uniform and equal to the ambient pressure. On the upper surface of the wing is a square inspection port 125 mm on a side. Calculate the load tending to lift the inspection panel and the air speed over the panel under the following conditions:

Mach number = 0.2; mean C_p over panel = −0.8
Mach number = 0.85; mean C_p over panel = −0.5

Under the first conditions, since the Mach number, 0.2, is small, it is a fair assumption that, although the speed over the panel is higher than the flight speed, it is still small enough for compressibility to be ignored. Then, using the definition of coefficient of pressure (see Section 1.6.3):

$$C_{p_1} = \frac{p_1 - p}{0.7pM^2}$$

$$p_1 - p = 0.7pM^2 C_{p_1} = 0.7 \times 46\,500 \times (0.2)^2 \times (-0.8)$$

$$= -1041\,\text{Nm}^{-2}$$

Load on the panel = pressure difference × area

$$= 1041 \times (0.15)^2$$

$$= 23.4\,\text{N}$$

Also,

$$C_{p_1} = 1 - \left(\frac{q}{v}\right)^2$$

that is,

$$-0.8 = 1 - \left(\frac{q}{v}\right)^2$$

whence

$$\left(\frac{q}{v}\right)^2 = 1.8 \text{ giving } \frac{q}{v} = 1.34$$

Now the speed of sound $= 20.05 \, (273 - 24.6)^{1/2} = 318 \text{ m s}^{-1}$, Therefore, the true flight speed $= 0.2 \times 318 = 63.6 \text{ m s}^{-1}$, and so the air speed over panel $q = 63.6 \times 1.34 = 85.4 \text{ m s}^{-1}$.

Under the second condition, with a Mach number of 0.85, the flow is definitely compressible. As before,

$$C_{p1} = \frac{p_1 - p}{0.7 p M^2}$$

and therefore

$$p_1 - p = 0.7 \times 46,500 \times (0.85)^2 \times (-0.5)$$

$$= -11,740 \text{ N m}^{-2}$$

Thus the load on the panel $= 11,740 \times (0.15)^2 = 264 \text{ N}$.

There are two ways to calculate the flow speed over the panel from Eq. (6.18). In one,

$$\frac{p_1}{p} = \left[1 - \frac{1}{5} \frac{q^2 - v^2}{a^2}\right]^{3.5}$$

where a is the speed of sound in the free stream:

$$\frac{p_1}{p} = \left[1 - \frac{1}{5}\left\{\left(\frac{q}{a}\right)^2 - M^2\right\}\right]^{3.5}$$

Now

$$p_1 - p = -11,740 \text{ N m}^{-2}$$

and therefore

$$p_1 = 46,500 - 11,740 = 34,760 \text{ N m}^{-2}$$

Thus substituting in the previous equation the known values $p = 46,500 \text{ N m}^{-2}$, $p_1 = 34,760$ N m^{-2}, and $M = 0.85$ leads to

$$\left(\frac{q}{a}\right)^2 = 1.124 \text{ giving } \frac{q}{a} = 1.06$$

Therefore,

$$q = 1.06a = 1.06 \times 318 = 338 \text{ m s}^{-1}$$

It is also possible to calculate the Mach number of the flow over the panel as follows. The local temperature T is found from

$$\frac{T_1}{T} = \left(\frac{p_1}{p}\right)^{1/3.5} = \left(\frac{34,760}{46,500}\right)^{1/3.5} = 0.920$$

giving

$$T_1 = 0.920\,T$$

and

$$a_1 = a(0.920)^{1/2} = 318(0.920)^{1/2} = 306\,\text{m s}^{-1}$$

Therefore, the Mach number over the panel $= 338/306 = 1.103$.

The alternative method of solution is as follows, with the total pressure of the flow denoted p_0:

$$\frac{p_0}{p} = \left[1 + \frac{1}{5}M^2\right]^{3.5} = \left[1 + \frac{(0.85)^2}{5}\right]^{3.5}$$

$$= (1.1445)^{3.5} = 1.605$$

Therefore,

$$p_0 = 46{,}500 \times 1.605 = 74{,}500\,\text{N m}^{-2}$$

As found in the first method,

$$p_1 - p = -11{,}740\,\text{N m}^{-2}$$

and

$$p_1 = 34{,}760\,\text{N m}^{-2}$$

Then

$$\frac{p_0}{p_1} = \frac{74{,}500}{34{,}760} = 2.15 = \left[1 + \frac{1}{5}M_1^2\right]^{3.5}$$

giving

$$M_1^2 = 1.22, M_1 = 1.103$$

which agrees with the result found in the first method.

The total temperature T_0 is given by

$$\frac{T_0}{T} = \left(\frac{p_0}{p}\right)^{1/3.5} = 1.1445$$

Therefore,

$$T_0 = 1.1445 \times 248.6 = 284\,\text{K}$$

Then

$$\frac{T_0}{T_1} = (2.15)^{1/3.5} = 1.244$$

giving

$$T_1 = \frac{284}{1.244} = 228\,\text{K}$$

and the local speed of sound over the panel a_1 is

$$a_1 = 20.05(228)^{1/2} = 305 \, \text{m s}^{-1}$$

Therefore, flow speed over the panel is

$$q = 305 \times 1.103 = 338 \, \text{m s}^{-1}$$

which also agrees with the answer obtained by the first method.

An interesting feature of this example is that, although the flight speed is subsonic ($M=0.85$), the flow over the panel is supersonic. This fact was used in the "wing-flow" method of transonic research, which dates from about 1940, when transonic wind tunnels were unsatisfactory. A small model was mounted on the upper surface of the wing of an airplane, which then dived at near maximum speed. As a result, the model experienced a flow that was supersonic locally. Though not very satisfactory, this was an improvement over other methods available at that time.

Example 6.8

A high-speed wind tunnel consists of a reservoir of compressed air that discharges through a convergent-divergent nozzle. The temperature and pressure in the reservoir are 400° F and 300 psi gauge, respectively. In the test section the Mach number is 2.5. If the test section is 25 in², what should the throat area be? Also calculate mass flow and pressure, temperature, and air speed in the test section.

$$\frac{A}{A^*} = \frac{1}{M}\left(\frac{5+M^2}{6}\right)^3 = \frac{1}{2.5}\left(\frac{5+6.25}{6}\right)^3 = 2.64$$

Therefore,

$$\text{Throat area} = \frac{25 \, \text{in}^3}{2.64} = 9.47 \, \text{in}^2$$

Since the throat is choked, the mass flow may be calculated from Eq. (6.24):

$$\text{Mass flow} = 0.0165 \frac{s\sqrt{°\text{R}}}{\text{ft}}\left(\frac{p_o}{\sqrt{T_o}}\right)A^*$$

Now the reservoir pressure is 300 psi gauge, or 314.7 psi absolute, while the reservoir temperature is 400°F = 857°R. Therefore,

$$\text{Mass flow} = 0.0165 \frac{s\sqrt{°\text{R}}}{\text{ft}} \times 314\,\text{psi}\left(\frac{12\,\text{in}}{\text{ft}}\right)^2 \times 9.47\,\text{in}^2\left(\frac{\text{ft}}{12\,\text{in}}\right)^2 \bigg/ \sqrt{857°\text{R}}$$

or $\dot{m} = 1.68 \, \text{slugs s}^{-1}$

In the test section,

$$1 + \frac{1}{5}M^2 = 1 + \frac{6.25}{5} = 2.25$$

Therefore,

$$p_0/p_1 = (2.25)^{3.5} = 17.1$$

and so

$$\text{Pressure in test section} = \frac{314\,\text{psi}}{17.1} = 18.4\,\text{psi}$$

Also,

$$\frac{T_0}{T_1} = 2.25$$

Therefore,

$$\text{Temperature in test section} = \frac{857°\text{R}}{2.25} = 381°\text{R} = -71°\text{F}$$

$$\text{Density in test section} = \frac{18.4\,\text{psi}\left(\frac{12\text{in}}{\text{ft}}\right)^2}{\left(1716\frac{\text{ft lb}}{\text{slug}\,°\text{R}}\right)(381°\text{R})} = 0.0041\,\text{slug ft}^{-3}$$

$$\text{Speed of sound in test section} = \sqrt{1.4\left(1716\frac{\text{ft lb}}{\text{slug}\,°\text{R}}\right)(381°\text{R})} = 957\,\text{ft s}^{-1}$$

$$\text{Air speed in test section} = 2.5 \times 957\,\text{ft s}^{-1} = 2390\,\text{ft s}^{-1}$$

As a check, the mass flow may be calculated from the above results. This gives

$$\text{Mass flow} = \rho v A = 0.0041\,\text{slug ft}^{-3} \times 2390\,\text{ft s}^{-1} \times 25\,\text{in}^2\left(\frac{\text{ft}}{12\,\text{in}}\right)^2 = 1.70\,\text{slug s}^{-1}$$

6.3 ONE-DIMENSIONAL FLOW: WEAK WAVES

To a certain extent, the results of this section were already assumed in that certain expressions for speed-of-sound propagation were used. Pressure disturbances in gaseous and other media propagate in longitudinal waves, and elementary physics provide an understanding of the phenomenon.

Consider the air in a stream tube to be initially at rest and, as a simplification, divided into layers 1, 2, 3, and so forth, normal to the possible direction of motion. A small pressure impulse felt on the face of the first layer moves the layer toward the right, and this first layer acquires a kinetic energy of uniform motion in doing so.

At the same time, since layers 1, 2, and 3 have inertia, layer 1 converts some kinetic energy of translational motion into molecular kinetic energy associated with heat (i.e., it becomes compressed). Eventually, all of the relative motion between layers 1 and 2 is absorbed in the pressure inequality between them and, in order to ease the pressure difference, the first layer acquires motion in the reverse direction. At the same time, the second layer acquires kinetic energy due to motion from left to right and proceeds to react on layer 3 in the same way. In the expansive condition, again because of inertia, layer 2 moves beyond the position it previously occupied. The necessary kinetic energy is acquired from internal conditions so that the pressure in the second layer falls below the original. Reversion to the *status quo* demands that the kinetic energy of motion to the left be transferred back to the conditions of pressure and temperature obtaining before the impulse was felt, with the fluid at rest and not displaced relative to its surroundings.

A first observation of this sequence of events is that the gas has no resultant mean displacement velocity or pressure different from that of the initial conditions; the gas serves only to transmit the pressure pulse throughout its length. Also, the displacement, and hence the velocity, pressure, and so forth, of an individual gas particle changes continuously while it is under the influence of the passing impulse.

A more graphic way of expressing the gas conditions in the tube is to plot the velocities of successive particles in the direction of movement of the impulse, at a given instant, while the impulse is passing. Another curve of the particles' velocities an instant later shows how individual particles behave.

Figure 6.5 is a typical set of curves for the passage of small pressure impulses. Immediately interesting is that an individual particle moves in the direction of wave propagation when its pressure is above the mean; it moves in the reverse direction in the expansive phase.

6.3.1 Speed of Sound (Acoustic Speed)

We now look at the changing conditions imposed on individual gas particles as the pressure pulse passes. As a first simple approach to defining the pulse and its speed of propagation, we consider the stream tube's velocity to be such that the pulse is stationary (see Fig. 6.6(a)). The flow upstream of the pulse has velocity u, density ρ, and pressure p, whereas the exit flow has these quantities changed by infinitesimal amounts to $u + \delta u, \rho + \delta \rho,$ and $p + \delta p$.

The flow is now quasi-steady, assumed inviscid, and adiabatic (since the very small pressure changes take place too rapidly for heat transfer to be significant). It takes place in the absence of external forces and is one-dimensional so that the differential equations of continuity and motion are, respectively,

$$u\frac{\partial \rho}{\partial x} + \rho\frac{\partial u}{\partial x} = 0 \qquad (6.31)$$

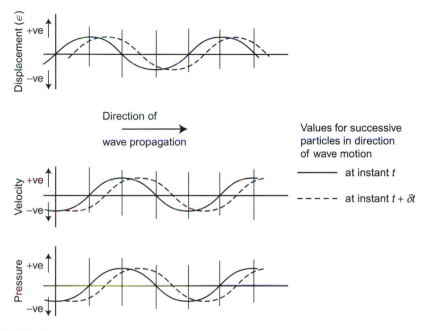

Direction of
→
wave propagation

Values for successive
particles in direction
of wave motion

——————— at instant t

- - - - - at instant $t + \delta t$

FIGURE 6.5

Relationship between displacement, mass velocity, and static pressure for simple wave motion.

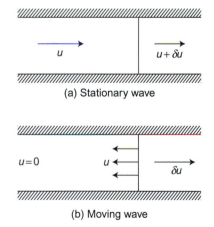

(a) Stationary wave

(b) Moving wave

FIGURE 6.6

Comparison of steady wave motion (a) and unsteady wave motion into still atmosphere (b). The two are related through a simple Newtonian transformation of the coordinate system.

and

$$u \frac{\partial u}{\partial x} = -\frac{1}{\rho} \frac{\partial p}{\partial x} \tag{6.32}$$

Eliminating $\partial u / \partial x$ from these equations leaves

$$u^2 = \frac{\partial p}{\partial \rho} \tag{6.33}$$

which implies that the speed of flow in the stream tube required to maintain a stationary pulse of weak strength is uniquely the speed given by $\sqrt{\partial p / \partial \rho}$ (see Eq. (1.6c) in Section 1.3.7).

The problem is essentially unaltered if the pulse advances at speed $u = \sqrt{\partial p / \partial \rho}$ through stationary gas and, since this is the (ideal) model of the propagation of weak pressure disturbances that are commonly sensed as sounds, the unique speed $\sqrt{\partial p / \partial \rho}$ is referred to as acoustic speed a. When the pressure-density relation is isentropic (as assumed previously), this velocity becomes (Eq. 1.6d)

$$a = \sqrt{\frac{\partial p}{\partial \rho}} = \sqrt{\frac{\gamma p}{\rho}} \tag{6.34a}$$

Recall that this is the speed the gas attains in the throat of a choked stream tube, and it follows that weak pressure disturbances do not propagate upstream into a flow where the velocity is greater than a (i.e., $u > a$ or $M > 1$).

For the usual case of air being an ideal gas, Eq. (6.34a) becomes

$$a = \sqrt{\gamma R T} \tag{6.34b}$$

MATLAB FUNCTIONS FOR COMPRESSIBLE FLOW

Many of the equations in Sections 6.1 through 6.3 are frequently used to produce numerical results. Additional equations from Sections 6.4 through 6.7 are also be used in this manner. At this point, we introduce a set of MATLAB functions, available for download from the companion site?

One beauty of MATLAB is its ease of computation with vectors of numbers, such as Mach numbers from 0 to 4 in steps of 0.1, which are defined as M = (0:0.1:4);. If we want to compute the ratio of stagnation to static temperature as a function of Mach number (Eq. 6.20a) and have defined gam as the ratio of specific heats, γ, we use the following statement:

```
Trat = 1+(gam-1)*0.5* M.^{2};
```

As this equation and others will be used numerous times in homework assignments, a set of MATLAB functions has been defined. All are listed here, and some will likely not make sense until the end of Section 6.7. An example use of the functions is given in the table, and more details can be obtained from the companion site?

Calling procedure	**Input Parameters** (Every input other than gamma and those for `OS_mt()` can be a vector.)	**Output**	**Eq. number**
`airconstants`		Printout of values	
`IF_p( )`	M vector [, gamma]	P_o/P vector	6.18a
`IF_r( )`	M vector [, gamma]	ρ_o/ρ vector	6.19a
`I_T( )`	M vector [, gamma]	T_o/T vector	6.20a
`IF_unp( )`	P_o/P vector [, gamma]	M vector	6.18a
`IF_unr( )`	ρ_o/ρ vector [, gamma]	M vector	6.19a
`IF_unT( )`	T_o/T vector [, gamma]	M vector	6.20a
`IF_Mstar( )`	M vector [, gamma]	M^* vector	6.28a
`IF_Munstar()`	M^* vector [, gamma]	M vector	6.28a
`NS_M1( )`	M_2 vector [, gamma]	M_1 vector*	6.49*
`NS_M2( )`	M_1 vector [, gamma]	M_2 vector*	6.49*
`NS_p2p1( )`	M_1 vector [, gamma]	P_2/P_1 vector	6.44
`NS_unp( )`	P_2/P_1 vector [, gamma]	M_1 vector	6.44
`NS_r2r1( )`	M_1 vector [, gamma]	ρ_2/ρ_1 vector	6.46
`NS_unr( )`	ρ_2/ρ_1 vector [, gamma]	M_1 vector	6.46
`NS_T2T1( )`	M_1 vector [, gamma]	T_2/T_1 vector	6.47
`NS_unT( )`	T_2/T_1 vector [, gamma]	M_1 vector	6.47
`NS_po2p1( )`	M_1 vector [, gamma]	P_{o2}/P_1 vector	6.57
`NS_po2po1( )`	M_1 vector [, gamma]	P_{o2}/P_{o1} vector	6.54a
`NS_s2s1( )`	M_1 vector [, gamma]	$(s_2 - s_1)/c_v$ vector	6.48a
`OS_tw( )`	Turning ,wave angle vectors [, gamma] in degrees	M_1 vector	6.83
`OS_mw( )`	M_1, wave angle in degrees vectors [, gamma]	Turning-angle vector in degrees	6.83
`OS_mt( )`	M_1, turning angle in degrees [, gamma] for the weak solution	Wave angle in degrees[†]	6.83
`OS_mt_str( )`	M_1, turning angle in degrees [, gamma] for the strong solution	Wave angle in degrees[†]	6.83
`OS_tmax( )`	Upstream Mach number vector [, gamma]	Maximum turning angle vector in degrees	None
`OS_wmax( )`	Upstream Mach number vector [, gamma]	Maximum wave angle in degrees	6.88
`PM_ang( )`	M [, gamma]	ν vector	6.68
`PM_Mach( )`	Prandtl-Meyer angle vector [, gamma]	M vector	6.68
`AreaMach( )`	M vector [, gamma]	A/A^* vector	6.25

* Returns a Mach number (not squared, to simplify use).

[†] Because an iterative solver is used in this function, vector inputs cannot be used.

Example

To find the wave angle of an oblique shock from M_1 and the turning angle:

```
turnang,
wvang = OS_mt(Mone, turnang)
```

To find the Mach number after a 12-degree Prandtl-Meyer expansion corner for Mach numbers from 1 to 2:

```
Mone = (1:0.1:2);
Mtwo = PM_Mach( 12+PM_ang(Mone))
```

To find the upstream Mach number and pressure ratio across a normal shock in helium at a downstream Mach number of 0.45:

```
Mone = NS_M1(0.45, 5/3)
Prat = NS_p2p1(Mone, 5/3)
```

To create an unadorned version of Fig. 6.24:

```
Mone = [1.4 1.8 2.2 2.6 3 3.4 3.8 100]; % Mach number
plot(OS_wmax(Mone), OS_tmax(Mone), '--k')
Machang = asin(1./Mone)*(180/pi); % Mach angle
hold on
for i = 1:length(M);
    % Create a wave-angle vector, degrees.
    wave = Machang(i) + [0:0.001:1]*(90-Machang(i))
    plot(wave, OS_mw(Mone(i), wave), '-k')
end
hold off
```

It is hoped that students of compressible flow will make frequent use of these functions to visualize the behavior of air by plotting the sometimes lengthy equations in this chapter or the results of analyses in assignments.

6.4 ONE-DIMENSIONAL FLOW: PLANE NORMAL SHOCK WAVES

In the previous section, the behavior of gas when transmitting waves of infinitesimal amplitude was considered, and the waves were shown to travel at an (acoustic) speed of $a = \sqrt{\partial p / \partial \rho}$ relative to the gas, while the gas properties of pressure, density, and similar vary continuously.

If a disturbance of large amplitude (e.g., a rapid pressure rise) is set up, there are almost immediate physical limitations to its continuous propagation. The accelerations of individual particles required for continuous propagation cannot be sustained, and a pressure front or discontinuity builds up. This front, known as a *shock wave*, travels through the gas at a speed always in excess of the acoustic speed a; together with the pressure jump, the density, temperature, and entropy of the gas increase suddenly while the normal velocity drops.

Useful and quite adequate expressions for the change in these flow properties across the shock can be obtained by assuming that the shock is of zero thickness. In fact, the shock wave is of finite thickness— a few molecular mean-free-path lengths in magnitude, the number depending on the initial gas conditions and shock intensity. For air-breathing flight, this approximation is very good; some reentry flows or flight in the Martian atmosphere have thicker shocks.

6.4.1 One-Dimensional Properties of Normal Shock Waves

Consider the flow model in Fig. 6.7(a) in which a plane shock advances from right to left with velocity u_1 into a region of still gas. Behind the shock, the velocity suddenly increases to some value u in the direction of the wave. It is convenient to superimpose on the system a velocity u_1 from left to right to bring the shock stationary relative to the walls of the tube through which gas flows undisturbed at u_1 (Fig. 6.7b). The shock becomes a stationary discontinuity into which gas flows with uniform conditions p_1, ρ_1, u_1, and so forth, and from which it flows with uniform conditions p_2, ρ_2, u_2,.... We assume that the gas is inviscid, and non-heat-conducting, so that the flow is adiabatic up to and beyond the discontinuity.

The equations of state and conservation for unit area of shock wave are, respectively,

$$\frac{p_1}{\rho_1 T_1} = \frac{p_2}{\rho_2 T_2} \tag{6.35}$$

and

$$\dot{m} = \rho_1 u_1 = \rho_2 u_2 \tag{6.36}$$

Momentum, in the absence of external and dissipative forces, is

$$p_1 + \rho_1 u_1^2 = p_2 + \rho_2 u_2^2 \tag{6.37}$$

Energy is

$$c_p T_1 + \frac{u_1^2}{2} = c_p T_2 + \frac{u_2^2}{2} = c_p T_0 \tag{6.38}$$

6.4.2 Pressure-Density Relations across the Shock

Equation (6.38) may be rewritten from, say, Eq. (6.27) as

$$\frac{u_1^2}{2} + \frac{\gamma}{\gamma - 1} \frac{p_1}{\rho_1} = c_p T_0 = \frac{u_2^2}{2} + \frac{\gamma}{\gamma - 1} \frac{p_2}{\rho_2}$$

which, on rearrangement, gives

$$\frac{\gamma}{\gamma - 1} \left(\frac{p_1}{\rho_1} - \frac{p_2}{\rho_2} \right) = \frac{1}{2} (u_2 - u_1)(u_2 + u_1) \tag{6.39}$$

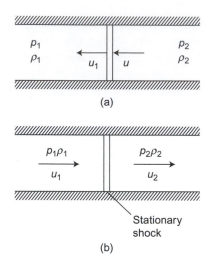

FIGURE 6.7

Comparison of unsteady normal shock (a) and steady normal shock (b). Because they differ only by a Newtonian transformation, all ratios of static thermodynamic quantities are unchanged; stagnation temperatures and densities are different.

From the continuity (Eq. 6.36),

$$u_2 + u_1 = \dot{m}\left(\frac{1}{\rho_2} + \frac{1}{\rho_1}\right)$$

and from the momentum Eq. (6.37),

$$u_2 - u_1 = \frac{1}{\dot{m}}(p_1 - p_2)$$

Substituting for both of these in the rearranged energy Eq. (6.39),

$$\frac{\gamma}{\gamma - 1}\left(\frac{p_1}{\rho_1} - \frac{p_2}{\rho_2}\right) = \frac{1}{2}(p_1 - p_2)\left(\frac{1}{\rho_1} + \frac{1}{\rho_2}\right) \tag{6.40}$$

and this, rearranged by isolating the respective pressure and density ratios, gives the *Rankine-Hugoniot relations*:

$$\frac{p_2}{p_1} = \frac{\dfrac{\gamma + 1}{\gamma - 1}\dfrac{\rho_2}{\rho_1} - 1}{\dfrac{\gamma + 1}{\gamma - 1} - \dfrac{\rho_2}{\rho_1}} \tag{6.41}$$

$$\frac{\rho_2}{\rho_1} = \frac{\dfrac{\gamma+1}{\gamma-1}\dfrac{p_2}{p_1}+1}{\dfrac{\gamma+1}{\gamma-1}+\dfrac{p_2}{p_1}} \tag{6.42}$$

Taking $\gamma = 1.4$ for air, these equations become

$$\frac{p_2}{p_1} = \frac{6\dfrac{\rho_2}{\rho_1}-1}{6-\dfrac{\rho_2}{\rho_1}} \tag{6.42a}$$

and

$$\frac{\rho_2}{\rho_1} = \frac{6\dfrac{p_2}{p_1}+1}{6+\dfrac{p_2}{p_1}} \tag{6.42b}$$

Eqs. (6.42) and (6.42a) show that, as the value of ρ_2/ρ_1 tends to $(\gamma+1)/(\gamma-1)$ (or 6 for air), p_2/p_1 tends to infinity, which indicates that the maximum possible density increase through a shock wave is about six times the undisturbed density.

6.4.3 Static Pressure Jump across a Normal Shock

From the momentum (Eq. 6.37) and using the continuity (Eq. 6.36),

$$\frac{p_2-p_1}{p_1} = \frac{\rho_1 u_1^2 - \rho_2 u_2^2}{p_1} = \frac{\rho_1 u_1^2}{p_1}\left[1-\frac{u_2}{u_1}\right]$$

or

$$\frac{p_2}{p_1}-1 = \gamma M_1^2\left[1-\frac{u_2}{u_1}\right]$$

but from continuity $u_2/u_1 = \rho_1/\rho_2$, and from the Rankine-Hugoniot relations, ρ_2/ρ_1 is a function of (p_2/p_1). Thus, by substitution,

$$\frac{p_2}{p_1}-1 = \gamma M_1^2\left[1-\left\{\left(\frac{\gamma+1}{\gamma-1}\frac{p_1}{p_2}+1\right)\bigg/\left(\frac{\gamma+1}{\gamma-1}+\frac{p_1}{p_2}\right)\right\}\right]$$

Isolating the ratio p_2/p_1 and rearranging gives

$$\frac{p_2}{p_1} = \frac{2\gamma M_1^2}{\gamma+1} - \frac{\gamma-1}{\gamma+1} \tag{6.43}$$

Note that, for air,

$$\frac{p_2}{p_1} = \frac{7M_1^2 - 1}{6} \tag{6.43a}$$

Expressed in terms of the downstream or exit Mach number M_2, the pressure ratio can be derived in a similar manner:

$$\frac{p_1}{p_2} = \frac{2\gamma M_2^2 - (\gamma - 1)}{\gamma + 1} \tag{6.44}$$

or

$$\frac{p_1}{p_2} = \frac{7M_2^2 - 1}{6} \quad \text{for air} \tag{6.44a}$$

6.4.4 Density Jump across the Normal Shock

Using the previous results, substituting for p_2/p_1 from Eq. (6.43) in the Rankine–Hugoniot relations, Eq. (6.42) gives

$$\frac{\rho_2}{\rho_1} = \frac{\left(\frac{\gamma+1}{\gamma-1} \left[\frac{2\gamma M_1^2 - (\gamma-1)}{\gamma+1} \right] + 1 \right)}{\left(\frac{\gamma+1}{\gamma-1} + \left[\frac{2\gamma M_1^2 - (\gamma-1)}{\gamma+1} \right] \right)}$$

or, rearranged,

$$\frac{\rho_2}{\rho_1} = \frac{(\gamma+1)M_1^2}{2 + (\gamma-1)M_1^2} \tag{6.45}$$

For air $\gamma = 1.4$ and

$$\frac{\rho_2}{\rho_1} = \frac{6M_1^2}{5 + M_1^2} \tag{6.45a}$$

Reversing to give the ratio in terms of the exit Mach number gives

$$\frac{\rho_1}{\rho_2} = \frac{(\gamma+1)M_2^2}{2 + (\gamma-1)M_2^2} \tag{6.46}$$

For air

$$\frac{\rho_1}{\rho_2} = \frac{6M_2^2}{5 + M_2^2} \tag{6.46a}$$

6.4.5 Temperature Rise across the Normal Shock

Directly from the equation of state and Eqs. (6.43) and (6.45),

$$\frac{T_2}{T_1} = \frac{p_2}{p_1} \bigg/ \frac{\rho_2}{\rho_1}$$

$$\frac{T_2}{T_1} = \left(\frac{2\gamma M_1^2 - (\gamma - 1)}{\gamma + 1}\right) \left(\frac{2 + (\gamma - 1)M_1^2}{(\gamma + 1)M_1^2}\right) \tag{6.47}$$

For air

$$\frac{T_2}{T_1} = \frac{7M_1^2 - 5/M_1^2 + 34}{36} \tag{6.47a}$$

Since the flow is non-heat-conducting, the total (or stagnation) temperature remains constant.

6.4.6 Entropy Change across the Normal Shock

Recalling the basic Eq. (1.32),

$$e^{\Delta S/c_V} = \left(\frac{T_2}{T_1}\right)^{\gamma} \left(\frac{p_1}{p_2}\right)^{\gamma - 1} = \left(\frac{p_2}{p_1}\right) \left(\frac{\rho_1}{\rho_2}\right)^{\gamma} \quad \text{from the equation of state}$$

which, on substituting for the ratios from the previous sections, may be written as a sum of the natural logarithms:

$$\frac{\Delta S}{c_V} = \ln \frac{2\gamma M_1^2 - (\gamma - 1)}{\gamma + 1} + \gamma \ln(2 + (\gamma - 1)M_1^2) - \gamma \ln M_1^2 \tag{6.48}$$

But what can be learned from this result? Rearrange Eq. (6.48) in terms of the variable $(M_1^2 - 1)$:

$$\frac{\Delta S}{c_V} = \ln \left[1 + \frac{2\gamma}{\gamma + 1}(M_1^2 - 1)\right] + \gamma \ln \left[1 + \frac{\gamma - 1}{\gamma + 1}(M_1^2 - 1)\right]$$
$$- \gamma \ln[1 + (M_1^2 - 1)]$$

On expanding these logarithms and collecting like terms, the first and second powers of $(M_1^2 - 1)$ vanish, leaving a converging series commencing with the term

$$\frac{\Delta S}{c_V} = \frac{2\gamma (\gamma - 1)}{(\gamma + 1)^2} \frac{(m_1^2 - 1)^3}{3} \tag{6.48a}$$

Inspection of this equation shows that (1) for the second law of thermodynamics to apply (i.e., for ΔS to be positive), M_1 must be greater than unity and an expansion shock is not possible; and (2) for values of M_1 close to (but greater than) unity, the values of the change in entropy are small and rise only slowly for increasing M_1. The appropriate curve in Fig. 6.9 shows that, for quite moderate supersonic Mach numbers (i.e., up to about $M_1 = 2$), we have a reasonable approximation to the flow conditions assuming an isentropic state.

6.4.7 Mach Number Change across the Normal Shock

Multiplying the previous pressure (or density) ratio equations together gives the Mach number relationship directly:

$$\frac{p_2}{p_1} \times \frac{p_1}{p_2} = \frac{2\gamma M_1^2 - (\gamma - 1)}{\gamma + 1} \times \frac{2\gamma M_2^2 - (\gamma - 1)}{\gamma + 1} = 1$$

Rearrangement gives the exit Mach number:

$$M_2^2 = \frac{(\gamma - 1)M_1^2 + 2}{2\gamma M_1^2 - (\gamma - 1)} \tag{6.49}$$

For air

$$M_2^2 = \frac{M_1^2 + 5}{7M_1^2 - 1} \tag{6.49a}$$

Inspection of these last equations shows that M_2 has upper and lower limiting values:

$$\text{For } M_1 \to \infty \quad M_2 \to \sqrt{\frac{(\gamma - 1)}{2\gamma}} = (1/\sqrt{7} = 0.378 \text{ for air})$$

$$\text{For } M_1 \to 1 \quad M_2 \to 1$$

Thus the exit Mach number from a normal shock wave is always subsonic and, for air, has values between 1 and 0.378.

6.4.8 Velocity Change across the Normal Shock

The velocity ratio is the inverse of the density ratio, since by continuity $u_2/u_1 = \rho_1/\rho_2$. Therefore, directly from Eqs. (6.45) and (6.45a),

$$\frac{u_2}{u_1} = \frac{2 + (\gamma - 1)M_1^2}{(\gamma + 1)M_1^2} \tag{6.50}$$

or for air

$$\frac{u_2}{u_1} = \frac{5 + M_1^2}{6M_1^2} \qquad (6.50a)$$

Of added interest is the following development. From the energy equations, with $c_p T$ replaced by $[\gamma/(\gamma - 1)]p/\rho$, p_1/ρ_1 and p_2/ρ_2 are isolated:

$$\frac{p_1}{\rho_1} = \frac{\gamma - 1}{\gamma}\left(c_p T_0 - \frac{u_1^2}{2}\right) \qquad \text{ahead of the shock}$$

and

$$\frac{p_2}{\rho_2} = \frac{\gamma - 1}{\gamma}\left(c_p T_0 - \frac{u_2^2}{2}\right) \qquad \text{downstream of the shock}$$

The momentum Eq. (6.37) is rearranged with $\rho_1 u_1 = \rho_2 u_2$ from the continuity Eq. (6.36) to

$$u_1 - u_2 = \frac{p_2}{\rho_2 u_2} - \frac{p_1}{\rho_1 u_1}$$

Substituting from the preceding line,

$$u_1 - u_2 = \frac{\gamma - 1}{\gamma}\left[(u_1 - u_2)\left(\frac{1}{2} + \frac{c_p T_0}{u_1 u_2}\right)\right]$$

Disregarding the uniform flow solution of $u_1 = u_2$, the conservation of mass, motion, and energy apply for this flow when

$$u_1 u_2 = \frac{2(\gamma - 1)}{\gamma + 1} c_p T_0 \qquad (6.51)$$

That is, the product of normal velocities through a shock wave is a constant that depends on the stagnation conditions of the flow and is independent of shock strength. Further, recall from Eq. (6.26) that

$$\frac{2(\gamma - 1)}{\gamma + 1} c_p T_0 = a^{*2}$$

where a^* is the critical speed of sound and an alternative parameter for expressing the gas conditions. Thus, in general, across the shock wave,

$$u_1 u_2 = a^{*2} \qquad (6.52)$$

Applying the definition of characteristic Mach number, Eq. (6.28a), alters Eq. (6.52) to be

$$M_1^* M_2^* = 1 \tag{6.52a}$$

Then, from Example 6.4 note that, when $M < 1$, $M^* < 1$, and when $M > 1$, $M^* > 1$. Therefore, Eq. (6.52a) states that a supersonic upstream flow results in a subsonic downstream flow. The other possibility, subsonic upstream and supersonic downstream, was ruled out in Section 6.4.6.

6.4.9 Total Pressure Change across the Normal Shock

From the previous sections, we see that a finite entropy increase occurs in the flow across a shock wave, implying a degradation of energy. Since, in the flow as a whole, no heat is acquired or lost, the total temperature (total enthalpy) is constant and the dissipation manifests itself as a loss in total pressure. Total pressure is defined as that obtained by bringing gas to rest isentropically.

Now the model flow of a uniform stream of gas of unit area flowing through a shock is extended upstream by assuming the gas to now have the conditions of suffix 1 by expansion from a reservoir of pressure p_{01} and temperature T_0, and extended downstream by bringing the gas to rest isentropically to a total pressure p_{02} (Fig. 6.8).

Isentropic flow from the upstream reservoir to just ahead of the shock gives, from Eq. (6.18a),

$$\frac{p_{01}}{p_1} = \left[1 + \frac{\gamma - 1}{2} M_1^2\right]^{\gamma/(\gamma - 1)} \tag{6.53}$$

and, from just behind the shock to the downstream reservoir,

$$\frac{p_{02}}{p_2} = \left[1 + \frac{\gamma - 1}{2} M_2^2\right]^{\gamma/(\gamma - 1)} \tag{6.54}$$

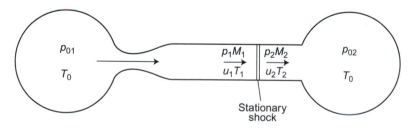

FIGURE 6.8

One geometry for the study of total pressure changes across a steady normal shock wave. The two reservoirs are considerably larger than as drawn.

Recall Eq. (6.43) is

$$\frac{p_2}{p_1} = \frac{2\gamma M_1^2 - (\gamma - 1)}{\gamma + 1}$$

and recall Eq. (6.49) is

$$M_2^2 = \frac{(\gamma - 1)M_1^2 + 2}{2\gamma M_1^2 - (\gamma - 1)}$$

These four expressions, by division and substitution, give successively

$$\frac{p_{01}}{p_{02}} = \frac{p_1}{p_2}\left[\frac{1 + \frac{\gamma-1}{2}M_1^2}{1 + \frac{\gamma-1}{2}M_2^2}\right]^{\gamma/(\gamma-1)} = \frac{p_1}{p_2}\left[\frac{2 + (\gamma - 1)M_1^2}{2 + (\gamma - 1)M_2^2}\right]^{\gamma/(\gamma-1)} \qquad (6.54a)$$

From entropy considerations, $M_1 > 1$ for a shock wave, so plotting Eq. (6.54a) shows that the total pressure always drops through a shock wave. The two phenomena, total pressure drop and entropy increase, are in fact related, as may be seen in the following (see Fig. 6.9):

Recalling Eq. (1.32) for entropy,

$$e^{\Delta S/c_v} = \frac{p_2}{p_1}\left(\frac{\rho_1}{\rho_2}\right)^\gamma = \frac{p_{02}}{p_{01}}\left(\frac{\rho_{01}}{\rho_{02}}\right)^\gamma$$

since

$$\frac{p_1}{\rho_1^\gamma} = \frac{p_{01}}{\rho_{01}^\gamma}, \text{ etc.}$$

But T_0 across the shock is constant; therefore, from the equation of state $p_{01}/\rho_{01} = p_{02}/\rho_{02}$, and entropy becomes

$$e^{\Delta S/c_v} = \left(\frac{p_{01}}{p_{02}}\right)^{\gamma-1}$$

$$\ln\frac{p_{01}}{p_{02}} = \frac{\Delta S}{c_V(\gamma - 1)} \qquad (6.55)$$

6.4.10 Pitôt Tube Equation

The pressure registered by a small open-ended tube facing a supersonic stream is effectively the "exit" (from the shock) total pressure p_{02}, since the bow shock wave is normal to the axial streamline and terminates in the tube's stagnation region. That is, the axial flow into the tube is assumed to be brought to rest at pressure p_{02} from the subsonic flow p_2 behind the wave, after it has been compressed from the supersonic region p_1 ahead of the wave (Fig. 6.10). In some applications, this is referred to as the *static pressure* of the free or undisturbed supersonic stream p_1, and is evaluated

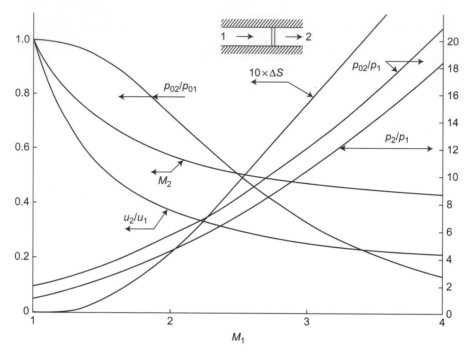

FIGURE 6.9

Property ratios versus upstream Mach number for a steady normal shock.

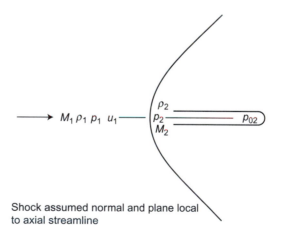

Shock assumed normal and plane local
to axial streamline

FIGURE 6.10

Operation of a Pitôt tube in a supersonic flow. We assume from the symmetry of the probe
tip that the shock wave upstream of the tip is normal to the free-stream flow.

in terms of the free-stream Mach number, thus providing a method of determining the undisturbed Mach number as follows.

From the normal shock static-pressure ratio Eq. (6.43),

$$\frac{p_2}{p_1} = \frac{2\gamma M_1^2 - (\gamma - 1)}{\gamma + 1}$$

From isentropic flow relations,

$$\frac{p_{02}}{p_2} = \left[1 + \frac{\gamma - 1}{2} M_2^2\right]^{\gamma/(\gamma-1)}$$

Dividing these expressions and recalling Eq. (6.64),

$$M_2^2 = \frac{(\gamma - 1)M_1^2 + 2}{2\gamma M_1^2 - (\gamma - 1)}$$

the required pressure ratio becomes

$$\frac{p_{02}}{p_1} = \frac{\left[\frac{\gamma+1}{2}M_1^2\right]^{\gamma/(\gamma-1)}}{\left[\frac{2\gamma M_1^2 - (\gamma-1)}{\gamma+1}\right]^{1/(\gamma-1)}} \tag{6.56}$$

and is sometimes called *Rayleigh's supersonic Pitôt tube equation.*

The observed curvature of the detached shock wave on supersonic Pitôt tubes was once thought sufficient to bring the plane-wave theory into question, but the agreement with theory reached in experimental work was well within the accuracy expected for that type of test and was held to support the assumption of a normal shock ahead of the wave [46].

6.4.11 Converging-Diverging Nozzle Operations

Now that we have studied the physics of normal shock waves and the mathematics that describe those physics, we can revisit the topic of converging-diverging nozzles and expand on our previous, isentropic, description of such flows. Normal shocks can be present in supersonic flow regions, which are only in the diverging section. Our assumption of quasi-one-dimensional flow limits our analysis to normal shocks. More detailed nozzle-flow analysis or design traditionally was by method of characteristics and boundary-layer estimates. New computational fluid dynamics tools provide even more detailed information.

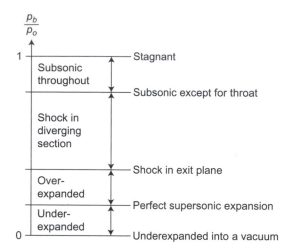

FIGURE 6.11

Illustration of several different cases of relevant motion of sound waves and the sound emitter.

One simple geometry, as shown in Fig. 6.11, suffices to describe all nozzle-flow conditions of interest here. A converging-diverging nozzle is supplied with air at stagnation pressure p_0. If exhausts into an environment (the atmosphere, a plumbing system, outer space, etc.) with an ambient pressure that we call "back pressure" p_b. Clearly, for air to flow from the inlet to the outlet, $p_b < p_0$. The throat area of the nozzle is A_T, and the exit area is A_e.

Textbooks provide a number of descriptions of the four flow types in converging-diverging nozzles. A direct, simple description is presented here. The four modes of nozzle operation are fully subsonic, normal shock in the diverging section, overexpanded, and underexpanded. They are defined by five values of back pressure p_b, which are often expressed as the ratio of back pressure to supply stagnation pressure $\frac{p_b}{p_0}$. A student preparing for an exam may consider these five pressures as the five fingers of a hand and the four modes of operation as the four spaces between them.

For all converging-diverging nozzles, the maximum value of $\frac{p_b}{p_0}$ is always 1 and the minimum is always 0. This leaves three other pressure ratios.

The two pressure ratios adjacent to the $\frac{p_b}{p_0} = 1$ and $\frac{p_b}{p_0} = 0$ limits (thumb and little finger for a hand placed palm up on the table) are subsonic and supersonic isentropic flow, respectively, through the nozzles. These pressures, represented respectively by the index and ring fingers, are found from

$$\frac{p_b}{p_0} = \left(\frac{p_e}{p_0}\right)^{-1} = \left[1 + \frac{\gamma - 1}{2}M_e^2\right]^{\frac{-\gamma}{\gamma+1}}$$

where M_e is the exit-plane Mach number, and M_e is found from the area-Mach number relation (Eq. 6.25). Tables or a numerical solver are commonly used for solving for the two M_e problems given the one exit throat-area ratio. Call these two Mach numbers $M_{e,sub}$ and $M_{e,super}$.

The final pressure to find is the central one. Simply take the supersonic isentropic flow solution from the previous paragraph and pass the air through a normal shock. The downstream static pressure after the shock is the value of back pressure we seek. That is, for the central value of important back pressures,

$$\frac{p_b}{p_0} = \frac{p_{b,super}}{p_0} \frac{p_2}{p_1}\bigg|_{M_{e,super}} = \frac{p_{b,super}}{p_0} \left[\frac{2\gamma M_{e,super}^2 - (\gamma - 1)}{\gamma + 1}\right]$$

Tabulating the results provides a structure for the student, as shown in Table 6.1.

Table 6.1 An ordered presentation of the vie back pressures that bound the four nozzle operating conditions.

Flow Description	Throat	$\frac{p_b}{p_0}$	$\frac{p_e}{p_b}$	Flow Mode	M_e	$\frac{p_e}{p_b}$
Upper limit—no flow	$M=0$	$=1$	$=1$			
				Subsonic in diverging section	<1	$=1$
Isentropic subsonic	$M=1$	<1	$=1$			
				Normal shock in diverging section	<1	$=1$
Normal shock in exit plane	$M=1$	<1	$=1$			
				Overexpanded	>1	<1
Isentropic supersonic	$M=1$	<1	$=1$			
				Underexpanded	>1	>1
Lower limit—no back pressure	$M=1$	$=0$	∞			

A review is appropriate here: Students are generally confused about exit pressure and back pressure. The two are different, but sometimes have the same value. Back pressure, as stated, is the static pressure of the atmosphere into which the nozzle exits; think of it as the pressure that exists when the nozzle flow is turned off. Exit pressure is the static pressure of the flow (not the atmosphere) in the nozzle's exit plane. When the exit-plane flow is subsonic, the exit and back pressures are equal. In addition, there is one specific value of back pressure for which a supersonic exit flow has $p_B = p_e$.

Example 6.9

A converging-diverging nozzle in a lab class having a throat area of $1\,\text{cm}^2$ and an exit area of $3.5\,\text{cm}^2$ exhausts into a plumbing system that provides a controllable back pressure. The supply pressure is 1 MPa. Find the four ranges of back pressure for the operating modes of the nozzle.

The solution is as follows:

The four ranges are bounded by five back pressure values. The maximum back pressure is the supply pressure, 1 MPa. The minimum back pressure is vacuum, or zero. This leaves three values of the back pressure to find. Two are found from isentropic expansion: one subsonic and one supersonic.

To start with, notice that the ratio of exit to throat area is $\frac{A_{exit}}{A*} = 3.5$. From the area-Mach number equation (Eq. 6.23 and Fig. 6.2) you see that there are two possible exit Mach numbers: 0.1682 and 2.800.

The ratios of stagnation to static pressures for isentropic flow at $M_{exit} = 0.1682$ is $\frac{p_o}{p_{exit}} = 1.020$ and at $M_{exit} = 2.800$ is $\frac{p_o}{p_{exit}} = 27.14$. These results are from using the MATLAB functions presented in this chapter.

For subsonic flows, the exit pressure is always equal to the back pressure, so $\frac{p_o}{p_{exit}} = 1.020$ leads to $p_b = 0.98$ MPa for the subsonic expansion.

For the supersonic case, this is called the perfect supersonic expansion because the exit plane pressure in the supersonic exit flow matches exactly the back pressure. Any change in back or supply pressures destroys this equality. Thus, $p_b = \frac{p_o}{27.14} = 0.0368$ MPa for the perfect supersonic expansion.

The one remaining back pressure to solve for is simply the perfect supersonic expansion case with the addition of a normal shock wave in the exit plane. The flow entering the normal shock wave is at M = 2.800 and thus there is a pressure ratio of 8.98 across the shock. So the back pressure to create this state will have to be 8.98 times greater than the exit pressure from perfect supersonic expansion, which we solved for above. Therefore, for a normal shock in the diverging section, we have $p_b = (0.0368\,\text{MPa})8.98 = 0.3305$ MPa.

Tabulating our results:

Subsonic expansion	1 MPa > p_b > 0.98 MPa
Normal shock in the divergent section	0.98 MPa > p_b > 0.3305 MPa
Overexpanded	0.3305 MPa > p_b > 0.0368 MPa
Underexpanded	0.0368 MPa > p_b > MPa

AERODYNAMICS AROUND US

Shock Waves

It is obvious by now that this chapter contains a wealth of equations that result from investigations. You might be asking yourself, "How am I going to remember all these equations?" The important point is, you need not memorize them as even the simpler ones will fade from your mind in a month if they are not used.

What is exceedingly important to memorize, however, is what a shock wave does to the flow of an ideal gas. For a steady flow of an ideal gas, the following table is important and you can always return to this book for the equations (which implies that you are wise enough to keep the book after this semester concludes):

Steady Normal Shock Wave in an Ideal Gas Flow

Quantity	Result	Note
M_2	< 1	Oblique shocks can have $M_2 > 1$
P_2/P_1	> 1	Compresses the gas
$T_2/T_1 = h_2/h_1 = e_2/e_1$	> 1	Heats the gas
$\rho_2/\rho_1 = u_1/u_2$	> 1	Gas takes up less volume
$s_2 - s_1$	> 0	Therefore, an irreversible process
P_{02}/P_{01}	< 1	Therefore, an irreversible process Eq. (6.55)
$T_{02}/T_{01} = h_{02}/h_{01} = e_{02}/e_{01}$	$= 1$	Adiabatic and steady
ρ_{02}/ρ_{01}	$= P_{02}/P_{01} < 1$	By ideal gas law
$s_{02} - s_{01}$	$= s_2 - s_1 > 0$	$= s_2 - s_1$ because of definition of stagnation conditions

Students often wonder how professors can be so quick to spot errors in their work. This table is one such "trick." For example, a computer-drawn plot showing the flow downstream of a steady shock as cooler than the upstream flow is obviously wrong. Similarly, an exam answer that reports supersonic flow after the normal shock is wrong, and there is no need to examine the work that led up to this conclusion. This table is an excellent tool for students to use in checking their work. If your professor permits you a formula sheet on exams, write this table on it and use it as a quick check. If you work in, or defend your PhD thesis in, supersonic aerodynamics, you can still lean on this table for quick error checks.

6.5 MACH WAVES AND SHOCK WAVES IN TWO-DIMENSIONAL FLOW

A small deflection in supersonic flow always takes place such that the flow properties are uniform along a front inclined to the flow direction. Also, their only change is in the direction normal to the front, which is known as a wave. For small flow changes, the wave sets itself up at the Mach angle (μ) appropriate to the upstream flow conditions.

For finite-positive or compressive-flow deflections—that is, when the downstream pressure is much greater than that upstream—the (shock) wave angle is greater than the Mach angle, causing characteristic changes in the flow (see Section 6.4). For finite-negative or expansive-flow deflections, where the downstream pressure is less, the turning power of a single wave is insufficient and a fan of waves is set up, each inclined to the flow direction by the local Mach angle and terminating in the wave whose Mach angle is appropriate to the downstream condition.

For small changes in supersonic flow deflection, both the compression-shock and expansion-fan systems approach the character and geometrical properties of a Mach wave. They retain only the algebraic sign of the change in pressure.

6.6 MACH WAVES

Figure 6.12 shows the wave pattern associated with a point source P of weak pressure disturbances: part (a) when stationary and parts (b) and (c) when moving in a straight line. In the stationary case (with the surrounding fluid at rest), the concentric circles mark the position of successive wave fronts at a particular instant in time. In three-dimensional flow, they are concentric spheres; however, a close analogy to the two-dimensional case is the appearance of ripples, on the still surface of a pond from a small disturbance. The wave fronts emanating from P advance at acoustic speed a, and consequently the radius of a wave t seconds after its emission is at. If t is large enough, the wave can traverse the whole of the fluid, which is thus made aware of the disturbance.

When the intermittent source moves at a speed u less than a in a straight line, the wave fronts adopt the pattern shown in part (b). The individual waves remain circular, with their centers on the line of motion of the source, and are eccentric but nonintersecting. The point source moves through a distance ut in the time the wave moves through the greater distance at. Once again, the waves propagating from the pressure disturbance move through the entire flow region, ahead of and behind the moving source.

If the steady speed of the source is increased beyond acoustic speed the individual sound waves (at any one instant) are seen in part (c) of Fig. 6.12 as eccentric intersecting circles with their centers on the line of motion. Further, the circles are tangential to two symmetrically inclined lines (a cone in three dimensions) with their apex at point source P.

While a wave has moved a distance at, point P has moved ut, and thus the semi-vertex angle is

$$\mu = \arcsin \frac{at}{ut} = \arcsin \frac{1}{M} \tag{6.57}$$

M, the Mach number of the speed of the point P relative to the undisturbed stream, is the ratio u/a, and the angle μ is known as the *Mach angle*. Were the disturbance continuous, the inclined lines (or cone), would be the envelope of all the waves produced and would then be known as *Mach waves* (or cones).

It is evident that the effect of the disturbance does not go beyond the Mach lines (or cone) into the surrounding fluid, which is thus unaffected by the disturbance. The region of fluid outside the Mach lines (or cone) is known as the zone of silence or, more dramatically, the zone of forbidden signals.

It is possible to project an image wedge (or cone) forward from apex P, (Fig. 6.12d). This extended wedge (or cone) contains the region of the flow where any disturbance ahead a point such as P_1 can have an effect on P. This is because a disturbance P_2 outside of the wedge (or cone) excludes P from its Mach wedge (or cone), assuming that P_1 and P_2 are moving at the same Mach number.

If a uniform supersonic stream M is superimposed from left to right on the flow in Fig. (6.12c) the system becomes a uniform stream of Mach number $M > 1$ flowing

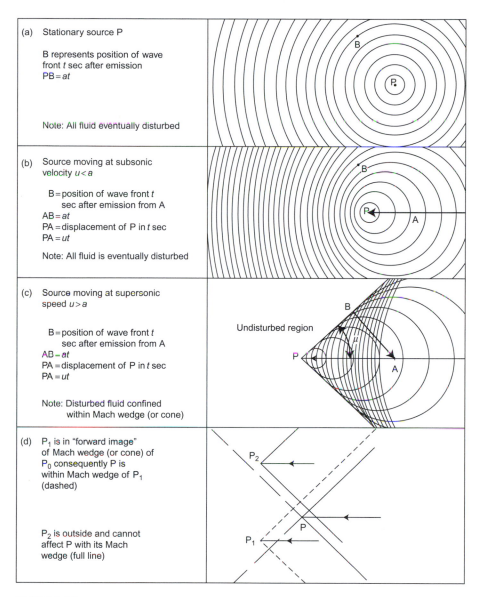

FIGURE 6.12

Sound and shock waves in subsonic to supersonic flows.

past a weak disturbance. Since the flow is symmetrical, the axis of symmetry may represent the surface of a flat plate along which an inviscid supersonic stream flows. Any small disturbance caused by, for example, a slight irregularity is communicated to the flow at large along a Mach wave. Figure 6.13 shows the Mach wave emanating

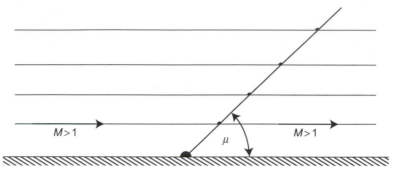

FIGURE 6.13

Infinitesimal, or Mach, wave in supersonic flow showing that straight streamlines persist through the wave.

from a disturbance, which has a net effect on the flow similar to a pressure pulse that leaves the downstream flow unaltered. If the pressure change across the Mach wave is to be permanent, the downstream flow direction must change. The converse is also true.

It was shown previously that a slight pressure change in a supersonic flow propagates along an oblique wave inclined at μ to the flow direction. The pressure difference is across, or normal to, the wave and, as a consequence, the gas velocity alters in its component perpendicular to the wave front. If the downstream pressure is less than the upstream pressure, the flow velocity component normal to the wave increases across it so that the resultant downstream flow inclines to the wave front at a greater angle (Fig. 6.14a). Thus the flow is expanded, accelerated, and deflected away from it. On the other hand, if the downstream pressure is greater (Fig. 6.14b), the flow component across the wave decreases, as does the net outflow velocity, which is now inclined at an angle less than μ to the wave front. The flow is compressed, retarded, and deflected toward the wave.

Quantitatively, the turning power of a wave may be obtained as described here. Figure 6.15 shows the slight expansion around a small deflection δv_p from flow conditions p, ρ, M, q, and so forth, across a Mach wave set at μ to the initial flow direction. Recall from the velocity components normal and parallel to the wave that the final velocity $q + \delta q$ changes only through a change in the normal velocity component u to $u + \delta u$ as it crosses the wave, since the tangential velocity remains uniform throughout the field. Then, from the velocity diagram after the wave,

$$(q + \delta q)^2 = (u + \delta u)^2 + v^2$$

On expanding,

$$q + 2q\delta q + (\delta q)^2 = u^2 + 2u\delta u + (\delta u)^2 + v^2$$

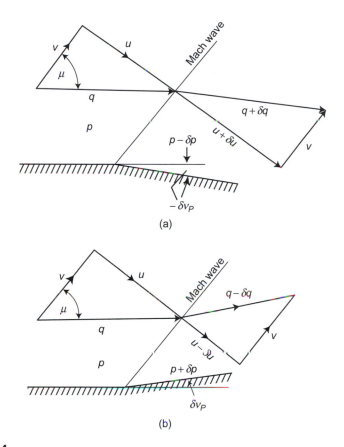

FIGURE 6.14

(a) Incremental expansion corner; (b) compression corner.

and in the limit, ignoring terms of the second order and putting $u^2 + v^2 = q^2$,

$$qdq = udu \tag{6.58}$$

Equally, from the definition of the velocity components,

$$\mu = \text{arc} \tan \frac{u}{t} \quad \text{and} \quad d\mu = \frac{1}{1 + (u/v)^2} \frac{du}{v} = \frac{v}{q^2} du$$

However, the change in deflection angle is the incremental change in Mach angle, so

$$dv_P = d\mu = \frac{v}{q^2} du \tag{6.59}$$

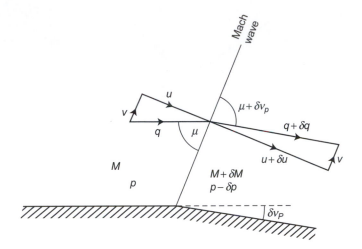

FIGURE 6.15

Expansion around an infinitesimal deflection through a Mach wave (labeling convention as used in Prandtl-Meyer flow).

Combining Eqs. (6.58) and (6.59) yields

$$\frac{dq}{dv_P} = q\frac{u}{v} \quad \text{since} \quad \frac{u}{v} = \arctan\mu\frac{1}{\sqrt{M^2-1}}$$

$$\frac{dq}{q} = \pm\frac{dv_P}{\sqrt{M^2-1}} = \pm dv_P\tan\mu \tag{6.60}$$

where q is the flow velocity inclined at v_P to some datum direction. It follows from Eq. (6.10), with q substituted for μ, that

$$\frac{dp}{p} = \pm dv_P\frac{\gamma M^2}{\sqrt{M^2-1}} \tag{6.61}$$

or, in pressure-coefficient form,

$$C_p = \pm\frac{2dv_P}{\sqrt{M^2-1}} \tag{6.62}$$

Flow behavior in the vicinity of a single weak wave due to a small pressure change can be used to study the effect of a larger pressure change that may be treated as the sum of a number of small pressure changes. We consider the expansive case first. Figure 6.16 shows expansion due to a pressure decrease equivalent to three incremental pressure reductions in a supersonic flow initially having pressure

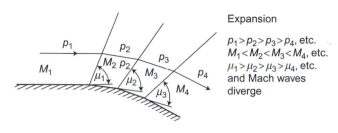

Expansion

$p_1 > p_2 > p_3 > p_4$, etc.
$M_1 < M_2 < M_3 < M_4$, etc.
$\mu_1 > \mu_2 > \mu_3 > \mu_4$, etc.
and Mach waves
diverge

FIGURE 6.16

Expansion over a curved wall represented as a series of Prandtl-Meyer expansions.

p_1 and Mach number M_1. On expansion through the wavelets, the Mach number of the flow successively increases because acceleration induced by the successive pressure reductions and the Mach angle ($\mu =$ arc sin $1/M$) successively decreases. Consequently, the Mach waves in such an expansive regime spread out or diverge and flow accelerates smoothly to the downstream conditions. It is evident that the number of steps shown in Figure 6.16 may be increased or that the generating wall may be continuous without the flow mechanism being altered except by the increased number of wavelets. In fact, the finite pressure drop can take place abruptly—for example, at a sharp corner—and the flow will continue to expand smoothly through a fan of expansion wavelets emanating from the corner. This case of two-dimensional expansive supersonic flow (i.e., around a corner) is known as the Prandtl-Meyer expansion. It has the same physical mechanism as the one-dimensional isentropic supersonic accelerating flow of Section 6.2. In the Prandtl-Meyer expansion, the streamlines turn through the wavelets as the pressure falls and the flow accelerates. The flow velocity, angular deflection (from some upstream datum), pressure, and so forth, at any point in the expansion may be obtained with reference to Fig. 6.17.

Algebraic expressions for the wavelets in terms of flow velocity can be obtained by further manipulation of Eq. (6.60), which, for convenience, is recalled in the form

$$\frac{1}{q}\frac{dq}{dv_\mathrm{P}} = +\tan\mu$$

If the velocity component $v = q\cos\mu$ is introduced along or tangential to the wave front (Fig. 6.14),

$$dv = dq\cos\mu - q\sin\mu\, d\mu = q\sin\mu\left(\frac{1}{q}\frac{dq}{\tan\mu} - d\mu\right) \qquad (6.63)$$

It is necessary to define the lower limiting or datum condition. This is most conveniently the sonic state where the Mach number is unity, $a = a^*$, $v_\mathrm{p} = 0$, and the wave angle is $\mu = \pi/2$. In the general case, the datum (sonic) flow may be inclined by some angle α to the coordinate in use. Substituting dv_P for $(1/q)dq/\tan\mu$ from Eq. (6.60) and, since $q\sin\mu = a$, Eq. (6.63) becomes $dv_\mathrm{P} - d\mu = dv/a$. But from the

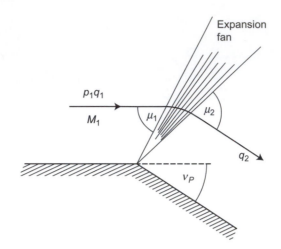

FIGURE 6.17

Prandtl-Meyer expansion with a finite deflection angle. The shape of a streamline passing through the expansion is a circular arc.

energy equation, with c=ultimate velocity, $a^2/(\gamma - 1) + (q^2/2) = (c^2/2)$, and with $q^2 = (v^2 + a^2)$ (Eq. 6.17),

$$a^2 = \frac{\gamma - 1}{\gamma + 1}(c^2 - v^2) \tag{6.64}$$

which gives the differential equation

$$d v_P - d\mu = \frac{dv}{\left(\frac{\gamma-1}{\gamma+1}(c^2 - v^2)\right)^{1/2}} \tag{6.65}$$

Equation (6.65) may now be integrated. Thus

$$\int_{v_P=\alpha}^{v_P} d v_P - \int_{\mu=\pi/2}^{\mu} d\mu = \left[\sqrt{\frac{\gamma+1}{\gamma-1}} \sin^{-1}\frac{v}{c}\right]_0^v$$

or

$$(v_P - \alpha) - \left(\mu - \frac{\pi}{2}\right) = \sqrt{\frac{\gamma+1}{\gamma-1}} \sin^{-1}\frac{v}{c} \tag{6.66}$$

From Eq. (6.64),

$$\sin^{-1}\frac{v}{c} = \tan^{-1}\sqrt{\frac{\gamma-1}{\gamma+1}}\frac{v}{a} = \tan^{-1}\sqrt{\frac{\gamma-1}{\gamma+1}}\cot\mu$$

which allows the flow deflection in Eq. (6.66) to be expressed as a function of Mach angle:

$$\nu_P - \alpha = \mu + \sqrt{\frac{\gamma - 1}{\gamma + 1}} \tan^{-1} \sqrt{\frac{\gamma - 1}{\gamma + 1}} \cot \mu - \frac{\pi}{2} \tag{6.67}$$

or

$$\nu_P - \alpha = f(\mu) \tag{6.67a}$$

The local velocity may also be expressed in terms of the Mach angle μ by rearranging the energy equation as follows:

$$\frac{q^2}{2} + \frac{a^2}{\gamma - 1} = \frac{c^2}{2}$$

but $a^2 = q^2 \sin^2 \mu$. Therefore,

$$q^2 \left(\frac{1}{2} + \frac{\sin^2 \mu}{\gamma - 1} \right) = \frac{c^2}{2}$$

or

$$q^2 = \frac{c^2}{\left(1 + \frac{2}{\gamma - 1} \sin^2 \mu \right)} \tag{6.68}$$

Equations (6.67) and (6.68) give expressions for flow velocity and direction at any point in a turning supersonic flow in terms of local Mach angle μ and thus local Mach number M.

We can compute values for the deflection angle from sonic conditions ($\nu_P - \alpha$), the deflection of the Mach angle from its position under sonic conditions φ, and the velocity ratio q/c for a given Mach number once and use them in tabular form thereafter. Numerous tables of these values exist, but most of them have the Mach number as a dependent variable. Recall that the turning power of a wave is a significant property; a more convenient tabulation has the angular deflection ($\nu_P - \alpha$) as the dependent variable, but usually, of course, α has the convenient value of zero.

Compression flow through three wavelets springing from the points of flow deflection is shown in Fig. 6.18. In this case, the flow velocity is decreasing, M is decreasing, the Mach angle is increasing, and the compression wavelets are converging toward a region away from the wall. If the curvature is continuous, the large number of wavelets reinforce each other in the convergence region, to become a finite disturbance forming the foot of a shock wave that propagates outward and through which the flow properties abruptly change. If the finite compressive deflection takes place abruptly at a point, the foot of the shock wave springs from that point and the initiating system of wavelets does not exist. In both cases, the boundary layers adjacent to real walls modify the flow locally, having a greater effect in the compressive case.

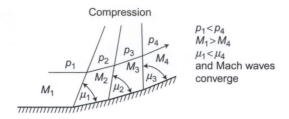

FIGURE 6.18

Flow over a compression corner with three small compressions.

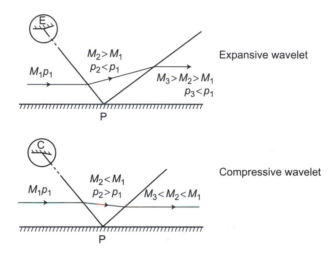

FIGURE 6.19

Impingement and reflection of plane wavelets on a plane surface. In both cases, a wave is reflected such that the velocity remains parallel to the wall.

6.6.1 Mach Wave Reflection

In certain situations, a Mach wave, generated somewhere upstream, may impinge on a solid surface. Unless the surface is bent at the point of contact, the wave is reflected as a wave of the same sign but at some other angle that depends on systems geometry. Figure 6.19 shows two wavelets, one expansive, the other compressive, each generated somewhere upstream, striking a plane wall at P along which the supersonic stream flows, at the Mach angle appropriate to the upstream flow.

Behind the wave, the flow is deflected away from the wave (and wall) in the expansive case and toward the wave (and wall) in the compressive case, with appropriate respective increase and decrease in the Mach number.

The physical requirement of the reflected wave is contributed by the wall downstream of point P that demands that the flow leave the reflected wave parallel to the wall. For this to be so, the reflected wave must turn the flow away from itself in the

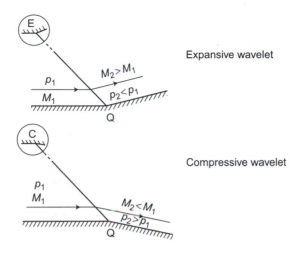

FIGURE 6.20

Impingement and absorption of plane wavelets at bent surfaces. Perfect matches between the downstream wall slope and the turning angle are shown.

expansive case, expanding it further to $M_3 < M_1$, and toward itself in the compressive case, thus additionally compressing and retarding its downstream flow.

If the wall is bent in the appropriate sense at the point of impingement at an angle sufficient for the exit flow from the impinging wave to be parallel to the wall, then the wave is absorbed and no reflection takes place (Fig. 6.20). If the wall is bent beyond this requirement, a wavelet of the opposite sign is generated.

A particular case arises in the impingement of a compressive wave on a wall if the upstream Mach number is not high enough to support supersonic flow after the two compressions through the impinging wave and its reflection. In this case, the impinging wave bends to meet the surface normally and the reflected wave forks from the incident wave above the normal part away from the wall (Fig. 6.21). The resulting wave system is Y-shaped.

Once reflected from an *open boundary*, the impinging wavelets change their sign as a consequence of the physical requirement of pressure equality with the free atmosphere through which the supersonic jet is flowing. A sequence of wave reflections is shown in Fig. 6.22, in which an adjacent solid wall serves to reflect the wavelets onto the jet boundary. As in a previous case, an expansive wavelet arrives from upstream and is reflected from the point of impingement P_1 while the flow behind it is expanded to the ambient pressure p and deflected away from the wall. Behind the reflected wave from P_1, the flow is further expanded to p_3 as discussed previously to bring the streamlines back parallel to the wall.

Once reflected from the free boundary in Q_1, the expansive wavelet P_1Q_1 must compress the flow from p_3 back to p along Q_1P_2. This compression deflects the flow toward the wall, where the compressive reflected wave from the wall (P_2Q_2) must

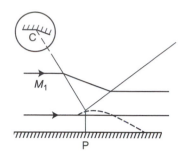

FIGURE 6.21

Mach reflection at a straight wall. Note the straight streamline near the wall.

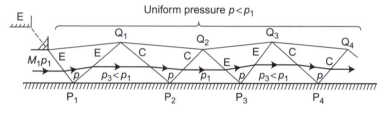

FIGURE 6.22

Wave reflection from an open boundary.

bring the flow back parallel to the wall. This process increases the air pressure to p_1 (greater than p). The requirement for reflection of P_2Q_2 in the open boundary is thus expansive wavelet Q_2P_3, which brings the pressure back, again, to the ambient value p. And so the cycle repeats.

The solid wall may be replaced by the axial streamline of a (two-dimensional) supersonic jet issuing into gas at a uniformly (slightly) lower pressure. If the ambient pressure were (slightly) greater than that in the jet, the system would start with a compressive wave and continue onward as before (Q_1P_2).

In the complete jet, the diamonds are regions where the pressure is alternately higher or lower than the ambient pressure but the streamlines are axial. When the streamlines are outside the diamonds, in the region of pressure equality with the boundary, they are alternately divergent or convergent.

The simple model discussed here is considerably different from the model of flow in a real jet, mainly because of jet entrainment of the ambient fluid, which affects the reflections from the open boundary. For a finite pressure difference between the jet and ambient conditions, the expansive waves are systems of fans and the compressive waves are shock waves.

6.6.2 Mach Wave Interference

Waves of the same character and strength intersect one another with the same configuration as those of reflections from the plane surface discussed previously, since the surface may be replaced by the axial streamline (Fig. 6.23a and 6.23b). When the intersecting wavelets are of opposite signs, the axial streamline is bent at the point of intersection in a direction away from the expansive wavelet. This is shown in Fig. 6.23(c). The streamlines also change direction at the intersection of waves of the same sign but of different turning power.

6.7 SHOCK WAVES

The generation of flow discontinuity called a shock wave was discussed in Section 6.4 for one-dimensional flow. Here the treatment is extended to plane-oblique and curved shocks in two-dimensional flows. Once again, the thickness of the shock wave is ignored, and the fluid is assumed to be inviscid and non-heat-conducting. In practice, the (thickness) distance in which the gas stabilizes its properties of state from initial to final conditions is small but finite. Treating a curved shock as consisting of small elements of a plane-oblique shock wave is reasonable only as long as radius of curvature is large compared to the thickness.

With these provisos, the following exact, but relatively simple, extension to one-dimensional shock theory will provide a deeper insight into problems of shock waves associated with aerodynamics.

6.7.1 Plane Oblique Shock Relations

Let a reference frame be fixed relative to the shock wave, and let angular displacements be measured from the free-stream direction. Then the model for general oblique flow through a plane shock wave may be taken, with the notation shown in Fig. 6.24, where V_1 is the incident flow from the shock wave and V_2 is the exit flow. The shock is inclined at an angle β to the direction of V_1, having components normal and tangential to the wave front of u_1 and v_1, respectively. The exit velocity V_2 (normal u_2 and tangential v_2 components) is also inclined to the wave but at some angle other than β. Relative to the incident-flow direction, the exit flow is deflected through δ. The equation of continuity for flow normal to the shock gives

$$\rho_1 u_1 = \rho_2 u_2 \tag{6.69}$$

Conservation of linear momentum parallel to the wave front yields

$$\rho_1 u_1 v_1 = \rho_2 u_2 v_2 \tag{6.70}$$

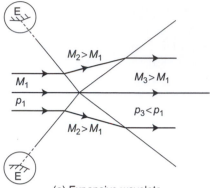

(a) Expansive wavelets

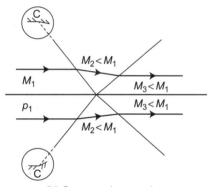

(b) Compressive wavelets

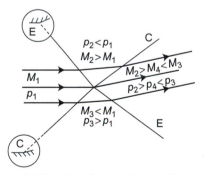

(c) Wavelets of opposite strength

FIGURE 6.23

Wavelet interference.

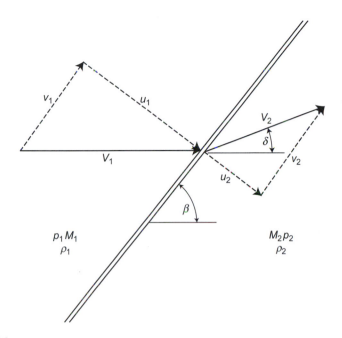

FIGURE 6.24

Geometry of flow through an oblique shock. This sketch can represent an infinitesimal section of a curved shock, so it is applicable to points on all three-dimensional shock wave shapes.

that is, since no tangential force is experienced along the wave front, the product of the mass entering the wave per unit second and its tangential velocity at entry must equal the product of the mass per second leaving the wave and the exit tangential velocity. From continuity, Eq. (6.70) yields

$$v_1 = v_2 \tag{6.71}$$

Thus the velocity component along the wave front is unaltered by the wave and the model reduces to one of one-dimensional flow (compare Section 6.4.1) on which a uniform velocity parallel to the wave front is superimposed.

Now, magnitude of the normal component of velocity decreases abruptly through the shock, and a consequence of the constant tangential component is that the exit-flow direction, as well as the magnitude, changes from that of the incident flow; the change in direction is toward the shock front. We see from this that the oblique shock is a mechanism for turning the flow inward as well as compressing it. In the expansive mechanism for turning a supersonic flow (Section 6.6), the angle of inclination to the wave increases.

Since the tangential-flow component is unaffected by the wave, the wave properties may be obtained from the one-dimensional flow case; however, they need to be referred to datum conditions and directions that are different from normal velocities and directions. In the present case,

$$M_1 = \frac{V_1}{a_1} = \frac{u_1}{a_1}\frac{V_1}{u_1}$$

or

$$M_1 = \frac{\frac{u_1}{a_1}}{\sin\beta} \tag{6.72}$$

Similarly,

$$M_2 = \frac{\frac{u_2}{a_2}}{\sin(\beta - \delta)} \tag{6.73}$$

The results from Section 6.4.2 may now be used directly, but with M_1 replaced by $M_1 \sin\beta$ and M_2 replaced by $M_2 \sin(\beta - \delta)$. The following ratios pertain.

The static pressure jump from Eq. (6.43) is

$$\frac{p_2}{p_1} = \frac{2\gamma M_1^2 \sin^2\beta - (\gamma - 1)}{\gamma + 1} \tag{6.74}$$

or, as inverted from Eq. (6.44),

$$\frac{p_1}{p_2} = \frac{2\gamma M_2^2 \sin^2(\beta - \delta) - (\gamma - 1)}{\gamma + 1} \tag{6.75}$$

The density jump from Eq. (6.45) is

$$\frac{\rho_2}{\rho_1} = \frac{(\gamma + 1)M_1^2 \sin^2\beta}{2 + (\gamma - 1)M_1^2 \sin^2\beta} \tag{6.76}$$

or, from Eq. (6.46),

$$\frac{\rho_1}{\rho_2} = \frac{(\gamma + 1)M_2^2 \sin^2(\beta - \delta)}{2 + (\gamma - 1)M_2^2 \sin^2(\beta - \delta)} \tag{6.77}$$

The static temperature change from Eq. (6.47) is

$$\frac{T_2}{T_1} = \frac{2\gamma M_1^2 \sin^2\beta - (\gamma - 1)}{\gamma + 1}\frac{2 + (\gamma - 1)M_1^2 \sin^2\beta}{(\gamma + 1)M_1^2 \sin^2\beta} \tag{6.78}$$

Finally, the Mach number change from Eq. (6.49) is

$$M_2^2 \sin^2(\beta - \delta) = \frac{(\gamma - 1)M_1^2 \sin^2 \beta + 2}{2\gamma M_1^2 \sin^2 \beta - (\gamma - 1)} \tag{6.79}$$

These equations contain one or both of the additional parameters β and δ that must be known for the appropriate ratios to be evaluated.

An expression relating the incident Mach number M_1, the wave angle β, and the flow deflection δ may be obtained by introducing the geometrical configuration of the flow components:

$$\frac{u_1}{v_1} = \tan \beta, \quad \frac{u_2}{v_2} = \tan(\beta - \delta)$$

but, by continuity,

$$v_1 = v_2 \text{ and } \frac{u_1}{u_2} = \frac{\rho_2}{\rho_1}$$

Thus

$$\frac{\rho_2}{\rho_1} = \frac{\tan \beta}{\tan(\beta - \delta)} \tag{6.80}$$

Equations (6.76) and (6.80) give the different expressions for ρ_2/ρ_1, so the right-hand sides may be set equal to give

$$\frac{\tan \beta}{\tan(\beta - \delta)} = \frac{(\gamma + 1)M_1^2 \sin^2 \beta}{2 + (\gamma - 1)M_1^2 \sin^2 \beta}$$

Algebraic rearrangement gives

$$\tan \delta = \cot \beta \frac{M_1^2 \sin^2 \beta - 1}{\frac{(\gamma + 1)}{2}M_1^2 - (M_1^2 \sin^2 \beta - 1)} \tag{6.81}$$

Plotting values of β against δ for various Mach numbers gives the carpet of graphs shown in Fig. 6.25.

We see that all of the curves are confined within the $M_1 = \infty$ curve and that, for a given Mach number, a certain value of deflection angle δ up to a maximum value δ_M may result in a smaller (weak) or larger (strong) wave angle β. To solve Eq. (6.81) algebraically (i.e., to find β for a given M_1 and δ) is difficult. However, any numerical package or spreadsheet has a solver that, with care, will produce good solutions.

Thus it is possible to obtain both physically possible values of the wave angle providing the deflection angle $\delta < \delta_{max}$. δ_{max} may be found in the normal way by differentiating Eq. (6.81) with reference to β, with M_1 constant and equating to zero.

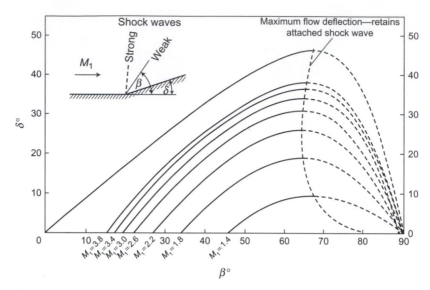

FIGURE 6.25

Plot of Eq. (6.81) as curves of a constant upstream Mach number. Note that each combination of Mach number and turning angle has two solutions for wave angle.

This gives, for the maximum value of tan δ,

$$\sin^2 \beta_{max} = \frac{1}{\gamma M_1^2} \left[\frac{\gamma+1}{4} M_1^2 - 1 + \sqrt{(\gamma+1) \left(1 + \frac{\gamma-1}{2} M_1^2 + \frac{\gamma+1}{16} M_1^4 \right)} \right]$$

(6.82)

Substituting back in Eq. (6.81) gives a value for tan δ_{max}.

6.7.2 Shock Polar

Although, in practice, data for plane shock waves are easily computed, numbers generally do not convey the trends and dependencies in the equations as well as a specialized plot can. The study of shock waves is still considerably helped by a traditional hodograph or velocity polar diagram set up for a given free-stream Mach number.

This section discusses construction and use of the shock polar, which is the exit velocity vector displacement curve for all possible exit flows downstream of an attached plane shock in a given undisturbed supersonic stream. To plot it requires rearrangement of the equations of motion in terms of the exit velocity components and the inlet flow conditions.

Figure 6.26 shows the exit component velocities to be used: q_t and q_n, the radial and tangential polar components with respect to the free stream V_1 direction taken

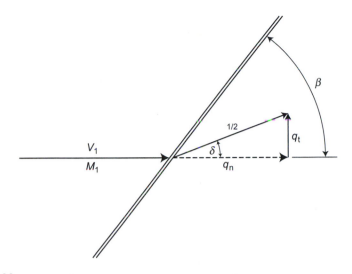

FIGURE 6.26

Variation on Fig. 6.24 that is useful for shock-polar analysis. Unlike Fig no new fluid phenomena are present here.

as a datum. It is immediately apparent that the exit-flow direction is given by arctan (q_t/q_n). For the wave angle β (recall the additional notation for Fig. 6.24), linear conservation of momentum along the wave front, from Eq. (6.71), gives $v_1 = v_2$, or, in terms of geometry,

$$V_1 \cos \beta = V_2 \cos(\beta - \delta) \tag{6.83}$$

Expanding the right-hand side and dividing through gives

$$V_1 = V_2[\cos \delta + \tan \beta \sin \delta] \tag{6.84}$$

or, in terms of the polar components,

$$V_1 = V_2 \left(\frac{q_n}{V_2} + \tan \beta \frac{q_t}{V_2} \right)$$

which, rearranged, gives the wave angle:

$$\beta = \arctan \frac{V_1 - q_n}{q_t} \tag{6.85}$$

To express the conservation of momentum normal to the wave in terms of polar velocity components, consider first the flow of unit area normal to the wave:

$$p_1 + \rho_1 u_1^2 = p_2 + \rho_2 u_2^2 \tag{6.86}$$

Then successively, using continuity and the geometric relations,

$$p_2 = p_1 + \rho_1 V_1 \sin\beta [V_1 \sin\beta - q_n \sin\beta + q_n \cos\beta \tan\delta]$$
$$p_2 = p_1 + \rho_1 V_1 \sin\beta [(V_1 - q_n)\sin\beta + q_t \cos\beta]$$

and, using Eq. (6.83),

$$p_2 = p_1 + \rho_1 V_1 (V_1 - q_n) \tag{6.87}$$

Again, from continuity (expressed in polar components),

$$\rho_1 V_1 \sin\beta = \rho_2 V_2 \sin(\beta - \delta) = \rho_2 q_n (\sin\beta - \cos\beta \tan\delta)$$

or

$$\rho_1 V_1 = \rho_2 q_n \left(1 - \frac{q_t}{q_n \tan\beta}\right) \tag{6.88}$$

Divide Eq. (6.87) by Eq. (6.88) to isolate pressure and density:

$$\frac{p_2}{\rho_2 q_n}\frac{1}{\left(1 - \frac{q_t}{q_n \tan\beta}\right)} = \frac{p_1}{\rho_1 V_1} + (V_1 - q_n)$$

Once more recalling Eq. (6.85) to eliminate the wave angle and rearranging,

$$\frac{p_2}{\rho_2}\frac{V_1 - q_n}{V_1 q_n - q_n^2 - q_t^2} = \frac{p_1}{\rho_1 V_1} + (V_1 - q_n) \tag{6.89}$$

Finally, from the energy equation expressed in polar velocity components: up to the wave

$$\frac{p_1}{\rho_1} = \frac{\gamma - 1}{2}\left(\frac{\gamma + 1}{\gamma - 1}a^{*2} - V_1^2\right) \tag{6.90}$$

and downstream from the wave

$$\frac{p_2}{\rho_2} = \frac{\gamma - 1}{2}\left(\frac{\gamma + 1}{\gamma - 1}a^{*2} - q_n^2 - q_t^2\right) \tag{6.91}$$

Substituting for these ratios in Eq. (6.87) and isolating the exit tangential velocity component gives the following equation:

$$q_t^2 = \frac{(V_1 - q_n)^2(V_1 q_n - a^{*2})}{\left(\frac{2}{\gamma+1}\right)V_1^2 - V_1 q_n + a^{*2}}$$

(6.92)

which is a basic form of the shock-polar equation.

To make Eq. (6.92) more amenable to graphical analysis, it may be made nondimensional. Any initial flow parameters, such as critical speed of sound a^{*2}, ultimate velocity c, and so forth, may be used, but here we follow the originator, A. Busemann [47], and divide through by the undisturbed acoustic speed a_1:

$$\hat{q}_t^2 = \frac{(M_1 - \hat{q}_n)^2 \left(M_1 \hat{q}_n - \frac{a^{*2}}{a_1^2}\right)}{\left(\frac{2}{\gamma+1}\right)M_1^2 - \left(M_1 \hat{q}_n - \frac{a^{*2}}{a_1^2}\right)}$$

(6.93)

where $\hat{q}_t^2 = (q_t/a_1)^2$, and so on. This may be further reduced to

$$\hat{q}_t^2 = (M_1 - \hat{q}_n)^2 \frac{\hat{q}_n - \overline{M}_1}{\frac{2}{\gamma+1}M_1 - (\hat{q}_n - \overline{M}_1)}$$

(6.94)

where

$$\overline{M}_1 = \frac{a^{*2}}{a_1^2 M_1} = \frac{\gamma-1}{\gamma+1}M_1 + \frac{2}{\gamma+1}\frac{1}{M_1}$$

$$= M_1 - \frac{2}{\gamma+2}\left(M_1 - \frac{1}{M_1}\right) = \overline{M}(M_1)$$

(6.95)

Equation (6.93) shows that the curve of the relationship between $\hat{q}_t$ and $\hat{q}_n$ is uniquely determined by the free-stream conditions (M_1); conversely, the shock polar curve is obtained for each free-stream Mach number. Further, since the nondimensional tangential component $\hat{q}_t$ appears in the expression as a squared term, the curve is symmetrical about the q_n axis.

Singular points are given by setting $\hat{q}_t = 0$ and ∞. For $\hat{q}_t = 0$,

$$(M_1 - \hat{q}_n)^2(\hat{q}_n^2 - \overline{M}_1) = 0$$

giving intercepts of the $\hat{q}_n$ axis at A:

$$\hat{q}_n = M_1 \quad \text{(twice)}$$

(6.96)

and at B:

$$\hat{q}_n = \overline{M}_1 = M_1 - \frac{2}{\gamma + 1}\left(M_1 - \frac{1}{M_1}\right) \tag{6.97}$$

For $\hat{q}_t = \infty$, at C:

$$\hat{q}_n = \frac{2}{\gamma + 1}M_1 + \overline{M}_1 = M_1 + \frac{2}{(\gamma + 1)M_1} \tag{6.98}$$

For a shock wave to exist, $M_1 > 1$. Therefore, the three points B, A, and C of the q_n axis referred to previously indicate values of $q_n < M_1, = M_1,$ and $> M_1$, respectively. Further, as the exit-flow velocity cannot be greater than the inlet-flow velocity for a shock wave, the region of the curve between A and C has no physical significance. We thus focus only on the curve between A and B.

Plotting Eq. (6.92) point by point confirms the values A, B, and C. Figure 6.27 shows the shock polar for the undisturbed-flow condition of $M_1 = 3$. The upper branch of the curve in Fig. 6.27 is plotted point by point for flow at a free stream of $M_1 = 3$. The lower branch, which is the image of the upper branch reflected in the $\hat{q}_n$ axis, shows the physically significant portion (i.e., the closed loop) obtained by a simple geometrical construction as follows:

(1) Find and plot points A, B, and C from the previous equations. All points are explicitly functions of M_1.
(2) Draw semicircles (for a half-diagram) with AB and CB as diameters.

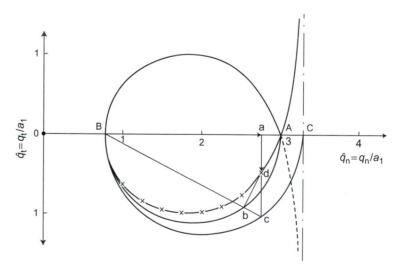

FIGURE 6.27

Construction of shock polar for Mach 3.

(3) At a given value $\hat{q}_n$ (Oa), erect ordinates to meet the larger semicircle in c.
(4) Join c to B, intersecting the smaller semi-circle at b.
The required point d is the intercept of bA and ac.

Geometrical proof Triangles Aad and acB are similar. Therefore,

$$\frac{ad}{aB} = \frac{aA}{ac}$$

that is,

$$(ad)^2 = (aA)^2 \left(\frac{aB}{ac}\right)^2 \tag{6.99}$$

Again, from the geometrical properties of circles,

$$(ac)^2 = aB \times aC$$

which, substituted in Eq. (6.100), gives

$$(ad)^2 = (aA)^2 \frac{aB}{aC} \tag{6.100}$$

Introducing the scaled values $ad = \hat{q}_t$,

$$aB = Oa - OB = \hat{q}_n - \overline{M}_1, \; aA = OA - Oa = M_1 - \hat{q}_n,$$
$$aC = OC - Oa = [2/(\gamma+1)]M_1 + \overline{M}_1 - \hat{q}_n$$

which reveals Eq. (6.94):

$$q_t^2 = \frac{(M_1 - \hat{q}_n)^2(\hat{q}_n - \overline{M}_1)}{\left(\frac{2}{\gamma+1}\right)M_1 - (\hat{q}_n - \overline{M}_1)}$$

Consider the physically possible flows represented by various points on the closed portion of the shock polar diagram in Fig. 6.28. Point A is the upper limiting value for the exit-flow velocity and is the case where the free stream is subjected to only an infinitesimal disturbance that produces a Mach wave inclined at μ to the free stream but no deflection of the stream and no change in exit velocity. The Mach wave angle is given by the inclination of the tangent of the curve at A to the vertical, and this is the limiting case of the construction required to find the wave angle in general.

Point D is the second point at which a general vector emanating from the origin cuts the curve (the first being point E). The representation means that, in going through a certain oblique shock, the inlet stream of direction and magnitude given by OA is deflected through an angle δ and has magnitude and direction given by vector

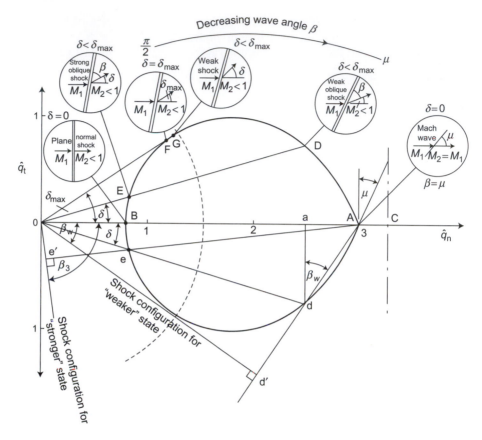

FIGURE 6.28

Shock polar for Mach 3.

OD (or Od in the lower half of Fig. 6.28). The ordinates of OD give the normal and tangential exit velocity components.

The appropriate wave angle β_w is determined by the geometrical construction shown in the lower half of the curve (i.e., by angle Ada). To establish this, recall Eq. (6.93):

$$\tan \beta_W = \frac{V_1 - q_n}{q_t} = \frac{M_1 - \hat{q}_n}{\hat{q}_t} = \frac{(OA) - (Oa)}{(ad)} = \frac{(aA)}{(ad)}$$

that is, $\beta_W = \widehat{adA}$.

The wave angle may be seen in better juxtaposition to the deflection δ by a small extension to the geometrical construction. Produce Ad to meet the perpendicular from O in d'. Since $\triangle aAd \, ||| \, \triangle Ad'O$,

$$\widehat{AOd'} = \widehat{Ada} = \beta_W$$

Of the two intercepts of the curve, point D yields the weaker shock wave—that is, the one whose inclination, characteristics, and so forth, are closer to the Mach wave at A. The other physically possible shock to produce deflection δ is represented by point E.

Regarding point E, the wave angle appropriate to this shock condition is found by a similar construction (see Fig. 6.28)—that is, by producing Ae to meet the perpendicular from O in e′.

We see that the wave is nearly normal to the flow; the velocity drop to OE from OA is much greater than the previous velocity drop OD for the same flow deflection; the shock is thus said to be the stronger one.

As drawn, OE is within the sonic line, which is an arc of center O and radius $[\hat{q}_n]_{M_2} = 1$, or of radius

$$\frac{a^*}{q_1}\sqrt{\frac{(\gamma-1)M_1^2+2}{\gamma+1}}$$

Point F is where the tangent to the curve through the origin meets the curve, and the angle so found by the tangent line and the $\hat{q}_n$ axis is the maximum possible flow deflection in the given supersonic stream that still retains an attached shock wave. For deflections less than this maximum, the curve is intersected in two physically real points, as shown previously in D and E. Of these two the exit flows OE, which correspond to the strong shock wave case, are always subsonic. The exit flows OD, due to the weaker shocks, are generally *supersonic*, but a few deflection angles close to δ_{max} allow weak shocks with subsonic exit flows. These are represented by point G. In practice, weaker waves occur in uniform flows with plane shocks. When curved detached shocks exist, their properties may be evaluated locally by reference to plane-shock theory; for the near-normal elements, the strong-shock representation OE may be used. Point B is the lower (velocity) limit to the polar curve and represents the normal (strongest) shock configuration in which the incident flow of velocity OA is compressed to the exit flow of velocity OB.

There is no flow deflection through a normal shock wave, which has the maximum reduction to subsonic velocity obtainable for the given undisturbed conditions.

6.7.3 Two-Dimensional Supersonic Flow Past a Wedge

Two-dimensional supersonic flow past a wedge can be described bearing the shock polar in mind. For an attached plane wave, the wedge semi-vertex angle Δ, say, Fig. 6.29, must be less than or equal to the maximum deflection angle δ_{max} given by point F of the polar. The shock wave then sets itself up at the angle given by the weaker shock. The exit flow is uniform and parallel at a lower, generally supersonic, Mach number, but has increased entropy compared with the undisturbed flow.

If the wedge angle Δ is increased, or if the free-stream conditions are altered to allow $\Delta > \beta_{max}$, the shock wave stands detached from the tip of the wedge and is

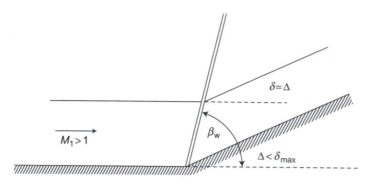

FIGURE 6.29

Geometry of an attached oblique shock at a compression corner. The corner angle must be less than the maximum turning angle for flow at M_1. See Fig. 6.30 for a corner angle greater than the maximum turning angle.

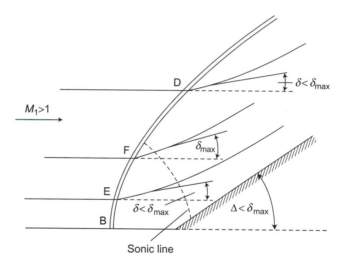

FIGURE 6.30

Geometry of a curved shock at a compression corner. The corner angle must be greater than the maximum turning angle for flow at M_1. See Fig. 6.29 for a corner angle less than the maximum turning angle.

curved from normal to the flow at the dividing streamline to an angle approaching the Mach angle a long way from the axial streamline (Fig. 6.30).

All of the conditions indicated by the closed loop of the shock polar can be identified:

B—on the axis, the flow is undetected but compressed through an element of normal shock, to a subsonic state.

E—a little way away from the axis, the stream deflection through the shock is less than the maximum possible, but the exit flow is still subsonic as a result of the stronger shock.

F—further out, the flow deflection through the shock wave reaches the maximum possible for the free-stream conditions and the exit flow is still subsonic. Beyond this point, elements of shock wave behave in the weaker fashion, giving a supersonic exit for streamlines meeting the shock wave beyond the intersection with the broken sonic line.

D—this point corresponds to the weaker shock wave. Further away from the axis, the inclination of the wave approaches the Mach angle (the case given by point A in Fig. 6.27).

It is evident that a significant variable along the curved wave front is the product $M_1 \sin \beta$, where β is the inclination of the wave locally to the incoming streamline. Uniform undisturbed flow is assumed for simplicity, but this is not a necessary restriction. Now $\mu < \beta < \pi/2$ and $M_1 \sin \beta$, the Mach number of the normal to the wave inlet component velocity, is the effective variable—that is, a maximum on the

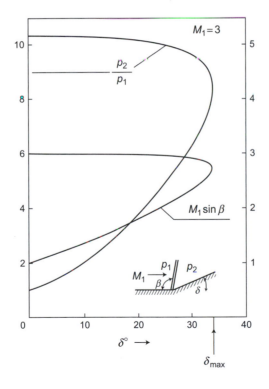

FIGURE 6.31

Curves of upstream normal Mach number (right scale) and pressure ratio (left scale) for points on the curved shock in the flow in Fig. 6.30.

axis—reducing to a minimum at the extremes of the wave (Fig. 6.31). Likewise, all other properties of the flow across the curved wave that are functions of $M_1 \sin \beta$ vary along the shock front. In particular, the entropy jump across the shock, which from Eq. (6.48) is

$$\Delta S = c_V \frac{2\gamma(\gamma-1)}{(\gamma+1)^2} \frac{(M_1^2 \sin^2 \beta - 1)^3}{3} \tag{6.101}$$

varies from streamline to streamline behind the shock wave. An entropy gradient in the flow is associated with rotational flow, and thus a curved shock wave produces a flow in which vorticity exists away from the surface of the associated solid body. At low initial Mach numbers, or with waves of small curvature, we can make the same approximations as those that result from assuming $\Delta S = 0$. For highly curved strong shock waves, such as may occur at hypersonic speed, the downstream flow contains shock-induced vorticity, or the entropy wake, as it is sometimes called, which forms a significant part of the flow in the immediate vicinity of the body.

6.8 EXERCISES

6.1   A convergent-divergent duct has a maximum diameter of 150 mm, and a Pitôt-static tube is placed in its throat. Neglecting the effect of the Pitôt-static tube on the flow, estimate the throat diameter under the following conditions:

   (a)   Air at the maximum section is of standard pressure and density, and the pressure difference across the pitôt-static tube $\equiv$ 127-mm water.

   (b)   Pressure and temperature in the maximum section are 100,300 N m^{-2} and 100°C, respectively, and the pressure difference across the pitôt-static tube $\equiv$ 127-mm mercury.

   (*Answer*: 123 mm; 66.5 mm)

6.2   In the wing-flow method of transonic research, an airplane dives at a Mach number of 0.87 at a height where the pressure and temperature are 46,500 N m^{-2} and –24.6°C, respectively. At the model's position, the pressure coefficient is –0.5. Calculate the speed, Mach number, $0.7\,pM^2$, and kinematic viscosity of the flow past the model.
   (*Answer*: 344 m s^{-1}; $M = 1.133$; $0.7\,pM^2 = 30{,}800 \text{ N m}^{-2}$; $v = 2.64 \times 10^{-3} \text{ m}^2\text{s}^{-1}$)

6.3   What is the indicated air speed and the true air speed of the airplane in Exercise 6.2 assuming that the air-speed indicator is calibrated for incompressible flow in standard conditions and that there are no instrument errors?
   (*Answer*: TAS $= 274 \text{ m s}^{-1}$; IAS $= 219 \text{ m s}^{-1}$)

6.4   On the basis of Bernoulli's equation, discuss the assumption that the compressibility of air may be neglected for low subsonic speeds.

A symmetric airfoil at zero lift has a maximum velocity that is 10% greater than the free-stream velocity. This maximum increases at the rate of 7% of the free-stream velocity for each degree of incidence. What is the free-stream velocity at which compressibility effects begin to become important (i.e., the error in pressure coefficient exceeds 2%) on the airfoil surface when the incidence is 5 degrees?

(*Answer*: Approximately 70 m s^{-1})

6.5   A closed-return wind tunnel with a large contraction ratio has air at standard conditions of temperature and pressure in the settling chamber upstream of the contraction to the working section. Assuming isentropic compressible flow in the tunnel, estimate the speed in the working section where the Mach number is 0.75. Take the ratio of specific heats for air as $\gamma = 1.4$.

(*Answer*: 242 m s^{-1})

6.6   Derive Eq. (6.28b).

6.7   Recreate Fig. 6.25 using `OS_mt.m` instead of `OS_mw.m`, as in the example in this chapter.

6.8   Traditionally, a second chart besides the wave angle versus turning angle chart is helpful for rapidly understanding the impact of oblique shocks on an airflow. This chart is pressure ratio across the shock $\frac{p_2}{p_1}$ versus turning angle for a constant Mach number. Use the MATLAB codes provided to create such a plot for upstream Mach numbers `Mone=[1.25:0.25:3]`.

6.9   At what Mach number does the pressure triple through a normal shock? Solve this three ways:

(a) Using the m-file `NS_unp.m`.
(b) To check the numerical result, solve Eq. (6.44) for Mach number and compute.
(c) Use the Normal Shock Table (NST) for the nearest value. That is, you will not find $\frac{p_2}{p_1} = 3$ in the NST, so use the row in the table (numerically) closest to it. That is the old-fashioned method, but it is still generally the in-class exam method.

6.10 Repeat Exercise 6.9 but for tripling of density.

6.11 Determine how downstream Mach number varies with upstream Mach number for a fixed turning angle. Specifically, for a 6-degree turning angle, plot the downstream versus the upstream Mach number from 1 to 10. Note that there are two solutions for each upstream Mach number and that some Mach numbers may be too low to have a solution for a 6-degree turning angle. It is assumed that you will make use of the compressible flow MATLAB functions as described above. Hand in the plot and the m-file.

6.12 Consider a Mach 2 flow that encounters a 15-degree compression corner and then a 15-degree expansion corner (Fig. Ex6.12a). What are the Mach number and pressure after the expansion corner?

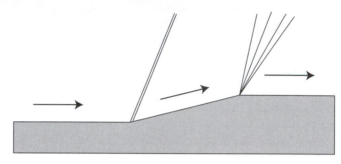

FIGURE Ex6.12a

Consider the case where the expansion corner is before the compression corner (Fig. Ex6.12b). What are the Mach number and pressure after the compression corner?

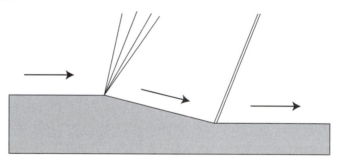

FIGURE Ex6.12b

6.13 A converging-diverging nozzle is supplied with air at a total pressure of 250 psi. The area of the throat is 0.1 in^2. The exit area is 1.6 in^2. Find the following (use of the nearest values in tables is acceptable):

(a) The greatest back pressure for which the throat is choked.
(b) The one value of the back pressure for perfect supersonic expansion.
(c) The lowest back pressure for which there is a shock in the nozzle.

Use these back pressures to determine:

(a) The range of back pressures for which a shock exists in the nozzle.
(b) The range of back pressures for which the flow is overexpanded.
(c) The range of back pressures for which the flow is underexpanded.

6.14 Consider how compressibility in subsonic flight affects the results of the lifting-line theory from Chapter 5. Derive an expression for how the induced drag coefficient for a thin rectangular wing of span b and chord c flying at a small angle of attack α is affected by compressibility in linearized small-perturbation subsonic compressible flow. Assume that you know whatever you need to know about the performance of the wing in incompressible flow; use the subscript o to denote those values when you write them. For example, induced drag coefficient at incompressible conditions is C_{Dio}. The flight Mach number in the subsonic compressible regime is M_∞. The pressure is p_∞, and the temperature is T_∞. There are two common answers to this problem, one is wrong. Be prepared to discuss why yours is correct.

Airfoils and Wings in Compressible Flow

LEARNING OBJECTIVES

- Apply the fluid physics knowledge from Chapter 6 to explore the elementary performance characteristics of airfoils and wings in compressible flow, including the concept of critical Mach number.

- Develop subsonic linearized compressible flow theory for extending the trusted results from incompressible flow into high subsonic flight.

- See how the different fluid physics for supersonic flow lead to a linearized solution different from that for subsonic compressible flow, and realize the remarkable simplicity of the linearized pressure coefficient for supersonic small-perturbation theory.

- Begin to explore the phenomenon of wave drag in supersonic flight and how it is predicted by both the shock-expansion method and linearized supersonic flow.

7.1 WINGS IN COMPRESSIBLE FLOW

In this section, the compressible-flow equations in their various forms are considered in order to predict the behavior of airfoil sections in high sub- and supersonic flows. Except in the descriptive portions of the chapter, the effects of viscosity are largely neglected.

7.1.1 Transonic Flow: The Critical Mach Number

When air flows past a body, or vice versa (e.g., a symmetrical airfoil section at low incidence) the local air speed adjacent to the surface just outside the boundary layer is higher or lower than the free-stream speed depending on whether local static pressure is less or greater than ambient pressure. In such a situation, the value of the velocity somewhere on the airfoil exceeds that of the free stream. Thus, as the free-stream flow speed rises, the Mach number at a point somewhere adjacent to the surface reaches sonic conditions before the free stream. This point is usually the minimum-pressure point, which in this case is on the upper surface. The value of the free-stream

Mach number (M_∞) at which the flow somewhere on the surface first reaches $M = 1$ is called the *critical* Mach number M_c. Typically for a slender wing section at low incidence, M_c may be about 0.75. Below that, the flow is subsonic throughout.

Above the critical Mach number, the flow is mixed—part supersonic part subsonic. As M_∞ increases progressively from low numbers to M_c, the aerodynamic characteristics of the airfoil section undergo progressive and generally smooth changes; for thin airfoil shapes at low incidences, these changes may be predicted by the small-perturbation or linearized theory due to Prandtl and Glauert, which we will discuss momentarily.

As M_∞ increases progressively beyond M_c, a limited region in which the flow is supersonic develops from the point where the flow first became sonic and grows outward and downstream, terminating in a shock wave that is at first approximately normal to the surface. With further increases in M_∞, the shock wave becomes stronger and longer and moves rearward. At some stage, at a value of $M_\infty > M_c$, the velocity somewhere on the lower surface approaches and passes the sonic value; a supersonic region terminating in a shock wave appears on the lower surface, which also grows stronger and moves back as the lower supersonic region increases.

Eventually, at a value of M_∞ close to unity, the upper and lower shock waves reach the trailing edge. In their rearward movement, they approach the trailing edge generally at different rates, the lower typically starting later and ahead of the upper, although moving more rapidly and overtaking the upper before reaching the trailing edge. When the free-stream Mach number reaches unity, a bow shock wave appears at a small standoff distance from the rounded leading edge. For higher Mach numbers, the extremes of the bow and trailing waves incline rearward to approach the Mach angle. For round-nosed airfoils or bodies, the bow wave is a "strong" one and always stands off; a small subsonic region exists around the front stagnation point. This sequence is shown in Fig. 7.1. For sharp leading edges, the bow shock waves are plane, and usually "weak," with the downstream flow still supersonic, at a lower Mach number. This case will be dealt with separately.

The effect on the airfoil characteristics of the flow sequence described previously is dramatic. The sudden loss of lift, increase in drag, and rapid movement in the center of pressure are similar in flight to those experienced at the stall; this flight regime is known as *shock stall*. Many of its effects can be minimized or delayed by design methods that are beyond the scope of the present volume.

To appreciate why the airfoil characteristics change so dramatically with supersonic flow, recall the properties of shock waves. Across a shock wave, the pressure rise is large and sudden. Moreover, the wave process is accompanied by an entropy increase that manifests itself as a total pressure loss, which is drag. In other words, an irreversible conversion of mechanical energy to heat (which is dissipated) takes place; sustaining this loss results in drag. The drag increase is directly related to wave strength, shock which in turn depends on the magnitude of the supersonic regions ahead. Another contribution to drag occurs if the boundary layer at the foot of the shock separates from the airfoil as a consequence of accommodating the large adverse pressure gradient.

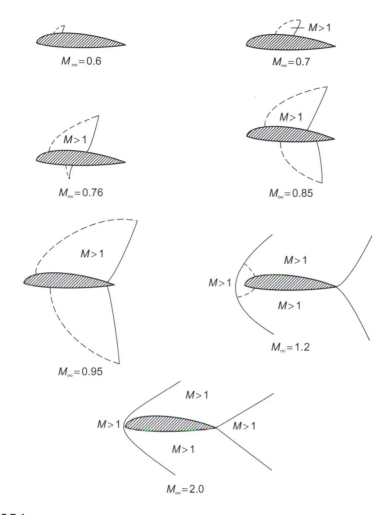

FIGURE 7.1

Flow development on a two-dimensional airfoil as M_∞ increases beyond M_c; in this case, $M_c = 0.58$.

In contrast, lift continues to rise smoothly with an increase in $M_\infty > M_c$ as a consequence of the increased low-pressure area on the upper surface. This sequence is seen in Fig. 7.2. Lift does not decrease significantly until the low-pressure area on the lower surface becomes appreciably large owing to the growth of the supersonic region there (Fig. 7.2(c)). The presence of the shock wave can be detected by the sharp vertical pressure recovery terminating the supersonic regions (the shaded areas in Fig. 7.2). It is apparent that the marked effect on lift is associated more with growth of the shock wave on the lower surface. Movement of the center of pressure also

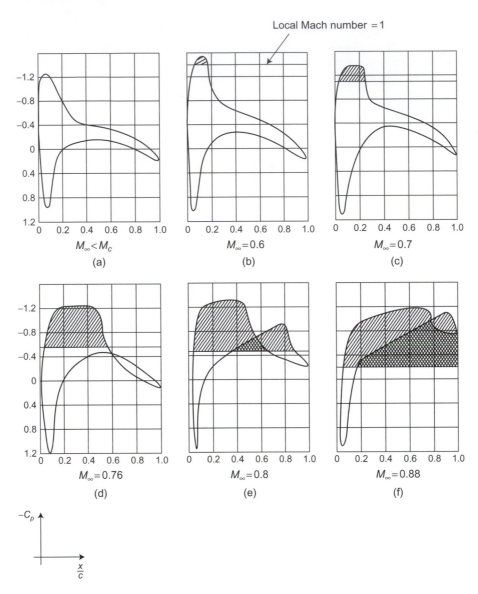

FIGURE 7.2

Pressure distribution on a two-dimensional airfoil ($M_c = 0.57$) as M_∞ increases through M_c. Important differences between the plots are discussed in the text.

follows because of the varying pressure distributions; it is particularly marked as the lower shock wave moves behind the upper at the higher Mach numbers approaching unity.

Note also from the pressure distributions (e) and (f) in Fig. 7.2 that the pressure recovery at the trailing edge is incomplete. This is due to flow separation at the feet of the shock waves, which leads to buffeting of any control surface near the trailing edge. It is worth noting as well that, even if the flow remains attached the pressure information that needs to propagate to the pressure distribution by a control movement (say) cannot propagate upstream through the supersonic region. Thus the effectiveness of a trailing-edge control surface is much reduced. As the free-stream Mach number M_∞ becomes supersonic, the flow over the airfoil, except for the small region near the stagnation point, is supersonic, and the shock system stabilizes to a form similar to the supersonic case shown in Fig. 7.1.

7.1.2 Subcritical Flow: The Small-Perturbation Theory (Prandtl-Glauert Rule)

In certain cases of compressible flow, notably in supersonic flow, exact solutions to the equations of motion may be found (always assuming the fluid to be inviscid). When these equations are applied to flow in the vicinity of airfoils, they are called *exact* theories. As described, airfoils in motion near the speed of sound, in the transonic region, have a mixed-flow regime where regions of subsonic and supersonic flow exist side by side around the airfoil. Mathematically, the analysis of this regime involves solving a set of nonlinear differential equations, a task that demands either advanced computational techniques or some form of approximation.

The most sweeping approximations, producing the simplest solutions, are made here and result in soluble linear differential equations. This leads to the expression *linearized theory* associated with airfoils in, for example, high-subsonic or low-supersonic flows. The approximations come about mainly from assuming that all disturbances are *small disturbances* or *small perturbations* to the free-stream flow conditions. As a consequence, these two terms are associated with the development of the theory.

Historically, H. Glauert was engaged in early theoretical treatment of compressibility effects on airfoils approaching the speed of sound, and he developed what are, in essence, the linearized equations for subsonic compressible flow, in *ARCR&M* (1927), in a note that had been previously published by the Royal Society [48]. In this, he mentions the same results seen by Prandtl in 1922. For this reason, the significant compressibility effect in subsonic flow is called the *Prandtl-Glauert rule* (or law).

Although the theory takes no account of viscous drag or the onset of shock waves in localized regions of supersonic flow, the relatively crude experimental results at the time (obtained from analysis of tests on an airscrew) did indicate the now well-investigated critical region of flight where the theory breaks down. Glauert suggested that the critical speed at which lift falls off depends on the shape and incidence of the airfoil, which is now well substantiated.

In what follows, we look at the approximate methods of satisfying the equations of motion for an inviscid compressible fluid. These depend on the simultaneous solution of the fundamental laws of conservation and state. Initially, a single equation is desired that combines all of the physical requirements. The complexity of this equation, and whether it is amenable to solution, depends on the nature of the problem and on quantities that may be conveniently minimized.

The Equations of Motion of a Compressible Fluid

The equation of continuity may be recalled in Cartesian coordinates for two-dimensional flow in the form

$$\frac{\partial \rho}{\partial t} + \frac{\partial (\rho u)}{\partial x} + \frac{\partial (\rho v)}{\partial y} = 0 \tag{7.1}$$

since, in what follows, analysis of two-dimensional conditions is sufficient to demonstrate the method and derive valuable equations. The equations of motion may also be recalled in similar notation as

$$\left.\begin{aligned} \frac{\partial u}{\partial t} + u\frac{\partial u}{\partial x} + v\frac{\partial u}{\partial y} &= -\frac{1}{\rho}\frac{\partial p}{\partial x} \\ \frac{\partial v}{\partial t} + u\frac{\partial v}{\partial x} + v\frac{\partial v}{\partial y} &= -\frac{1}{\rho}\frac{\partial p}{\partial y} \end{aligned}\right\} \tag{7.2}$$

and, for steady flow,

$$\left.\begin{aligned} -\frac{1}{\rho}\frac{\partial p}{\partial x} &= u\frac{\partial u}{\partial x} + v\frac{\partial u}{\partial y} \\ -\frac{1}{\rho}\frac{\partial p}{\partial y} &= u\frac{\partial v}{\partial x} + v\frac{\partial v}{\partial y} \end{aligned}\right\} \tag{7.3}$$

For adiabatic flow (since the assumption of negligible viscosity has already been made, further stipulations of adiabatic compression and expansion imply isentropic flow),

$$p = k\rho^\gamma, \quad \frac{\partial p}{\partial \rho} = a^2 = \frac{\gamma p}{\rho} \tag{7.4}$$

For steady flow, Eq. (7.1) may be expanded to

$$u\frac{\partial \rho}{\partial x} + v\frac{\partial \rho}{\partial y} + \rho\frac{\partial u}{\partial x} + \rho\frac{\partial v}{\partial y} = 0 \tag{7.5}$$

but

$$\frac{\partial \rho}{\partial x} = \frac{\partial p}{\partial x}\frac{\partial \rho}{\partial p} = \frac{1}{a^2}\frac{\partial p}{\partial x}, \text{ etc.}$$

so Eq. (7.5) becomes

$$\frac{u}{a^2}\frac{\partial p}{\partial x} + \frac{v}{a^2}\frac{\partial p}{\partial y} + \rho\frac{\partial u}{\partial x} + \rho\frac{\partial v}{\partial y} = 0 \tag{7.6}$$

Substituting in Eq. (7.6) for $\partial p/\partial x$, $\partial p/\partial y$ from Eq. (7.3) and cancelling ρ gives

$$-\frac{u^2}{a^2}\frac{\partial u}{\partial x} - \frac{uv}{a^2}\frac{\partial u}{\partial y} - \frac{vu}{a^2}\frac{\partial v}{\partial x} - \frac{v^2}{a^2}\frac{\partial v}{\partial y} + \frac{\partial u}{\partial x} + \frac{\partial v}{\partial y} = 0$$

or, collecting like terms,

$$\left(1 - \frac{u^2}{a^2}\right)\frac{\partial u}{\partial x} - \frac{uv}{a^2}\left(\frac{\partial v}{\partial x} + \frac{\partial u}{\partial y}\right) + \left(1 - \frac{v^2}{a^2}\right)\frac{\partial v}{\partial y} = 0 \qquad (7.7)$$

For irrotational flow $\partial v/\partial x = \partial u/\partial y$, and a velocity potential φ_1 (say) exists, so

$$\left(1 - \frac{u^2}{a^2}\right)\frac{\partial u}{\partial x} - \frac{2uv}{a^2}\frac{\partial u}{\partial y} + \left(1 - \frac{v^2}{a^2}\right)\frac{\partial v}{\partial y} = 0 \qquad (7.8a)$$

and since $u = \partial\varphi_1/\partial x, v = \partial\varphi_1/\partial y$, Eq. (7.8a) can be written as

$$\left(1 - \frac{u^2}{a^2}\right)\frac{\partial^2\phi_1}{\partial x^2} - \frac{2uv}{a^2}\frac{\partial^2\phi_1}{\partial x\partial y} + \left(1 - \frac{v^2}{a^2}\right)\frac{\partial^2\phi_1}{\partial y^2} = 0 \qquad (7.8b)$$

Finally, the energy equation provides the relation between a, u, v, and acoustic speed. Thus

$$\frac{u^2 + v^2}{2} + \frac{a^2}{\gamma - 1} = \text{constant} \qquad (7.9)$$

or

$$\left(\frac{\partial\phi_1}{\partial x}\right)^2 + \left(\frac{\partial\phi_1}{\partial y}\right)^2 + \frac{2}{\gamma - 1}a^2 = \text{constant} \qquad (7.9a)$$

Combining Eqs. (7.8b) and (7.9a) gives an expression in terms of the local velocity potential.

Even without continuing the algebra beyond this point, we note that the resulting nonlinear differential equation in φ_1 is not amenable to a simple closed solution and that further restrictions on the variables are required. Since all possible restrictions on the generality of the flow properties have been made, it is necessary to consider the component velocities themselves.

Small Disturbances

So far we have tacitly assumed that the flow is steady at infinity and that the local-flow velocity has components u and v parallel to coordinate axes x and y, respectively, with the origin of the coordinate axes furnishing the necessary datum. Let the equations now refer to a class of flows in which the velocity changes only slightly from its steady value at infinity and in which the velocity gradients themselves are small (thin

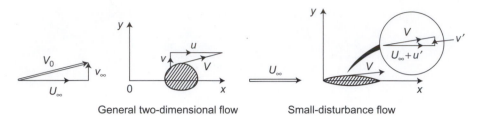

General two-dimensional flow　　**Small-disturbance flow**

FIGURE 7.3

Velocity vector, components, and perturbations over typical blunt *(left)* and slender *(right)* bodies.

wings at low incidence, etc.). Further, identify the *x*-axis with the undisturbed flow direction (see Fig. 7.3). The local velocity components *u* and *v* can now be written as

$$u = U_\infty + u', \ v = v'$$

where u' and v' are small compared to the undisturbed stream velocity and are termed the perturbation or disturbance velocities. These may be expressed nondimensionally in the form

$$\frac{u'}{U_\infty} \ll 1, \quad \frac{v'}{U_\infty} \ll 1$$

Similarly, $\partial u'/\partial x, \partial v'/\partial y$ are small.

Making this substitution, Eq. (7.9) becomes

$$\frac{(U_\infty + u')^2 + v'^2}{2} + \frac{a^2}{\gamma - 1} = \frac{U_\infty^2}{2} + \frac{a_\infty^2}{\gamma - 1}$$

When the squares of small quantities are neglected, this equation simplifies to

$$U_\infty u' = \frac{a_\infty^2 - a^2}{\gamma - 1}$$

Similarly,

$$a^2 = a_\infty^2 - (\gamma - 1)U_\infty u'$$

Thus the coefficient terms of Eq. (7.8a) become

$$1 - \frac{u^2}{a^2} = 1 - \left(\frac{U_\infty + u'}{a}\right)^2 = 1 - \frac{U_\infty^2 + 2U_\infty u'}{a_\infty^2 - (\gamma - 1)\,U_\infty u'}$$

Putting $U_\infty/a_\infty = M_\infty$, the free-stream Mach number is

$$1 - \left(\frac{u}{a}\right)^2 = 1 - M_\infty^2 \left(\frac{1 + \frac{2u'}{U_\infty}}{1 - (\gamma - 1)M_\infty^2(u'/U_\infty)}\right)$$

$$= 1 - M_\infty^2 \left(\frac{u'[2 + (\gamma - 1)M_\infty^2}{U_\infty[1 - (\gamma - 1)M_\infty^2(u'/U_\infty)]}\right)$$

and

$$\frac{2uv}{a^2} = \frac{2(U_\infty + u')}{a^2}v' = \frac{2U_\infty v'}{a_\infty^2 - (\gamma - 1)U_\infty u'}$$

$$= M_\infty^2 \frac{\frac{2v'}{U_\infty}}{1 - [(\gamma - 1)M_\infty^2(u'/U_\infty)]}$$

Also, $1 - (v'/U_\infty)^2 = 1$ from the small disturbance assumption.

Now, if the velocity potential φ_1 is expressed as the sum of a velocity potential due to the flow at infinity plus a velocity potential due to the disturbance (i.e., $\varphi_1 = \varphi_\infty + \varphi$), Eq. (7.8b) becomes, with slight rearrangement,

$$\left(1 - M_\infty^2\right)\frac{\partial^2 \phi}{\partial x^2} + \frac{\partial^2 \phi}{\partial y^2} = \frac{M_\infty^2}{1 - (\gamma - 1)M_\infty^2 u'/U_\infty}$$

$$\times \left[[2 + (\gamma - 1)M_\infty^2]\frac{u'}{U_\infty}\frac{\partial^2 \phi}{\partial x^2} + \frac{2v'}{U_\infty}\frac{\partial^2 \phi}{\partial x \partial y}\right] \quad (7.10)$$

where φ is the disturbance potential and $u' = \partial\varphi/\partial x, v' = \partial\varphi/\partial y$, and so on.

The right-hand side of Eq. (7.10) vanishes when $M_\infty = 0$, and the coefficient of the first term becomes unity so that the equation reduces to the Laplace equation—that is, when $M_\infty = 0$, Eq. (7.10) becomes

$$\frac{\partial^2 \phi}{\partial x^2} + \frac{\partial^2 \phi}{\partial y^2} = 0 \quad (7.11)$$

Since velocity components and their gradients are of the same small order their products can be neglected and the bracketed terms on the right-hand side of Eq. (7.10) become negligibly small. This controls the magnitude of the right-hand side, which can therefore be assumed essentially zero unless the remaining quantity outside the bracket becomes large or indeterminate. This occurs when $M_\infty^2(\gamma - 1)u'/U_\infty \to 1$,

$$\frac{M_\infty^2}{\left[1 - (\gamma - 1)M_\infty^2\frac{u'}{U_\infty}\right]} \to \infty$$

We see that, by assigning reasonable values to u'/U_∞ and γ, the equality is made when $M_\infty = 5$. In other words, put $u'/U_\infty = 0.1, \gamma = 1.4$; then $M_\infty^2 = 25$.

Within the limitations given, the equation of motion reduces to the linear equation

$$(1 - M_\infty^2)\frac{\partial^2\phi}{\partial x^2} + \frac{\partial^2\phi}{\partial y^2} = 0 \qquad (7.12)$$

A further limitation in the application of Eq. (7.10) occurs when M_∞ has a value in the vicinity of unity—that is, where the flow regime may be described as transonic. Inspection of Eq. (7.10) also shows a fundamental change in form as M_∞ approaches and passes unity: the quantity $(1 - M_\infty^2)$ changes sign, the equation changing from elliptic to hyperbolic.

Because of these restrictions, the further application of the equations is most useful in the high subsonic region, where $0.4 < M_\infty < 0.8$, and in the supersonic region, where $1.2 < M_\infty < 5$.

To extend our theoretical investigation to transonic or hypersonic Mach numbers requires further development of the equations, which we do not consider here.

Prandtl-Glauert Rule: The Application of Linearized Theories of Subsonic Flow

Consider Eq. (7.12) in the subsonic two-dimensional form:

$$(1 - M_\infty^2)\frac{\partial^2\phi}{\partial x^2} + \frac{\partial^2\phi}{\partial y^2} = 0$$

For a given Mach number M_∞, this equation can be written as

$$B^2\frac{\partial^2\phi}{\partial x^2} + \frac{\partial^2\phi}{\partial y^2} = 0 \qquad (7.13)$$

where B is a constant. This bears a superficial resemblance to the Laplace equation:

$$\frac{\partial^2\Phi}{\partial \xi^2} + \frac{\partial^2\Phi}{\partial \eta^2} = 0 \qquad (7.14)$$

Moreover, if the problem expressed by Eq. (7.12)—that of finding φ for the subsonic compressible flow around a thin airfoil, say—can be transformed into an equation such as Eq. (7.14), its solution is possible by standard methods.

Figure 7.4 shows the thin airfoil occupying the Ox-axis because it is thin, in the definitive sense, in the *real* or *compressible* plane, where the velocity potential φ exists in the region defined by the (x,y) ordinates. The corresponding airfoil in the *Laplace* or *incompressible* (ξ,η) plane has a velocity potential Φ. If the simple relations

$$\Phi = A\phi, \quad \xi = Cx, \quad \text{and} \quad \eta = Dy \qquad (7.15)$$

are assumed, where A, C, and D are constants, the transformation can proceed.

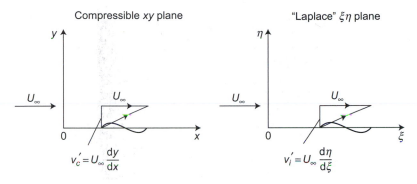

Compressible *xy* plane   "Laplace" $\xi\eta$ plane

FIGURE 7.4

Correspondence between the real plane (*left*) and the transformed, or solution, plane (*right*).

The boundary conditions on the airfoil surface demand that the flow be locally tangential to it so in each plane, respectively,

$$v'_c = U_\infty \frac{dy}{dx} = \left(\frac{\partial\phi}{\partial y}\right)_{y=0} \tag{7.16}$$

and

$$v'_i = U_\infty \frac{d\eta}{d\xi} = \left(\frac{\partial\Phi}{\partial\eta}\right)_{\eta=0} \tag{7.17}$$

where the subscripts c and i denote the compressible and incompressible planes.

Using the simple relationships of Eq. (7.15) gives

$$\frac{\partial\Phi}{\partial\xi} = \frac{\partial(A\phi)}{\partial(Cx)} = \frac{A}{C}\frac{\partial\phi}{\partial x}, \quad \frac{\partial^2\Phi}{\partial\xi^2} = \frac{A}{C^2}\frac{\partial^2\phi}{\partial x^2}$$

and

$$\frac{\partial\Phi}{\partial\eta} = \frac{A}{D}\frac{\partial\phi}{\partial y}, \quad \frac{\partial^2\Phi}{\partial\eta^2} = \frac{A}{D^2}\frac{\partial^2\phi}{\partial y^2}$$

Thus Eq. (7.14), by substitution and rearrangement of constants, becomes

$$\frac{\partial^2\Phi}{\partial\xi^2} + \frac{\partial^2\Phi}{\partial\eta^2} = \frac{A}{D^2}\left[\left(\frac{D^2}{C^2}\right)\frac{\partial^2\phi}{\partial x^2} + \frac{\partial^2\phi}{\partial y^2}\right] = 0 \tag{7.18}$$

Comparison of Eqs. (7.18) and (7.13) shows that a solution to Eq. (7.14) (the left-hand part of Eq. 7.18) provides a solution to Eq. (7.13) (the right-hand part of Eq. 7.13) if the bracketed constant can be identified as B^2—that is, when

$$\frac{D}{C} = B = \sqrt{1 - M_\infty^2} \tag{7.19}$$

Without generalizing further, two simple procedures emerge from Eq. (7.19): making C or D unity when $D = B$ or when $1/C = B$, respectively. Since C and D control the spatial distortion in the Laplace plane, the two procedures reduce to the distortion of one or the other of the two ordinates.

Constant Chordwise Ordinates

If the airfoil is thin—and by definition this must be so for the small-disturbance conditions of the theory from which Eq. (7.12) is derived—the implication is that the airfoils are of similar shape in both planes. Take the case of $C = 1$ (i.e., $\xi = x$), which gives $D = B = \sqrt{1 - M_\infty^2}$ from Eq. (7.19). A solution to Eq. (7.12) or Eq. (7.13) is found by applying the transformation $\eta = \sqrt{1 - M^2}\, y$ (see Fig. 7.5). Then Eqs. (7.16) and (7.17) give

$$v_c' = U_\infty \frac{dy}{dx} = \left[\frac{\partial \phi}{\partial y}\right]_{y=0} = \frac{B}{A}\left[\frac{\partial \Phi}{\partial \eta}\right]_{\eta=0} = \frac{B}{A} U_\infty \frac{d\eta}{d\xi}$$

but $D = B$, since $D/C = B$ and $C = 1$.

For similar airfoils, it is required that $dy/dx = d\eta/d\xi$ at corresponding points. For this to be so,

$$A = B = \sqrt{1 - M_\infty^2} \tag{7.20}$$

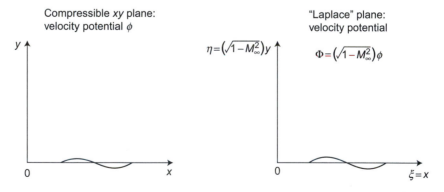

Compressible xy plane: velocity potential ϕ

"Laplace" plane: velocity potential

$\eta = \left(\sqrt{1 - M_\infty^2}\right) y$

$\Phi = \left(\sqrt{1 - M_\infty^2}\right)\phi$

$\xi = x$

FIGURE 7.5

Transformation for the choice of constant chordwise ordinates in linearized subsonic flow theory.

The transformed potential is thus

$$\Phi = \sqrt{1 - M_\infty^2}\, \phi \tag{7.21}$$

The horizontal flow perturbations are now easily found:

$$u_c' = \frac{\partial \phi}{\partial x} = \frac{1}{\sqrt{1 - M_\infty^2}} \frac{\partial \Phi}{\partial x} = \frac{u_i'}{\sqrt{1 - M_\infty^2}} \tag{7.22}$$

$$C_{p_c} = \frac{-2u_c'}{U_\infty} = \frac{-1}{\sqrt{1 - M_\infty^2}} \frac{-2u_i'}{U_\infty} = \frac{C_{pi}}{\sqrt{1 - M_\infty^2}} \tag{7.23}$$

Since

$$C_L = \frac{1}{c} \oint C_p dx$$

the relationship between compressible and incompressible lift coefficients follows that of the pressure coefficients:

$$C_L = \frac{C_{L_i}}{\sqrt{1 - M_\infty^2}} \tag{7.24}$$

This simple use of the factor $\sqrt{1 - M_\infty^2}$ is known as the Prandtl-Glauert rule (or law), and $\sqrt{1 - M_\infty^2}$ is known as Glauert's factor.

Constant Normal Ordinates
Glauert developed the affine transformation implicit in taking a transformed plane distorted in the x direction. The consequence is that, for a thin airfoil, the transformed section about which the potential Φ exists has its chordwise lengths altered by the factor $1/C$ (Fig. 7.6).

With $D = 1$ (i.e., $\eta = y$), Eq. (7.25) gives

$$C = \frac{1}{B} = \frac{1}{\sqrt{1 - M_\infty^2}} \tag{7.25}$$

Thus the solution to Eq. (7.12) or Eq. (7.13) is found by applying the transformation

$$\xi = \frac{x}{\sqrt{1 - M_\infty^2}}$$

For this and the geometrical condition of Eqs. (7.16) and (7.17) to apply, A can be found. Eq. (7.16) gives

$$v_c' = U_\infty \frac{dy}{dx} = \left[\frac{\partial \phi}{\partial y} \right]_{y=0} = \frac{1}{A} \left[\frac{\partial \Phi}{\partial \eta} \right]_{\eta=0}$$

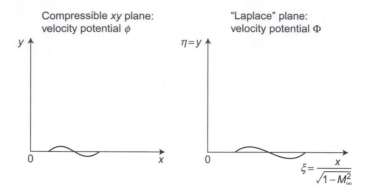

FIGURE 7.6

Transformation for the choice of constant normal ordinates in linearized subsonic flow theory.

By substituting $\Phi = A\varphi$ (Eq. 7.15), $y = \eta$ (Eq. 7.25). However, from Eq. (7.16),

$$v_c' = \frac{U_\infty}{A} \frac{d\eta}{d\zeta} = \frac{U_\infty}{A} \frac{dy}{dx} \sqrt{1 - M_\infty^2}$$

To preserve the identity, $A = \sqrt{1 - M_\infty^2}$ and the transformed potential $\Phi = \sqrt{1 - M_\infty^2}\phi$, as shown in Eq. (7.21). The horizontal flow perturbations, pressure coefficients, and lift coefficients follow as before.

Glauert explained the latter transformation in physical terms by showing that the flow at infinity in both the original compressible plane and the transformed ideal, or Laplace plane is the same, and hence the overall lifts to the systems are the same. But the chord of the ideal airfoil is greater (because of the $\xi = x/\sqrt{1 - M_\infty^2}$ distortion) than that of the equivalent compressible airfoil, and thus, for an identical airfoil in the compressible plane, the lift is greater than that in the ideal (or incompressible) case. The ratio L_c/L_i is as before $(1 - M_\infty^2)^{-1/2}$.

Critical Pressure Coefficient

The pressure coefficient of the point of minimum pressure on an airfoil section, using the notation of Fig. 7.7(b), is

$$C_{p_{min}} = \frac{p_{min} - p_\infty}{\frac{1}{2}\rho_\infty V_\infty^2} \tag{7.26}$$

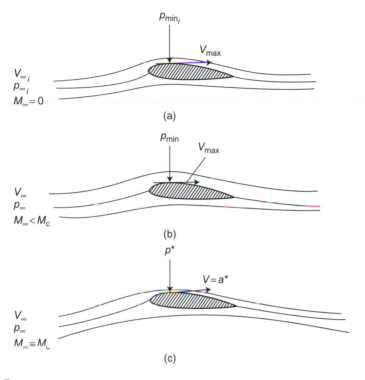

FIGURE 7.7

(a) "Incompressible" flow. (b) Compressible subcritical flow. (c) Critical flow.

but since $\frac{1}{2}\rho_\infty V_\infty^2 = \frac{1}{2}\gamma p_\infty M_\infty^2$, Eq. (7.26) may be written as

$$C_{p_{min}} = \left[\frac{p_{min}}{p_\infty} - 1\right]\frac{2}{\gamma M_\infty^2} \tag{7.27}$$

The critical condition obtains when p_{min} first reaches the sonic pressure p^* and M_∞ becomes M_c (Fig. 7.7(c)). $C_{p_{min}}$ is then the critical pressure coefficient of the airfoil section. Thus

$$C_{p_{crit}} = \left[\frac{p^*}{p_\infty} - 1\right]\frac{2}{\gamma M_c^2} \tag{7.28}$$

An expression for p^*/p_∞ in terms of M_c may be readily found by recalling the energy equation applied to isentropic flow along a streamline (see Section 6.2), which in the present notation gives

$$\frac{V_\infty^2}{2} + \frac{a_\infty^2}{\gamma - 1} = \frac{V^2}{2} + \frac{a^2}{\gamma - 1}$$

Dividing through by a_∞^2 for the condition when $V = a = a^*, M_\infty = M_c$, and so on,

$$\frac{M_c^2}{2} + \frac{1}{\gamma - 1} = \left(\frac{a^*}{a_\infty}\right)\left(\frac{1}{2} + \frac{1}{\gamma - 1}\right)$$

Rearranging,

$$\left(\frac{a^*}{a_\infty}\right)^2 = M_c^2 \frac{\gamma - 1}{\gamma + 1} + \frac{2}{\gamma + 1}$$

However,

$$\left(\frac{p^*}{p_\infty}\right) = \left(\frac{a^*}{a_\infty}\right)^{2\gamma/(\gamma-1)}$$

and substituting p^*/p_∞ in Eq. (7.28),

$$C_{p_{\text{crit}}} = \left[\left(\frac{\gamma - 1}{\gamma + 1}M_c^2 + \frac{2}{\gamma + 1}\right)^{\gamma/(\gamma-1)} - 1\right]\frac{2}{\gamma M_c^2} \tag{7.29}$$

In this expression, M_c is the critical Mach number of the wing and is the parameter often required to be found. $C_{p_{\text{crit}}}$ is the pressure coefficient at the point of maximum velocity on the wing when locally sonic conditions are just attained, and it is usually unknown in practice. It must be predicted from the corresponding minimum pressure coefficient (C_{p_i}) in incompressible flow. C_{p_i} may be obtained from pressure-distribution data from low-speed models or, as previously, from the solution of the Laplace equation of a potential flow.

The approximate relationship between $C_{p_{\text{crit}}}$ and C_{p_i} was discussed earlier for two-dimensional wings. The Prandtl-Glauert rule gives

$$C_{p_{\text{crit}}} = \frac{C_{p_i}}{\sqrt{1 - M_c^2}} \tag{7.30}$$

A simultaneous solution of Eqs. (7.29) and (7.30) with a given C_{p_i} yields values of the critical Mach number M_{crit}.

Application to Swept Wings

In the same way as for the incompressible case (see Section 5.7), compressible flow over an infinite-span swept (or sheared) wing can be considered the superposition of two flows. One is the flow perpendicular to the swept leading edge, the other is the flow parallel to the leading edge. The free-stream velocity now consists of two components (see Fig. 7.8). For the component perpendicular to the leading edge, Eq. (7.12) becomes

$$(1 - M_\infty^2 \cos^2 \Lambda)\frac{\partial^2 \phi}{\partial x^2} + \frac{\partial^2 \phi}{\partial y^2} \tag{7.31}$$

Only the perpendicular component affects the pressure, so Eqs. (7.22) and (7.23) become

$$(u_c')_n = \frac{\partial \phi}{\partial x} = \frac{1}{\sqrt{1 - M_\infty^2 \cos^2 \Lambda}}\frac{\partial \Phi}{\partial x} = \frac{(u_i')_n}{\sqrt{1 - M_\infty^2 \cos^2 \Lambda}} \tag{7.32}$$

$$C_{p_c} = -2\frac{\cos \Lambda\,(u_c')_n}{U_\infty} = -2\frac{\cos \Lambda\,(u_i')_n}{U_\infty\sqrt{1 - M_\infty^2 \cos^2 \Lambda}} \tag{7.33}$$

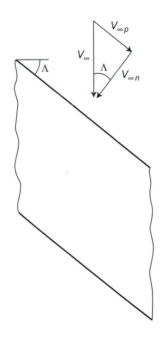

FIGURE 7.8

Definitions for wing sweep of a constant-chord wing.

It follows directly that

$$C_L = \frac{C_{Li}}{\sqrt{1 - M_\infty^2 \cos^2 \Lambda}} \tag{7.34}$$

Example 7.1

For the NACA 4-digit series of symmetrical airfoil sections in incompressible flow, the maximum disturbance velocity $(u'/V_\infty)_{max}$ (corresponding to $(C_p)_{min}$) varies, as shown in the table, with the thickness-to-chord ratio t/c.

NACA Airfoil Designation	t/c	$(u'/V_\infty)_{max}$
NACA0006	0.06	0.107
NACA0008	0.08	0.133
NACA0010	0.10	0.158
NACA0012	0.12	0.188
NACA0015	0.15	0.233
NACA0018	0.18	0.278
NACA0021	0.21	0.325
NACA0024	0.24	0.374

Use this data to determine the critical Mach number for

- A straight wing of infinite span with a NACA0010 wing section
- An infinite-span wing with a 45-degree sweep-back with the same wing section perpendicular to the leading edge.

All 4-digit NACA wing sections are essentially the same shape, but have different thickness-to-chord ratios, as denoted by the last two digits. Thus a NACA0010 wing section at a given free-stream Mach number M_∞ is equivalent to a 4-digit NACA series in incompressible flow having a thickness of

$$(t/c)_i = 0.10\sqrt{1 - M_\infty^2}$$

The maximum disturbance velocity, $[(u'/V_\infty)_{max}]_i$, for such a wing section is obtained by linear interpolation on the data in the previous table. The maximum perturbation velocity in the actual compressible flow at M_∞ is given by

$$(u'/V_\infty)_{max} = \frac{1}{1 - M_\infty^2}\left[(u'/V_\infty)_{max}\right]_i \tag{a}$$

The maximum local Mach number is given approximately by

$$M_{max} \simeq \frac{V_\infty + (u_c)_{max}}{a_\infty} = \frac{V_\infty}{a_\infty}\left(1 + \frac{(u_c)_{max}}{V_\infty}\right) = M_\infty\left(1 + \frac{(u_c)_{max}}{V_\infty}\right) \tag{b}$$

Equations (a) and (b) and linear interpolation can be used to determine M_{max} for a specified M_∞. The results are set out in the following table:

M_∞	$\sqrt{1-M_\infty^2}$	$\left[\left(\frac{u'}{V_\infty}\right)_{max}\right]_i$	$\left[\left(\frac{u'}{V_\infty}\right)_{max}\right]_c$	M_{max}
0.5	0.866	0.141	0.188	0.594
0.6	0.08	0.133	0.2078	0.725
0.7	0.0714	0.120	0.2353	0.865
0.75	0.066	0.114	0.2606	0.945
0.8	0.06	0.107	0.2972	1.038

Linear interpolation between $M_\infty = 0.75$ and $M_\infty = 0.8$ gives the critical value of $M_\infty = 0.78$ (i.e., corresponding to $M_{max} = 1.0$).

For the 45-degree swept-back wing,

$$(t/c)_i = 0.10\sqrt{1 - M_\infty^2 \cos^2 \Lambda} = 0.10\sqrt{1.0 - 0.5M_\infty^2}$$

V_∞ must be replaced by $(V_\infty)_n$ (i.e., $V_\infty \cos \Lambda$), so the maximum disturbance velocity is given by

$$\left[\left(\frac{u'}{V_{\infty n}}\right)_{max}\right]_c = \frac{1}{1 + 0.5M_\infty^2}\left[\left(\frac{u'}{V_{\infty n}}\right)_{max}\right]_i$$

The maximum local Mach number is then obtained from

$$M_{max} \simeq M_\infty \left\{1 + \left[\left(\frac{u'}{V_{\infty n}}\right)_{max}\right]_c\right\}$$

Thus, in a similar way as for the straight wing, the following table is obtained. Linear interpolation gives a critical Mach number of about 0.87.

M_∞	$\sqrt{1-M_\infty^2}$	$\left[\left(\frac{u'}{V_\infty}\right)_{max}\right]_i$	$\left[\left(\frac{u'}{V_\infty}\right)_{max}\right]_c$	M_{max}
0.5	0.0935	0.144	0.116	0.558
0.6	0.0906	0.143	0.123	0.674
0.7	0.0869	0.141	0.132	0.792
0.8	0.0825	0.136	0.141	0.913
0.85	0.0799	0.133	0.147	0.975
0.9	0.0771	0.128	0.152	1.037

Note that, although the wing section is the same in both cases, the critical Mach number is much higher for the swept-back wing principally because $V_{\infty n}$ is considerably less than V_∞.

For this reason, aircraft that are designed to cruise at high subsonic Mach numbers, such as airliners, invariably have swept-back wings in order to keep wave drag to a minimum.

7.1.3 Supersonic Linearized Theory (Ackeret's Rule)

Before considering a solution to the supersonic form of the simplified (small-perturbation) equation of motion (Eq. 7.12)—that is, where the Mach number is everywhere greater than unity—we review the early work of Ackeret [49] in this field. Notwithstanding the intrinsic historical value of Ackeret's achievements, a fresh reading is interesting for its general development of first-order theory.

Making obvious simplifications—such as thin sharp-edged wings of small camber at low incidence in two-dimensional frictionless shock-free supersonic flow—Ackeret argued that the flow in the vicinity of the airfoil may be likened directly to that of the Prandtl-Meyer expansion around a corner. With the restrictions imposed previously, any leading-edge effect produces two Mach waves issuing from the sharp leading edge (Fig. 7.9), ahead of which the flow is undisturbed. Over the upper surface of the airfoil, the flow may expand according to the two-dimensional solution to the flow equations originated by Prandtl and Meyer (see p. 401). If the same restrictions apply to the leading edge and lower surface, providing that the inclinations are gentle and that no shock waves exist, the Prandtl-Meyer solution may still be used if the following device is employed. Since the undisturbed flow is supersonic, we may assume that it reaches that condition by expanding through the appropriate angle v_p from sonic conditions. Then any isentropic compressive deflection δ leads to flow conditions equivalent to an expansion of $(v_p - \delta)$ from sonic flow conditions.

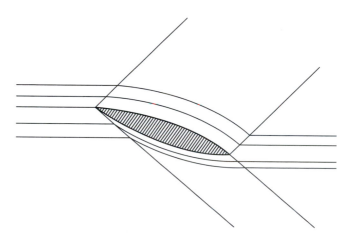

FIGURE 7.9

Example airfoil in supersonic flight. Note the presence of four shock waves, curved streamlines over and under the airfoil, and the straight, undisturbed streamlines ahead of the airfoil.

Therefore, if nowhere on the surface are any compressive deflections large, the Prandtl-Meyer values of pressure may be found by reading off the values [50] appropriate to the flow deflection caused by the airfoil surface and the aerodynamic forces and so forth, obtained from pressure integration.

Referring to Eq. (7.12) with $M_\infty > 1$,

$$(M_\infty^2 - 1)\frac{\partial^2 \phi}{\partial x^2} - \frac{\partial^2 \phi}{\partial y^2} = 0 \tag{7.35}$$

This wave equation has a general solution:

$$\phi = F_1\left(x - \sqrt{M_\infty^2 - 1}\, y\right) + F_2\left(x + \sqrt{M_\infty^2 - 1}\, y\right)$$

where F_1 and F_2 are independent functions with forms that depend on the flow boundary conditions. In the present case, physical considerations show that each function exists separately in well-defined flow regions (Figs 7.10 through 7.12).

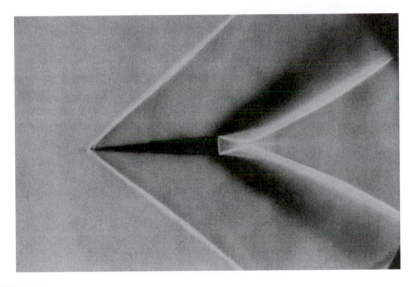

FIGURE 7.10

Supersonic flow over a wedge: The Schlieren method was used to obtain this flow visualization. A parallel light beam is refracted by the density differences in the flow field and is then focused on a knife edge to give a flow visualization the image plane. This takes the form of bright or dark patterns, depending on the direction in which the beam is bent by refraction. The main features of the flow field are the oblique bow shock wave, which is slightly rounded at the nose (see Fig. 8.46); the expansion fans at the trailing edge; and the recompression shock waves, which form downstream in the wake. These shock waves are slightly curved owing to the interaction with the expansion waves from the trailing edge. *Source: Photograph by D.J. Buckingham School of Engineering, University of Exeter, United Kingdom.*

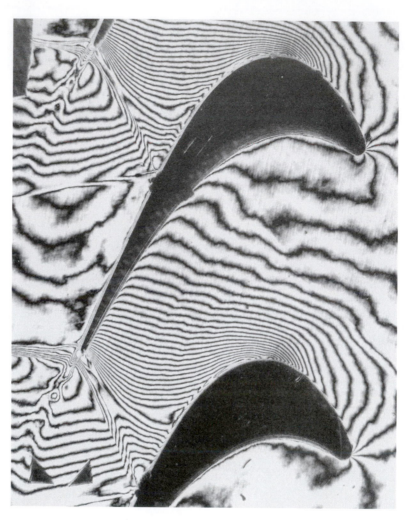

FIGURE 7.11

Transonic flow (from right to left) through a turbine cascade: The holographic interferogram shows fringes corresponding to lines of constant density. The flow enters from the right and exits at a Mach number of about 1.3 from the left. The convex and concave surfaces of the turbine blades act as suction and pressure surfaces, respectively. Various features of the flow field may be discerned from the interferogram; for example the gradual drop in density from inlet to outlet until the formation of a sharp density gradient marking a shock wave where the constant-density lines fold together. The shock formation at the trailing edge may be compared with Fig. 8.49. *Source: Photogtraph by P.J. Bryanston-Cross, Engineering Department, University of Warwick, United Kingdom.*

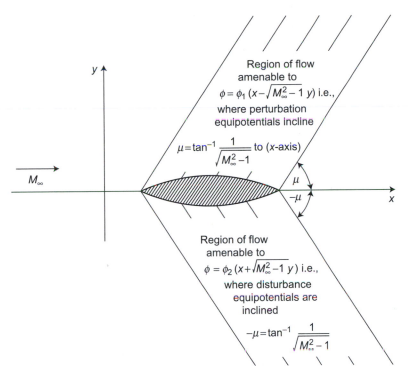

FIGURE 7.12

Definitions for linearized supersonic theory applied to a thin airfoil. Take special note of the differences on the upper and lower surfaces described in the text.

By inspection, the solution $\phi = F_1(x - \sqrt{M_\infty^2 - 1}\, y)$ allows constant values of φ along the lines $x - \sqrt{M_\infty^2 - 1}\, y = C$—that is, along the straight lines with an inclination of arc tan $1/\sqrt{M_\infty^2 - 1}$ to the x-axis (Fig. 7.12). This means that the disturbance originating on the airfoil shape (as shown) propagates into the flow at large along the straight lines $x = \sqrt{M_\infty^2 - 1}\, y + C$. Similarly, the solution $\phi = F_2(x + \sqrt{M_\infty^2 - 1}\, y)$ allows constant values of φ along the straight lines $x = -\sqrt{M_\infty^2 - 1}\, y + C$ with inclinations of the following to the axis:

$$\arctan\left(-\frac{1}{\sqrt{M_\infty^2 - 1}}\right)$$

Remember that Mach lines incline at an angle

$$\mu = \arctan\left(\pm\frac{1}{\sqrt{M_\infty^2 - 1}}\right)$$

to the free-stream direction, and thus the lines along which the disturbances propagate coincide with Mach lines.

Since disturbances cannot propagate forward into supersonic flow, the appropriate solution is such that the lines incline downstream. In addition, the effect of the disturbance is felt only in the region between the first and last Mach lines, and any flow conditions away from it replicate those adjacent to the body. Within the region in which the disturbance potential exists, then, taking the positive solution, for example,

$$u' = \frac{\partial \phi}{\partial x} = \frac{\partial F_1}{\partial \left(x - \sqrt{M_\infty^2 - 1}\, y\right)} \frac{\partial \left(x - \sqrt{M_\infty^2 - 1}\, y\right)}{\partial x}$$

$$u' = F_1'$$

(7.36)

and

$$v' = \frac{\partial \phi}{\partial y} = \frac{\partial F_1}{\partial (x - \sqrt{M_\infty^2 - 1}\, y)} \frac{\partial (x - \sqrt{M_\infty^2 - 1}\, y)}{\partial y}$$

$$v' = -\sqrt{M_\infty^2 - 1}\, F_1'$$

(7.37)

Now the boundary conditions for the problem require that the velocity on the body's surface be tangential to it. This gives an alternative value for v':

$$v' = U_\infty \frac{df(x)}{dx}$$

(7.38)

where $df(x)/dx$ is the local surface slope, $f(x)$ is the shape of the disturbing surface, and U_∞ is the undisturbed velocity. Equating Eqs. (7.37) and (7.38) on the surface where $y = 0$,

$$[F_1']_{y=0} - \frac{U_\infty}{\sqrt{M_\infty^2 - 1}} f'(x)$$

or

$$\left[\frac{\partial F_1}{\partial (x - \sqrt{M_\infty^2 - 1}\, y)} \right]_{y=0} = -\frac{U_\infty}{\sqrt{M_\infty^2 - 1}} \left[\frac{df(x - \sqrt{M_\infty^2 - 1}\, y)}{d(x - \sqrt{M_\infty^2 - 1}\, y)} \right]_{y=0}$$

Integrating gives

$$\phi = (F_1 =) \frac{-U_\infty}{\sqrt{M_\infty^2 - 1}} f\left(x - \sqrt{M_\infty^2 - 1}\, y\right)$$

(7.39)

With the value of φ (the disturbance potential) found, the horizontal perturbation velocity on the surface becomes, from Eq. (7.36),

$$[u']_{y=0} = \frac{-U_\infty}{\sqrt{M_\infty^2 - 1}}\left[f'(x - \sqrt{M_\infty^2 - 1}y)\right]_{y=0} = \frac{-U_\infty}{\sqrt{M_\infty^2 - 1}}\frac{\mathrm{d}f x}{\mathrm{d}x}$$

The local pressure coefficient, which in the linearized form is $C_p = -(2u')/U_\infty$, gives for this flow:

$$C_p = \frac{2}{\sqrt{M_\infty^2 - 1}}\frac{\mathrm{d}f x}{\mathrm{d}x}$$

However, $\mathrm{d}f x/\mathrm{d}x$ is the local inclination of the surface to the direction of motion $(\mathrm{d}f x/\mathrm{d}x = \varepsilon$, say). Thus

$$C_p = \frac{2\varepsilon}{\sqrt{M_\infty^2 - 1}} \tag{7.40}$$

Example 7.2

A shallow irregularity of length l in a plane wall along which a two-dimensional supersonic stream $M_0 = u_0/a_0$ is flowing is given approximately by the expression $y = kx[1 - (x/l)]$, where $0 < x < l$ and $k \ll 1$ (see Fig. E7.2). Using Ackeret's theory, prove that the velocity potential due to disturbance in the flow is

$$\phi = \frac{-u_0}{\sqrt{M_0^2 - 1}}k(x - \sqrt{M_0^2 - 1}y)\left[1 - \frac{x - \sqrt{M_0^2 - 1}y}{l}\right]$$

and obtain a corresponding expression for the local pressure coefficient anywhere on the irregularity:

$$\left(M_0^2 - 1\right)\frac{\partial^2 \phi}{\partial x^2} - \frac{\partial^2 \phi}{\partial y^2} = 0$$

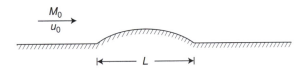

FIGURE E7.2

Supersonic flow over a single small bump.

This has the solution, applicable here, of $\phi = \phi(x - \sqrt{M_0^2 - 1}y)$, where φ is the disturbance potential function. Local perturbation velocity components on the wall are

$$u = \frac{\partial\phi}{\partial x} = \left[\frac{\partial\phi}{\partial(x - \sqrt{M_0^2 - 1}y)} \right]_{y=0}$$

$$v = \frac{\partial\phi}{\partial x} = -\sqrt{M_0^2 - 1} \left[\frac{\partial\phi}{\partial(x - \sqrt{M_0^2 - 1}y)} \right]_{y=0}$$

At x from the leading edge, the boundary conditions require the flow velocity to be tangential to the surface:

$$v = (u_0 - u)\frac{dy}{dx} \simeq u_0\frac{dy}{dx} = u_0 k\left[1 - \frac{2x}{l} \right]$$

Equating Eqs. (7.37) and (7.38) on the surface, where $y = 0$, Eq. (7.39) gives

$$[\partial\phi/\partial y]_{y=0} = \left[\frac{-u_0}{\sqrt{M_0^2 - 1}}k\left(1 - \frac{2x}{l}\right)d(x - \sqrt{M_0^2 - 1}y) \right]_{y=0}$$

$$\phi = \frac{-u_0}{\sqrt{M_0^2 - 1}}k(x - \sqrt{M_0^2 - 1}\,y)\left[1 - \frac{x - \sqrt{M_0^2 - 1}\,y}{l} \right]$$

Also on the surface, with the same assumptions,

$$C_p = -\frac{2u}{U_0} = \left[-\frac{2\frac{\partial\phi}{\partial x}}{u_0} \right]_{y=0} = \left[-\frac{2\frac{\partial\phi}{\partial(x - \sqrt{M_0^2-1}y)}}{u_0} \right]_{y=0}$$

$$C_p = \frac{2}{\sqrt{M_0^2 - 1}}k\left(l - \frac{2x}{l}\right)$$

It is now much more convenient to use the pressure coefficient (dropping the suffix ∞):

$$C_p = \frac{2\varepsilon}{\sqrt{M^2 - 1}} \tag{7.41}$$

where ε may be taken as +ve or −ve according to whether the flow is respectively compressed or expanded. Some care is necessary in designating the sign in a particular case; in the use of this result, the angle ε is always measured from the undisturbed stream direction where the Mach number is M, not from the previous flow direction if different.

Symmetrical Double Wedge Airfoil in Supersonic Flow

The linearized supersonic theory is applicable to many thin airfoils and wings plus slender bodies. One traditional shape, still in use, is the symmetric double-wedge airfoil. From the early days of supersonic flight this geometry has persisted; one recent use is the roll-control vanes on the successful Armadillo Aerospace STIG-A rocket launched to 137,500 feet on December 4, 2011.

Example 7.3

Plot the pressure distribution over the symmetrical double-wedge, 10% thick supersonic airfoil shown in Fig. E7.3(a) when the Mach 2.2 flow meets the upper surface (a) tangentially and (b, c) at an incidence 2 degrees above and below this. Also estimate the lift, drag, and pitching moment coefficients for these incidences.

For the semi-wedge angle $\varepsilon_0 = \arctan 0.1 = 5.72° = 0.1$ radians and angle of attack $\alpha = 5.72$ degrees $= 0.1$ radians,

$$M = 2.2; \quad M^2 = 4.84; \quad \sqrt{M^2 - 1} = 1.96$$

And for the incidence $\alpha = \varepsilon_0 = 0.1$ degrees. Using Eq. (7.40), the distribution is completed in tabular and graphical forms in Fig. E7.3(b).

For lift, drag, and moment, a more general approach can be adopted. If a chordwise element δx, located x from the leading edge is taken, the net force normal to the chord is

$$(p_L - p_U)\delta x = (C_{pL} - p_{pU})\frac{1}{2}\rho V^2 \delta x$$

Total normal force = lift (since α is small):

$$L = C_L \frac{1}{2}\rho V^2 c = \int_0^c (C_{pL} - p_{pU})\frac{1}{2}\rho V^2 dx \tag{7.42}$$

In this case, Eq. (7.42) integrates to give

$$C_L c = \frac{c}{2}(C_{p3} - C_{p1}) + \frac{c}{2}(C_{p4} - C_{p2})$$

and, on substituting $C_{p2} = 2\varepsilon_1\sqrt{M^2 - 1}$, and so forth,

$$C_L = \frac{1}{\sqrt{M^2 - 1}}[\varepsilon_3 - \varepsilon_1 + \varepsilon_4 - \varepsilon_2] \tag{7.43}$$

However, for the present configuration,

$$\varepsilon_1 = \varepsilon_0 - \alpha, \quad \varepsilon_2 = -\varepsilon_0 - \alpha, \quad \varepsilon_3 = \varepsilon_0 + \alpha, \quad \varepsilon_4 = -\varepsilon_0 + \alpha$$

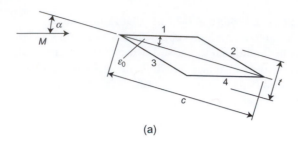

(a)

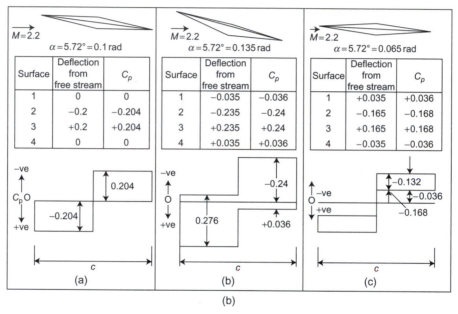

(b)

FIGURE E7.3

(a) Symmetrical double-wedge airfoil in supersonic flight. The symmetry is both top to bottom about the chord line and left to right about $x/c = 0.5$. (b) Solution to Example 7.3.

and Eq. (7.43) becomes

$$C_L = \frac{1}{\sqrt{M^2-1}}[(\varepsilon_0 + \alpha) - (\varepsilon_0 - \alpha) + (-\varepsilon_0 + \alpha) + (\varepsilon_0 - \alpha)]$$

$$C_L = \frac{4\alpha}{\sqrt{M^2-1}}$$

(7.44)

In the present example,

$$C_{L3.42} = 0.132, \quad C_{L5.72} = 0.204, \quad C_{L7.72} = 0.275$$

The contribution to drag due, say, to a chordwise element of lower surface is

$$p_L \varepsilon_L \delta x = C_{pL} \frac{1}{2} \rho V^2 \varepsilon_L \delta x + p_0 \varepsilon_L \delta x$$

where p_0 is the free-stream static pressure, which integrates to zero and may be neglected throughout. Again using $C_{pL} = 2\varepsilon_L^2 \sqrt{M^2 - 1}$, the elemental contribution to drag becomes

$$\frac{2\varepsilon_L^2}{\sqrt{M^2 - 1}} \frac{1}{2} \rho V^2 \delta x$$

The corresponding contribution from the upper surface is

$$\frac{2\varepsilon_U^2}{\sqrt{M^2 - 1}} \frac{1}{2} \rho V^2 \delta x$$

The total wave drag becomes

$$D = \frac{2}{\sqrt{M^2 - 1}} \frac{1}{2} \rho V^2 \int_0^c \left(\varepsilon_L^2 + \varepsilon_U^2 \right) dx - C_D \frac{1}{2} \rho V^2 c$$

$$C_D = \frac{2}{\sqrt{M^2 - 1}} \int_0^1 \left(\varepsilon_L^2 - \varepsilon_U^2 \right) d\left(\frac{x}{c} \right)$$

In the present case, with $\varepsilon_1 = \varepsilon_0 - \alpha, \varepsilon_2 = -(\alpha + \varepsilon_0), \varepsilon_3 = \varepsilon_0 + \alpha$, and $\varepsilon_4 = \alpha - \varepsilon_0$,

$$C_D = \frac{2}{\sqrt{M^2 - 1}} \left\{ \left[(\alpha - \varepsilon_0)^2 + (\varepsilon_0 - \alpha)^2 \right] \frac{1}{2} + \left[(\alpha + \varepsilon_0)^2 + (\varepsilon_0 + \alpha)^2 \right] \frac{1}{2} \right\}$$

$$= \frac{4}{\sqrt{M^2 - 1}} \left[\alpha^2 + \varepsilon_0^2 \right]$$

However, $\varepsilon_0^2 = (t/c)^2$; therefore,

$$C_D = \frac{4}{\sqrt{M^2 - 1}} \left[\alpha^2 + \left(\frac{t}{c} \right)^2 \right] \tag{7.45}$$

We now see that airfoil thickness contributes to wave drag, which is a minimum for a wing of zero thickness (i.e., a flat plate). Alternatively, for airfoils other than the flat plate, minimum wave drag occurs at zero incidence. This is generally true for symmetrical sections, although the magnitude of the minimum wave drag varies. In the present example, the required values are

$$C_{D_{3.72}} = 0.029, \quad C_{D_{5.72}} = 0.0408, \quad C_{D_{7.72}} = 0.0573$$

Find the lift wave-drag ratio directly from Eqs. (7.44) and (7.45):

$$\frac{L}{D} = \frac{4\alpha}{\sqrt{M^2 - 1}} \left\{ 4 \left[\alpha^2 + \left(\frac{t}{c}\right)^2 \right] \frac{1}{\sqrt{M^2 - 1}} \right\}^{-1} = \frac{\alpha}{\alpha^2 + (t/c)^2} \qquad (7.46)$$

Now L/D is a maximum when $D/L = \alpha + (t/c)^2/\alpha$ is a minimum; this occurs when the two terms involved are numerically equal—in this case, when $\alpha = t/c$. Substituting back gives the maximum L/D ratio as

$$\left[\frac{L}{D} \right]_{\max} = \frac{t/c}{2\left(\frac{t}{c}\right)^2} = \frac{1}{2\left(\frac{t}{c}\right)}$$

For the present example, with $t/c = 0.1, [L/D >]_{\max} = 5$ occurring at 5.72 degrees of incidence, and

$$\left[\frac{L}{D} \right]_{3.72} = 4.55, \quad \left[\frac{L}{D} \right]_{5.72} = 5.0, \quad \left[\frac{L}{D} \right]_{7.72} = 4.8$$

Moment about the Leading Edge   Directly from the lift case, the force normal to the chord from an element δx on chord x from the leading edge is $(C_{P_L} - C_{P_U}) \frac{1}{2} \rho V^2 \delta x$. This produces the negative increment of pitching moment:

$$\Delta M = -(C_{P_L} - C_{P_U}) x \frac{1}{2} \rho V^2 \delta x$$

Integrating gives the total moment:

$$-M = \int_0^c (C_{P_L} - C_{P_U}) \frac{1}{2} \rho V^2 x dx = -C_M \frac{1}{2} \rho V^2 c^2$$

Making the appropriate substitution for C_p gives

$$-C_M = \frac{2}{c^2 \sqrt{M^2 - 1}} \int_0^c M(\varepsilon_L - \varepsilon_U) x dx$$

which, for the profile of the present example, gives

$$-C_M = \left(\frac{2}{\sqrt{M^2 - 1}}\right)\frac{1}{c^2}\left\{[\alpha - \varepsilon_0 - (\varepsilon_0 - x)]\frac{c^2}{4} + [(\varepsilon_0 + \alpha) + (\alpha + \varepsilon_0)]\frac{c^2}{4}\right\}$$

$$= \frac{1}{\sqrt{M^2 - 1}}[2x]$$

that is,

$$-C_M = \frac{2\alpha}{\sqrt{M^2 - 1}} = \frac{C_L}{2} \tag{7.47}$$

Hence

$$C_{M_{3.72}} = -0.066, \quad C_{M_{6.72}} = -0.102, \quad C_{M_{7.72}} = -0.138$$

The center-of-pressure coefficient k_{CP} is

$$k_{CP} = \frac{-C_M}{C_L} = 0.5 \tag{7.48}$$

which is independent of both Mach number and (for symmetrical sections) incidence.

Supersonic Biconvex Circular Arc Airfoil in Supersonic Flow

While still dealing with symmetrical sections it is useful to consider another class of profile, one made up of biconvex circular arcs. Much early experimental work was done on these sections, with both symmetrical and cambered profiles; this work is readily available to compare with the theory.

Consider the thin symmetrical airfoil section shown in Fig. 7.13. On the upper surface, the deflection of the flow from the free-stream direction at a distance x

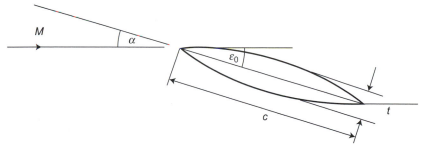

FIGURE 7.13

Angles and dimensions for the thin biconvex airfoil in supersonic flight. In general, $\varepsilon_0 \neq \alpha$, unlike the special case shown here.

behind the leading edge is ε_U. Because $\varepsilon_U = -\alpha + \varepsilon_0[1 - (2x/c)]$,[1] the local pressure coefficient is

$$C_{p_U} = \frac{2}{\sqrt{M^2 - 1}}\left[-\alpha + \varepsilon_0\left(1 - \frac{2x}{c}\right)\right] \qquad (7.49)$$

For the lower surface,

$$C_{p_L} = \frac{2}{\sqrt{M^2 - 1}}\left[\alpha + \varepsilon_0\left(1 - \frac{2x}{c}\right)\right]$$

So the contribution to lift from the upper and lower surfaces x from the leading edge is

$$\delta L = \frac{2}{\sqrt{M^2 - 1}} 2\alpha \frac{1}{2}\rho V^2 \delta x = \delta C_L \frac{1}{2}\rho V^2 c$$

Integrating over the chord gives, as before,

$$C_L = \frac{4\alpha}{\sqrt{M^2 - 1}} \qquad (7.50)$$

The contributions to wave drag of each of the surfaces x from the leading edge are together

$$\delta D = \frac{2}{\sqrt{M^2 - 1}}\frac{1}{2}\rho V^2\left(\varepsilon_U^2 + \varepsilon_L^2\right)\delta x = \delta C_D \frac{1}{2}\rho V^2 c$$

Integrating gives, after substituting for ε_U^2 and ε_L^2,

$$C_D = \frac{4}{\sqrt{M^2 - 1}}\left[\alpha^2 + \frac{\varepsilon_0^2}{3}\right]$$

Now, by geometry, and since ε_0 is small, $\varepsilon_0 = 2(t/c)$, giving

$$C_D = \frac{4}{\sqrt{M^2 - 1}}\left[\alpha^2 + \frac{4}{3}\left(\frac{t}{c}\right)^2\right]$$

---

[1]This approximate form of equation for a circular arc is justified for shallow concavities (i.e., large radii of curvature) and follows from simple geometry (from Fig. 7.14):

$$\varepsilon_1 = \varepsilon_0 - \theta = \varepsilon_0\left(1 - \frac{\theta}{\varepsilon_0}\right), \qquad \theta = \frac{s}{R} = \frac{x}{R}, \qquad \varepsilon_0 = \sin^{-1}\frac{c}{2R} \simeq \frac{c}{2R}$$

Therefore,

$$\frac{\theta}{\varepsilon_0} = \frac{2x}{c} \quad \text{and} \quad \varepsilon = \varepsilon_0 = \left(1 - \frac{2x}{c}\right)$$

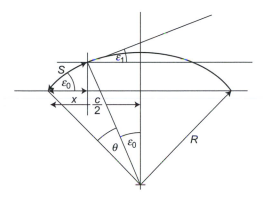

FIGURE 7.14

Angles and dimensions used in the solutions for the geometry of the thin biconvex airfoil.

The lift/drag ratio is a maximum when, by division, $D/L = \alpha + \left[\frac{4}{3}(t/c)^2\, 1/\alpha\right]$ is a minimum. This occurs when

$$\alpha = \frac{4}{3}\left(\frac{t}{c}\right)^2 \frac{1}{\alpha}$$

Then

$$\left[\frac{L}{D}\right]_{\max} = \frac{\alpha}{\alpha^2 + \alpha^2} = \frac{1}{2\alpha} = \frac{\sqrt{3}}{4}\frac{c}{t} = \frac{0.433}{t/c}$$

For a 10% thick section, $(L/D)_{\max} = 4\frac{1}{3}$ at $\alpha = 6.5$ degrees.

Moment Coefficient and k_{CP} Directly from previous work—that is, taking the moment of δL about the leading edge—

$$M = C_M \frac{1}{2}\rho V^2 c^2 = -\int_0^c \frac{4\alpha}{\sqrt{M^2 - 1}} x \frac{1}{2}\rho V^2 dx$$

$$C_{M_0} = \frac{-2\alpha}{\sqrt{M^2 - 1}} \qquad\qquad (7.51)$$

and the center-of-pressure coefficient $k_{CP} = -(C_M/C_L) = 0.5$ as before. Test results on supersonic airfoil sections published by A. Ferri [51] compare with the theory. The set chosen here is for a symmetrical biconvex airfoil section of $t/c = 0.1$ set in an airflow of Mach number 2.13. The incidence varies from -10 to 28 degrees. Plotted on the graphs of Fig. 7.15 are the theoretical values of Eqs. (7.50) and (7.51).

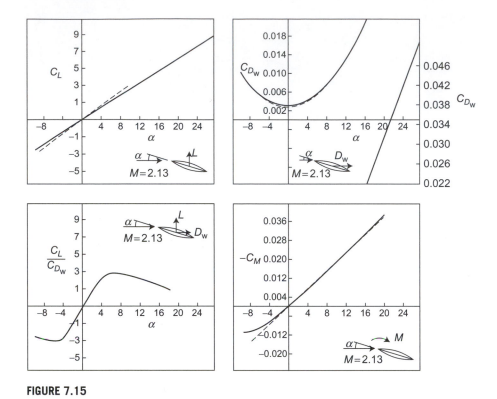

FIGURE 7.15

Performance results for the thin biconvex airfoil, comparing theory (dashed lines) and measurement (solid lines).

Figure 7.15 shows the close approximation of the theoretical values to experimental results. The lift coefficient varies linearly with incidence but at some slightly smaller value than predicted. No significant reduction in C_L, as is common at high incidences in low-speed tests, was found even with an incidence > 20 degrees .

The measured drag values are slightly higher than predicted, which is understandable because the theory accounts for wave drag only. The difference between the two may be attributed to skin-friction drag or, more generally, to viscosity and boundary-layer behavior. It is unwise, however, to expect the excellent agreement of these particular results to extend to more general airfoil sections or, indeed, to other Mach numbers for the same section, as severe limitations on the theory appear at extreme Mach numbers. Nevertheless, these and other published data amply justify the theory's continued use.

General Airfoil Section
Retaining the major assumptions of the theory that airfoil sections must be slender and sharp-edged permits us to assess the overall aerodynamic properties as the sum

of contributions due to thickness, camber, and incidence. From previous sections, we know that the local pressure at any point on the surface is due to the magnitude and sense of the angular deflection of the flow from the free-stream direction. This deflection in turn can be resolved into components arising from the separate geometric quantities of the section—that is, from thickness, camber, and chord incidence.

This principle is shown in Fig. 7.16, where pressure p acting on the airfoil at a point where the flow deflection from the free stream is ε represents the sum of $p_t + p_c + p_\alpha$. If, more conveniently, we consider the pressure coefficient, then care must be taken to evaluate the sum algebraically. With the notation shown in Fig. 7.16,

$$C_p = C_{p_t} + C_{p_c} + C_{p_\alpha} \tag{7.52}$$

or

$$\frac{2}{\sqrt{M^2 - 1}}\varepsilon = \frac{2}{\sqrt{M^2 - 1}}(\varepsilon_t + \varepsilon_c + \varepsilon_\alpha) \tag{7.53}$$

Lift   The lift coefficient due to the element of surface is

$$\delta C_L = \frac{-2}{\sqrt{M^2 - 1}}(\varepsilon_t + \varepsilon_c + \varepsilon_\alpha)\frac{\delta x}{c}$$

which is made up of terms due to thickness, camber, and incidence. Integrating around the surface of the airfoil, the contributions due to thickness and camber vanish, leaving only that due to incidence. This can be easily shown by isolating, say, the camber

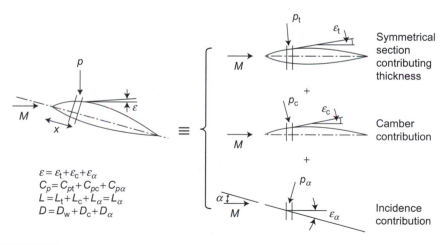

FIGURE 7.16

An airfoil decomposed into the three contributions to shape, and hence, surface slope.

contribution for the upper surface. From Eq. (7.42),

$$C_{L_{\text{camber}}} = \frac{-2}{\sqrt{M^2 - 1}} \int_0^c \varepsilon_c \frac{dx}{c}$$

but

$$\int_0^c \varepsilon_c dx = \int_0^c \left(\frac{dy}{dx}\right)_c dx = \int_0^c dy_c = [y]_0^c = 0$$

Therefore,

$$C_{L_{\text{camber}}} = 0$$

Similar treatment of the lower surface gives the same result; likewise for the contribution to lift due to thickness.

This result is borne out by the values of C_L found in the previous examples:

$$C_L = \frac{-2}{\sqrt{M^2 - 1}} \left(\int_0^c \varepsilon_\alpha \frac{dx}{c} - \int_0^c \varepsilon_\alpha \frac{dx}{c} \right)$$

(upper surface, lower surface)

Now ε_α (upper surface) $= -\alpha$ and ε_α (lower surface) $= +\alpha$:

$$C_L = \frac{-2}{\sqrt{M^2 - 1}} \frac{1}{c} \left([-\alpha^2]_0^c + [\alpha]_c^0 \right)$$

$$C_L = \frac{4\alpha}{\sqrt{M^2 - 1}} \tag{7.54}$$

Drag (Wave) The drag coefficient due to the element of surface shown in Fig. 7.16 is

$$\delta C_D = C_p \varepsilon^2 \frac{\delta x}{c}$$

which, on putting $\varepsilon = \varepsilon_t + \varepsilon_c + \varepsilon_\alpha$ and so on, becomes

$$\delta C_D = \frac{2}{\sqrt{M^2 - 1}} (\varepsilon_t + \varepsilon_c + \varepsilon_\alpha)^2 \frac{\delta x}{c}$$

By integrating this expression around the contour to find the overall drag, only the integration of the squared terms contributes, since integration of other products vanishes for the same reason given previously for the development leading to

Eq. (7.54). Thus

$$C_D = \frac{2}{\sqrt{M^2 - 1}} \oint \left(\varepsilon_t^2 + \varepsilon_c^2 + \varepsilon_\alpha^2\right)^2 \frac{dx}{c} \tag{7.55}$$

Now

$$2 \oint \varepsilon_\alpha^2 dx = 4\alpha^2 c$$

and, for a particular section,

$$2 \oint \varepsilon_t^2 dx = k_t \left(\frac{t}{c}\right)^2 c$$

and

$$2 \oint \varepsilon_c^2 dx = k_c \beta^2 c$$

so, for a given airfoil profile, the drag coefficient becomes in general

$$C_D = \frac{2}{\sqrt{M^2 - 1}} \left(4\alpha^2 + k_t \left(\frac{t}{c}\right)^2 + k_c \beta^2\right) \tag{7.56}$$

where t/c and β are the thickness-chord ratio and camber, respectively, and k_t, k_c are geometric constants.

Lift/Wave Drag Ratio   It follows from Eqs. (7.54) and (7.55) that

$$\frac{D}{L} = \alpha + \frac{k_t(t/c)^2 + k_t \beta^2}{4\alpha}$$

which is a minimum when

$$\alpha^2 = \frac{k_t(t/c)^2 + k_c \beta^2}{4}$$

Moment Coefficient and Center-of-Pressure Coefficient   Once again, the moment about the leading edge is generated from the normal contribution, and for the general element of surface x from the leading edge:

$$\delta C_M = -\left(\frac{2}{\sqrt{M^2 - 1}}\right) \frac{x}{c} \varepsilon \frac{dx}{c}$$

$$C_M = \frac{-2}{\sqrt{M^2 - 1}} \oint (\varepsilon_\alpha + \varepsilon_t + \varepsilon_c) \frac{x}{c} \frac{dx}{c}$$

Now

$$\oint \varepsilon_t \frac{x}{c} \frac{dx}{c}$$

is zero for the general symmetrical thickness, since the pressure distribution due to the section (which, by definition, is symmetrical about the chord) provides neither lift nor moment. In other words, the net lift at any chordwise station is zero. However, the effect of camber is not zero in general, although overall lift is (since the integral of the slope is zero), and the influence of camber is to exert a negative pitching moment (nose down for positive camber) (i.e., concave downward). Thus

$$ C_M = \frac{-2}{\sqrt{M^2 - 1}} \left(2\frac{\alpha}{2} + k_c\beta \right) $$

The center-of-pressure coefficient follows from

$$ k_{CP} = \frac{-C_M}{C_L} = \frac{\frac{2}{\sqrt{M^2-1}}(\alpha + k_c\beta)}{\frac{4\alpha}{\sqrt{M^2-1}}} $$

$$ k_{CP} = 0.5 \left(1 + \frac{k_c\beta}{\alpha} \right) $$

and this is no longer independent of incidence, although it is still independent of Mach number.

Airfoil Section Made Up of Unequal Circular Arcs
As a first example, consider the biconvex airfoil used by Stanton [52] in some early work on airfoils at speeds near the speed of sound. In his experimental work Stanton used a conventional (round-nosed) airfoil (RAF 31a) in addition to the biconvex sharp-edged section at subsonic as well as supersonic speeds. However, the only results used for comparison here are those for the biconvex section at the supersonic speed $M = 1.72$.

Example 7.4
Made up of two unequal circular arcs, a profile has the dimensions shown in Fig. E7.4(a). Compare the values of lift, drag, moment, and center-of-pressure coefficients found by Stanton with those predicted by Ackeret. From the geometric data given, the tangent angles at the leading and trailing edges are 16 degrees = 0.28 radians and 7 degrees = 0.12 radians for upper and lower surfaces, respectively. Measuring x from the leading edge, the local deflections from the free-stream direction are

$$ \varepsilon_U = 0.28 \left(1 - 2\frac{x}{c} \right) - \alpha $$

and

$$ \varepsilon_L = 0.12 \left(1 - 2\frac{x}{c} \right) + \alpha $$

for the upper and lower surfaces, respectively.

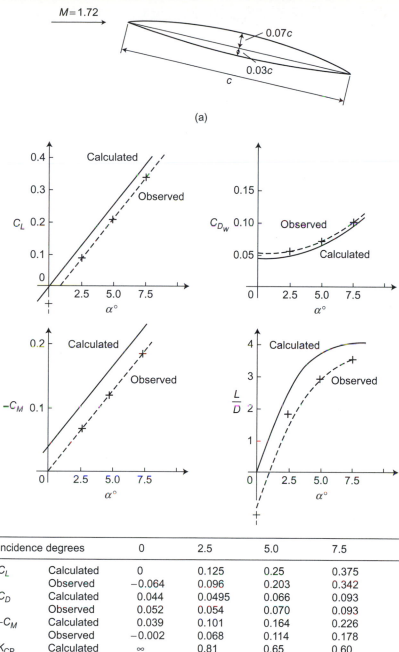

Incidence degrees		0	2.5	5.0	7.5
C_L	Calculated	0	0.125	0.25	0.375
	Observed	−0.064	0.096	0.203	0.342
C_D	Calculated	0.044	0.0495	0.066	0.093
	Observed	0.052	0.054	0.070	0.093
$-C_M$	Calculated	0.039	0.101	0.164	0.226
	Observed	−0.002	0.068	0.114	0.178
K_{CP}	Calculated	∞	0.81	0.65	0.60
	Observed	0.03	0.69	0.54	0.49
	Calculated	0	2.5	3.8	4.0

(b)

FIGURE E7.4

(a) Stanton's biconvex airfoil section $t/c = 0.1$. (b) Performance results for Stanton's biconvex airfoil. Comparison between theory (solid lines) and measurement (dashed lines) are made.

Lift Coefficient

$$C_L = \frac{-2}{c\sqrt{M^2-1}}\left[\int_0^c \varepsilon_U dx - \int_0^c \varepsilon_L dx\right]$$

$$= \frac{-2}{c\sqrt{M^2-1}}\left[\int_0^c 2\alpha + 0.16\left(1-2\frac{x}{c}\right)dx\right]$$

$$C_L = \frac{4\alpha}{\sqrt{M^2-1}}$$

For $M = 1.72$,

$$C_L = 2.86\alpha$$

Drag (Wave) Coefficient

$$C_D = \frac{-2}{c\sqrt{M^2-1}}\int_0^c\left[\left(0.28\left(1-\frac{2x}{c}\right)-\alpha\right)^2 + \left(0.12\left(1-2\frac{x}{c}\right)-\alpha\right)^2\right]dx$$

$$C_D = \frac{(4\alpha^2+0.0619)}{\sqrt{M^2-1}}$$

For $M = 1.72$ (as in Stanton's case),

$$C_D = 2.86\alpha^2 + 0.044$$

Moment Coefficient (about Leading Edge)

$$M_{LE} = C_{M_{LE}}\frac{1}{2}\rho_0 V^2 c^2 = -\int_0^c \frac{\rho_0 V^2}{\sqrt{M^2-1}}\left[2\alpha - 0.16\left(1-2\frac{x}{c}\right)\right]x dx$$

or

$$C_{M_{LE}} = \frac{2}{\sqrt{M^2-1}}[\alpha + 0.0271]$$

For $M = 1.72$,

$$-C_{M_{LE}} = 1.43\alpha + 0.039$$

Center-of-Pressure Coefficient

$$k_{CP} = \frac{-C_{M_{LE}}}{C_L} = \frac{2\alpha + 0.054}{4\alpha}$$

$$k_{CP} = \frac{0.5 + 0.0135}{\alpha}$$

Lift/Drag Ratio

$$\frac{L}{D} = \frac{\dfrac{4\alpha}{\sqrt{M^2-1}}}{\dfrac{4\alpha^2+0.0619}{\sqrt{M^2-1}}} = \frac{\alpha}{\alpha^2 - 0.0155}$$

This is a maximum when $\alpha = \sqrt{0.0155} = 0.125$ radians $= 8.4$ degrees, giving $(L/D)_{max} = 4.05$.

Again, note that the calculated and observed values are close in shape but the latter are lower in value (Fig. E7.4(b)). The differences between theory and experiment are probably explained by the fact that the theory neglects viscous drag.

AERODYNAMICS AROUND US

Shock-Expansion Shapes

Students may find the double-wedge and other supersonic airfoil shapes used in shock-expansion models "too simple" to be relevant to modern flight. Certainly, airfoil design for supersonic flight has progressed from the 1950s-style ultra-thin diamonds and the F-104 Starfighter era. Combining good supersonic cruise performance with good handling at lower (safer) take-off and landing speeds was made possible through hard work, decades, and CFD. Regardless of progress, the basic performance results—lift, drag, and pitching coefficients—show the representative behavior of thin supersonic airfoils.

These "simple" designs have certainly been put to use in flight. Students taking an aerodynamics class who are anticipating a future in rockets rather than aircraft should pay close attention to aerodynamics. The image shows a fin on a Saturn 1-B rocket. Of course, this rocket is no longer flying, but the lesson is that fins are simple low-aspect-ratio wings that exhibit lift, drag, pitching moment, stall, and other wing-like behaviors. Note too that the successful STIG-A liquid-fueled

Lowermost section of the Saturn 1-B on display at the NASA Kennedy Space Center.
Source: Photograph courtesy of Steven Collicott.

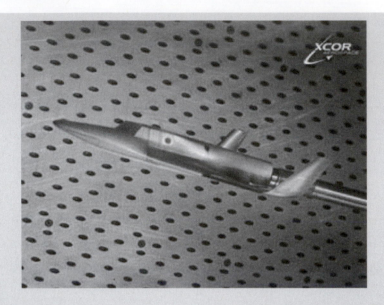

Wind-tunnel model of the XCOR Lynx suborbital rocket plane being developed for tourism and science flights. Used by permission of XCOR.
Source: Photograph courtesy of XCOR.

low-cost reusable commercial sounding rocket launched by Armadillo Aerospace on December 4, 2011 at Spaceport America in New Mexico used a double-edge airfoil on the two roll-control vanes. These vanes controlled toe roll of the vehicle from Mach zero up to perhaps nearly Mach 3 in this flight.

Modern supersonic "airfoil" design may not even be airfoil design but rather vehicle design. The image above of the wind-tunnel model for the XCOR Lynx suborbital rocket highlights that the wing, forebody, and vertical surfaces are coupled design problems. Modern CFD and modern wind-tunnel testing both supplement old-fashioned experience, creative thinking, and hard work, to produce functional and safe aerodynamic designs—in this case for suborbital space tourism and research.

Double-Wedge Airfoil Section

Various measures of performance of airfoil sections are of common interest. Maximum lift-to-drag ratio is one important measure. Ackeret's, or linearized supersonic, theory permits comparison of airfoil performance measures such as $[L/D]_{max}$.

Example 7.5

Using Ackeret's theory, obtain expressions for the lift and drag coefficients of the cambered double-wedge airfoil shown in Fig. E7.5(a). Show that the maximum lift/drag ratio for the uncambered double-wedge airfoil is $\sqrt{2}$ times that for a cambered airfoil with $h = t/2$.

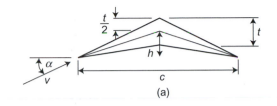

(a)

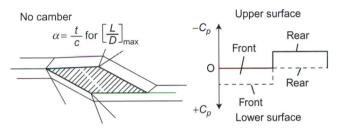

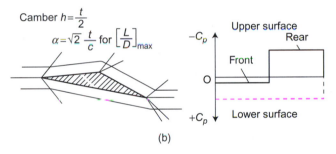

(b)

FIGURE E7.5

(a) Cambered double-wedge airfoil. Fore–aft symmetry exists. (b) Flow patterns and pressure distributions around both airfoils at an incidence of $[L/D]_{max}$.

Sketch the flow patterns and pressure distributions around both airfoils at the incidence for $(L/D)_{max}$.

Lift   Eq. (7.54) showed that

$$C_L = \frac{4\alpha}{\sqrt{M^2 - 1}}$$

Drag (Wave)   From Eq. (7.55) on the general airfoil,

$$C_D = \frac{2}{\sqrt{M^2 - 1}} \oint \left(\varepsilon_t^2 + \varepsilon_c^2 + \varepsilon_\alpha^2 \right) \frac{dx}{c}$$

Here, as before,

$$2 \oint \varepsilon_\alpha^2 \frac{dx}{c} = 4\alpha^2 c$$

For the given geometry,

$$\oint \varepsilon_t^2 \frac{dx}{c} = 4\left[\frac{1}{2}\left(\frac{t}{c}\right)^2\right]$$

that is, one equal contribution from each of four flat surfaces, and

$$\oint \varepsilon_c^2 \frac{dx}{c} = 4\left[\frac{1}{2}\left(\frac{h}{c/2}\right)^2\right]$$

also one equal contribution from each of four flat surfaces. Therefore,

$$C_D = \frac{2}{\sqrt{M^2-1}}\left[\alpha^2 + \left(\frac{t}{c}\right)^2 + 4\left(\frac{h}{c}\right)^2\right]$$

Lift–Drag Ratio

$$\frac{L}{D} = \frac{C_L}{C_D} = \frac{\alpha}{\left[\alpha^2 + \left(\frac{t}{c}\right)^2 + 4\left(\frac{h}{c}\right)^2\right]}$$

For the uncambered airfoil $h = 0$,

$$\left[\frac{L}{D}\right]_{max} = \left[\frac{\alpha}{\alpha^2 + (t/c)^2}\right]_{max} \quad \overset{\alpha = t/c}{\Rightarrow} \quad \left[\frac{L}{D}\right]_{max} = \frac{1}{2(t/c)}$$

For the cambered section, given $h = t/c$,

$$\left[\frac{L}{D}\right]_{max} = \left[\frac{\alpha}{\alpha^2 + 2(t/c)^2}\right]_{max} \quad \overset{\alpha = \sqrt{2}t/c}{\Rightarrow} \quad \left[\frac{L}{D}\right]_{max} = \frac{1}{2\sqrt{2}(t/c)}$$

The results are plotted in Figure E7.5(b). Note the substantially greater lift produced on the rear half of the triangle airfoil also adds significant drag.

7.1.4 Other Aspects of Supersonic Wings

Wing shapes are nearly limitless. No one text can discuss a majority of the options. Consistent with the introductory purpose of text, we will explore a few examples to illustrate fundamental concepts in supersonic lift and drag.

The Shock-Expansion Approximation

The supersonic linearized theory has the advantage of giving relatively simple formulae for an airfoil's aerodynamic characteristics. However, as shown in Example 7.5, the exact pressure distribution for a double-wedge airfoil can be readily found, which means that the coefficients of lift and drag can be obtained.

Example 7.6

Consider a symmetrical double-wedge airfoil at zero incidence, similar in shape to that shown in Fig. E7.3(a) except that the semi-wedge angle $\varepsilon_0 = 10$ degrees. Sketch the wave pattern for $M_\infty = 2.0$, calculate the Mach number and pressure on each face of the airfoil, and thus determine C_D. Compare the results with those obtained by the linear theory. Assume that the free-stream stagnation pressure $p_{0\infty} = 1$ bar.

The wave pattern is shown in Fig. E7.6(a). The flow properties in the various regions can be obtained using isentropic flow and oblique shock tables [53]. In region 1, $M = M_\infty = 2.0$ and $p_{0\infty} = 1$ bar. From the isentropic flow tables, $p_{01}/p_1 = 7.83$, leading to $p_1 = 0.1277$ bar.

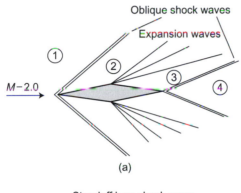

Oblique shock waves

Expansion waves

① $M-2.0$ ② ③ ④

(a)

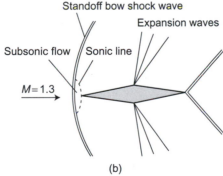

Standoff bow shock wave

Expansion waves

Subsonic flow / Sonic line

$M=1.3$

(b)

FIGURE E7.6

(a) Effects of the differing magnitudes of leading-edge turning angle and maximum turning angle for the flight Mach number. Solutions for the flow behind the detached curved bow shock, such as for pressures on the front half of the airfoil surfaces as in (b), are far more complex than for the attached oblique shock.

In region 2, the oblique shock-wave tables give $p_2/p_1 = 1.7084$ (leading to $p_2 = 0.2182$ bar), $M_2 = 1.6395$, and shock angle = 39.33 degrees. Therefore,

$$C_{p2} = \frac{p_2 - p_\infty}{\frac{1}{2}\rho_\infty V_\infty^2} = \frac{(p_2/p_\infty) - 1}{\frac{1}{2}\gamma(\rho_\infty/\gamma\rho_\infty)V_\infty^2} = \frac{(p_2/p_\infty) - 1}{\frac{1}{2}\gamma M_\infty^2}$$

$$= \frac{(0.2182/0.1277) - 1}{0.5 \times 1.4 \times 2^2} = 0.253$$

Using the linear theory, Eq. (7.39) gives

$$C_{p2} = \frac{2\varepsilon}{\sqrt{M_\infty^2 - 1}} = \frac{2 \times (10\pi/180)}{\sqrt{2^2 - 1}} = 0.202$$

To continue the calculation into region 3 it is first necessary to determine the Prandtl-Meyer angle and the stagnation pressure in region 2. These can be obtained as follows using the isentropic flow tables: $p_{02}/p_2 = 4.516$, giving $p_{02} = 4.516 \times 0.2182 = 0.9853$ bar; Mach angle, $\mu_2 = 37.57$ degrees; and Prandtl-Meyer angle, $v_2 = 16.01$ degrees.

Between regions 2 and 3, the flow expands isentropically through 20 degrees so $v_3 = v_2 + 20$ degrees $= 36.01$ degrees. From the isentropic flow tables, this value of v_3 corresponds to $M_3 = 2.374$, $\mu_3 = 24.9$ degrees, and $p_{03}/p_3 = 14.03$. Since the expansion is isentropic, $p_{03} = p_{02} = 0.9853$ bar, so $p_3 = 0.9853/14.03 = 0.0702$ bar. Thus

$$C_{p3} = \frac{(0.0702/0.1277) - 1}{0.7 \times 2^2} = -0.161$$

Using the linear theory, Eq. (7.39) gives

$$C_{p3} = \frac{2\varepsilon}{\sqrt{M_\infty^2 - 1}} = \frac{-2 \times (10\pi/180)}{\sqrt{2^2 - 1}} = -0.202$$

There is an oblique shock wave between regions 3 and 4. The oblique shock tables give $p_4/p_3 = 1.823$ and $M_4 = 1.976$ giving $p_4 = 1.823 \times 0.0702 = 0.128$ bar and a shock angle of 33.5 degrees.

The drag per unit span acting on the airfoil is given by resolving the pressure forces:

$$D = 2(p_2 - p_3) \times \frac{(c/2)}{\cos(10°)} \times \sin(10°)$$

so

$$C_D = (C_{p2} - C_{p3})\tan(10°) = 0.0703$$

Using the linear theory, Eq. (7.45) with $\alpha = 0$ gives

$$C_D = \frac{4(t/c)^2}{\sqrt{M_\infty^2 - 1}} = \frac{4\tan^2(10°)}{\sqrt{2^2 - 1}} = 0.072$$

These calculations show that, although the linear theory does not accurately approximate the value of C_p, it does accurately estimate C_D.

When $M_\infty = 1.3$, it can be seen from the oblique shock tables that the maximum compression angle is less than 10 degrees. This implies that, in this case, the flow can negotiate the leading edge only by being compressed through a shock wave that stands off from the leading edge and is normal to the flow where it intersects the extension of the chord line. This causes the formation of a region of subsonic flow between the standoff shock wave and the leading edge. The corresponding flow pattern is shown in Fig. E7.6(b).

A similar procedure to that in Example 7.6 can be followed for airfoils with curved profiles. In this case, though, it is approximate because it ignores the effect of the Mach waves reflected from the bow shock wave (see Fig. 7.17). The so-called shock-expansion approximation is made clearer by the following example.

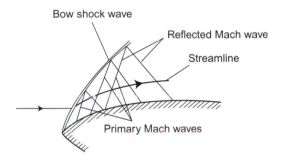

FIGURE 7.17

Primary and reflected Mach lines between the airfoil and the shock wave showing how the flow is similar to Prandtl-Meyer flow.

Example 7.7

Consider a biconvex airfoil at zero incidence in supersonic flow at $M_\infty = 2$, similar in shape to that shown in Fig. 7.13 so that, as before, the shape of the upper surface is given by

$$y = x\varepsilon_0 \left(1 - \frac{x}{c}\right) \text{ giving local flow angle } \theta(=\varepsilon) = \arctan\left[\varepsilon_0\left(1 - \frac{2x}{c}\right)\right]$$

Calculate the pressure and Mach number along the surface as functions of x/c for the case of $\varepsilon_0 = 0.2$. Compare the result with those obtained with linear theory. Take the free-stream stagnation pressure to be 1 bar.

Region 1 is, as in Example 7.6, $M_1 = 2.0$, $p_{01} = 1$ bar, and $p_1 = 0.1277$ bar.

At $x = 0$, $\theta = \arctan(0.2) = 11.31$ degrees. Hence, the flow is initially turned by the bow shock through an angle of 11.31 degrees, so the oblique shock tables give $p_2/p_1 = 1.827$ and $M_2 = 1.59$. Thus $p_2 = 1.827 \times 0.1277 = 0.233$ bar. From the isentropic flow tables, it is found that $M_2 = 1.59$ corresponds to $p_{02}/p_2 = 4.193$, giving $p_{02} = 0.977$ bar.

Thereafter the pressures and Mach numbers around the surface can be obtained using the isentropic flow tables, as shown in the table.

$\frac{x}{c}$	$\tan\theta$	θ	$\Delta\theta$	v	M	$\frac{p_0}{p}$	p (bar)	C_p	$(C_p)_{\text{lin}}$
0.0	0.2	11.31°	0°	14.54°	1.59	4.193	0.233	0.294	0.228
0.1	0.16	9.09°	2.22°	16.76°	1.666	4.695	0.208	0.225	0.183
0.2	0.12	6.84°	4.47°	19.01°	1.742	5.265	0.186	0.163	0.138
0.3	0.08	4.57°	6.74°	21.28°	1.820	5.930	0.165	0.104	0.092
0.5	0.0	0.0	11.31°	25.85°	1.983	7.626	0.128	0.0008	0
0.7	−0.08	−4.57°	15.88°	30.42°	2.153	9.938	0.098	−0.0831	−0.098
0.8	−0.12	−6.84°	18.15°	32.69°	2.240	11.385	0.086	−0.1166	−0.138
0.9	−0.16	−9.09°	20.40°	34.94°	2.330	13.104	0.075	−0.1474	−0.183
1.0	−0.20	−11.31°	22.62°	37.16°	2.421	15.102	0.065	−0.1754	−0.228

Wings of Finite Span

When the component of free-stream velocity perpendicular to the leading edge is greater than the local speed of sound the wing is said to have a *supersonic leading edge*. In this case, as illustrated in Fig. 7.18, there is two-dimensional supersonic flow over much of the wing, which can be calculated using supersonic airfoil theory. For the rectangular wing shown in Fig. 7.18, the presence of a wingtip can be communicated only within the Mach cone apex, which is located at the wingtip. The same consideration applies to any inboard three-dimensional effects, such as the "kink" at the centerline of a swept-back wing.

In the opposite case, the component of free-stream velocity perpendicular to the leading edge is less than the local speed of sound, and the term *subsonic leading edge* is used. A typical example is the swept-back wing shown in Fig. 7.19. In this case, the Mach cone generated by the leading edge of the center section subtends the

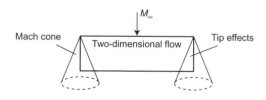

FIGURE 7.18

Typical wing with a supersonic leading edge.

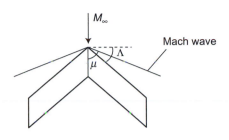

FIGURE 7.19

Example of a wing with a subsonic leading edge.

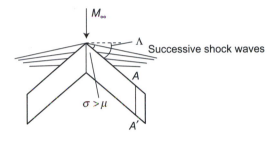

FIGURE 7.20

A real wing with nonzero thickness creates a more complex shock structure than the zero-thickness wing in Fig. 7.19 creates.

whole wing. This implies that the leading edge of the outboard portions of the wing influences the oncoming flow just as it does for subsonic flow. Wings having finite thickness and incidence actually generate a shock cone rather than a Mach cone, as shown in Fig. 7.20. Additional shocks are generated by other points on the leading edge, and the associated shock angles tend to increase because each successive shock wave leads to a reduction in Mach number. These shock waves progressively decelerate the flow so that, at some section such as AA', the flow approaching the leading edge is subsonic. Thus subsonic wing sections are used over most of the wing.

Wings with subsonic leading edges have lower wave drag than those with supersonic edges. Consequently, highly swept wings (e.g., slender deltas) are the preferred configuration at supersonic speeds. On the other hand, swept wings with supersonic leading edges tend to have a greater wave drag than do straight wings.

Computational Methods

Computational methods for compressible flows, particularly transonic flow over wings, have been the subject of a considerable research effort over the past three decades. Substantial progress has been made, although much still remains to be done. A discussion of these methods is beyond our scope save to note that for linearized compressible potential flow (Eq. 7.12), panel methods have been developed for both subsonic and supersonic flow (see Sections 4.10, and 5.8). These can be used to obtain approximate numerical solutions in cases with exceedingly complex geometries. A review of the computational methods developed for the full inviscid and viscous equations of motion is given by Jameson [54].

7.2 EXERCISES

7.1 Consider a two-dimensional flat plate inclined at a positive angle of attack of 10 degrees in a supersonic air stream of Mach 2 with pressure of 50 kPa and temperature of 250 K. Assume that the length of the plate is 1 meter.

(a) What is the angle of the oblique shock?
(b) What is the pressure acting on the lower surface of the plate?
(c) What is the pressure acting on the upper surface of the plate?
(d) What is the lift? What is the drag? (Provide the formulae for lift and drag, and then substitute the numerical values into them to obtain the numerical predictions.)
(e) What are the lift and drag coefficients?
(f) Solve the same problem for angles of attack of 5 and 15 degrees. What is the slope of the C_L versus angle-of-attack curve for angles in radians?

7.2 Examine the flow over the supersonic airfoil illustrated in Fig. E7.2, using shock-expansion theory. Assume that the Mach number $M_\infty = 2.5$, the chord length $c = 1$ m, the camber ratio $h/c = 0.05$, and the thickness ratio $t/c = 0.02$.

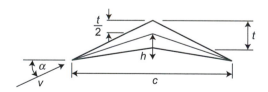

FIGURE Ex7.2

(a) What are the lift and drag coefficients on this airfoil at angles of attack α, from -5 to 5 degrees? Display your results graphically.
(b) Compare your results with those predicted by applying the linearized theory.
(c) What is the lift coefficient at a zero angle of attack for this airfoil?

7.3 Consider a very thin, flat-plate airfoil of chord c acting as a fin at the aft end of a rocket. If the rocket veers by an angle of attack α, compare the location of the center of pressure for the airfoil for subsonic and supersonic flight. For subsonic flight, review thin-airfoil theory in Chapter 4.

7.4 Use the shock-expansion method to solve for the lift and drag coefficients of the cross-section of the Saturn 1-B fin shown in Fig. E7.4 as functions of angle of attack from −5 to 5 degrees at Mach 2. Representative dimensions (the chord of the fin varies with spanwise position, so consider just the airfoil shape) are presented in the figure below. Assume that the pressure on the base of the fin is zero.

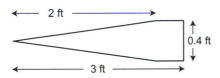

FIGURE Ex7.4

Viscous Flow and Boundary Layers

LEARNING OBJECTIVES

- Learn more about the boundary layer.

- Recall how the Navier-Stokes equations can be simplified for analyzing boundary-layer flows.

- Learn the details of Blasius's solution to the boundary-layer equations.

- Learn about the phenomena of flow separation and transition from laminar to turbulent flow.

- Learn how to derive the momentum-integral form of the boundary-layer equations. Explore how it is used for obtaining approximate solutions for laminar, turbulent, and mixed laminar-turbulent boundary layers. Learn how it is applied for estimating profile drag.

- Learn how commercially available computational fluid dynamics tools are applied to solve approximate forms of the Navier-Stokes equations in aerodynamics.

8.1 INTRODUCTION

The aerodynamics problem of interest in this chapter is illustrated in Fig. 8.1 (note that the figure exaggerates the thickness of the boundary layers and the wake). The problem is to determine the lift and drag components of force acting on an airfoil in a uniform stream. We are also interested in the moment of force acting about, for example, the quarter-chord location as measured from the leading edge. Hence, the aerodynamic problem is to determine the surface distribution of pressure p (the normal component of the force per unit area) and the shear stress $\tau_w = \mu \partial u / \partial y$ (the tangential component of the force per unit area) acting on the airfoil surface S_B. The velocity in the direction of flow parallel to the surface of the airfoil, which is the x direction, is denoted u. The direction y is perpendicular to the surface and hence perpendicular to the velocity near the wall. Once the force per unit area is known, the net force and moment are determined by integration over S_B.

How do we solve this engineering problem? We will begin with the well-known and well-established real-flow formulas.

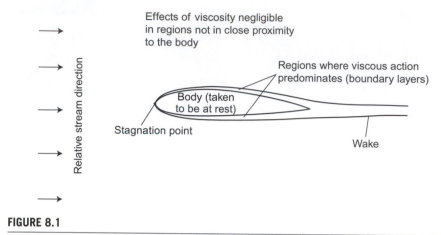

FIGURE 8.1

Complete airfoil problem.

The equations of motion (as described in Chapter 2) for real, viscous Newtonian fluids such as water and air are the Navier-Stokes (NS) equations. For incompressible flows (of air at low Mach number, $M < 0.3$, and water), the equations in three dimensions in dimensional form are

$$u'_{t'} + u'u'_{x'} + v'u'_{y'} + w'u'_{z'} = -\frac{1}{\rho}p'_{x'} + v\left(u'_{x'x'} + u'_{y'y'} + u'_{z'z'}\right) \tag{8.1}$$

$$v'_{t'} + u'v'_{x'} + v'v'_{y'} + w'v'_{z'} = -\frac{1}{\rho}p'_{y'} + v\left(v'_{x'x'} + v'_{y'y'} + v'_{z'z'}\right) \tag{8.2}$$

$$w'_{t'} + u'w'_{x'} + v'w'_{y'} + w'w'_{z'} = -\frac{1}{\rho}p'_{z'} + v\left(w'_{x'x'} + w'_{y'y'} + w'_{z'z'}\right) \tag{8.3}$$

$$u'_{x'} + v'_{y'} + w'_{z'} = 0 \tag{8.4}$$

where the subscripts imply differentiation with respect to the variable used and the primes (′) indicate that the unknowns are dimensional. This system of equations is complete because the number of unknowns, u', v', w', and p', equal the number of equations available.[1]

It is useful to scale the NS equations to help interpret experimental data and to help guide the engineer in selecting approximations of this system of equations to facilitate solutions to real problems (e.g., of the airfoil). We divide all components of the velocity vector by U, the free-stream speed, and scale all distances by the length scale L. (Typically we select L to be equal to the chord of the aerodynamic object.) The time scale is thus L/U, which is the time it takes a fluid particle moving at U to

---

[1]They are *not* the Reynolds-averaged Navier-Stokes (RANS) equations that can be solved by "advanced computational methods"; the RANS equations are introduced and discussed later in this chapter.

move a distance L. Let $u = u'/U$, $v = v'/U$, $w = w'/U$, $x = x'/L$, $y = y'/L$, $z = z'/L$, $t = t'U/L$, and $p = p'/\rho U^2$. Applying this to the four NS equations, we get

$$u_t + uu_x + vu_y + wu_z = -p_x + Re^{-1}\left(u_{xx} + u_{yy} + u_{zz}\right) \tag{8.5}$$

$$v_t + uv_x + vv_y + wv_z = -p_y + Re^{-1}\left(v_{xx} + v_{yy} + v_{zz}\right) \tag{8.6}$$

$$w_t + uw_x + vw_y + ww_z = -p_z + Re^{-1}\left(w_{xx} + w_{yy} + w_{zz}\right) \tag{8.7}$$

$$u_x + v_y + w_z = 0 \tag{8.8}$$

where $Re = UL/\nu$ is the Reynolds number based on a length scale in the direction of flow. This system of dimensionless equations indicates, for example, that the drag coefficient

$$C_D = \frac{D}{\frac{1}{2}\rho U^2 L(1)} = f(Re_L) \tag{8.9}$$

that is, C_D is a function of Reynolds number only as long as the dimensionless geometry of the object under examination is the same.

Now we examine the drag data available for flat plates exposed to free-stream flows U parallel to the surface of the plate. The data for all aerodynamically smooth flat plates are given in Fig. 8.2. The curves in the figure have been verified experimentally [55]. The drag is the force acting on the plate in the direction of flow. The curves are labeled laminar, transition, and turbulent to distinguish the three possible flow regimes that can occur at intermediate Reynolds numbers. At low Reynolds numbers, $\log_{10} Re < 5.5$, the flow in the boundary layers is laminar. This means that a velocity probe placed in the boundary layer indicates a steady-state condition—that is, a measurement that does not change with time. For large Reynolds numbers, $\log_{10} Re > 7$, the flow in the boundary layer aft of transition is turbulent. A velocity probe placed in this region measures a highly unsteady apparently random velocity. In this region a time average can be found. Because of the kinetic energy lost to random fluctuations, the wall shear stress is significantly larger if the boundary layer is turbulent as compared with a laminar boundary layer. Note that in both the laminar and turbulent cases, the drag coefficient decreases as the Reynolds number increases. This does not mean that the drag decreases. In fact, it increases, although not as fast as it would if the drag coefficient were constant. There is an increase in drag coefficient when transition occurs somewhere between the leading edge and the end of the plate. The Reynolds number for transition varies between $3 \times 10^5 \leq Re_T \leq 3 \times 10^6$. This uncertainty is due partly to the turbulence level in the outer flow and partly to deviations in aerodynamic smoothness. The main point here is the uncertainty in drag prediction in the Reynolds number range $5.5 < \log_{10} Re < 7$.

The flow over a flat plate is sketched in Fig. 8.3. For a plate in a uniform stream, the boundary layer grows as a laminar boundary layer for $Re_x < Re_T$, where $Re_x = Ux/\nu$. After reaching the transition point, it transitions from laminar to turbulent. This occurs over a relatively small but finite distance in the flow direction, after which the boundary-layer flow is turbulent.

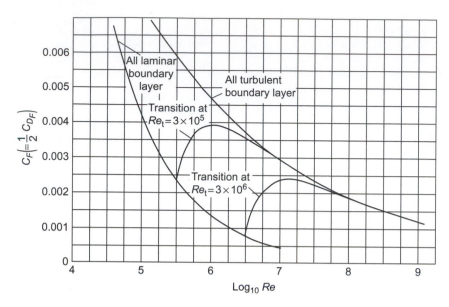

FIGURE 8.2

Skin-friction drag coefficients for a flat plate. The flat-plate Reynolds number is $Re = UL/v$. The transition Reynolds number is $Re_T = Ux_T/v$, where x_T is the location from the leading edge at which transition from laminar to turbulent flow occurs. The skin-friction force coefficient is $C_F = F/(\frac{1}{2})\rho U^2 S_w$, where F is the skin-friction drag acting on the plate and S_w is the entire surface area of the plate exposed to shear. Twice this coefficient is the drag coefficient based on the plan area, which is one-half the plate surface area because the surface of the plate includes both sides of it.

Another sketch of the velocity profile illustrating the definition of boundary-layer thickness is given if Fig. 8.4. In part (a) of this figure, the boundary-layer thickness δ is the distance from the wall where the velocity is $0.99U$. In part (b) the dimensionless form of the boundary layer is shown. In the latter coordinates, we can compare the laminar velocity profile with the time-averaged turbulent velocity profile. This is done in Fig. 8.5, which illustrates that the turbulent boundary layer is "fuller" than the laminar boundary layer; hence the time-averaged wall shear stress is larger because $\partial \bar{u}/\partial y$ is significantly greater compared with the laminar profile of the same thickness.

The data for flat-plate skin-friction drag is useful for two reasons: (1) they provide a method to estimate the drag on an airfoil because an airfoil is almost a flat plate and hence its drag is of the order 95% flat-plate skin-friction drag; (2) analytical procedures developed to examine the drag on a plate can be checked with the data and so can be more comfortably extended to examine the flows around airfoils where pressure variations in the outer flow (outside the boundary layer) near the leading and

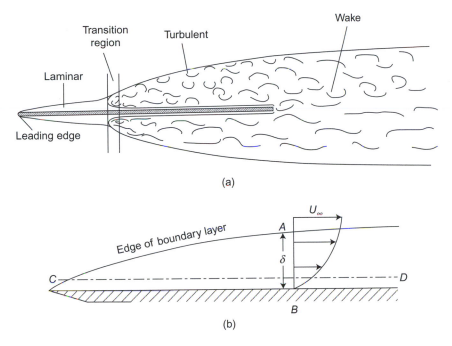

FIGURE 8.3

Boundary layer on a flat plate. (a) Boundary-layer development at intermediate Reynolds numbers. (b) Velocity profile in the boundary layer illustrating the no-slip condition at the plate and the approach to the outer flow speed at the boundary layer edge. Note that the scale normal to the surface of the plate is greatly exaggerated.

trailing edges are important. In addition, flat-plate boundary layer problems help us gain insight into the physical behavior of boundary layers.

The fact that transitional and turbulent boundary-layer flows are highly unsteady and, inherently, three-dimensional, makes it difficult to solve the NS equations for them. Even if the flows are laminar, the NS equations are formidable. The engineering question is, then, can we simplify the problem? Can we segregate the outer-potential-flow problem from the inner-boundary-layer flow problem? Since airfoils are thin and Reynolds numbers are relatively high, the answer is yes. Prandtl, in 1904, showed how this could be done.

Prandtl observed that the effect of viscous shear for high Reynolds-number flows is confined to thin layers adjacent to the surface of airfoils and flat plates. Outside the boundary layer, the flow behaves as if it were inviscid. Hence the outer-potential-flow–inner-viscous-boundary-layer-flow theory was born. We already investigated the outer-potential-flow problem. By imposing a circulation around the airfoil such that the Kutta condition is satisfied at the trailing edge, we were able to predict the lift reasonably accurately for a well-designed airfoil. However, the drag for

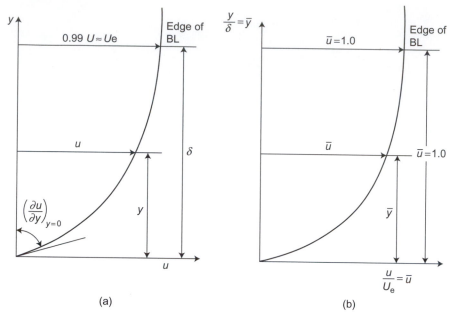

FIGURE 8.4

Velocity profile in a boundary layer.

two-dimensional potential flow is identically zero. To determine the drag, we need to take into account viscous effects. Prandtl's boundary-layer theory provides a reasonable approach to this that leads to useful results.

8.2 BOUNDARY-LAYER THEORY

Here we discuss ideas for simplifying the NS equations to come up with Prandtl's boundary-layer equations. We begin by considering the boundary layer shown in Fig. 8.3(b). Note that the direction of flow is x, and hence u is in the direction of the flow; v is perpendicular to the direction of flow, and hence, normal to the wall; and $w = 0$. (If the boundary-layer flow is turbulent, it is the mean values of the velocity components we are talking about.) In the boundary layer, as the flow moves horizontally from the leading edge, where the profile is uniform at speed U just before point C, toward point D somewhere downstream, the speed changes from U to $u \rightarrow 0$ near the surface (i.e., near the no-slip wall). Therefore, the size of various terms and factors in the NS equations are $u_x \sim U/L$, $u_y \sim U/\delta$, $u_{xx} \sim U/L^2$, $u_{yy} \sim U/\delta^2$, and $(u_x + v_y) \sim U/L + v/\delta$. Thus $v \sim U\delta/L$ and $u_{xx} \ll u_{yy}$. With these observations of term and factor size, we can reduce the NS equations to the following

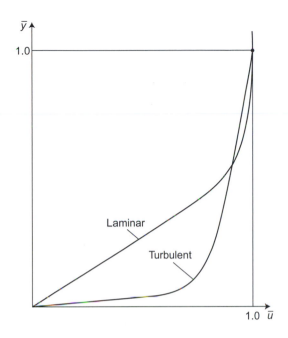

FIGURE 8.5

Comparison of dimensionless laminar and time-averaged turbulent velocity profiles in the boundary layer.

thin boundary-layer equations of Prandtl:

$$u_t + uu_x + vu_y = -P_x + Re^{-1}u_{yy} \qquad (8.10)$$

$$u_x + v_y = 0 \qquad (8.11)$$

Across the boundary layer, $p_y = 0$, so $p = P(x)$ is a function of x only and can be determined at the outer edge of the boundary layer via potential-flow theory.

8.2.1 Blasius's Solution

For a flat plate, $P_x = 0$. For steady laminar flows, $u_t = 0$. Therefore, the laminar boundary-layer flow over a flat plate is described by

$$uu_x + vu_y = Re^{-1}u_{yy} \qquad (8.12)$$

$$u_x + v_y = 0 \qquad (8.13)$$

Note that the dimensionless free-stream speed and the overall length of the plate are $U = 1$ and $L = 1$, respectively. This system of equations is subject to the following boundary conditions:

$$u = 1, \quad x = 0, \quad \text{for all } y$$

$$u = v = 0, \quad 0 < x < 1, \quad y = 0 \quad \text{(no-slip)}$$

$$u_y = 0, \quad 0 < x < 1, \quad y = \delta/L \quad \text{(no-shear)}$$

Now we examine the size of each term in Equation (8.12) for the flow in the boundary layer between the leading edge and a location x on the plate downstream. The size of the inertia term is $1/x$. The size of the viscous diffusion term on the right-hand side is $1/(Re\delta^2)$. Equating the two measures, we get

$$\delta \propto \frac{1}{\sqrt{Re_x}} \tag{8.14}$$

where $Re = Ux'/\nu$. Note that $\delta = \delta'/x'$, and so δ is the dimensionless boundary-layer thickness scaled by its horizontal distance from the leading edge.

The dimensional value of the boundary-layer thickness is thus

$$\delta' \propto \sqrt{x'} \tag{8.15}$$

That is, the laminar boundary layer grows like the square root of the distance from the leading edge, where it starts with zero thickness.

The results of this scaling argument provide the method for transforming the boundary-layer equations from a system of partial differential equations to an ordinary differential equation. It is known as the *method of similarity*, and it works if such a transformation is found.

We search for a solution of the form $u = f'(\eta)$, where $\eta = Cy/\sqrt{x}$. The similarity variable was selected based on the previous scaling arguments that suggest how the laminar boundary layer grows with x. (This is what Blasius did under the direction of Prandtl.) If $u = f'(\eta)$, then the stream function must take the form $\psi = a(x)f(\eta)$. The stream function is introduced here to provide a way to find v in terms of f. The horizontal component of the velocity is

$$u = \frac{\partial \psi}{\partial y} = a(x)f'\frac{\partial \eta}{\partial y} = a(x)f'\frac{C}{\sqrt{x}}$$

Therefore, we set $a(x) = \sqrt{x}/c$ so that $u = f'$:

$$\psi = \frac{\sqrt{x}}{C}f(\eta)$$

The vertical component of the velocity is thus

$$v = -\frac{\partial \psi}{\partial x} = -\frac{1}{2C\sqrt{x}}f + \frac{y}{2x}f'$$

Also,

$$u_y = \frac{\partial u}{\partial y} = \frac{C}{\sqrt{x}}f'', \quad u_{yy} = \frac{\partial^2 u}{\partial y^2} = \frac{C^2}{x}f'''$$

$$u_x = \frac{\partial u}{\partial x} = \frac{\partial \eta}{\partial x}f'' = -\frac{Cy}{2x\sqrt{x}}$$

Substituting these relationships into the boundary-layer equation, we get

$$-ff'' = 2C^2 Re^{-1}f'''$$

Next, we set the constant $C^2 = Re/2$ to get the ordinary differential equation

$$f''' + ff'' = 0, \quad \eta = y\sqrt{\frac{U}{2\nu x}} \qquad (8.16)$$

subject to the boundary conditions

$$f' \to 1, \eta \to \infty, \quad f = f' = 0, \eta = 0 \qquad (8.17)$$

Equation (8.16) was solved numerically by Blasius. The velocity profile he predicted is illustrated in Fig. 8.6. Experimental shapes for laminar flows match this result (see Fig. 8.7).

8.2.2 Definitions of Boundary-Layer Thickness

In deriving the boundary-layer equations presented in Section 2.12, we showed how boundary-layer thickness varies with Reynolds number in a laminar boundary layer, $\delta \propto Re^{-1/2}$ (see Eq. 2.137). This is another example of obtaining useful practical information from an equation without solving it. Its practical use will be illustrated in Example 8.1. Notwithstanding such practical applications, however, we saw that the boundary-layer thickness is a rather imprecise concept—it is difficult to give it a precise numerical value. To do so in Section 2.11.3 it was necessary, rather arbitrarily, to identify the edge of the boundary layer as corresponding to the point where $u = 0.99 U_e$. Partly owing to this rather unsatisfactory vagueness, several more precise definitions of boundary-layer thickness are given next. As will become clear, each definition also has a useful and significant physical interpretation relating to boundary-layer characteristics.

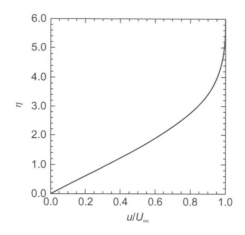

FIGURE 8.6

Blasius's laminar boundary-layer similarity solution.

FIGURE 8.7

Velocity profile in a boundary layer: The profile that forms along a flat wall is made visible by lines of aluminium powder dropped from a trough onto the flowing fluid surface. The fluid is a dilute solution of wallpaper paste in water. The Reynolds number based on distance along the wall is about 50,000 (see also Fig. 8.4). *Source: Photograph by D.J. Buckingham, School of Engineering, University of Exeter, United Kingdom.*

Displacement Thickness (δ*)

Consider the flow past a flat plate (Fig. 8.8(a)). Because of the buildup of the boundary layer on the plate surface, a stream tube that is close to the surface at the leading edge becomes entrained into the boundary layer. As a result, the mass flow in the stream tube decreases from ρU_e in the mainstream to some value ρu, and—to satisfy continuity—the tube cross-section increases. In the two-dimensional flows considered here, this means that the widths, normal to the plate surface, of the boundary-layer stream tubes increase and stream tubes that are in the mainstream are displaced slightly away from the surface. The effect on the mainstream flow is then as if, with no boundary layer present, the solid surface had been displaced a small distance into the stream. The amount by which the surface is displaced under such conditions is termed the boundary-layer displacement thickness (δ^*) and may be calculated as follows, provided the velocity profile $\bar{u} = f(\bar{y})$ (see Fig. 8.3) is known.

At station x (Fig. 8.8(c)), owing to the presence of the boundary layer, the mass flow rate is reduced by an amount equal to

$$\int_0^\infty (\rho U_e - \rho u)\,dy$$

corresponding to area OABR. This must equate to the mass flow rate deficiency that occurs at uniform density ρ and velocity U_e through the thickness δ^*, corresponding to area OPQR. Equating these mass flow rate deficiencies gives

$$\int_0^\infty (\rho U_e - \rho u)\,dy = \rho U_e \delta^*$$

that is,

$$\delta^* = \int_0^\infty \left(1 - \frac{u}{U_e}\right) dy \tag{8.18}$$

The idea of displacement thickness is put forward on the basis of two-dimensional flow past a flat plate purely so that the concept may be considered in its simplest form. The definition just given may be used for any incompressible two-dimensional boundary layer without restriction and is also largely true for boundary layers over three-dimensional bodies, provided the curvature, in planes normal to the free-stream direction, is not large. In other words, the local radius of curvature should be much greater than the boundary-layer thickness. If the curvature is large, a displacement thickness may still be defined but the form of Eq. (8.18) is slightly modified. The use of displacement thickness will be treated in Examples 8.2 and 8.3.

Arguments similar to those given previously will be used to define other boundary-layer thicknesses, using either momentum rates or energy flow rates.

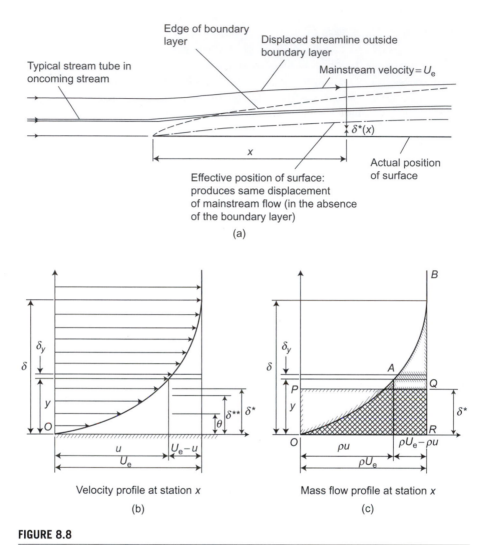

FIGURE 8.8

Measures of boundary-layer thickness.

Momentum Thickness (θ)

Momentum thickness is defined in relation to the momentum flow rate within the boundary layer. This rate is less than the rate that would occur if no boundary layer existed, when the velocity in the vicinity of the surface, at the station considered, would be equal to the mainstream velocity U_e.

For the typical stream tube within the boundary layer (Fig. 8.8(b)), the rate of momentum defect (relative to the mainstream) is $\rho u(U_e - u)\delta y$. Note that the mass flow rate ρu actually within the stream tube must be used here, because the momentum defect of this mass is the difference between its momentum based on mainstream velocity and its actual momentum at position x in the boundary layer.

The rate of momentum defect for the thickness θ (the distance through which the surface must be displaced so that, with no boundary layer, the total flow momentum at the station considered is the same as that actually occurring) is given by $\rho U_e^2 \theta$. Thus

$$\int_0^\infty \rho u (U_e - u) \mathrm{d}y = \rho U_e^2 \theta$$

that is,

$$\theta = \int_0^\infty \frac{u}{U_e}\left(1 - \frac{u}{U_e}\right)\mathrm{d}y \tag{8.19}$$

The momentum thickness concept is used in the calculation of skin-friction losses.

Kinetic-Energy Thickness (δ^{**})

Kinetic-energy thickness is defined with reference to kinetic energies of the fluid similarly to momentum thickness. The rate of kinetic-energy defect within the boundary layer at any station x is given by the difference between the energy that the element would have at mainstream velocity U_e and that it actually has at velocity u, being equal to

$$\int_0^\infty \frac{1}{2}\rho u \left(U_e^2 - u^2\right)\mathrm{d}y$$

while the rate of kinetic-energy defect in the thickness δ^{**} is $\frac{1}{2}\rho U_e^3 \delta^{**}$. Thus

$$\int_0^\infty \rho u \left(U_e^2 - u^2\right)\mathrm{d}y = \rho U_e^2 \delta^{**}$$

that is,

$$\delta^{**} = \int_0^\infty \frac{u}{U_e}\left[1 - \left(\frac{u}{U_e}\right)^2\right]\mathrm{d}y \tag{8.20}$$

8.2.3 Skin-Friction Drag

The shear stress between adjacent layers of fluid in a laminar flow is given by $\tau = \mu(\partial u/\partial y)$, where $\partial u/\partial y$ is the transverse velocity gradient. Adjacent to the solid surface at the base of the boundary layer, the shear stress in the fluid is due entirely to viscosity and is given by $\mu(\partial u/\partial y)_w$. This statement is true for both laminar and turbulent boundary layers because a viscous sublayer exists at the surface even if the main boundary-layer flow is turbulent. The shear stress in the fluid layer in contact with the surface is essentially the same as the shear stress between that layer and the surface; thus, for all boundary layers, the shear stress at the wall, due to the presence of the boundary layer, is given by

$$\tau_w = \mu\left(\frac{\partial u}{\partial y}\right)_w \tag{8.21}$$

where τ_w is the wall shear stress or surface friction stress, usually known as skin friction.

Once the velocity profile (laminar or turbulent) of the boundary layer is known, the surface (or skin) friction can be calculated. The skin-friction stress can be defined in terms of a nondimensional local skin-friction coefficient C_f as follows:

$$\tau_w = C_f \frac{1}{2}\rho U_e^2 \tag{8.22}$$

Of particular interest is the total skin-friction force F on the surface under consideration. This force is obtained by integrating the skin-friction stress over the surface. For a two-dimensional flow, the force F per unit width of surface may be evaluated, with reference to Fig. 8.9, as follows. The skin-friction force per unit width on an elemental length (δx) of surface is

$$\delta F = \tau_w \delta x$$

Therefore, the total skin-friction force per unit width on length L is

$$F = \int_0^L \tau_w \mathrm{d}x \tag{8.23}$$

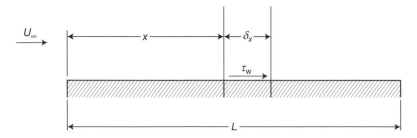

FIGURE 8.9

Direction of wall shear of fluid acting on flat plate.

The skin-friction force F may be expressed in terms of a nondimensional coefficient C_F, defined by

$$F = C_F \frac{1}{2} \rho U_\infty^2 S_w \tag{8.24}$$

where S_w is the wetted area of the surface under consideration. Similarly for a flat plate or airfoil section, the total skin-friction drag coefficient C_{D_F} is defined by

$$D_F = C_{D_F} \frac{1}{2} \rho U_\infty^2 S \tag{8.25}$$

where D_F = total skin-friction force on both surfaces resolved in the direction of the free stream, and S = plan area of the plate or airfoil. For a flat plate or symmetrical airfoil section, at zero incidence, when the top and bottom surfaces behave identically, $D_F = 2F$ and $S = S_w$ (the wetted area for each surface). Thus

$$C_{D_F} = \frac{2F}{\frac{1}{2}\rho U_\infty^2 S} = 2C_F \tag{8.26}$$

When flat-plate flows (at constant pressure) are considered, $U_e = U_\infty$. Except where a general definition is involved, U_e is used throughout.

Subject to the previous condition, Eqs. (8.22) and (8.24) in Eq. (8.21) lead to

$$C_F = \int_0^1 C_f \mathrm{d}\left(\frac{x}{L}\right) \tag{8.27}$$

Equation (8.27) is strictly applicable only to a flat plate, but on a thin airfoil, for which U_e does not vary greatly from U_∞ over most of the surface, the expression gives a good approximation to C_F.

We saw in Eq. (8.14) how the boundary-layer thickness varies with Reynolds number. This result can be used to show how skin friction and skin-friction drag vary with Re_L. It can be seen that

$$\tau_w = \mu \frac{\partial u}{\partial y} \propto \frac{\mu U_e}{\delta} \propto \frac{\mu U_\infty}{L}\sqrt{Re_L}$$

But, by definition, $Re_L = \rho U_\infty L/\mu$, so the previous equation becomes

$$\tau_w \propto \rho U_\infty^2 \frac{\mu}{\rho U_\infty L}\sqrt{Re_L} = \rho U_\infty^2 \frac{1}{\sqrt{Re_L}} \tag{8.28}$$

It therefore follows from Eqs. (8.22) and (8.27) that the relationships between the coefficients of skin friction and skin-friction drag and Reynolds number are identical and given by

$$C_f \propto \frac{1}{\sqrt{Re_L}} \quad \text{and} \quad C_{D_f} \propto \frac{1}{\sqrt{Re_L}} \tag{8.29}$$

Example 8.1

Some engineers want a good estimate of drag and boundary-layer thickness at the trailing edge of a miniature wing. The chord and span of the wing are 6 mm and 30 mm, respectively and a typical flight speed is 5 m/s in air (kinematic viscosity $= 15 \times 10^{-6}$ m^2/s; density $= 1.2$ kg/m^3). An engineer may decide to make a superseding model with chord and span of 150 mm and 750 mm, respectively. Measurements on the model in a water channel flowing at 0.5 m/s (kinematic viscosity $= 1 \times 10^{-6}$ m^2/s, density $= 1000$ kg/m^3) give a drag of 0.19 N and a boundary-layer thickness of 3 mm. Estimate the corresponding values for the prototype.

The Reynolds numbers of both model and prototype are given by

$$(Re_L)_m = \frac{0.15 \times 0.5}{1 \times 10^{-6}} = 75\,000 \quad \text{and} \quad (Re_L)_p = \frac{0.006 \times 5}{15 \times 10^{-6}} = 2000$$

Evidently, the Reynolds numbers are not the same for the model and the prototype, so the flows are not dynamically similar. But, as a streamlined body is involved, we can use Eqs. (8.14) and (8.29). From Eq. (8.14),

$$\delta_p = \delta_m \times \frac{L_p}{L_m} \times \left(\frac{(Re_L)_m}{(Re_L)_p}\right)^{1/2}$$

$$= 3 \times \frac{6}{150} \times \left(\frac{75\,000}{2000}\right)^{1/2} = 0.735\,\text{mm} = 735\,\mu\text{m}$$

and from Eq. (8.29),

$$(C_{Df})_p = (C_{Df})_m \times \left(\frac{(Re_L)_m}{(Re_L)_p}\right)^{1/2}$$

But $D_f = \frac{1}{2}\rho U_\infty^2 S C_{Df}$, so, if we assume that skin-friction drag is the dominant type of drag and that it scales in the same way as total drag, the prototype drag is given by

$$D_p = D_m \frac{(\rho U_\infty^2 S)_p}{(\rho U_\infty^2 S)_m} \left(\frac{(Re_L)_m}{(Re_L)_p}\right)^{1/2} = 0.19 \times \frac{1.2 \times 5^2 \times 6 \times 30}{1000 \times 0.5^2 \times 150 \times 750} \times \left(\frac{75\,000}{2000}\right)^{1/2}$$

$$= 0.00022\,\text{N} = 220\,\mu\text{N}$$

8.2.4 Laminar Boundary-Layer Thickness along a Flat Plate

From Blasius, there is an exact solution to the boundary-layer equations. This was examined in Section 2.12. Integrating the velocity profile determined by Blasius, the displacement, momentum, and energy thicknesses can be determined.

The displacement thickness is

$$\delta^* = \int_0^\infty \left(1 - \frac{u}{U_\infty}\right) dy = \int_0^\infty \left(1 - \frac{u}{U_\infty}\right) \underbrace{\frac{dy}{d\eta}}_{\sqrt{\frac{2vx}{U_\infty}}} d\eta$$

$$= \sqrt{\frac{2vx}{U_\infty}} \underbrace{\int_0^\infty \left(1 - \frac{df}{d\eta}\right) d\eta}_{1.7208} = 1.7208 \sqrt{\frac{vx}{U_\infty}} \qquad (8.30)$$

The momentum thickness is

$$\theta = \int_0^\infty \frac{u}{U_\infty}\left(1 - \frac{u}{U_\infty}\right) dy = \sqrt{\frac{2vx}{U_\infty}} \int_0^\infty \frac{df}{d\eta}\left(1 - \frac{df}{d\eta}\right) d\eta - 0.664 \sqrt{\frac{vx}{U_\infty}} \qquad (8.31)$$

The energy thickness is

$$\delta^{**} = \int_0^\infty \frac{u}{U_\infty}\left(1 - \frac{u^2}{U_\infty^2}\right) dy = \sqrt{\frac{2vx}{U_\infty}} \int_0^\infty \frac{df}{d\eta}\left\{1 - \left(\frac{df}{d\eta}\right)^2\right\} d\eta = 1.0444 \sqrt{\frac{vx}{U_\infty}} \qquad (8.32)$$

The local wall shear stress and thus the skin-friction drag can also be calculated readily from function $f(\eta)$—that is, the similarity solution described in detail in Section 2.12, as follows:

$$\tau_w(x) = \mu \left(\frac{\partial u}{\partial y}\right)_{y=0} = \mu \underbrace{\frac{\partial \eta}{\partial y}}_{\sqrt{\frac{U_\infty}{2vx}}} \times \underbrace{\left(\frac{\partial u}{\partial \eta}\right)_{\eta=0}}_{U_\infty\left(\frac{d^2 f}{d\eta^2}\right)_{\eta=0}}$$

$$= \mu U_\infty \sqrt{\frac{U_\infty}{2vx}} \underbrace{\left(\frac{d^2 f}{d\eta^2}\right)_{\eta=0}}_{0.332\sqrt{2}} = 0.332 \mu U_\infty \sqrt{\frac{U_\infty}{vx}} \qquad (8.33)$$

Thus the skin-friction coefficient is

$$C_f(x) = \frac{\tau_{w(x)}}{\frac{1}{2}\rho U_\infty^2} = \frac{0.664}{\sqrt{Re_x}} \quad \text{where} \quad Re_x = \frac{\rho U_\infty x}{\mu} \tag{8.34}$$

The drag of one side of the plate (spanwise breadth B and length L) is given by

$$D_F = B \int_0^L \tau_w(x)dx \tag{8.35}$$

Thus, combining Eqs. (8.33) and (8.35), we find that the drag of one side of the plate is given by

$$D_F = 0.332\,\mu BU_\infty \sqrt{\frac{U_\infty}{\nu}} \int_0^L \frac{dx}{\sqrt{x}} = 0.664\,\mu BU_\infty \sqrt{\frac{LU_\infty}{\nu}} = 0.664\,\mu BU_\infty \sqrt{Re_L} \tag{8.36}$$

So the coefficient of skin-friction drag is given by

$$C_{DF} = \frac{D_F}{\frac{1}{2}\rho U_\infty^2 BL} = \frac{1.328}{\sqrt{Re_L}} \tag{8.37}$$

Example 8.2

We use the Blasius solution for the laminar boundary layer over a flat plate to estimate the boundary-layer thickness and skin-friction drag for the miniature wing from Example 8.1.

The Reynolds number is based on length $Re_L = 2000$; according to Eqs. (2.142) and (8.30), then the boundary-layer thicknesses at the trailing edge are given by

$$\delta_{0.99} = \frac{5.0L}{Re_L^{1/2}} = \frac{5 \times 6}{\sqrt{2000}} = 0.67\,\text{mm} \quad \delta^* = \frac{1.7208}{5} \times 0.67 = 0.23\,\text{mm}$$

Remember that the wing has two sides, so we get an estimate for its skin-friction drag by

$$D_F = 2 \times C_{DF} \times \frac{1}{2}U_\infty^2 BL = \frac{1.328}{\sqrt{2000}} \times 1.2 \times 5^2 \times 30 \times 6 \times 10^{-6} = 160\,\mu\text{N}$$

8.2.5 Solving the General Case

The solution to the boundary-layer equations for the flat plate described in Section 8.2.2 is a very special case. Although other *similarity* solutions exist (i.e., where the boundary-layer equations reduce to an ordinary differential equation), they are of limited practical value. In general, it is necessary to solve Eqs. (8.10) and (8.11) as partial differential equations.

To fix ideas, consider the flow over an airfoil, as shown in Fig. 8.10, in which the boundary-layer thickness is greatly exaggerated. The first step is to determine the potential flow around the airfoil. This is done computationally using the panel method described in Section 3.6 for nonlifting airfoils and in Section 4.10 in the case where lift is generated. From this solution for the potential flow, the velocity U_e along the surface of the airfoil can be determined. This is assumed to be the velocity at the edge of the boundary layer. The location of the fore stagnation point F can also be determined from the solution for U_e. Plainly, it corresponds to $U_e = 0$. (For the nonlifting case of a symmetric airfoil at zero angle of attack, the location of the fore stagnation point is known in advance from symmetry.) This point corresponds to $x = 0$. The development of the boundary layers over the top and bottom of the airfoil must be calculated separately unless they are identical, as in symmetric airfoils at zero angle of attack.

Mathematically, the boundary-layer equations are *parabolic*, which means that their solution (i.e., the boundary-layer velocity profile) at an arbitrary point P_1, say where $x = x_1$, on the airfoil depends only on the solutions upstream (i.e., at $x < x_1$). This property allows special efficient numerical methods to be used whereby we begin with the solution at the fore stagnation point and march step by step around the airfoil, solving the boundary-layer equations at each value of x in turn. This is much easier than solving the Navier–Stokes equations that, in subsonic steady flow, are *elliptic* (like the Laplace equation). The term *elliptic* implies that the solution (i.e., the velocity field) at a particular point depends on the solutions at all other points. For elliptic equations, the flow field upstream does depend on conditions downstream. How else would the flow approaching the airfoil sense its presence and begin to gradually deflect from uniform flow in order to flow smoothly around the airfoil? Still, numerical solution of the boundary-layer equations is not simple. To avoid numerical

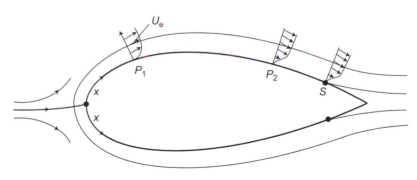

FIGURE 8.10

Boundary layer developing around an airfoil.

instability, so-called implicit methods are usually required. These are largely beyond our scope, but are described in a simple treatment given in Section 8.10.3.

For airfoils and other bodies with rounded leading edges, the stagnation flow field determined in Section 2.10.3 gives the initial boundary-layer velocity profile in the vicinity of $x = 0$. The velocity U_e along the edge of the boundary layer increases rapidly away from the fore stagnation point F. We find the evolving velocity profile in the boundary layer by solving the boundary-layer equations step by step, marching around the airfoil surface. At some point, U_e reaches a maximum at the point of minimum pressure. From this point onward, the pressure gradient along the surface changes sign to become adverse and begins to slow the boundary-layer flow (as explained earlier in Section 2.11.6 and to be discussed in Section 8.3). A point of inflexion develops in the velocity profile (e.g., at point P_2 in Fig. 8.10) that moves toward the wall as x increases. Eventually, the inflexion point reaches the wall itself, the shear stress at the wall falls to zero, reverse flow occurs, and the boundary layer separates from the airfoil surface at point S. The boundary-layer equations cease to be valid just before separation (where $\tau_w = \mu(\partial u/\partial y)_w = 0$), and the calculation is terminated.

Overall, the same procedures are involved in the approximate methods described in Section 8.10. A more detailed account of the computation of the boundary layer around an airfoil is presented there.

8.3 BOUNDARY-LAYER SEPARATION

The behavior of a boundary layer in a positive pressure gradient (i.e., pressure increasing with distance downstream) may be considered with reference to Fig. 8.11, which shows a length of surface with a gradual but steady convex curvature, such as that of an airfoil beyond the point of maximum thickness. In such a flow region, because of the retardation of the mainstream flow, the pressure in the mainstream rises (Bernoulli's equation). The variation in pressure along a normal to the surface through the boundary-layer thickness is essentially zero, so the pressure at any point in the mainstream, adjacent to the edge of the boundary layer, is transmitted unaltered through the layer to the surface. In light of this, consider the small element of fluid (Fig. 8.11) marked ABCD. On face AC, the pressure is p; on face BD, the pressure has increased to $p + (\partial p/\partial y)\delta x$. Thus the net pressure force on the element tends to retard the flow. This retarding force is in addition to the viscous shears that act along AB and CD, and it continuously slows the element as it progresses downstream.

This slowing effect is more pronounced near the surface where the elements are remote, via shearing actions, from the accelerating effect of the mainstream, so successive profile shapes in the streamwise direction change in the manner shown.

Ultimately, at a point S on the surface, the velocity gradient $(\partial u/\partial y)_w$ becomes zero. Apart from the change in profile shape, it is evident that the boundary layer must thicken rapidly under these conditions in order to satisfy continuity within the boundary layer. Downstream of point S, the flow adjacent to the surface is in an upstream direction so that a circulatory movement, in a plane normal to the surface,

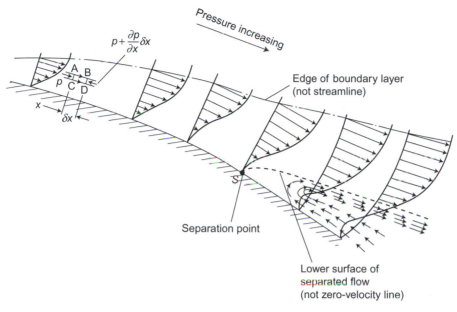

FIGURE 8.11

Boundary-layer separation,

takes place near it. A line (dotted in Fig. 8.11) may be drawn from point S such that the mass flow above it corresponds to the mass flow ahead of point S. This line represents the continuation of the lower surface of the upstream boundary layer so that, in effect, the original boundary layer separates from the surface at point S. This is termed the separation point.

The velocity profiles for laminar and turbulent layers in Fig. 8.5 make it clear that, owing to the greater extent of lower-energy fluid near the surface in the laminar boundary layer, the effect of a positive pressure gradient causes separation of the flow much more rapidly than if the flow were turbulent. A turbulent boundary layer is said to stick to the surface better than a laminar one.

The result of separation on the rear half of an airfoil is an increase in thickness of the wake flow, with a consequent reduction in pressure rise that should occur near the trailing edge. This rise means that the forward-acting pressure-force components on the rear part of the airfoil do not develop to offset the rearward-acting pressures near the front stagnation point. In consequence, the pressure drag of the airfoil increases. In fact, if there were no boundary layers, there would be a stagnation point at the trailing edge and the boundary-layer pressure drag, as well as the skin-friction drag, would be zero. If the airfoil incidence is sufficiently large, separation takes place not far downstream of the maximum suction point, and a very large wake develops. This causes such a marked redistribution of the flow over the airfoil that the large area of low pressure near the upper-surface leading edge is seriously reduced, with the result that the lift force is also greatly reduced. This condition is referred to as *stall*.

A negative pressure gradient obviously has the reverse effect, since the streamwise pressure forces cause energy to be added to the slower-moving air near the surface, decreasing any tendency for the layer adjacent to the surface to come to rest.

8.3.1 Separation Bubbles

On many airfoils with relatively large upper-surface curvatures, high local curvature over the forward part of the chord may initiate a laminar separation when the airfoil is at a moderate angle of incidence (Fig. 8.12).

Small disturbances grow much more readily and at low Reynolds numbers in separated, as compared to attached, boundary layers. Consequently, the separated laminar boundary layer may undergo transition to turbulence with characteristic rapid thickening. This thickening may be sufficient for the lower edge of the now turbulent shear layer to come back into contact with the surface and reattach as a turbulent boundary layer on the surface. In this way, a bubble of fluid is trapped under the separated shear layer between the separation and reattachment points. Within the bubble, the boundary of which is usually the streamline that leaves the surface at the separation point, two regimes exist. In the upstream region, a pocket of stagnant fluid at constant pressure extends back some way; behind this, a circulatory motion develops, as shown in Fig. 8.12, with the pressure in this latter region increasing rapidly towards the reattachment point.

Two distinct types of bubble are observed:

- A short bubble of the order of 1% of the chord in length (or 100 separation-point displacement thicknesses[2]) that exerts negligible effect on the peak suction value just ahead of it.
- A long bubble that may be of almost any length from a few percent of the chord (10,000 separation-point displacement thicknesses) to almost the entire chord, which exerts a large effect on the value of the peak suction near the airfoil leading edge.

A useful criterion for whether a short or long bubble is formed is the value at the separation point of the displacement-thickness Reynolds number $Re_{\delta*} = U_e\delta^*/v$. If $Re_{\delta*} < 400$, a long bubble almost certainly forms; for values > 550, a short bubble is almost certain. Between these values either type is possible. This is the Owen-Klanfer [56] criterion.

Short bubbles exert very little influence on the pressure distribution over the airfoil surface and remain small, with increasing incidence, right up to the stall. In general, they move slowly forward along the upper surface as incidence increases. The final stall may be caused by forward movement of the rear turbulent separation point (trailing-edge stall) or by breakdown of the small bubble at the leading edge caused by the failure of the separated shear flow to reattach at high incidence (leading-edge stall).

---

[2] Displacement thickness δ^* is defined in Section 8.2.2.

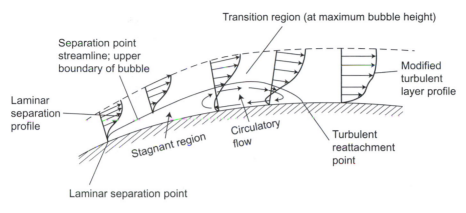

FIGURE 8.12

Laminar separation and turbulent reattachment points.

If a long bubble forms at moderate incidence, its length rapidly increases with increasing incidence, causing a continuous reduction of the leading-edge suction peak. The bubble may ultimately extend to the trailing edge or even into the wake downstream. This condition results in a low lift coefficient and effective stalling of the airfoil. Known as progressive stall, this usually occurs with thin airfoils and is often referred to as thin-airfoil stall. There are thus three alternative mechanisms that may produce subsonic stalling of airfoil sections.

8.4 FLOW PAST CYLINDERS AND SPHERES

Some of the properties of boundary layers just discussed help to explain the behavior, under certain conditions, of a cylinder or sphere immersed in a uniform free stream. So far, our discussion has been restricted to the flow over bodies of reasonably streamline form, behind which a relatively thin wake is formed. In such cases, the drag forces are largely due to surface friction (i.e., shear stresses at the base of the boundary layer). When dealing with nonstreamlined or bluff bodies, because of the adverse effect of a positive pressure gradient on the boundary layer, the flow usually separates somewhere near points at the maximum cross-section and forms a broad wake. As a result, the skin-friction drag is small, and total drag now consists mostly of form drag due to the large area at the rear of the body acted on by a reduced pressure in the wake region. Experimental observation of the flow past a sphere or cylinder indicates that the drag of the body is markedly influenced by the cross-sectional area of the wake; a broad wake is accompanied by a relatively high drag and vice versa.

We can understand the way the flow pattern around a bluff body can change dramatically as the Reynolds number changes by considering the flow past a circular cylinder. For the most part, the flow past a sphere behaves in a similar way. At very low Reynolds numbers[3] (i.e., less than unity), the flow behaves as if it were purely viscous with negligible inertia. This is known as *creeping* or *Stokes* flow, for which there are no boundary layers and the effects of viscosity extend an infinite distance from the body. The streamlines are completely symmetrical fore and aft, as depicted in Fig. 8.13(a). The streamline pattern is superficially similar in appearance to that for potential flow. For creeping flow, however, the influence of the cylinder on the streamlines extends to much greater distances than for potential flow. Skin-friction drag is the only force generated by the fluid flow on the cylinder. Consequently, the body with the lowest drag for a fixed volume is the sphere. (Perhaps this is the reason that microscopic swimmers such as protozoa, bacteria, and spermatazoa tend to be nearly spherical.) In the range $1 < Re < 5$, the streamline pattern remains fairly similar to that in Fig. 8.13(a), except that as *Re* increases within this range an increasingly pronounced asymmetry develops between the fore and aft directions. Nevertheless, the flow remains attached.

When *Re* exceeds about 5, a much more profound change in flow pattern occurs. The flow separates from the cylinder surface to form a closed wake of recirculating flow (see Fig. 8.13(b)). The wake grows progressively longer as *Re* increases from 5 up to about 41. The flow pattern is symmetrical about the horizontal axis and is *steady*—that is, it does not change with time. At these comparatively low Reynolds numbers, the effects of viscosity still extend a considerable distance from the surface, so the concept of the boundary layer is not valid here; nevertheless, the explanation for flow separation is substantially the same as that given in Section 8.3.

When *Re* exceeds about 41, another profound change occurs: steady flow becomes impossible. In some respects, this is similar to the early stages of laminar-turbulent transition (see Section 8.8), in that the steady recirculating wake flow, seen in Fig. 8.13(b), becomes unstable to small disturbances. In this case, though, the small disturbances develop as vortices rather than waves. Also in this case, the small disturbances do not develop into turbulent flow; rather, a steady laminar wake becomes unsteady, but stable. The vortices are generated periodically on alternate sides of the horizontal axis through the wake and the center of the cylinder, and in this way a row of vortices is formed. The row persists a very considerable distance downstream. This phenomenon was first explained theoretically by von Kármán in the first decade of the twentieth century.

For Reynolds numbers between just above 40 and up to about 100, the vortex street develops from amplified disturbances in the wake. However, as the Reynolds number rises, an identifiable thin boundary layer begins to form on the cylinder surface and the disturbance develops increasingly closer to the cylinder. Finally, above about $Re = 100$, eddies are shed alternately from the laminar separation points on either side of the cylinder (see Fig. 8.14). Thus a vortex is generated in the region behind the separation point on one side, and a corresponding vortex on the other

[3]The Reynolds number here is defined as $U_\infty D/\nu$, where U_∞ is the free-stream velocity, ν is the kinematic viscosity in the free stream, and D is the cylinder diameter.

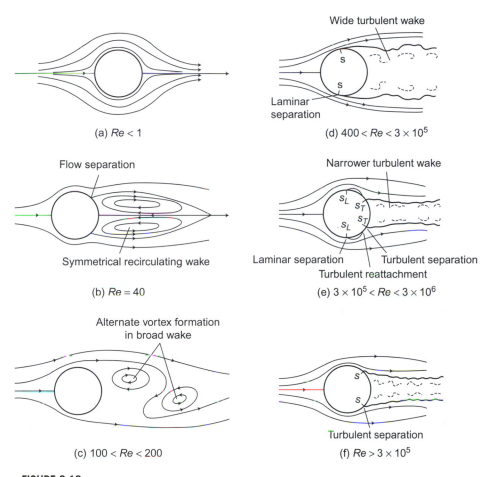

Wide turbulent wake

s
s

Laminar
separation

(a) Re < 1

(d) 400 < Re < 3 × 10⁵

Flow separation

Symmetrical recirculating wake

(b) Re = 40

Narrower turbulent wake

s_L s_T
s_T
s_L

Laminar separation   Turbulent separation
Turbulent reattachment

(e) 3 × 10⁵ < Re < 3 × 10⁶

Alternate vortex formation
in broad wake

(c) 100 < Re < 200

s
s

Turbulent separation

(f) Re > 3 × 10⁵

FIGURE 8.13

Illustration of flows over circular cylinders. (Note: The Reynolds number limits are approximate, depending appreciably on the free-stream turbulence level.)

side breaks away from the cylinder and moves downstream in the wake. When the attached vortex reaches a particular strength, it in turn breaks away and a new vortex begins to develop, again on the second side, and so on.

The wake thus consists of a procession of equal-strength vortices, equally spaced but alternating in sign. This wake type, which can occur behind all long cylinders of bluff cross-section, including flat plates normal to the flow direction, is termed a von Kármán *vortex street* or *trail* (see Fig. 8.15(a)). In a uniform stream flowing past a cylinder, the vortices move downstream at a speed somewhat less than the free-stream velocity; this reduction in speed is inversely proportional to the streamwise distance separating alternate vortices.

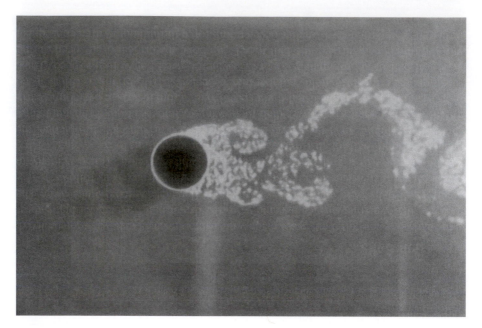

FIGURE 8.14

Wake of a circular cylinder at $Re_D = 5000$. Vortices are formed when flow passes over circular cylinders for a wide range of Reynolds numbers. The flow is from left to right. A laminar boundary layer is formed on the upstream surface of the cylinder. It separates just ahead of the maximum thickness and breaks up into a turbulent wake, which is dominated by large-scale vortices. Flow visualization is obtained by using aluminum particle tracers on water flow. *Source: Photograph by D.J. Buckingham, School of Engineering, University of Exeter, United Kingdom.*

During the formation of any single vortex while it is bound to the cylinder, an increasing circulation exists about the cylinder and generates a transverse (lift) force. With the development of each successive vortex, this force changes sign, giving rise to an alternating transverse force on the cylinder at the same frequency as that of the vortex shedding. If the frequency happens to coincide with the natural frequency of the cylinder's oscillation, however it may be supported, appreciable vibration may result. This phenomenon is what causes, for example, the singing of telegraph wires in the wind (Aeolian tones).

A unique relationship exists between the Reynolds number and a dimensionless parameter involving shedding frequency. Known as the Strouhal number, this parameter is defined by the expression $S = nD/U_\infty$, where n is the vortex shedding frequency. Figure 8.15(b) shows the typical variation of S with Re in the vortex street range.

Despite the many other changes, to be described momentarily, that occur in flow pattern as Re increases, markedly periodic vortex shedding remains a characteristic flow around the circular cylinder and other bluff bodies up to the highest

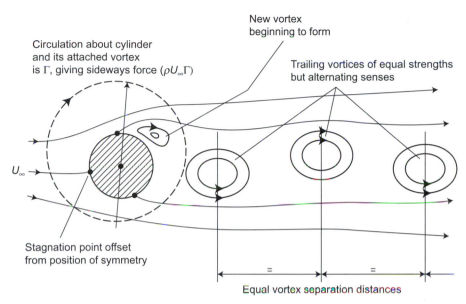

Circulation about cylinder
and its attached vortex
is Γ, giving sideways force ($\rho U_\infty \Gamma$)

New vortex
beginning to form

Trailing vortices of equal strengths
but alternating senses

U_∞

Stagnation point offset
from position of symmetry

Equal vortex separation distances

Flow pattern and sideways force reverse when vortex detaches and new vortex
begins to form on opposite side

(a)

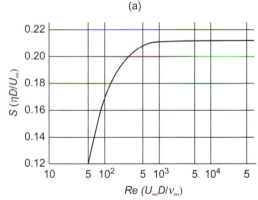

Approximate relation between Strouhal number and
Reynolds number for circular cylinder

(b)

FIGURE 8.15

Periodic vortex shedding and Strouhal number versus Reynolds number for flow around
circular cylinders.

Reynolds numbers. This phenomenon can have important consequences in engineering applications—for example, the Tacoma Narrows Bridge (in Washington State). A natural frequency of the bridge deck was close to its shedding frequency, causing resonant behavior in moderate winds. The bridge's collapse in 1940 was due to torsional aeroelastic instability excited by stronger winds.

For two ranges of Reynolds number—$200 < Re < 400$ and $3 \times 10^5 < Re < 3 \times 10^6$—the regularity of vortex shedding is greatly diminished. In the former range, considerable scatter occurs in the Strouhal number; in the latter range, all periodicity disappears except very close to the cylinder. The Reynolds numbers marking the limits of these two ranges are associated with pronounced changes in flow pattern. For $Re \simeq 400$ and 3×10^6, the transitions in flow pattern are such as to restore periodicity.

Below $Re \simeq 200$, the vortex street persists to great distances downstream. Above it, transition to turbulent flow occurs in the wake and so destroys the periodic vortex wake far downstream. At $Re \simeq 200$, the vortex street also becomes unstable to three-dimensional disturbances, leading to greater irregularity.

At $Re \simeq 400$, a further change occurs. Transition to turbulence is now close to the separation points on the cylinder. Curiously, this has a stabilizing effect on shedding frequency, even though the vortices themselves develop considerable irregular fluctuations. This pattern of laminar boundary-layer separation and a turbulent vortex wake persists until $Re \simeq 3 \times 10^5$; it is illustrated in Figs. 8.13(d) and 8.14. Note that with laminar separation, the flow separates at points on the front half of the cylinder, forming a large wake and producing high-level form drag. In this case, the contribution of skin-friction drag is negligible. When the Reynolds number reaches the vicinity of 3×10^5, the laminar boundary layer undergoes transition to turbulence almost immediately after separation. The increased mixing reenergizes the separated flow, causing it to reattach as a turbulent boundary layer, thereby forming a separation bubble, as described in Section 8.3.1 (see Fig. 8.13(e)).

At this critical stage, the second and final point of separation, which now takes place in a turbulent layer, moves suddenly downstream because of the better sticking property of the turbulent layer, and the wake width appreciably decreases. This stage is therefore accompanied by a sudden decrease in total cylinder drag. For this reason, the Re at which this transition in flow pattern occurs is often called the *critical Reynolds* number. The wake vorticity remains random with no clearly discernible frequency. Further increase in Reynolds number causes the wake width to gradually increase to begin with, as the turbulent separation points slowly move upstream around the rear surface. The total drag continues to increase steadily because of increases in both pressure and skin-friction drag, although the drag coefficient, defined by

$$C_D = \frac{\text{drag per unit span}}{\frac{1}{2}\rho_\infty U_\infty^2 D}$$

tends to become constant, at about 0.6, for values of $Re > 1.3 \times 10^6$. The final change in the flow pattern occurs at $Re \simeq 3 \times 10^6$, when the separation bubble disappears (see Fig. 8.13(f)). This transition has a stabilizing effect on shedding frequency, which becomes discernible again. C_D rises slowly as the Reynolds number increases beyond 3×10^6.

The actual value of the Reynolds number at the critical stage, when dramatic drag decrease occurs, depends, for a smooth cylinder, on the small-scale turbulence level in the oncoming free stream. Increased turbulence or, alternatively, increased surface roughness provokes turbulent reattachment, with its accompanying drag decrease, at a lower Reynolds number. The behavior of a smooth sphere under similarly varying conditions exhibits the same characteristics as the cylinder (see Fig. 8.16), although the Reynolds numbers corresponding to changes in flow regime are somewhat different. One marked difference is that the eddying vortex street, typical of bluff cylinders, does not develop as regularly behind a sphere. Graphs showing the variations in drag coefficient with Reynolds number for circular cylinders and spheres are given in Fig. 8.17.

AERODYNAMICS AROUND US

Viscous Flow

Perhaps the most common aerodynamic flow separation you will encounter is blunt-body flow—your car. While this differs in numerous ways from the flow over low-drag bodies like airfoils, it is sometimes possible to observe flow structures on your car. A long sloping hatchback can provide a good view of surface streamlines on a rainy day. Streaks of water flowing down the rear window can show if the flow is attached, streaming constantly, or separated. A separated flow produces largely stagnant droplets on the window. Of course, you should observe these flows while you are the passenger, not the driver.

If you are standing at the side of a road and a truck speeds past you, you feel another result of viscous flow and, on a blunt body, of flow separation—the wake. A streamlined body may also have a substantial wake. One aerospace engineer who is also a diver described the wake from the left-to-right motion of the tail of a 50-foot whale shark he was photographing as sufficiently strong to send him tumbling. The shark is shaped for efficient motion in the water (the higher-drag fish are outcompeted or eaten), but the propulsive motion of its tail relies on pressure differences across it, and this pressure difference results in motion of the water. For the shark to generate thrust, it pushes back on the water, generating a wake in which large vortices are present.

8.4.1 Turbulence on Spheres

The effect of free-stream turbulence on the Reynolds number at which critical drag decreases was widely used many years ago to ascertain the turbulence level in the airstream of a wind-tunnel working section. In this application, a smooth sphere was mounted in the working section, and its drag, for a range of tunnel speeds, was read off on the drag balance. The speed, and hence the Reynolds number, at which the drag suddenly decreased was recorded. Experiments in air with virtually zero small-scale turbulence indicated that the highest critical-sphere Reynolds number attainable is 385,000. A turbulence factor for the tunnel under test was thus defined as the ratio of 385,000 to the critical Reynolds number of the test tunnel.

A major difficulty in this application was the necessity for extreme accuracy in the manufacture of the sphere, as small variations from the true spherical shape could cause appreciable differences in behavior at the critical stage. As a result, this technique for turbulence measurement is no longer in favor, and more recent methods, such as hot-wire anemometry, took its place some time ago.

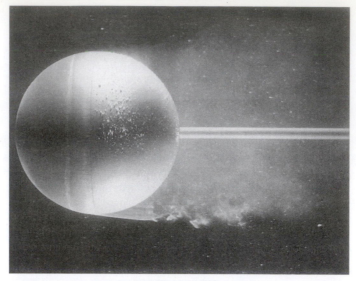

(a) $Re_D = 15,000$

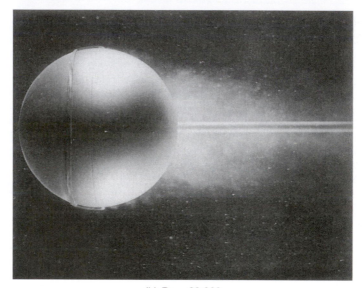

(b) $Re_D = 30,000$

FIGURE 8.16

Flow past a sphere, in both cases from left to right. $Re_D = 15,000$ for (a), which uses dye in water to show a laminar boundary layer separating ahead of the equator and remaining laminar for almost one radius before becoming turbulent. Air bubbles in water provide the flow visualization in (b). For this case $Re_D = 30,000$ and a wire hoop on the downstream surface trips the boundary layer to ensure that transition occurs in the separation bubble leading to reattachment and a final turbulent separation much further rearward. The much reduced wake in (b) as compared with (a) leads to dramatically reduced drag. The use of a wire hoop to promote transition artificially produces the drag reduction at a much lower Reynolds number than for the smooth sphere. *Source: Photographs by H. Werlé, ONERA, France.*

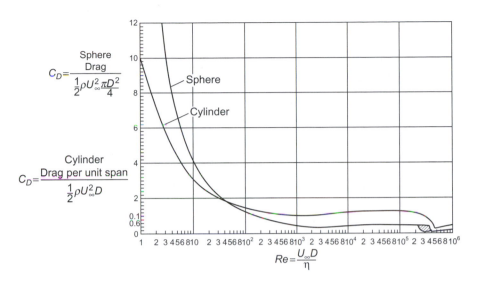

$$C_D = \frac{\text{Sphere Drag}}{\frac{1}{2}\rho U_\infty^2 \frac{\pi D^2}{4}}$$

$$C_D = \frac{\text{Cylinder Drag per unit span}}{\frac{1}{2}\rho U_\infty^2 D}$$

$$Re = \frac{U_\infty D}{\eta}$$

FIGURE 8.17

Approximate values of C_D with Re for spheres and circular cylinders.

8.4.2 Golf Balls

In the early days of golf, balls were made with a smooth surface. It was soon realized, however, that when its surface became worn the driver ball traveled farther. Subsequently, golf balls were manufactured with a dimpled surface to simulate wear. The reason for the increase in driven distance with the rough surface is as follows.

The diameter of a golf ball is about 42 mm, which gives a critical velocity in air, for a smooth ball, of just over 135 m s^{-1} (corresponding to $Re = 3.85 \times 10^5$). This is much higher than the average flight speed of a driven ball. In practice, the critical speed is somewhat lower than this owing to imperfections in manufacture, but it is still higher than the usual flight speed. With a rough surface promoting early transition, the critical Reynolds number may be as low as 10^5, giving a critical speed for a golf ball of approximately 35 m s^{-1}, which is well below flight speed. Thus, with the roughened surface, the ball travels at above the critical drag speed during flight and so experiences a smaller decelerating force throughout, with a consequent increase in range.

8.4.3 Cricket Balls

The art of the seam bowler in cricket is also explainable by boundary-layer transition and separation. The idea is to align the seam at a small angle to the flight path (see Fig. 8.18) by spinning the ball about an axis perpendicular to the plane of the seam and using gyroscopic inertia to stabilize the seam position during the trajectory. On the side of the front stagnation point where the boundary layer passes over the seam, the boundary layer is induced to become turbulent before reaching the point of laminar separation. The boundary layer remains attached to a greater angle from

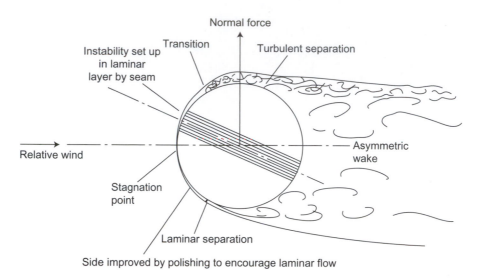

FIGURE 8.18

Flow around a cricket ball.

the fore stagnation point on this side than it does on the other side, where no seam is present to trip it. The flow past the ball thus becomes asymmetric with a larger area of low pressure on the turbulent side, producing a lateral force that moves the ball in a direction normal to its flight path.

The range of flight speeds over which this phenomenon can be employed corresponds to speeds achieved by the medium to medium-fast pace bowler. The diameter of a cricket ball is between 71 and 72.5 mm. In air, the critical speed for a smooth ball is about 75 m s^{-1}. In practice, however, transition to turbulence for the seam-free side occurs at speeds of 30 to 35 m s^{-1} because of inaccuracies in the ball's shape and minor surface irregularities. The critical speed for a rough ball with early transition ($Re \approx 10^5$) is about 20 m s^{-1}; below this, flow asymmetry tends to disappear because laminar separation occurs before the transition, even on the seam side.

Thus, from 20 to about 30 m s^{-1}, very approximately, a cricket ball may be made to swing by a skillful bowler. A very fast bowler produces a flight speed in excess of the upper critical, so no swing is possible. A bowler may make the ball swing late by bowling at a speed just too high for the asymmetric condition to exist so that, as the ball loses speed in flight, the asymmetry develops later in the trajectory. It is obvious that considerable skill and experience are required to know at just what speed the delivery must be to do this.

The surface condition, apart from the seam, also affects the possibility of swinging the ball. For example, a new, smooth-surfaced ball tends to maintain laminar layers up to separation even on the seam side, while a badly worn ball tends to induce turbulence on the side remote from the seam. The slightly worn ball is best, especially if one side can be kept reasonably polished to help maintain laminar flow on that side only.

8.5 THE MOMENTUM-INTEGRAL EQUATION

Accurately evaluating most of the quantities defined in Sections 8.2.2 and 8.2.3 requires numerical solution of the differential equations of motion. This will be discussed in Section 8.10. Here an integral form of the equations of motion is derived that allows practical solutions to be found fairly easily for certain engineering problems.

We derive the required momentum-integral equation considering mass and momentum balances on a thin slice of boundary layer of length δx (illustrated in Fig. 8.19). Remember that quantities generally vary with x (i.e., along the surface), so it follows from elementary differential calculus that the value of a quantity f, say, on CD (where the distance from the origin is $x + \delta x$) is related to its value on AB (where the distance from the origin is x) in the following way:

$$f(x + \delta x) \simeq f(x) + \frac{df}{dx}\delta x \tag{8.38}$$

We consider first the conservation of mass for an elemental slice of boundary layer (see Fig. 8.19(b)). Since the density is assumed to be constant, the mass flow balance for a slice ABCD is, in words,

$$\begin{array}{l}\text{Volumetric flow rate into} \\ \text{the slice across AB}\end{array} = \begin{array}{l}\text{Volumetric flow rate out across CD} \\ + \text{ Volumetric flow rate out across AD} \\ + \text{ Volumetric flow rate out across BC}\end{array}$$

The last item allows for the possibility of flow due to suction passing through a porous wall. In the usual case of an impermeable wall, $V_s = 0$. Expressed mathematically, this equation becomes

$$\underbrace{Q_i}_{\text{across AB}} = Q_i + \underbrace{\frac{dQ_i}{dx}\delta x}_{\text{across CD}} + \underbrace{V_e \delta x - U_e \frac{d\delta}{dx}\delta x}_{\text{across AD}} + \underbrace{V_s \delta x}_{\text{across BC}} \tag{8.39}$$

Note that Eq. (8.38) has been used, Q_i replacing f where

$$Q_i \equiv \int_0^\delta u \, dy$$

Cancelling common factors, rearranging Eq. (8.39), and taking the limit $\delta x \to dx$ leads to an expression for the perpendicular velocity component at the edge of the boundary layer:

$$V_e = -\frac{d}{dx}\int_0^\delta u \, dy + U_e \frac{d\delta}{dx} - V_s$$

The definition of displacement thickness, Eq. (8.18), is now introduced to give

$$V_e = -\frac{d}{dx}(\delta^* U_e) - V_s \tag{8.40}$$

Now we turn to the y momentum balance for slice ABCD of the boundary layer, which is illustrated in Fig. 8.19(c). In this case, noting that the y component of momentum can be carried by the flow across side AD only, and that the only force in the y direction is pressure,[4] the momentum theorem states that the

Rate at which y component of momentum crosses AD	$=$	Net pressure force in y direction acting on slice ABCD

or, in mathematical terms,

$$\rho V_e^2 \delta x = (p_w - p_1)\delta x$$

Cancelling the common factor δx thus leads to

$$p_w - p_1 = \rho V_e^2$$

We easily see from this result that the net pressure difference across the boundary layer is negligible (i.e., $p_w \simeq p_1$), as it should be according to boundary-layer theory. For simplicity, the case of the boundary layer along an impermeable flat plate when $U_e = U_\infty (=\text{const.})$ and $V_s = 0$ is considered so that, from Eq. (8.40),

$$V_e = U_\infty \frac{d\delta^*}{dx} \Rightarrow p_w - p_1 = \rho U_\infty^2 \left(\frac{d\delta^*}{dx}\right)^2$$

Remember, however, that the boundary layer is very thin compared with the length of the plate; thus $d\delta^*/dx \ll 1$ so that its square is negligibly small. This argument can be extended to the more general case where U_e varies along the edge of the boundary layer. Thus it can be demonstrated that the assumption of a thin boundary layer implies that the pressure does not vary appreciably across it. This is one of the major features of boundary-layer theory (see Section 8.2). It also implies that within the boundary layer pressure p is a function of x only.

Finally, we look at the x momentum balance for slice ABCD. This case is more complex since there are both pressure and surface friction forces to be considered; furthermore, the x component may be carried across AB, CD, and AD. The forces involved are depicted in Fig. 8.19(d) while the momentum fluxes are shown in

---

[4]The force of gravity is usually ignored in aerodynamics.

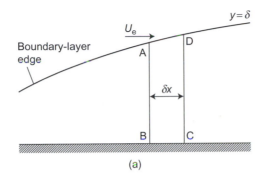

(a)

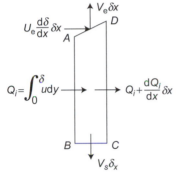

$$Q_i = \int_0^\delta u\,dy$$

Mass balance

(b)

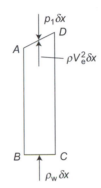

y-momentum balance

(c)

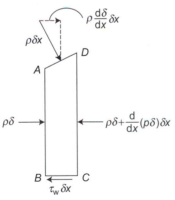

Forces in x direction

(d)

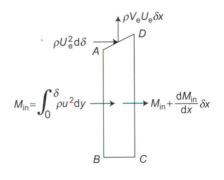

$$M_{in} = \int_0^\delta \rho u^2\,dy$$

Momentum fluctuations in x direction

(e)

FIGURE 8.19

Mass and momentum balances on a thin slice of boundary layer.

Fig. 8.19(e). In this case, the momentum theorem states that

$$\begin{pmatrix} \text{Rate at which} \\ \text{momentum leaves} \\ \text{across CD and AD} \end{pmatrix} - \begin{pmatrix} \text{Rate at which} \\ \text{momentum enters} \\ \text{across AB} \end{pmatrix}$$

$$= \begin{pmatrix} \text{Net pressure force} \\ \text{in } x \text{ direction acting} \\ \text{on ABCD} \end{pmatrix} - \begin{pmatrix} \text{Surface friction} \\ \text{force acting} \\ \text{on BC} \end{pmatrix}$$

Using Eq. (8.38), this can be expressed mathematically as

$$\underbrace{M_{\text{in}} + \frac{dM_{\text{in}}}{dx}}_{\text{out across CD}} + \underbrace{\rho V_e U_e \delta x - \rho U_e^2 \frac{d\delta}{dx}\delta x}_{\text{out across AD}} \underbrace{-M_{\text{in}}}_{\text{in across AB}}$$

$$= \underbrace{p\delta}_{\text{on AB}} - \underbrace{\left(p\delta + \frac{d}{dx}(p\delta)\delta x \right)}_{\text{on CD}} + \underbrace{p\frac{d\delta}{dx}\delta x}_{\text{on AD}} - \underbrace{\tau_w \delta x}_{\text{on BC}} \tag{8.41}$$

where

$$M_{\text{in}} = \int_0^\delta \rho u^2 dy.$$

After cancelling common factors, taking the limit $\delta x \to dx$, and simplifying, Eq. (8.41) becomes

$$\frac{d}{dx}\int_0^\delta \rho u^2 dy - \rho U_e^2 \frac{d\delta}{dx} + \rho U_e V_e = -\delta\frac{dp}{dx} - \tau_w \tag{8.42}$$

The Bernoulii's equation can be used at the edge of the boundary layer so that

$$p + \rho U_e^2 = \text{const.}, \quad \frac{dp}{dx} = -\rho U_e \frac{dU_e}{dx}$$

After substituting Eq. (8.40) for V_e, introducing the definition of momentum thickness, Eq. (8.19), and using the result just given, Eq. (8.42) reduces to

$$\frac{d}{dx}\left(U_e^2 \theta \right) + \delta^* U_e \frac{dU_e}{dx} - \rho U_e V_s = \frac{\tau_w}{\rho} \tag{8.43}$$

This is the momentum-integral equation first derived by von Kármán. Since no assumption is made at this stage about the relationship between τ_w and the velocity gradient at the wall, the equation applies equally well to laminar and turbulent flow.

When suitable forms are selected for the velocity profile, the momentum-integral equation can be solved to provide variations in δ, δ^*, θ, and C_f along the surface. A suitable approximate form for the velocity profile in the laminar boundary layer is derived in Section 8.5.1. To solve Eq. (8.43) in the turbulent case, additional semi-empirical relationships must be introduced. In the simple case of the flat plate, the solution to Eq. (8.43) can be found in closed form, as shown in Section 8.6. In the general case with a nonzero pressure gradient, computational methods are needed to solve Eq. (8.43). Such methods are discussed in Section 8.10.

8.5.1 An Approximate Velocity Profile for the Laminar Boundary Layer

As explained in the previous subsection, an approximate expression is required for the velocity profile in order to use the momentum-integral equation. A reasonably accurate approximation can be obtained using a cubic polynomial in the form

$$\bar{u}(\equiv u/U_e) = a + b\bar{y} + c\bar{y}^2 + d\bar{y}^3 \tag{8.44}$$

where $\bar{y} = y/\delta$. To evaluate the coefficients a, b, c, and d, four conditions are required: two at $\bar{y} = 0$ and two at $\bar{y} = 1$. Of these conditions, two are are readily available:

$$\bar{u} = 0 \quad \text{at} \quad \bar{y} = 0 \tag{8.45a}$$

$$\bar{u} = 1 \quad \text{at} \quad \bar{y} = 1 \tag{8.45b}$$

In real boundary-layer velocity profiles (see Fig. 8.6), velocity varies smoothly to reach U_e; there is no kink at the edge of the boundary layer. It follows then that the velocity gradient is zero at $y = \delta$, giving a third condition:

$$\frac{\partial \bar{u}}{\partial \bar{y}} = 0 \quad \text{at} \quad \bar{y} = 1 \tag{8.45c}$$

To obtain the fourth and final condition, it is necessary to return to the boundary-layer Eq. (8.10). At the wall $y = 0$, $u = v = 0$, so both terms on the left-hand side are zero at $y = 0$. Noting that $\tau = \mu \partial u / \partial y$, the required condition is thus given by

$$\frac{dp}{dx} = \frac{\partial \tau}{\partial y} \quad \text{at} \quad y = 0$$

Since $y = \bar{y}\delta$ and $p + \rho U_e^2 = \text{const.}$, this equation can be rearranged to read

$$\frac{\partial^2 \bar{u}}{\partial \bar{y}^2} = -\frac{\delta^2}{\nu} \frac{dU_e}{dx} \quad \text{at} \quad \bar{y} = 0 \tag{8.45d}$$

In terms of coefficients a, b, c, and d, the four conditions (8.45a, b, c, d) become

$$a = 0 \tag{8.46a}$$

$$b + c + d = 1 \tag{8.46b}$$

$$b + 2c + 3d = 0 \tag{8.46c}$$

$$2c = -\Lambda \quad \text{where} \Lambda \equiv \frac{\delta^2}{\nu} \frac{dU_e}{dx} \tag{8.46d}$$

Equations (8.46b,c,d) are easily solved for b, c, and d to give the following approximate velocity profile:

$$\bar{u} = \frac{3}{2}\bar{y} - \frac{1}{2}\bar{y}^3 + \frac{\Lambda}{4}(\bar{y} - 2\bar{y}^2 + \bar{y}^3) \tag{8.47}$$

In Eq. (8.47), Λ is often called the Pohlhausen parameter. It determines the effect of an external pressure gradient on the shape of the velocity profile. $\Lambda > 0$ and < 0 correspond, respectively, to favorable and unfavorable pressure gradients. For $\Lambda = -6$, the wall shear stress $\tau_w = 0$; for more negative values of Λ, flow reversal at the wall develops. Thus $\Lambda = -6$ corresponds to boundary-layer separation. Velocity profiles corresponding to various values of Λ are plotted in Fig. 8.20, in which the flat-plate profile corresponds to $\Lambda = 0$; $\Lambda = 6$ for the favorable pressure gradient; $\Lambda = -4$ for the mild adverse pressure gradient; $\Lambda = -6$ for the strong adverse pressure gradient; and $\Lambda = -9$ for the reversed-flow profile.

For the flat-plate case $\Lambda = 0$, we compare the approximate velocity profile of Eq. (8.47) with two other approximate profiles in Fig. 8.20. The profile labeled Blasius is the accurate solution of the differential equations of motion given in Section 8.2.1 and in Fig. 8.6.

The various quantities introduced in Sections 8.2.2 and 8.2.3 can be evaluated using the approximate velocity profile (Eq. 8.47). For example, if Eq. (8.47) is substituted in turn into Eqs. (8.18), (8.19) and (8.22), the following are obtained using Eq. (8.21):

$$I_1 = \frac{\delta^*}{\delta} = \int_0^1 \left[1 - \left(\frac{3}{2} + \frac{\Lambda}{4} \right) \bar{y} + \frac{\Lambda}{2} \bar{y}^2 + \left(\frac{1}{2} - \frac{\Lambda}{4} \right) \bar{y}^3 \right] d\bar{y}$$

$$= \left[\bar{y} - \left(\frac{3}{2} + \frac{\Lambda}{4} \right) \frac{\bar{y}^2}{2} + \frac{\Lambda}{2} \frac{\bar{y}^3}{3} + \left(\frac{1}{2} - \frac{\Lambda}{4} \right) \frac{\bar{y}^4}{4} \right]_0^1 \qquad (8.48a)$$

$$= 1 - \left(\frac{3}{4} + \frac{\Lambda}{8} \right) + \frac{\Lambda}{6} + \left(\frac{1}{8} - \frac{\Lambda}{16} \right)$$

$$= \frac{3}{8} - \frac{\Lambda}{48}$$

$$I = \frac{\theta}{\delta} = \int_0^1 \left[\left(\frac{3}{2} + \frac{\Lambda}{4} \right) \bar{y} - \frac{\Lambda}{2} \bar{y}^2 - \left(\frac{1}{2} - \frac{\Lambda}{4} \right) \bar{y}^3 \right]$$

$$\times \left[1 - \left(\frac{3}{2} + \frac{\Lambda}{4} \right) \bar{y} + \frac{\Lambda}{2} \bar{y}^2 + \left(\frac{1}{2} - \frac{\Lambda}{4} \right) \bar{y}^3 \right] d\bar{y}$$

$$= \left[\left(\frac{3}{2} + \frac{\Lambda}{4} \right) \frac{\bar{y}^2}{2} - \left\{ \left(\frac{3}{2} + \frac{\Lambda}{4} \right)^2 + \frac{\Lambda}{2} \right\} \frac{\bar{y}^3}{3} \right.$$

$$+ \left\{ 2 \left(\frac{3}{2} + \frac{\Lambda}{4} \right) \frac{\Lambda}{2} - \left(\frac{1}{2} - \frac{\Lambda}{4} \right) \right\} \frac{\bar{y}^4}{4} \qquad (8.48b)$$

$$+ \left\{ 2 \left(\frac{3}{2} + \frac{\Lambda}{4} \right) \left(\frac{1}{2} - \frac{\Lambda}{4} \right) - \frac{\Lambda^2}{4} \right\} \frac{\bar{y}^5}{5}$$

$$\left. - \left\{ 2 \frac{\Lambda}{2} \left(\frac{1}{2} - \frac{\Lambda}{4} \right) \right\} \frac{\bar{y}^6}{6} - \left(\frac{1}{2} - \frac{\Lambda}{4} \right)^2 \frac{\bar{y}^7}{7} \right]_0^1$$

$$= \frac{1}{280} \left(39 - \frac{\Lambda}{2} - \frac{1}{6} \Lambda^2 \right)$$

$$C_f = \frac{\tau_w}{\frac{1}{2} \rho U_e^2} = \frac{2\mu}{\rho U_e^2} \left(\frac{\partial u}{\partial y} \right)_w = \frac{2\mu}{\rho U_e \delta} \left(\frac{d\bar{u}}{d\bar{y}} \right)_{\bar{y}=0}$$

$$= \frac{\mu}{\rho U_e \delta} \left(3 + \frac{\Lambda}{2} \right) \qquad (8.48c)$$

Quantities I_1 and I depend only on the shape of the velocity profile; for this reason, they are usually known as *shape parameters*. If the more accurate differential form of the boundary-layer equations were used, rather than the momentum-integral equation

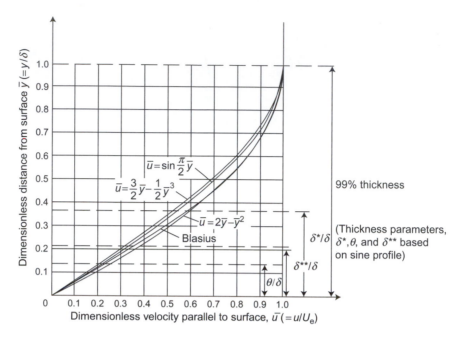

FIGURE 8.20

Laminar velocity profile.

with approximate velocity profiles, the boundary-layer thickness δ would become rather less precise. For this reason, it is more common to use the shape parameter $H = \delta^*/\theta$. H is frequently referred to simply as the shape parameter.

For the numerical methods discussed in Section 8.10, which are used in the general case with an external pressure gradient, it is preferable to employ a somewhat more accurate quartic polynomial as the approximate velocity profile, particularly for predicting the transition point. The quartic velocity profile is derived in a very similar way to that given earlier, with the main differences being the addition of another term $e\bar{y}^4$ on the right-hand side of Eq. (8.44) and the need for an additional condition at the edge of the boundary layer. This latter difference requires that

$$\frac{d^2\bar{u}}{d\bar{y}^2} = 0 \quad \text{at} \quad \bar{y} = 1$$

which has the effect of making the velocity profile even smoother at the edge of the boundary layer, thereby improving the approximation. The resulting quartic velocity profile takes the form

$$\bar{u} = 2\bar{y} - 2\bar{y}^3 + \bar{y}^4 + \frac{\Lambda}{6}(\bar{y} - 3\bar{y}^2 + 3\bar{y}^3 - \bar{y}^4) \qquad (8.49)$$

Using this profile and following procedures similar to those outlined previously leads to the following expressions:

$$I_1 = \frac{3}{10} - \frac{\Lambda}{120} \tag{8.48a'}$$

$$I = \frac{1}{63}\left(\frac{37}{5} - \frac{\Lambda}{15} - \frac{\Lambda^2}{144}\right) \tag{8.48b'}$$

$$C_f = \frac{\mu}{\rho U_e \delta}\left(4 + \frac{\Lambda}{3}\right) \tag{8.48c'}$$

Note that it follows from Eq. (8.48c') that, with the quartic velocity profile, the separation point where $\tau_w = 0$ now corresponds to $\Lambda = -12$.

8.6 APPROXIMATE METHODS FOR A BOUNDARY LAYER ON A FLAT PLATE WITH ZERO PRESSURE GRADIENT

In this section, we solve the momentum-integral Eq. (8.43) to give approximate expressions for skin-friction drag and for variation in δ, δ^*, θ, and C_f along a flat plate with laminar, turbulent, and mixed laminar/turbulent boundary layers. This may seem a rather artificial and restrictive case to study in depth. Note, however, that these results can be used to provide rough but reasonable estimates for any streamlined body. The equivalent flat plate for a specific streamlined body has the same surface area and total streamwise length as the body. In this way, reasonable estimates can be obtained, especially for skin-friction drag, provided that the transition point is correctly located using the guidelines given at the end of Section 8.8.

8.6.1 Simplified Form of the Momentum-Integral Equation

For the flat plate, $dp/dx = 0$ and $U_e = U_\infty = $ const, so $dU_e/dx = 0$. Accordingly, the momentum-integral Eq. (8.43) reduces to the simple form

$$\tau_w = \rho U_e^2 \frac{d\theta}{dx} \tag{8.50}$$

Now the shape factor $I = \theta/\delta$ is simply a numerical quantity that depends only on the shape of the velocity profile. Thus Eq. (8.50) may be expressed in the alternative form

$$C_f = 2I\frac{d\delta}{dx} \tag{8.51}$$

where I is assumed to be independent of x. Equations (8.50) and (8.51) are forms of the simple momentum-integral equation.

8.6.2 Rate of Growth of a Laminar Boundary Layer on a Flat Plate

The rate of increase in boundary-layer thickness δ may be found by integrating Eq. (8.51), after setting $\Lambda = 0$ in Eqs. (8.48b,c) and substituting for I and C_f. Thus Eq. (8.51) becomes

$$\frac{d\delta}{dx} = \frac{C_f}{2I} = \frac{140}{13} \frac{\mu}{\rho U_\infty \delta}$$

Therefore,

$$\delta d\delta = \frac{140}{13} \frac{\mu}{\rho U_\infty}$$

whence

$$\frac{\delta^2}{2} = \frac{140}{13} \frac{\mu x}{\rho U_\infty}$$

The integration constant is zero if x is measured from the fore stagnation point where $\delta = 0$—that is,

$$\delta = 4.64x/(Re_x)^{1/2} \tag{8.52}$$

The other thickness quantities may now be evaluated using Eqs. (8.48a,b) with $\Lambda = 0$. Thus

$$\delta^* = 0.375\delta = 1.74x/(Re_x)^{1/2} \tag{8.53}$$

$$\theta = 0.139\delta = 0.646x/(Re_x)^{1/2} \tag{8.54}$$

8.6.3 Drag Coefficient for a Flat Plate of Streamwise Length *L* with a Wholly Laminar Boundary Layer

Note that

$$C_F = \frac{1}{L} \int_0^L C_f dx = \frac{2}{L} \int_0^L \frac{d\theta}{dx} = \frac{2\theta(L)}{L} \tag{8.55}$$

where $\theta(L)$ is the value of the momentum thickness at $x = L$. Thus using Eq. (8.54) in Eq. (8.55) gives

$$C_F = 1.293/Re^{1/2}$$

and

$$C_{D_F} = 2.586/Re^{1/2} \tag{8.56}$$

These expressions are plotted in Fig. 8.2 (lower curve).

Example 8.3

Consider a flat plate of 0.6-m chord at zero incidence in a uniform airstream of 45 m s^{-1}. Estimate (1) the displacement thickness at the trailing edge, and (2) the overall drag coefficient of the plate.

At the trailing edge, $x = 0.6$ m and

$$Re_x = \frac{45 \times 0.6}{14.6 \times 10^{-6}} = 1.85 \times 10^6$$

Therefore, using Eq. (8.53),

$$\delta^* = \frac{1.74 \times 0.6}{\sqrt{1.85 \times 10^3}} = 0.765 \times 10^{-3}\,\text{m} = 0.8\,\text{mm} \tag{a}$$

Re has the same value as Re_x at the trailing edge, so Eq. (8.56) gives

$$C_{D_F} = \frac{2.54}{\sqrt{1.85 \times 10^3}} = 0.0019 \tag{b}$$

8.6.4 Turbulent Velocity Profile

A commonly employed turbulent-boundary-layer profile is the seventh-root profile, which was proposed by Prandtl on the basis of friction-loss experiments with turbulent flow in circular pipes correlated by Blasius. Blasius investigated experimental results on the resistance to flow and proposed the following empirical relationships between the local skin-friction coefficient at the walls, $\overline{C}_f(= \tau_w / \frac{1}{2}\rho \overline{U}^2)$, and the Reynolds number of the flow $\overline{Re}$ (based on average flow velocity $\overline{U}$ in the pipe and diameter D). Blasius proposed the relationship

$$\overline{C}_f = \frac{0.0791}{\overline{Re}^{1/4}} \tag{8.57}$$

which is in reasonably good agreement with experiment for values of $\overline{Re}$ up to about 2.5×10^5.

Assuming that the velocity profile in the pipe may be written in the form

$$\frac{u}{U_m} = \left(\frac{y}{D/2}\right)^n = \left(\frac{y}{a}\right)^n \tag{8.58}$$

where u is the velocity at distance y from the wall and $a =$ pipe radius $D/2$, it remains to determine the value of n. From Eq. (8.58), writing $U_m = C\overline{U}$, where C is a constant to be determined,

$$u = C\overline{U}\left(\frac{y}{a}\right)^n$$

that is,

$$\overline{U} = \frac{u}{C}\left(\frac{a}{y}\right)^n \tag{8.59}$$

Substituting for $\overline{C}_f$ in the expression for surface-friction stress at the wall,

$$\tau_w = \overline{C}_f \frac{1}{2}\rho\overline{U}^2 = \frac{0.0791\nu^{1/4}}{D^{1/4}\overline{U}^{1/4}}\frac{1}{2}\rho\overline{U}^2 = 0.03955\rho\overline{U}^{7/4}\left(\frac{\nu}{D}\right)^{1/4} \tag{8.60}$$

From Eq. (8.59),

$$\overline{U}^{7/4} = \frac{u^{7/4}}{C^{7/4}}\left(\frac{a}{y}\right)^{7n/4}$$

so Eq. (8.60) becomes

$$\tau_w = \frac{0.03955}{C^{7/4}}\rho u^{7/4}\frac{a^{[(7n/4)-(1/4)]}}{y^{7n/4}}\nu^{1/4}\left(\frac{1}{2}\right)^{1/4}$$

that is,

$$\tau_w = \frac{0.0333}{C^{7/4}}\rho u^{7/4}\frac{\nu^{1/4}}{y^{7n/4}}a^{[(7n/4)-(1/4)]} \tag{8.61}$$

Now, we may argue that very close to the wall, in the viscous sublayer ($u \neq 0$), velocity u does not depend on the overall size of the pipe (i.e., $u \neq f(a)$). If this is so, it immediately follows that τ_w, which is $\mu(\partial u/\partial y)_w$, cannot depend on the pipe diameter, and therefore the term $a^{[(7n/4)-(1/4)]}$ in Eq. (8.61) must be unity in order not to affect the expression for τ_w. For this to be so, $7n/4 - 1/4 = 0$, which immediately gives $n = \frac{1}{7}$. Substituting this back into Eq. (8.58) gives $u/U_m = (y/a)^{1/7}$, which thus relates the velocity u at distance y from the surface to the centerline velocity U_m at distance a from the surface. Assuming that this holds for very large pipes, we can argue that the flow at a section along a flat, two-dimensional plate is similar to that along a small peripheral length of pipe. Therefore, replacing a by δ gives the profile for the free boundary layer on the flat plate:

$$\frac{u}{U_\infty} = \left(\frac{y}{\delta}\right)^{1/7} \quad \text{or} \quad \overline{u} = \overline{y}^{1/7} \tag{8.62}$$

This is Prandtl's seventh-root law, and it gives surprisingly good overall agreement with practice for moderate Reynolds numbers ($Re_x < 10^7$). However, it breaks down at the wall where the profile is tangential to the surface and gives an infinite value of $(\partial \bar{u}/\partial \bar{y})_w$. To find the wall shear stress, we use Eq. (8.61). The constant C may be evaluated by equating expressions for the total volume flow through the pipe—that is, using Eqs. (8.59) and (8.62),

$$\pi a^2 \bar{U} = 2\pi \int_0^a ur\,dr = 2\pi \bar{U} C \int_0^a \left(\frac{y}{a}\right)^{1/7} (a-y)dy = \frac{49}{60}\pi \bar{U} C a^2$$

giving $C = \frac{60}{49} = 1.224$. Substituting for C and n in Eq. (8.61) then gives

$$\tau_w = 0.0234\rho u^{7/4} \left(\frac{\nu}{y}\right)^{1/4}$$

which, on substituting for u from Eq. (8.62), gives

$$\tau_w = 0.0234\rho U_\infty^{7/4} \left(\frac{\nu}{\delta}\right)^{1/4} \tag{8.63}$$

Finally, since

$$C_f = \frac{\tau_w}{\frac{1}{2}\rho U_\infty^2}$$

for a free boundary layer

$$C_f = 0.0468 \left(\frac{\nu}{U_\infty \delta}\right)^{1/4} = \frac{0.0468}{Re_\delta^{1/4}} \tag{8.64}$$

Using Eqs. (8.62) and (8.64) in the momentum-integral equation enables investigation of turbulent boundary-layer growth on a flat plate.

8.6.5 Rate of Growth of a Turbulent Boundary Layer on a Flat Plate

The differential equation for the rate of growth of a turbulent boundary layer is

$$d\delta/dx = C_f/2I$$

where

$$C_f = 0.0468(\nu/U_\infty \delta)^{1/4}$$

and

$$I = \int_0^1 \bar{u}(1-\bar{u})d\bar{y} = \int_0^1 \bar{y}^{1/7}\left(1-\bar{y}^{1/7}\right)d\bar{y}$$

(a)

$$= \left[\frac{7}{8}\bar{y}^{8/7} - \frac{7}{9}\bar{y}^{9/7}\right]_0^1 = \frac{63-56}{72} = \frac{7}{72}$$

Therefore,

$$\frac{d\delta}{dx} = \frac{72 \times 0.0468\nu^{1/4}}{2 \times 7 \times (U_\infty\delta)^{1/4}}$$

that is,

$$\delta^{1/4}d\delta = 0.241\left(\frac{\nu}{U_\infty}\right)^{1/4}dx$$

Therefore,

$$\frac{4}{5}\delta^{5/4} = 0.241\left(\frac{\nu}{U_\infty}\right)^{1/4}x$$

$$\delta = \left(\frac{5 \times 0.241}{4}\right)^{4/5}\left(\frac{\nu}{U_\infty}\right)^{1/5}x^{4/5}$$

or, in terms of Reynolds number Re_x,

$$\delta = 0.383\frac{x}{(Re_x)^{1/5}}$$

(8.65)

Development of laminar and turbulent layers for a given stream velocity is plotted in Fig. 8.21.

To estimate the other thickness quantities for the turbulent layer, we evaluate the following integrals:

$$\int_0^1 (1-\bar{u})d\bar{y} = \int_0^1\left(1-\bar{y}^{1/7}\right)d\bar{y} = \left[\bar{y} - \frac{7}{8}\bar{y}^{8/7}\right]_0^1 = 1 - \frac{7}{8} = 0.125$$

(b)

$$\int_0^1 \bar{u}\left(1-\bar{u}^2\right)d\bar{y} = \int_0^1\left(\bar{y}^{1/7} - \bar{y}^{3/7}\right)d\bar{y} = \left[\frac{7}{8}\bar{y}^{8/7} - \frac{7}{10}\bar{y}^{10/7}\right]_0^1$$

(c)

$$= \frac{7}{8} - \frac{7}{10} = 0.175$$

Using the value for I in Eq. (a) ($I = \frac{7}{72} = 0.0973$) and substituting appropriately for δ from Eq. (8.65) and for the integral values from Eqs. (b) and (c) in Eqs. (8.18),

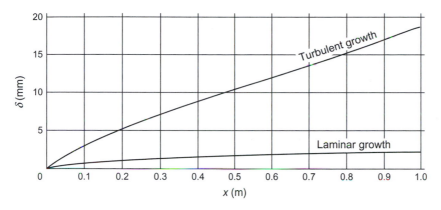

FIGURE 8.21

Boundary-layer growth on a flat plate at a free-stream speed of 60 m s^{-1}.

(8.19), and (8.20), we have

$$\delta^* = 0.125\delta = \frac{0.0479x}{(Re_x)^{1/5}} \tag{8.66}$$

$$\theta = 0.0973\delta = \frac{0.0372x}{(Re_x)^{1/5}} \tag{8.67}$$

$$\delta^{**} = 0.175\delta = \frac{0.0761x}{(Re_x)^{1/5}} \tag{8.68}$$

The seventh-root profile with the previous thickness quantities indicated is plotted in Fig. 8.22.

Example 8.4

A wind-tunnel working section is being designed to work with no streamwise pressure gradient when running empty at an airspeed of 60 m s^{-1}. The working section is 3.6 m long and has a rectangular cross-section 1.2 m wide by 0.9 m high. An approximate allowance for boundary-layer growth is made by allowing the side walls of the working section to diverge slightly. It is assumed that, at the upstream end of the working section, the turbulent boundary layer is equivalent to one that has grown from zero thickness over a length of 2.5 m; the wall divergence is determined on the assumption that the net area of flow is correct at the entry and exit sections of the working section. What must the width between the walls at the exit section be if the width at the entry section is exactly 1.2 m?

For the seventh-root profile, the displacement thickness is

$$\delta^* = \frac{0.0479x}{(Re_x)^{1/5}}$$

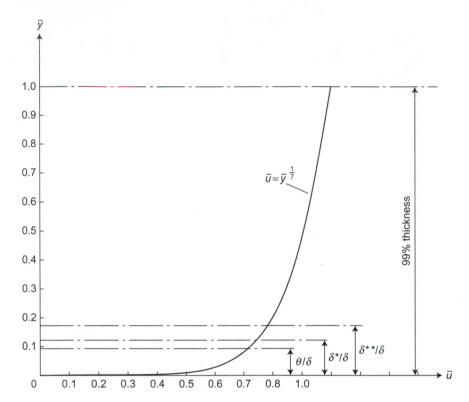

FIGURE 8.22

Turbulent velocity profile.

At entry, $x = 2.5$ m. Therefore,

$$Re_x = \frac{U_\infty x}{\nu} = \frac{60 \times 2.5}{14.6 \times 10^{-6}} = 102.7 \times 10^5$$

$$Re_x^{1/5} = 25.2$$

that is,

$$\delta^* = \frac{0.0479 \times 2.5}{25.2} = 0.00475 \, \text{m}$$

At exit, $x = 6.1$ m. Therefore,

$$Re_x = \frac{60 \times 6.1}{14.6 \times 10^{-6}} = 251 \times 10^5$$

$$Re_x^{1/5} = 30.2$$

that is,

$$\delta^* = \frac{0.0479 \times 6.1}{30.2} = 0.00968 \, \text{m}$$

Thus δ^* increases by $(0.00968 - 0.00475) = 0.00493\,\text{m}$. This increase in displacement thickness occurs on all four walls—that is, total displacement area at exit (relative to entry) $= 0.00493 \times 2(1.2 + 0.9) = 0.0207\,\text{m}^2$.

The allowance is made on the two side walls only so that the displacement area on the side walls $= 2 \times 0.9 \times \Delta^* = 1.8\Delta^*\,\text{m}^2$, where Δ^* is the exit displacement per wall. Therefore,

$$\Delta^* = \frac{0.0207}{1.8} = 0.0115\,\text{m}$$

This is the displacement for each wall, so the total width between side walls at the exit section $= 1.2 + 2 \times 0.0115 = 1.223\,\text{m}$.

8.6.6 Drag Coefficient for a Flat Plate with a Wholly Turbulent Boundary Layer

The local friction coefficient C_f may now be expressed in terms of x by substituting from Eq. (8.65) in Eq. (8.64). Thus

$$C_f = 0.0468 \left(\frac{v}{U_\infty}\right)^{1/4} \frac{(Re_x)^{1/20}}{(0.383x)^{1/4}} = \frac{0.0595}{(Re_x)^{1/5}} \tag{8.69}$$

whence

$$\tau_w = \frac{0.0595}{2(U_\infty x)^{1/5}} v^{1/5} \rho U_\infty^2 = 0.02975 \rho v^{1/5} U_\infty^{9/5} x^{-1/5} \tag{8.70}$$

The total surface-friction force and drag coefficient for a wholly turbulent boundary layer on a flat plate follows:

$$C_F = \int_0^1 C_f d\left(\frac{x}{L}\right) = \int_0^1 0.0595 \left(\frac{v}{U_\infty}\right)^{1/5} x^{-1/5} d\left(\frac{x}{L}\right)$$

$$= \left(\frac{v}{U_\infty L}\right)^{1/5} \times 0.0595 \left[\frac{5}{4}\left(\frac{x}{L}\right)^{4/5}\right]_0^1 = 0.0744 Re^{-1/5} \tag{8.71}$$

and

$$C_{D_F} = 0.1488 Re^{-1/5} \tag{8.72}$$

These expressions are plotted in Fig. 8.2 (upper curve). It should be clearly understood that these last two coefficients refer to a flat plate for which the boundary layer is turbulent over the entire streamwise length.

In practice, for Reynolds numbers (Re) up to at least 3×10^5, the boundary layer is entirely laminar. If the Reynolds number increases further (by an increase in flow speed), transition to turbulence in the boundary layer may be initiated (depending on free-stream and surface conditions) at the trailing edge, with the transition point moving forward with increasing Re (such that Re_x at transition remains approximately constant at a specific value, say Re_t). However large the value of Re, there is inevitably a short length of boundary layer near the leading edge that remains laminar as far back on the plate as the point corresponding to $Re_x = Re_t$. Thus, for a large range of practical Reynolds numbers, the boundary-layer flow on the plate is partly laminar and partly turbulent. The next stage is to investigate the conditions at transition in order to evaluate the overall drag coefficient for the plate with mixed boundary layers.

8.6.7 Conditions at Transition

For boundary-layer calculations, it is usually assumed that the transition from laminar to turbulent flow within the boundary layer occurs instantaneously. Obviously, this is not exactly true, but observations of the transition process do indicate that the transition region (streamwise distance) is fairly small; thus, as a first approximation, the assumption is reasonably justified. An abrupt change in momentum thickness at the transition point implies that $d\theta/dx$ is infinite, which, as the simplified momentum-integral Eq. (8.50) shows, in turn implies that the local skin-friction coefficient C_f is infinite. This is plainly unacceptable on physical grounds, so it follows that the momentum thickness remains constant across the transition position. Thus

$$\theta_{L_t} = \theta_{T_t} \tag{8.73}$$

where the subscripts L and T refer to laminar and turbulent boundary-layer flows, respectively, and t indicates that these are particular values at transition. Thus

$$\delta_{L_t} \left(\int_0^1 \overline{u}(1 - \overline{u}) d\overline{y} \right)_L = \delta_{T_t} \left(\int_0^1 \overline{u}(1 - \overline{u}) d\overline{y} \right)_T$$

The integration performed in each case uses the appropriate laminar or turbulent profile. The ratio of turbulent to laminar boundary-layer thicknesses is then given directly by

$$\frac{\delta_{T_t}}{\delta_{L_t}} = \frac{\left(\int_0^1 \overline{u}(1 - \overline{u}) d\overline{y} \right)_L}{\left(\int_0^1 \overline{u}(1 - \overline{u}) d\overline{y} \right)_T} = \frac{I_L}{I_T} \tag{8.74}$$

Using the values of I previously evaluated for the cubic and seventh-root profiles (Sections 8.5.1 and 8.6.4),

$$\frac{\delta_{T_t}}{\delta_{L_t}} = \frac{0.139}{0.0973} = 1.43 \tag{8.75}$$

This indicates that the boundary layer on a flat plate increases in thickness by about 40% at transition.

We then assume that the turbulent layer, downstream of transition, grows as if it had started from zero thickness at some point ahead of transition and developed along the surface so that its thickness reaches the value δ_{T_t} at the transition position.

8.6.8 Mixed Boundary-Layer Flow on a Flat Plate with Zero Pressure Gradient

Figure 8.23 indicates the symbols employed to denote the various physical dimensions used. At the leading edge, a laminar layer begins to develop, thickening with distance downstream, until transition to turbulence occurs at some Reynolds number $Re_t = U_\infty x_t / \nu$. At transition, the thickness increases suddenly from δ_{L_t} in the laminar layer to δ_{T_t} in the turbulent layer, and the latter then continues to grow as if it had started from some point on the surface distant x_{T_t} ahead of transition. This distance is given by the relationship, for the seventh-root profile,

$$\delta_{T_t} = \frac{0.383 x_{T_t}}{(Re_x)_{T_t}^{1/5}}$$

The total skin-friction force coefficient C_F for one side of the plate of length L may be found by adding the skin-friction force per unit width for the laminar boundary layer of length x_t to that for the turbulent boundary layer of length $(L - x_t)$, and dividing by $\frac{1}{2}\rho U_\infty^2 L$, where L here is the wetted surface area per unit width. Working

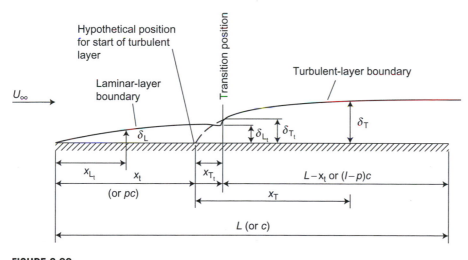

FIGURE 8.23

Engineering analysis of mixed boundary layers.

in terms of Re_t, the transition position is given by

$$x_t = \frac{v}{U_\infty} Re_t \tag{8.76}$$

The laminar boundary-layer momentum thickness at transition is then obtained from Eq. (8.54):

$$\theta_{L_t} = \frac{0.646 x_t}{(Re_t)^{1/2}} = 0.646 \left(\frac{v}{U_\infty} \right)^{1/2} x_t^{1/2}$$

which, on substituting for x_t from Eq. (8.76), gives

$$\theta_{L_t} = 0.646 \frac{v}{U_\infty} (Re_t)^{1/2} \tag{8.77}$$

The corresponding turbulent boundary-layer momentum thickness at transition follows directly from Eq. (8.67):

$$\theta_{T_t} = \frac{0.037 x_{T_t}}{(Re_x)_{T_t}^{1/5}} \tag{8.78}$$

The equivalent length of the turbulent layer (x_{T_t}) to give this thickness is obtained from setting $\theta_{L_t} = \theta_{T_t}$; using Eqs. (8.77) and (8.78), this gives

$$0.646 x_t \left(\frac{v}{U_\infty x_t} \right)^{1/2} = 0.037 x_{T_t} \left(\frac{v}{U_\infty x_{T_t}} \right)^{1/5}$$

leading to

$$x_{T_t}^{4/5} = \frac{0.646}{0.037} \left(\frac{v}{U_\infty} \right)^{4/5} Re_t^{1/2}$$

Thus

$$x_{T_t} = 35.5 \frac{v}{U_\infty} Re_t^{5/8} \tag{8.79}$$

Now, on a flat plate with no pressure gradient, the momentum thickness at transition is a measure of the momentum defect produced, by the surface-friction stresses only, in the laminar boundary layer between the leading edge and the transition position. As we also assume here that the momentum thickness through transition is constant, it is clear that the actual surface-friction force under the laminar boundary layer of length x_t must be the same as the force that exists under a turbulent boundary layer of length x_{T_t}. It then follows that total skin-friction force for the whole plate may be found simply by calculating the skin-friction force under a turbulent boundary layer acting over a length from the point a distance x_{T_t} ahead of transition to the trailing edge. Figure 8.23 shows that the total effective length of the turbulent boundary layer is therefore $L - x_t + x_{T_t}$.

Now, from Eq. (8.23),

$$F = \int_0^{L-x_t+x_{T_t}} \tau_w dx = \frac{1}{2}\rho U_\infty^2 \int_0^{L-x_t+x_{T_t}} C_f\, dx$$

where C_f is given from Eq. (8.69) as

$$\frac{0.0595}{(Re_x)^{1/5}} = 0.0595\left(\frac{\nu}{U_\infty}\right)^{1/5} x^{-1/5}$$

Thus

$$F = \frac{1}{2}\rho U_\infty^2 \times 0.0595 \left(\frac{\nu}{U_\infty}\right)^{1/5} \frac{5}{4}\left[x^{4/5}\right]^{L-xt+xT_t}$$

Now $C_F = F/\frac{1}{2}\rho U_\infty^2 L$, where L is the total chordwise length of the plate, so

$$C_F = 0.0744\left(\frac{\nu}{U_\infty}\right)^{1/5} \frac{(L-x_t+x_{T_t})^{4/5}}{L}$$

$$= 0.0744\left(\frac{\nu}{U_\infty L}\right)\left(\frac{U_\infty L}{\nu} - \frac{U_\infty x}{\nu} + \frac{U_\infty x_{T_t}}{\nu}\right)^{4/5}$$

that is,

$$C_F = \frac{0.0744}{Re}\left(Re - Re_t + 35.5Re_t^{5/8}\right)^{4/5} \tag{8.80}$$

This result could have been obtained by direct substitution of the appropriate value of Re in Eq. (8.71), making the necessary correction for effective chord length (see Example 8.5).

The expression enables the curve of either C_F or C_{D_F} for the flat plate to be plotted against plate Reynolds number $Re = (U_\infty L/\nu)$ for a known value of the transition Reynolds number Re_t. Two such curves for extreme values of Re_t of 3×10^5 and 3×10^6 are plotted in Fig. 8.2.

Note that Eq. (8.80) is not applicable to Re values less than Re_t when Eqs. (8.55) and (8.56) are used. For Re values greater than about 10^8, the appropriate all-turbulent expressions should be used. However, Eqs. (8.69) and (8.72) are inaccurate for $Re > 10^7$. At higher Reynolds numbers, the semi-empirical expressions due to Prandtl and Schlichting should be employed:

$$C_f = [2\log_{10}(Re_x) - 0.65]^{-2.3} \tag{8.81a}$$

$$C_F = \frac{0.455}{(\log_{10} Re)^{2.58}} \tag{8.81b}$$

For the lower transition Reynolds number of 3×10^5, the corresponding Re value, above which the all-turbulent expressions are reasonably accurate, is 10^7.

Example 8.5
(1) Develop an expression for the drag coefficient of a flat plate of chord c and infinite span at zero incidence in a uniform stream of air, when transition occurs at distance pc from the leading edge. Assume the following relationships for laminar and turbulent boundary-layer velocity profiles, respectively:

$$\bar{u}_L = \frac{3}{2}\bar{y} - \frac{1}{2}\bar{y}^3, \quad \bar{u}_T = \bar{y}^{1/7}$$

(2) On a thin two-dimensional airfoil of 1.8-m chord in an airstream of 45 m s^{-1}, estimate the required position of transition to give a drag-per-meter span 4.5 N less than that for transition at the leading edge.

For (1), refer to Fig. 8.23 for notation. From Eq. (8.79), setting $x_t = pc$,

$$x_{T_t} = 35.5(pc)^{(5/8)}\left(\frac{\nu}{U_\infty}\right)^{(3/8)} \tag{a}$$

Equation (8.72) gives the drag coefficient for an all-turbulent boundary layer as $C_{D_F} = 0.1488/Re^{1/5}$. For the mixed boundary layer, drag is obtained as for an all-turbulent layer of length $[x_{T_t} + (1-p)c]$. The corresponding drag coefficient (defined with reference to length $[x_{T_t} + (1-p)c]$) is then obtained directly from the all-turbulent expression where Re is based on the same length $[x_{T_t} + (1-p)c]$. To relate the coefficient to the whole plate length c, then, requires that the quantity obtained be factored by the ratio

$$\frac{[x_{T_t} + (1-p)c]}{c}$$

Thus

$$C_{D_F} = \frac{D_F}{\frac{1}{2}\rho U_\infty^2 c} = \frac{0.1488}{Re_{[x_{T_t}+(1-p)c]}^{1/5}} \times \frac{[x_{T_t} + (1-p)c]}{c} = \frac{0.1488}{\left(\frac{U_\infty}{\nu}\right)^{1/5}}\frac{[x_{T_t} + (1-p)c]^{4/5}}{c}$$

$$= \frac{0.1488[x_{T_t} + (1-p)c]^{4/5}}{\frac{U_\infty}{\nu}c\left(\frac{\nu}{U_\infty}\right)^{4/5}} = \frac{0.1488}{Re}\left[\frac{U_\infty}{\nu}x_{T_t} + (1-p)Re\right]^{4/5}$$

Note that Re here is based on total plate length c. Substituting from Eq. (a) for x_{T_t} gives

$$C_{D_F} = \frac{0.1488}{Re}\left[35.5p^{5/8}Re^{5/8} + (1-p)Re\right]^{4/5}$$

This form of expression (as an alternative to Eq. (8.80)) is convenient for enabling a quick approximation of skin-friction drag when the position of transition is likely fixed, rather than the transition Reynolds number (e.g., by position of maximum thickness), although, strictly, the profile shapes are not unchanged with length under these conditions, nor is U_e over the length.

For (2), with transition at the leading edge,

$$C_{D_F} = \frac{0.1488}{Re^{1/5}}$$

In this case,

$$Re = \frac{Uc}{\nu} = \frac{45 \times 1.8}{14.6 \times 10^{-6}} = 55.5 \times 10^5$$

$$Re^{1/5} = 22.34$$

and

$$C_{D_F} = \frac{0.1488}{22.34} = 0.00667$$

The corresponding airfoil drag is thus $D_F = 0.00667 \times 0.6125 \times (45)^2 \times 1.8 = 14.88$ N. With transition at pc, $D_F = 14.86 - 4.5 = 10.36$ N:

$$C_{D_F} = \frac{10.36}{14.88} \times 0.00667 = 0.00465$$

Using this value in (a), with $Re^{5/8} = 16480$, gives

$$0.00465 = \frac{0.1488}{55.8 \times 10^5}\left[35.5p^{5/8} \times 16480 + 55.8 \times 10^5 - 55.8 \times 10^5 p\right]^{4/5}$$

That is,

$$5.84 \times 10^5 p^{5/8} - 55.8 \times 10^5 p = \left(\frac{55.8 \times 465}{0.1488}\right)^{5/4} - 55.8 \times 10^5 = (35.6 - 55.8)10^5$$

or

$$55.8p - 5.84p^{5/8} = 20.2$$

The solution to this (by successive approximation) is $p = 0.423$:

$$pc = 0.423 \times 1.8 = 0.671 \text{ m behind leading edge}$$

Example 8.6

A light aircraft has a tapered wing with root and tip chord lengths of 2.2 m and 1.8 m, respectively, and a wingspan of 16 m. Estimate the skin-friction drag of the wing when the aircraft is traveling at 55 m/s. On the upper surface, the point of minimum pressure is located at 0.375 chord length from the leading edge. Dynamic viscosity and air density may be taken as 1.8×10^{-5} kg/s/m and 1.2 kg/m^3, respectively.

The average wing chord is given by $\bar{c} = 0.5(2.2 + 1.8) = 2.0$ m, so the wing is equivalent to a flat plate measuring 2.0 m $\times$ 16 m. The overall Reynolds number based on average chord is given by

$$Re = \frac{1.2 \times 55 \times 2.0}{1.8 \times 10^{-5}} = 7.33 \times 10^6$$

Since this is below 10^7, the guidelines at the end of Section 8.6 suggest that the transition point is very shortly after the point of minimum pressure, so $x_t \simeq 0.375 \times 2.0 = 0.75$ m; Eq. (8.80) may also be used:

$$Re_t = 0.375 \times Re = 2.75 \times 10^6$$

Thus Eq. (8.80) gives

$$C_F = \frac{0.0744}{7.33 \times 10^6} \{7.33 \times 10^6 - 2.75 \times 10^6 + 35.5(2.75 \times 10^6)^{5/8}\}^{4/5} = 0.0023$$

Therefore, the skin-friction drag of the upper surface is given by

$$D = \frac{1}{2}\rho U_\infty^2 \bar{c} s C_F = 0.5 \times 1.2 \times 55^2 \times 2.0 \times 16 \times 0.0023 = 133.8\,\text{N}$$

Finally, assuming that the drag of the lower surface is similar, the estimate for total skin-friction drag for the wing is $2 \times 133.8 \simeq 270$ N.

8.7 ADDITIONAL EXAMPLES OF THE MOMENTUM-INTEGRAL EQUATION

For the general solution of the momentum-integral equation, computational methods, as described in Section 8.10, are needed. It is possible, however, in certain cases with external pressure gradients, to find engineering solutions using the momentum-integral equation without a computer. Two examples are given here: one involves the use of suction to control the boundary layer; the other determines the boundary-layer properties at the leading-edge stagnation point of an airfoil. For such applications, Eq. (8.43) can be written in the alternative form with $H = \delta^*/\theta$:

$$\frac{C_f}{2} = \frac{V_s}{U_e} + \frac{\theta}{U_e}\frac{dU_e}{dx}(H+2) + \frac{d\theta}{dx} \qquad (8.82)$$

In addition, when there is no pressure gradient and no suction, this further reduces to the simple momentum-integral equation previously obtained (Section 8.6.1, Eq. (8.50)); that is, $C_f = 2(d\theta/dx)$.

Example 8.7

A two-dimensional divergent duct has a total included angle, between the plane diverging walls, of 20 degrees. To prevent separation from the walls and to maintain a laminar boundary-layer flow, the walls are to be constructed of porous material so that suction may be applied to them. At entry to the diffuser duct, where the flow velocity is 48 m s^{-1}, the section is square with a side length of 0.3 m, and the laminar boundary layers have a general thickness (δ) of 3 mm. If the boundary-layer thickness is to be maintained constant at this value, obtain an expression in terms of x for the value of the suction velocity required along the diverging walls. Assume that the laminar velocity profile for the diverging walls remains constant and is given approximately by $\bar{u} = 1.65\bar{y}^3 - 4.30\bar{y}^2 + 3.65\bar{y}$.

The momentum equation for steady flow along the porous walls is given by Eq. (8.82):

$$\frac{V_s}{U_e} = \frac{C_f}{2} - \frac{1}{U_e}\frac{dU_e}{dx}(H+2)\theta - \frac{d\theta}{dx}$$

If the thickness δ and the profile are to remain constant, then $\theta =$ constant and $d\theta/dx = 0$. Also,

$$\frac{C_f}{2} = \frac{\tau_w}{\rho U_e^2} = \frac{\mu\frac{U_e}{\delta}\left(\frac{\partial\bar{u}}{\partial\bar{y}}\right)_w}{\rho U_e^2} = \frac{\nu}{U_e\delta}\left(\frac{\partial\bar{u}}{\partial\bar{y}}\right)_w$$

that is,

$$V_s = \frac{\nu}{\delta}\left(\frac{\partial\bar{u}}{\partial\bar{y}}\right)_w - \frac{dU_e}{dx}(H+2)\times\delta\frac{\theta}{\delta}$$

$$\frac{\partial\bar{u}}{\partial\bar{y}} = 4.95\bar{y}^2 - 8.60\bar{y} + 3.65$$

$$\left(\frac{\partial\bar{u}}{\partial\bar{y}}\right)_w = 3.65$$

Equation (8.18) gives

$$\frac{\delta^*}{\delta} = \int_0^1 (1-\bar{u})d\bar{y} = \int_0^1 (1 - 1.65\bar{y}^3 + 4.30\bar{y}^2 - 3.65\bar{y})d\bar{y} = 0.1955$$

Equation (8.19) gives

$$\frac{\theta}{\delta} = \int_0^1 \bar{u}(1-\bar{u})d\bar{y} = \int_0^1 (3.65\bar{y} - 17.65\bar{y}^2 + 33.05\bar{y}^3 - 30.55\bar{y}^4 + 14.2\bar{y}^5 - 2.75\bar{y}^6)d\bar{y} = 0.069$$

$$H = \frac{\delta^*}{\theta} = \frac{0.1955}{0.069} = 2.83$$

Also, $\delta = 0.003$ m.

The diffuser duct cross-sectional area $= 0.09 + 0.06x \tan 10$ degrees, where $x =$ distance from the entry section:

$$A = 0.09 + 0.106x$$

and

$$A/A_e = 1 + 1.178x$$

Let the suffix i denotes the value at the entry section. Thus,

$$A_e U_{e_i} = A U_e$$

$$U_e = \frac{A_e}{A} U_{e_i} = \frac{48}{1 + 1.178x}$$

Then

$$\frac{dU_e}{dx} = -48 \times 1.178(1 + 1.178x)^{-2}$$

Finally,

$$V_s = \frac{14.6 \times 10^{-6}}{0.003} \times 3.65 + \frac{48 \times 1.178 \times 4.83 \times 0.003 \times 0.069}{(1 + 1.178x)^2}$$

$$= 0.0178 + \frac{0.0565}{(1 + 1.178x)^2} \, \mathrm{m\,s^{-1}}$$

Thus the maximum suction is required at entry, where $V_s = 0.0743 \, \mathrm{m\,s^{-1}}$.

For bodies with sharp leading edges, such as flat plates, the boundary layer grows from zero thickness. However, in most engineering applications, (e.g., conventional airfoils), the leading edge is rounded. Under these circumstances, the boundary layer has a finite thickness at the leading edge, as shown in Fig. 8.24(a). To estimate the initial boundary-layer thickness, we can assume that the flow in the vicinity of the stagnation point is similar to that approaching a flat plate oriented perpendicularly to the free stream, Fig. 8.24(b). For this flow, $U_e = cx$ (where c is a constant) and the boundary-layer thickness does not change with x. In the example given next, the momentum-integral equation is used to estimate the initial boundary-layer thickness for the flow depicted in Fig. 8.24(b). An exact solution to the Navier-Stokes equations can be found for this stagnation-point flow (see Section 2.10.3). Here the momentum-integral equation is used to obtain an approximate solution.

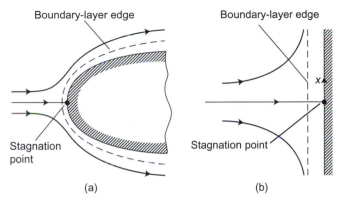

FIGURE 8.24

Boundary-layer flow in the vicinity of the fore stagnation point.

Example 8.8

Use the momentum-integral Eq. (8.43) and the results in Eqs. (8.48a', b', c') to obtain expressions for δ, δ^*, θ, and C_f. Assume that the boundary-layer thickness does not vary with x and that $U_e = cx$.

$$\Lambda = \frac{\delta^2}{\nu}\frac{dU_e}{dx} = \frac{\delta^2}{\nu}c = \text{const.}$$

Therefore, $\theta = \text{const.}$ also, and Eq. (8.43) becomes

$$\theta\frac{d}{dx}(c^2x^2) + \delta^* c^2 x = \frac{\tau_w}{\rho}$$

Substituting Eqs. (8.48a', b', c') leads to

$$2\delta c^2 x \frac{1}{63}\left(\frac{37}{5} - \frac{\Lambda}{15} - \frac{\Lambda^2}{144}\right) + c^2 x\delta\left(\frac{3}{10} - \frac{\Lambda}{210}\right) = \nu cx\frac{1}{\delta}\left(2 + \frac{\Lambda}{6}\right)$$

Multiplying both sides by $\delta/\nu cx$, and using the result for Λ, gives

$$\frac{2}{63}\left(\frac{37}{5} - \frac{\Lambda}{15} - \frac{\Lambda^2}{144}\right)\Lambda + \left(\frac{3}{10} - \frac{\Lambda}{120}\right)\Lambda = 2 + \frac{\Lambda}{6}$$

After rearrangement, this simplifies to

$$\frac{1}{4536}\Lambda^3 + \left(\frac{1}{120} + \frac{2}{945}\right)\Lambda^2 + \left(\frac{1}{6} - \frac{3}{10} - \frac{74}{315}\right)\Lambda + 2 = 0$$

or

$$0.00022\Lambda^3 + 0.01045\Lambda^2 - 0.3683\Lambda + 2 = 0$$

We know that Λ lies somewhere between 0 and 12, so it is relatively easy to solve this equation by trial and error to obtain

$$\Lambda = 7.052 \quad \Rightarrow \quad \delta = \sqrt{\frac{v\Lambda}{c}} = 2.655\sqrt{\frac{v}{c}}$$

Equations. (8.48a′, b′, c′) then give

$$\delta^* = 0.641\sqrt{\frac{v}{c}}, \quad \theta = 0.278\sqrt{\frac{v}{c}}, \quad C_f = 2.392\sqrt{\frac{v}{c}}\frac{1}{x}$$

Once the value of $c = (dU_e/dx)_{x=0}$ is specified (see Example 2.4), the results given previously can be used to supply initial conditions for boundary-layer calculations over airfoils.

8.8 LAMINAR-TURBULENT TRANSITION

We saw in Section 2.11.4 that transition from laminar to turbulent flow usually occurs at some point along the surface. This process is exceedingly complex and remains an active area of research. Owing to very rapid changes in both space and time, the simulation of transition is arguably the most challenging problem in computational fluid dynamics. Still, despite the formidable difficulties considerable progress has been made, and transition can now be reliably predicted in simple engineering applications. The theoretical treatment of transition is beyond our scope, although a physical understanding of it is vital for many engineering applications of aerodynamics. Accordingly, we present a brief account of the underlying physics of transition in a boundary layer on a flat plate.

Transition occurs because of the growth of small disturbances in the boundary layer. In many respects, the boundary layer can be regarded as a complex nonlinear oscillator that under certain circumstances has an initially linear wavelike response to external stimuli (or inputs), which is illustrated schematically in Fig. 8.25. In free flight or in high-quality wind-tunnel experiments, several stages in the process can be discerned. The first stage is the conversion of external stimuli or disturbances into low-amplitude waves. These disturbances may arise from a variety of sources: free-stream turbulence, sound waves, surface roughness, vibration, and the like. The conversion process is not well understood. One of the main difficulties is that the wavelength of a typical external disturbance is invariably much larger than that of the wavelike response of the boundary layer. Once the low-amplitude wave is generated, it propagates downstream in the boundary layer and, depending on local conditions, grows or decays. It eventually develops into turbulent flow.

While their amplitude remains small, the waves are predominantly two-dimensional (see Figs. 8.25 and 8.26). This phase of transition is well understood and was first explained theoretically by Tollmien [57] with later extensions by Schlichting [58] and many others. For this reason, the growing waves in the early so-called linear phase of transition are known as *Tollmien-Schlichting waves*. The linear phase extends for some 80% of the total transition region. The more advanced engineering predictions are, in fact, based on modern versions of Tollmien's theory, which is linear because it assumes that the wave amplitudes are so small that their products can be neglected. In the later nonlinear stages of transition, the disturbances become increasingly three-dimensional and develop rapidly. In other words, as the amplitude of the disturbance increases, the response of the boundary layer becomes more and more complex.

This view of transition originated with Prandtl [59] and his research team at Göttingen, which included Tollmien and Schlichting. Earlier theories, based on neglecting viscosity, seemed to suggest that small disturbances could not grow in

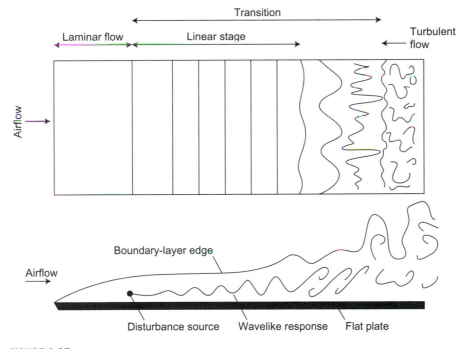

FIGURE 8.25

Transition in a boundary layer over a flat plate with disturbances generated by a harmonic line source.

the boundary layer. One effect of viscosity was well known: its so-called dissipative action in removing energy from a disturbance, thereby causing it to decay. Prandtl realized that, in addition to its dissipative effect, viscosity played a subtle but essential role in promoting the growth of wavelike disturbances by causing a transfer of energy to them. His explanation is illustrated in Fig. 8.27. Consider a small-amplitude wave passing through a small element of fluid within the boundary layer, as shown in Fig. 8.27(a). The instantaneous velocity components of the wave are (u', v') in the (x, y) directions; u' and v' are much smaller than the velocity u in the boundary layer in the absence of the wave. The instantaneous rate of increase in kinetic energy in the small element is given by the difference between the rates at which kinetic energy leaves the top of the element and enters the bottom:

$$-\rho u'v'\frac{\partial u}{\partial y} + \text{higher-order terms}$$

In the absence of viscosity, u' and v' are exactly 90 degrees out of phase and the average of their product over a wave period, denoted $\overline{u'v'}$, is zero (see Fig. 8.27(b)).

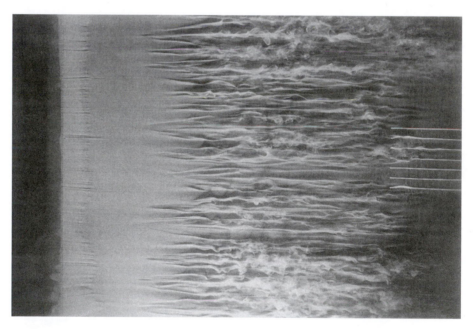

FIGURE 8.26

Laminar-turbulent transition in a flat-plate boundary layer: This is a planform view of a dye sheet emitted upstream parallel to the wall into water flowing from left to right. Successive stages of transition are revealed (i.e., laminar flow on the upstream side), then the two-dimensional tollmien-schlichting waves, followed by the formation of turbulent spots, and finally fully developed turbulent flow. The reynolds number based on distance along the wall is about 75,000. *Source: The photograph taken by H. Werle, ONERA, France.*

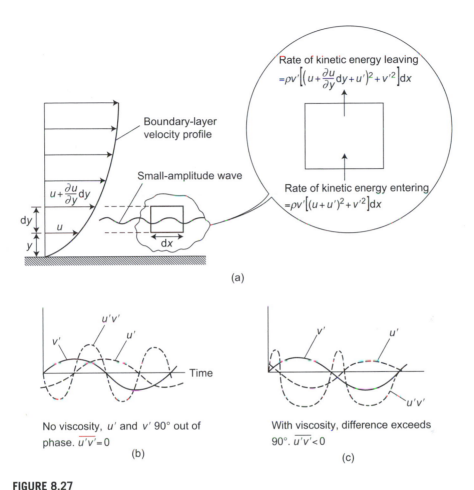

Rate of kinetic energy leaving
$$= \rho v' \left[\left(u + \frac{\partial u}{\partial y} dy + u' \right)^2 + v'^2 \right] dx$$

Boundary-layer
velocity profile

Small-amplitude wave

Rate of kinetic energy entering
$$= \rho v' \left[(u + u')^2 + v'^2 \right] dx$$

$u + \frac{\partial u}{\partial y} dy$

u

(a)

$u'v'$

v'

u'

Time

No viscosity, u' and v' 90° out of
phase. $\overline{u'v'} = 0$

(b)

v'

u'

$u'v'$

With viscosity, difference exceeds
90°. $\overline{u'v'} < 0$

(c)

FIGURE 8.27

Prandtl's explanation for disturbance growth.

However, as Prandtl realized, the effects of viscosity are to increase the phase difference between u' and v' to slightly more than 90 degrees. Consequently, as shown in Fig. 8.27(c), $\overline{u'v'}$ is now negative, resulting in a net energy transfer to the disturbance. The quantity $-\rho \overline{u'v'}$ is, in fact, the Reynolds stress referred to in Section 2.11.4. Accordingly, this energy transfer is usually referred to as *energy production by the Reynolds stress*. The mechanism is active throughout the transition process and plays a key role in sustaining fully turbulent flow (see Section 8.9).

Tollmien was able to verify Prandtl's hypothesis theoretically and so lay the foundations of modern transition theory. It was some time, however, before the ideas of the Göttingen group were accepted by the aeronautical community. This was in part because experimental corroboration was lacking. No sign of Tollmien-Schlichting waves could at first be found in experiments on natural transition. Schubauer and Skramstadt did succeed in seeing them but realized that, in order to study such

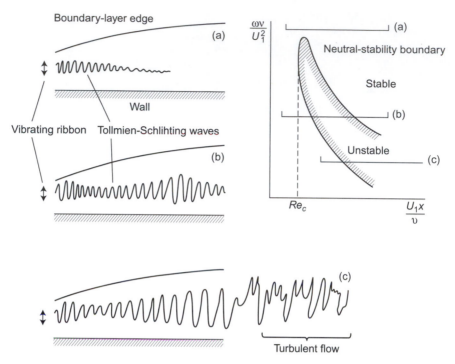

FIGURE 8.28

Schematic of Schubauer and Skramstadt's experiment.

waves systematically, they would have to create them artificially in a controlled manner. Thus they placed a vibrating ribbon having a controlled frequency ω within the boundary layer to act as a wave maker, rather than rely on natural sources of disturbance. The results are illustrated schematically in Fig. 8.28. Schubauer and Skramstadt found that, for high ribbon frequencies, (part (a) in the figure), the waves always decay. For intermediate frequencies (part (b)), they attenuate just downstream of the ribbon and then, at a greater distance downstream, begin to grow; finally, at still greater distances downstream, decay resumes. For low frequencies, the waves grow until their amplitude is sufficiently large for the nonlinear effects, referred to previously, to set in, with complete transition to turbulence occurring shortly afterward.

As shown in Fig. 8.28, Schubauer and Skramstadt were able to map out a curve of nondimensional frequency versus $Re_x (= U_\infty x/v)$, separating the disturbance frequencies that grow at a given position along the plate from those that decay. When disturbances grow, the boundary-layer flow is said to be *unstable* to small disturbances; conversely, when they decay, the flow is said to be *stable*. When the disturbances neither grow nor decay, the flow is in a state of *neutral stability*. Thus the curve shown in Fig. 8.28 is known as the *neutral-stability boundary* or curve. Inside the neutral-stability curve, production of energy by the Reynolds stress exceeds viscous dissipation and vice versa outside.

Note that a *critical Reynolds number Re_c* and a *critical frequency ω_c* exist. The Tollmien–Schlichting waves cannot grow at Reynolds numbers below Re_c or at frequencies above ω_c. However, since the disturbances causing the transition to turbulence are considerably lower than the critical frequency, the transitional Reynolds number is generally considerably greater than Re_c.

The shape of the neutral-stability curve obtained by Schubauer and Skramstadt [60] agreed well with Tollmien's theory, especially at the lower frequencies of interest for transition. Moreover, Schubauer and Skramstadt were able to measure the growth rates of the waves, and these too agreed well with Tollmien and Schlichting's theoretical calculations. Publication of Schubauer and Skramstadt's results finally led to general acceptance of the Göttingen "small-disturbance" theory of transition.

It was mentioned earlier that Tollmien-Schlichting waves can not be easily observed in experiments on natural transition. This is because natural sources of disturbance tend to generate wave packets in an almost random fashion in time and space. Thus, at any given instant, there is a great deal of "noise," tending to obscure the wavelike response of the boundary layer, and disturbances of a wide range of frequencies are continually being generated. In contrast, the Tollmien-Schlichting theory is based on disturbances with a single frequency. Nevertheless, providing that the initial level of the disturbances is low, it seems that the boundary layer responds preferentially so that waves of a certain frequency grow most rapidly and are primarily responsible for transition. These rapidly growing waves are those predicted by modern versions of the Tollmien-Schlichting theory, allowing the theory to predict, approximately at least, the onset of natural transition.

We explained previously that, provided the initial level of the external disturbances is low, as in typical free-flight conditions, there is a considerable difference between the critical and transitional Reynolds number. In fact, the latter is about 3×10^6, whereas $Re_c \simeq 3 \times 10^5$. However, if the initial level of disturbances rises, for example because of increased free-stream turbulence or surface roughness, the downstream distance required for the disturbance amplitude to grow sufficiently to produce nonlinear effects becomes shorter. Therefore, the transitional Reynolds number is reduced to a value closer to Re_c. In fact, for high-disturbance environments, such as those encountered in turbomachinery, the linear transition phase is bypassed completely and laminar flow abruptly breaks down into fully developed turbulence.

The Tollmien-Schlichting theory can also successfully predict how transition is affected by an external pressure gradient. The neutral-stability boundaries for the flat plate and for typical adverse and favorable pressure gradients are plotted schematically in Fig. 8.29. In accordance with theoretical treatment, Re_δ is used as the abscissa in place of Re_x. However, since the boundary layer grows with passage downstream, Re_δ can still be regarded as a measure of distance along the surface. From Fig. 8.29, we see that, for adverse pressure gradients, not only is $(Re_\delta)_c$ smaller than for a flat plate, but a much wider band of disturbance frequencies are unstable and will grow.

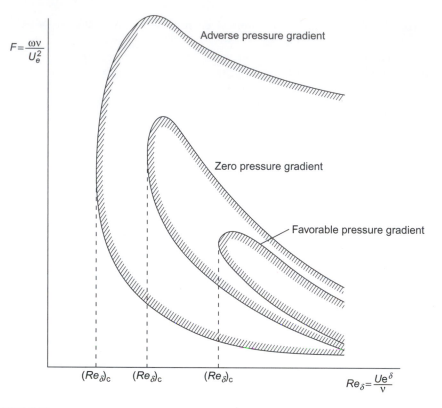

$$F = \frac{\omega v}{U_e^2}$$

Adverse pressure gradient

Zero pressure gradient

Favorable pressure gradient

$(Re_\delta)_c$ $(Re_\delta)_c$ $(Re_\delta)_c$

$$Re_\delta = \frac{U_e \delta}{v}$$

FIGURE 8.29

Plot of the effect of external pressure gradients on neutral stability boundaries.

Recall that the boundary-layer thickness also grows more rapidly in an adverse pressure gradient, thereby reaching a given critical value of Re_δ sooner. Thus we easily see that transition is promoted under these circumstances.

Exactly the converse is found for the favorable pressure gradient. This circumstance allows rough and ready predictions for the transition point on bodies and wings, especially in the case of more classic streamlined shapes. These guidelines may be summarized as follows:

1. If $10^5 < Re_L < 10^7$ (where $Re_L = U_\infty L/v$ is based on the total length or chord of the body or wing), transition occurs very shortly downstream of the point of minimum pressure. For airfoils at zero incidence or for streamlined bodies of revolution, the point of minimum pressure often, but not invariably, coincides with the point of maximum thickness.

2. If, for an airfoil, Re_L is kept constant, increasing the angle of incidence advances the point of minimum pressure toward the leading edge on the upper surface, which causes transition to move forward. The opposite occurs on the lower surface.

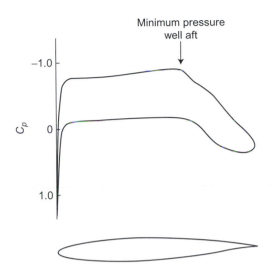

Minimum pressure
well aft

FIGURE 8.30

Modern laminar-flow airfoil and its pressure distribution.

3. At constant incidence, an increase in Re_L tends to advance transition.
4. For $Re_L > 10^7$, the transition point may slightly precede the point of minimum pressure.

The effects of external pressure gradient on transition also explain how transition may be postponed by designing airfoils with points of minimum pressure further aft. A typical modern airfoil of this type is shown in Fig. 8.30. The problem here is that, although the onset of the adverse pressure gradient is postponed, it tends to be more severe, giving rise to boundary-layer separation. This necessitates boundary-layer suction aft of the point of minimum pressure to prevent separation and maintain laminar flow.

8.9 THE PHYSICS OF TURBULENT BOUNDARY LAYERS

Here we give a brief account of the physics of turbulent boundary layers. This is still very much a developing subject and an active topic of research. However, some classic empirical knowledge, results, and methods have stood the test of time and are worth describing in a general textbook on aerodynamics. Moreover, turbulent flows are so important for engineering applications that some understanding of relevant flow physics is essential for predicting and controlling flows.

8.9.1 Reynolds Averaging and Turbulent Stress

Turbulent flow is a complex motion, fundamentally three-dimensional and highly unsteady. Figure 8.31(a) depicts a typical variation in a flow variable f, such as velocity or pressure, with time at a fixed point in a turbulent flow. The usual approach in engineering, originating with Reynolds [61], is to take a time average. Thus the instantaneous velocity is given by

$$f = \bar{f} + f' \tag{8.83}$$

where the time average is denoted $(^{-})$ and $(\ )'$ denotes fluctuation (or deviation from the time average). The strict mathematical definition of the time average is

$$\bar{f} \equiv \lim_{T \to \infty} \frac{1}{T} \int_0^T f(x,y,z,t = t_0 + t')dt' \tag{8.84}$$

where t_0 is the time at which measurement notionally begins. For practical measurements, T is merely taken as suitably large rather than infinite. The basic approach is often known as *Reynolds averaging*.

We use the Reynolds averaging approach on the continuity Eq. (8.8) and the x momentum Navier-Stokes Eq. (8.5). When Eq. (8.83), with u for f and similar expressions for v and w, is substituted into Eq. (8.8), we obtain

$$\frac{\partial \bar{u}}{\partial x} + \frac{\partial \bar{v}}{\partial y} + \frac{\partial \bar{w}}{\partial z} + \frac{\partial u'}{\partial x} + \frac{\partial v'}{\partial y} + \frac{\partial w'}{\partial z} = 0 \tag{8.85}$$

Taking a time average of fluctuation gives zero by definition, so taking a time average of Eq. (8.85) gives

$$\frac{\partial \bar{u}}{\partial x} + \frac{\partial \bar{v}}{\partial y} + \frac{\partial \bar{w}}{\partial z} = 0 \tag{8.86}$$

Subtracting Eq. (8.86) from Eq. (8.85) gives

$$\frac{\partial u'}{\partial x} + \frac{\partial v'}{\partial y} + \frac{\partial w'}{\partial z} = 0 \tag{8.87}$$

This result will be used in further discussions.

We now substitute Eq. (8.83) to give expressions for u, v, w, and p in Eq. (8.5) to obtain

$$\rho \left(\frac{\partial (\bar{u} + u')}{\partial t} + (\bar{u} + u')\frac{\partial (\bar{u} + u')}{\partial x} + (\bar{v} + v')\frac{\partial (\bar{u} + u')}{\partial y} + (\bar{w} + w')\frac{\partial (\bar{u} + u')}{\partial z} \right)$$

$$= -\frac{\partial (\bar{p} + p')}{\partial x} + \mu \left(\frac{\partial^2 (\bar{u} + u')}{\partial x^2} + \frac{\partial^2 (\bar{u} + u')}{\partial y^2} + \frac{\partial^2 (\bar{u} + u')}{\partial z^2} \right) \tag{8.88}$$

Next we take a time average of each term, noting that, although the time average of a fluctuation is zero by definition (see Fig. 8.31(b)), the time average of a

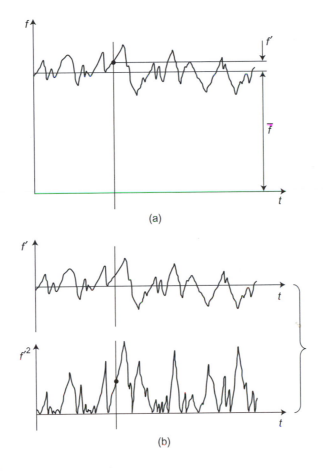

FIGURE 8.31

Reynolds averaging.

product of fluctuations is not, in general, equal to zero (e.g., plainly $\overline{u'u'} = \overline{u'^2} > 0$; see Fig. 8.31(b)). We also assume that the turbulent boundary-layer flow is two-dimensional when time-averaged so that no time-averaged quantities vary with z and $\overline{w} = 0$. Therefore, if we take the time average in each term in Eq. (8.88), we simplify to

$$\rho\left(\overline{u}\frac{\partial\overline{u}}{\partial x} + \overline{v}\frac{\partial\overline{u}}{\partial y}\right) + \underbrace{\overline{\left(u'\frac{\partial u'}{\partial y} + v'\frac{\partial u'}{\partial y} + w'\frac{\partial u'}{\partial z}\right)}}_{*}$$

$$= -\frac{\partial\overline{p}}{\partial x} + \mu\left(\frac{\partial^2\overline{u}}{\partial x^2} + \frac{\partial^2\overline{u}}{\partial y^2}\right)$$

(8.89)

The term marked with an asterisk can be written as

$$\overline{\left(\frac{\partial u'^2}{\partial x} - u'\frac{\partial u'}{\partial y} + \frac{\partial (u'v')}{\partial y} - u'\frac{\partial v'}{\partial y} + \frac{\partial (u'w')}{\partial z} - u'\frac{\partial w'}{\partial z}\right)}$$

$$= \frac{\partial \overline{u'^2}}{\partial x} + \frac{\partial \overline{u'v'}}{\partial y} + \frac{\partial \overline{u'w'}}{\partial z} - \underbrace{\overline{u'\left(\frac{\partial u'}{\partial x} + \frac{\partial v'}{\partial y} + \frac{\partial w'}{\partial z}\right)}}_{=0 \text{ from Eq. (8.87)}}$$

$$= \frac{\partial \overline{u'^2}}{\partial x} + \frac{\partial \overline{u'v'}}{\partial y} + \underbrace{\frac{\partial \overline{u'w'}}{\partial z}}_{=0 \text{ no variation with } z}$$

So Eq. (8.89) becomes

$$\rho\left(\overline{u}\frac{\partial \overline{u}}{\partial x} + \overline{v}\frac{\partial \overline{u}}{\partial y}\right) = -\frac{\partial \overline{p}}{\partial x} + \mu\left(\frac{\partial \overline{\sigma}_{xx}}{\partial x} + \frac{\partial \overline{\sigma}_{xy}}{\partial y}\right) \qquad (8.90)$$

where we write

$$\overline{\sigma}_{xx} = \mu\frac{\partial \overline{u}}{\partial x} - \rho\overline{u'^2}, \quad \overline{\sigma}_{xy} = \mu\frac{\partial \overline{u}}{\partial y} - \rho\overline{u'v'}$$

This notation makes it evident that, when the turbulent flow is time-averaged, $-\rho\overline{u'^2}$ and $-\rho\overline{u'v'}$ take on the character of a direct and a shear stress, respectively. For this reason, the quantities are known as *Reynolds stresses* or *turbulent stresses*. In fully turbulent flows, the Reynolds stresses are usually much greater than the viscous stresses. If the time-averaging procedure is applied to Eqs. (8.5), (8.6) and (8.7), the full three-dimensional Navier-Stokes equations, a Reynolds stress tensor is generated:

$$-\rho\begin{pmatrix} \overline{u'^2} & \overline{u'v'} & \overline{u'w'} \\ \overline{u'v'} & \overline{v'^2} & \overline{v'w'} \\ \overline{u'w'} & \overline{v'w'} & \overline{w'^2} \end{pmatrix} \qquad (8.91)$$

We can see that there are generally nine components of the Reynolds stress that comprise six distinct quantities.

8.9.2 Boundary-Layer Equations for Turbulent Flows

For the applications considered here—two-dimensional boundary layers (more generally two-dimensional shear layers)—only one of the Reynolds stresses is significant, and that is the Reynolds shear stress $-\rho\overline{u'v'}$. Thus, for two-dimensional turbulent boundary layers, the time-averaged boundary-layer equations can be written in the form

$$\frac{\partial \bar{u}}{\partial x} + \frac{\partial \bar{v}}{\partial y} = 0 \qquad (8.92a)$$

$$\bar{u}\frac{\partial \bar{u}}{\partial x} + \bar{v}\frac{\partial \bar{u}}{\partial y} = -\frac{d\bar{p}}{dx} + \frac{\partial \bar{\tau}}{\partial y} \qquad (8.92b)$$

The chief difficulty with turbulence is that there is no way to determine the Reynolds stresses from first principles, apart from solving the unsteady three-dimensional Navier-Stokes equations. We must formulate semi-empirical approaches for modeling the Reynolds shear stress before we can solve Eqs. (8.92a,b).

The momentum-integral form of the boundary-layer equations from Section 8.5.1 is equally applicable to laminar or turbulent boundary layers, if we recognize that the time-averaged velocity should be used to define momentum and displacement thicknesses. This is the basis of the approximate methods described in Section 8.6, which assume a one-seventh power velocity profile and use semi-empirical formulae for the local skin-friction coefficient.

8.9.3 Eddy Viscosity

Away from the immediate influence of the wall, which has a damping effect on turbulent fluctuations, we expect the Reynolds shear stress to be much greater than the viscous shear stress. This can be seen by comparing rough order-of-magnitude estimates of the Reynolds and viscous shear stress:

$$-\rho\overline{u'v'} \quad \text{c.f.} \quad \mu\frac{\partial \bar{u}}{\partial y}$$

Assume that $\overline{u'v'} \simeq CU_\infty^2$ (where C is a constant); then

$$\mu\frac{\partial \bar{u}}{\partial y} = O\left(\frac{\mu U_\infty}{\delta}\right) = O\left(\frac{1}{C^2}\underbrace{\frac{\mu}{\rho U_\infty \delta}}_{1/Re}\right)\rho\overline{u'v'}$$

where δ is the shear-layer width. Provided $C = O(1)$, then

$$\frac{\mu\partial \bar{u}/\partial y}{-\rho\overline{u'v'}} = O\left(\frac{1}{Re}\right)$$

showing that, for large values of Re (recall that turbulence occurs only at high Reynolds numbers), the viscous shear stress is negligible compared with the

Reynolds shear stress. Boussinesq [62] drew an analogy between viscous and Reynolds shear stresses with his introduction of *eddy viscosity* ε_T:

$$\underbrace{\tau = \mu \frac{\partial \bar{u}}{\partial y}}_{\text{viscous shear stress}} \quad \text{cf.} \quad \underbrace{-\rho \overline{u'v'} = \rho \varepsilon_T \frac{\partial \bar{u}}{\partial y}}_{\text{Reynolds shear stress}} : \quad \varepsilon_T \gg \nu \, (= \mu/\rho) \tag{8.93}$$

Boussinesq himself merely assumed that eddy viscosity is constant everywhere in the flow field, like molecular viscosity but much larger, and until comparatively recently his approach was widely used by oceanographers for modeling turbulent flows. In fact, though, a constant eddy viscosity is a very poor approximation for wall shear flows such as boundary layers and pipe flows. For simple turbulent free shear layers, such as the mixing layer and jet (see Fig. 8.32) and wake, it is a reasonable assumption that eddy viscosity varies in the streamwise direction but not across a particular cross-section. Thus, using simple dimensional analysis, Prandtl [64] and Reichardt [65] proposed that

$$\varepsilon_T = \underbrace{\kappa}_{\text{const.}} \times \underbrace{\Delta U}_{\text{velocity difference across shear layer}} \times \underbrace{\delta}_{\text{shear-layer width}} \tag{8.94}$$

κ is often called the *exchange coefficient*, and it varies somewhat from one type of flow to another. Equation (8.94) gives excellent results and can be used to determine the variation of the overall flow characteristics in the streamwise direction (see Example 8.9).

The outer 80% or so of the turbulent boundary layer is largely free from the effects of the wall and, in this respect, is quite similar to a free turbulent shear layer. In this outer region, it is commonly assumed, following Laufer (1954) [63], that eddy viscosity can be determined by a version of Eq. (8.94) whereby

$$\varepsilon_T = \kappa U_e \delta^* \tag{8.95}$$

Example 8.9

The spreading rate of a mixing layer, illustrated in Fig. 8.32 shows the mixing layer in the initial region of a jet. To a good approximation, the external mean pressure field for a free shear layer is atmospheric and therefore constant. Furthermore, the Reynolds shear stress is much larger than the viscous stress so that, after substituting Eqs. (8.93) and (8.94), the turbulent boundary-layer Eq. (8.92b) becomes

$$\bar{u} \frac{\partial \bar{u}}{\partial x} + \bar{v} \frac{\partial \bar{u}}{\partial y} = \varepsilon_T \frac{\partial^2 \bar{u}}{\partial y^2} \quad \left(= \kappa U_J \delta \frac{\partial^2 \bar{u}}{\partial y^2} \right)$$

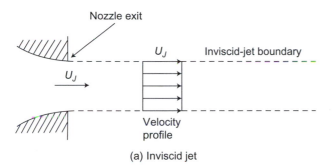

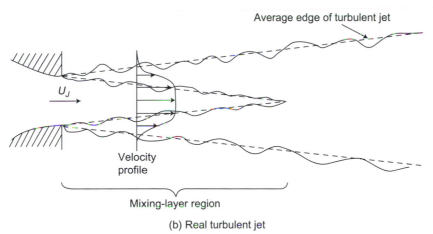

(b) Real turbulent jet

FIGURE 8.32

Ideal inviscid jet compared with a real turbulent jet near the nozzle exit.

The only length scale is the mixing-layer width $\delta(x)$, which increases with x, so dimensional arguments suggest that the velocity profile does not change shape when expressed in terms of dimensionless y:

$$\frac{\bar{u}}{U_J} = F \underbrace{\left(\frac{y}{\delta}\right)}_{\eta,\,\text{say}}$$

This is known as a *similarity* assumption. The assumed form of the velocity profile implies that

$$\frac{\partial \bar{u}}{\partial x} = \frac{\partial}{\partial x}\left(U_J F(\eta)\right) = \underbrace{\frac{dU_J}{dx}}_{=0} F(\eta) + U_J F'(\eta) \underbrace{\left(-\frac{\eta}{\delta}\frac{d\delta}{dx}\right)}_{\partial \eta/\partial x}$$

where $F'(\eta) \equiv dF/d\eta$.

We integrate Eq. (8.92a) to get

$$\bar{v} = -\int \frac{\partial \bar{u}}{\partial x} dy = U_J \frac{\partial \delta}{\partial x} \int \eta F'(\eta) d\eta$$

so

$$\bar{v} = U_J \frac{\partial \delta}{\partial x} G(\eta) \quad \text{where} \quad G = \int \eta F'(\eta) d\eta$$

The derivatives with respect to y are given by

$$\frac{\partial \bar{u}}{\partial y} = \frac{\partial \eta}{\partial y} \frac{d\bar{u}}{d\eta} = \frac{U_j}{\delta} F'(\eta)$$

$$\frac{\partial^2 \bar{u}}{\partial y^2} = \frac{\partial \eta}{\partial y} \frac{d}{d\eta}\left(\frac{\partial \bar{u}}{\partial y}\right) = \frac{U_j}{\delta^2} F''(\eta)$$

The results are substituted into the reduced boundary-layer equation to obtain, after removing common factors,

$$\underbrace{-\frac{1}{\delta}\frac{d\delta}{dx}}_{\text{fn. of } x \text{ only}} \underbrace{\left(\eta FF' + GF'\right)}_{\text{fn. of } \eta \text{ only}} = \underbrace{\kappa \frac{1}{\delta}}_{\text{fn. of } x \text{ only}} \underbrace{F''}_{\text{fn. of } \eta \text{ only}}$$

The braces indicate which terms are functions of x only or η only. We separate the variables and thereby see that, for the similarity form of the velocity to be viable, we must require

$$\frac{\frac{1}{\delta}\frac{d\delta}{dx}}{\frac{\kappa}{\delta}} = -\frac{F''}{\eta FF' + GF'} = \text{const.}$$

After simplification, the term on the left-hand side implies

$$\frac{d\delta}{dx} = \text{const.} \quad \text{or} \quad \delta \propto x$$

Setting the term equal to a constant, depending on η and with F'' as numerator, leads to a differential equation for F that can be solved to give the velocity profile. In fact, it is easy to derive a good approximation to the velocity profile, so this is a less valuable result.

When a turbulent (or laminar) flow is characterized by only one length scale—as in the present case—the term *self-similarity* is commonly used. Solutions found this way are called *similarity* solutions. Similar methods can be used to determine the overall flow characteristics of other turbulent free shear layers.

8.9.4 Prandtl's Mixing-Length Theory of Turbulence

Equation (8.95) is not a good approximation in the region of the turbulent boundary layer or pipe flow near the wall. Eddy viscosity varies with distance from the wall in this region. A common approach in the near-wall region is based on Prandtl's *mixing-length* theory [66]. This approach to modeling turbulence is loosely based on the kinetic theory of gases. We give a brief account of this and illustrate it in Fig. 8.33.

Imagine that a blob of fluid is transported upward by a fluctuating turbulent velocity v' through an average distance ℓ_m—the mixing length (analogous to the mean free path in molecular dynamics). In the new position, assuming that the streamwise velocity of the blob remains unchanged at the value in its original position, the fluctuation in velocity is generated by the difference in the blob's velocity and that of its new surroundings. Thus

$$u' \simeq \underbrace{\left(\bar{u} + \frac{\partial \bar{u}}{\partial y}\ell_m\right)}_{(i)} - \underbrace{\bar{u}}_{(ii)} = \frac{\partial \bar{u}}{\partial y}\ell_m$$

Term (i) is the mean flow speed in the new environment. In writing the term in this form, we assume that $\ell_m \ll \delta$ so that, in effect, (i) is the first two terms in a Taylor series expansion.

Term (ii) is the mean velocity of the blob.

If it is also assumed that $v' \simeq (\partial \bar{u}/\partial y)\ell_m$, then

$$-\overline{u'v'} \simeq \ell_m^2 \left(\frac{\partial \bar{u}}{\partial y}\right)\underbrace{|\frac{\partial \bar{u}}{\partial y}|}_{(iii)} \quad \text{implying} \quad \varepsilon_T = \text{const} \times \ell_m^2 |\frac{\partial \bar{u}}{\partial y}| \tag{8.96}$$

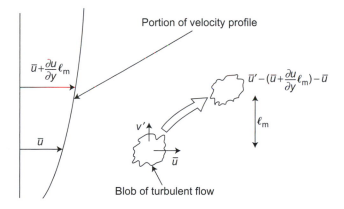

FIGURE 8.33

Prandtl's notion of mixing length.

Term (iii) is written with an absolute value sign so that the Reynolds stress changes sign with $\partial \bar{u}/\partial y$, just as the viscous shear stress does.

8.9.5 Regimes of Turbulent Wall Flow

As the wall is approached, it has a damping effect on the turbulence so that, very close to the wall, the viscous shear stress greatly exceeds the Reynolds shear stress. This region next to the wall where viscous effects dominate is usually known as the *viscous sublayer*. Beyond it is a *transition* or *buffer* layer where the viscous and Reynolds shear stresses are roughly equal. This region blends into the fully turbulent region where the Reynolds shear stress is much larger than the viscous shear stress. It is in this fully turbulent near-wall region that the mixing-length theory can be used. The outer part of the boundary layer is more like a free shear layer, and there the Reynolds shear stress is given by Eq. (8.95).

A major assumption is that the fully turbulent layer begins at a height above the wall of $y \ll \delta$, so

$$\tau = \tau_{\mathrm{w}} + \underbrace{\frac{\mathrm{d}\tau}{\mathrm{d}y}y}_{\ll \tau_{\mathrm{w}}} + \cdots \simeq \tau_{\mathrm{w}} \tag{8.97}$$

Near the wall in the viscous sublayer, the turbulence is almost completely damped, so only molecular viscosity is important; thus

$$\tau = \mu \frac{\mathrm{d}\bar{u}}{\mathrm{d}y} = \tau_{\mathrm{w}} \quad \text{therefore} \quad \bar{u} = \frac{\tau_{\mathrm{w}}}{\mu}y \tag{8.98}$$

In the fully turbulent region, the Reynolds shear stress is much greater than the viscous shear stress, so

$$\tau = -\rho \overline{u'v'}$$

Therefore, if Eq. (8.96) is used and it is assumed that $\ell_{\mathrm{m}} \propto y$,

$$\left(\frac{\mathrm{d}\bar{u}}{\mathrm{d}y}\right)^2 \propto \frac{\tau_{\mathrm{w}}}{\rho}\frac{1}{y^2} \quad \text{implying} \quad \frac{\mathrm{d}\bar{u}}{\mathrm{d}y} \propto \frac{V_*}{y}$$

where we introduce the *friction velocity*,

$$V_* = \sqrt{\tau_{\mathrm{w}}/\rho} \tag{8.99}$$

as the reference velocity subsequently used to render the velocity in the near-wall region nondimensional.

We integrate the equation just before Eq. (8.99) and divide by V_* to obtain the nondimensional velocity profile in the fully turbulent region, and we rewrite Eq. (8.98) to obtain the same in the viscous sublayer. Thus the fully turbulent flow is

$$\frac{\bar{u}}{V_*} = C_1 \ell n\left(\frac{yV_*}{v}\right) + C_2 \tag{8.100}$$

and the viscous sublayer is

$$\frac{\bar{u}}{V_*} = \frac{yV_*}{v} \tag{8.101}$$

where C_1 and C_2 are constants of integration to be determined by comparison with experimental data, and η or $y^+ = yV_*/v$ is the dimensionless distance from the wall. The length $\ell^+ = v/V_*$ is usually known as the *wall unit*.

Figure 8.34 compares (8.100) and (8.101) with experimental data for a turbulent boundary layer; we can thereby deduce that the viscous sublayer is

$$\frac{yV_*}{v} < 5 \quad \mu\frac{\partial\bar{u}}{\partial y} \gg -\rho\overline{u'v'}$$

the buffer layer is

$$5 < \frac{yV_*}{v} < 50 \quad \mu\frac{\partial\bar{u}}{\partial y} \simeq -\rho\overline{u'v'} \tag{8.102}$$

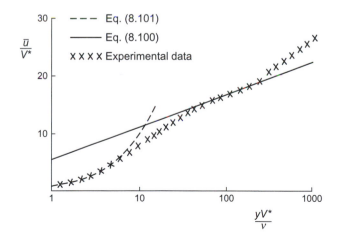

FIGURE 8.34

Average velocity profile near the wall of a turbulent boundary layer.

and the fully developed turbulence is

$$\frac{yV_*}{\nu} > 50 \quad \mu\frac{\partial\overline{u}}{\partial y} \ll -\rho\overline{u'v'}$$

The constants C_1 and C_2 can be determined from comparison with the experimental data so that Eq. (8.100) becomes the logarithmic velocity profile:

$$\frac{\overline{u}}{V_*} = 2.54\,\ell n\left(\frac{yV_*}{\nu}\right) + 5.56 \qquad (8.103)$$

C_1 is often written as $1/\kappa$, where $\kappa = 0.41$ is known as the von Kármán constant [67] because von Kármán was the first to derive the logarithmic velocity profile. Equation (8.100) is often known as the *law of the wall*. It applies equally well to the near-wall region of turbulent pipe and channel flows for which better agreement with experimental data is found for slightly different constant values. Note that it is not essential to evoke Prandtl's mixing-length theory to derive the law of the wall. The logarithmic form of the velocity profile can also be derived purely by means of dimensional analysis [68].

Outer Boundary Layer

The outer part of the boundary layer that extends for 70% or 80% of the total thickness is unaffected by the direct effect of the wall. In Fig. 8.34 the velocity profile deviates considerably from the logarithmic form in this outer part of the boundary layer. In many respects, this is analogous to a free shear layer, especially a wake and it is sometimes referred to as the *defect layer* or *wake region*. Here inertial effects dominate and viscous effects are negligible, so the appropriate reference velocity and length scales for nondimensionalization are U_e (the streamwise flow speed at the boundary-layer edge) and δ (the boundary-layer thickness) or some similar length scale. Thus the so-called *outer variables* are

$$\frac{\overline{u}}{U_e} \quad \text{and} \quad \frac{y}{\delta}$$

8.9.6 Formulae for Local Skin-Friction Coefficient and Drag

Although it is not valid in the outer part of the boundary layer, Eq. (8.100) can be used to obtain the following more accurate semi-empirical formulae for the local skin-friction coefficient and the corresponding drag coefficient for turbulent boundary layers over flat plates:

$$c_f \equiv \frac{\tau_w}{\frac{1}{2}\rho U_\infty^2} = (2\log_{10} Re_x - 0.65)^{-2.3} \qquad (8.104)$$

$$C_{Df} \equiv \frac{D_f}{\frac{1}{2}\rho U_\infty^2 BL} = \frac{0.455}{(\log_{10} Re_L)^{2.58}} \qquad (8.105)$$

where B and L are the breadth and length of the plate. The *Prandtl-Schlichting formula* (8.105) is more accurate than Eq. (8.72) when $Re_L > 10^7$.

Effects of Wall Roughness

Turbulent boundary layers, especially at high Reynolds numbers, are very sensitive to wall roughness. This is because any roughness element that protrudes through the viscous sublayer modifies the law of the wall. The effect of wall roughness on the boundary layer depends on the size, shape, and spacing of the elements. To bring a semblance of order, Nikuradze [70] matched each "type" of roughness against an *equivalent sand-grain roughness* having roughness of height k_s. Three regimes of wall roughness, corresponding to the three sections of the near-wall region, can be defined as follows:

Hydraulically smooth If $k_s V_*/\nu \leq 5$, the roughness elements lie wholly within the viscous sublayer; the roughness therefore has no effect on the velocity profile or on the value of skin friction or drag.

Completely rough If $k_s V_*/\nu \geq 50$, the roughness elements protrude into the region of fully developed turbulence. This has the effect of displacing the logarithmic profile downwards—that is, reducing the value of C_2 in Eq. (8.100). In such cases the local skin-friction and drag coefficients are independent of Reynolds number and are given by

$$c_f = [2.87 + 1.58 \log_{10}(x/k_s)]^{-2.5} \qquad (8.106)$$

$$C_{Df} = [1.89 + 1.62 \log_{10}(L/k_s)]^{-2.5} \qquad (8.107)$$

Transitional roughness If $5 \leq k_s V_*/\nu \leq 50$, the effect of roughness is more complex and the local skin-friction and drag coefficients depend on both Reynolds number and relative roughness k_s/δ.

Relative roughness plainly varies along the surface, but the viscous sublayer increases slowly and, although its maximum thickness is located at the trailing edge, the trailing-edge value represents most of the rest of the surface. The degree of roughness considered *admissible* in engineering practice is that for which the surface remains hydraulically smooth throughout—that is, the roughness elements remain within the viscous sublayer all the way to the trailing edge. Thus

$$k_{adm} = 5\nu/(V_*)_{TE} \qquad (8.108)$$

In the case of a flat plate, Eq. (8.108) is approximately equivalent to

$$k_{adm} \propto \frac{\nu}{V_\infty} \propto \frac{L}{Re_L} \qquad (8.109)$$

Thus, for plates of similar length, the admissible roughness diminishes with increasing Re_L. For ships' hulls, admissible roughness ranges from 7 (large fast ships) to 20 μm (small slow ships); such values are utterly impossible to achieve in practice, and it is always neccessary to allow for considerable increase in drag due to roughness. For aircraft, admissible roughness ranges from 10 to 25 μm, which is just about attainable in practice. Model aircraft and compressor blades require the same order of admissible roughness, and hydraulically smooth surfaces can be obtained without undue difficulty. At the other extreme are steam-turbine blades that combine a small chord (L) with a fairly high Reynolds number (5×10^6) owing to the high velocities involved and comparatively high pressures. In this case, admissible roughness values are very small, ranging from 0.2 to 2 μm. This degree of smoothness can barely be achieved on newly manufactured blades, and certainly the admissible roughness is exceeded after a period of operation because of corrosion and scaling.

The description of the aerodynamic effects of surface roughness just given was in terms of equivalent sand-grain roughness. It is important to remember that the aerodynamic effects of a particular type of roughness may differ greatly from that of sand-grain roughness of the same size. It is even possible (see Section 9.4.3) for special forms of wall "roughness," such as riblets, to reduce drag [69].

8.9.7 Distribution of Reynolds Stresses and Turbulent Kinetic Energy across the Boundary Layer

Figure 8.35 plots the variation in Reynolds shear stress and kinetic energy (per unit mass), $k = (\overline{u'^2} + \overline{v'^2} + \overline{w'^2})/2s$, across the boundary layer. Immediately striking is how comparatively high the levels are in the near-wall region. The Reynolds shear stress reaches a maximum at about $y^+ \simeq 100$, while the turbulence kinetic energy appears to reach its maximum not far above the edge of the viscous sublayer.

Figure 8.36 plots the distributions of the so-called turbulence intensities of the velocity components: the square roots of the direct Reynolds stresses $\overline{u'^2}$, $\overline{v'^2}$, and $\overline{w'^2}$. Note that, in the outer part of the boundary layer, the three turbulent intensities tend to be the same (they are "isotropic"), but they diverge widely as the wall is approached (i.e., they become "anisotropic"). The distribution of eddy viscosity across the turbulent boundary layer is plotted in Fig. 8.37. For engineering calculations of turbulent boundary layers this quantity is important. Note that the form adopted in Eq. (8.96) for the near-wall region according to the mixing-length theory, with $\ell_m \propto y$, is borne out by the behavior shown in the figure.

A probe placed in the outer region of a boundary layer shows that the flow is turbulent only for part of the time. This proportion of time is called the *intermittency* (γ). The intermittency distribution is also plotted in Fig. 8.37.

8.9.8 Turbulence Structures in the Near-Wall Region

The dominance of the near-wall region in terms of turbulence kinetic energy and Reynolds shear stress motivated engineers to study it in more detail to identify its time-varying flow structures. Kline *et al.* [71] carried out a seminal study of this

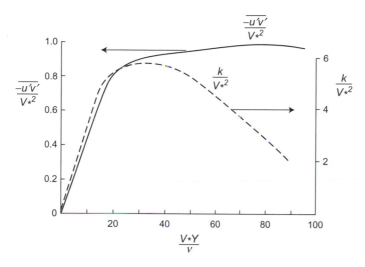

FIGURE 8.35

Variations in Reynolds shear stress and turbulence kinetic energy across the near-wall region of the turbulent boundary layer.

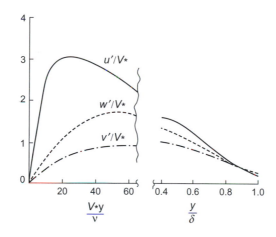

FIGURE 8.36

Variations in the root mean squares of $\overline{u'^2}, \overline{v'^2}$, and $\overline{w'^2}$ across a turbulent boundary layer.

kind, obtaining hydrogen-bubble flow visualizations for the turbulent boundary layer that revealed streak-like structures developing within the viscous sublayer. Depicted schematically in Fig. 8.38. These structures are continuously changing with time, and observations over time reveal that there are low- and high-speed streaks. The structures become less noticeable further away from the wall and apparently disappear in

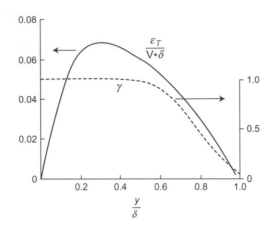

FIGURE 8.37

Distributions of eddy viscosity and intermittency across a turbulent boundary layer.

the law-of-the-wall region. Experiments reveal the turbulence to be intermittent and of larger scale in the outer region (Fig. 8.38).

The conventional view is that the streaks are a manifestation of developing hairpin vortices. See Figs. 8.38 and 8.39 for a schematic illustration of the conceptual burst cycle of these structures, which are responsible for generating transient high levels of wall shear stress. The development of these vortices tends to be quasi-periodic with the following sequence of events:

1. *Formation of the low-speed streaks*   During this process the legs of the vortices lie close to the wall.
2. *Lift-up* or *ejection*   This is stage E in Fig. 8.38. The velocity induced by the vortex legs cause the vortex head to lift away from the wall.
3. *Oscillation* or *instability*   This is the first part of stage B in Fig. 8.38. A local point of inflexion develops in the velocity profile, and the flow becomes susceptible to Helmholtz instability, locally causing the head of the vortex to oscillate violently.
4. *Bursting* or *break-up*   This is the latter part of stage B in Fig. 8.38. The oscillation culminates in the vortex head bursting.
5. *High-speed sweep*   This is stage S in Fig. 8.38. After a period of quiescence, the bursting event is followed by a high-speed sweep toward the wall. During this process the shear stress at the wall is at its greatest and new hairpin vortices are generated.

Understand that Fig. 8.38 corresponds to a frame moving downstream with the evolving vortex structure. Thus a constant streamwise velocity is superimposed on the ejections and sweeps.

The events just described are quasi-periodic in a statistical sense. The mean values of their characteristics are as follows: the spanwise spacing of streaks $= 100\nu/V_*$; the

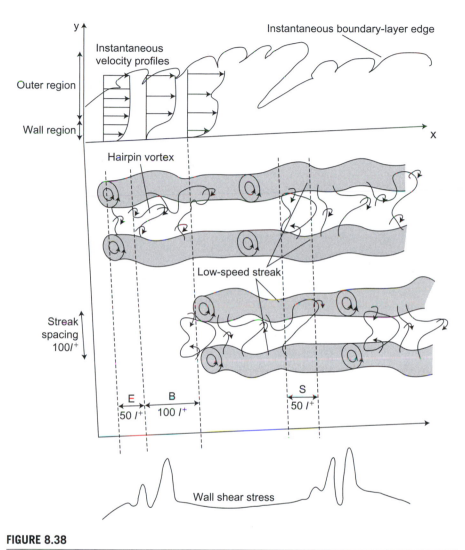

FIGURE 8.38

Flow structures in a turbulent boundary layer showing the conceptual burst cycle.
E, ejection stage; B, breakup stage; S, sweep stage.

streaks reach a vertical height of $50\nu/V_*$; their streamwise extent is about $1000\nu/V_*$; and their bursting frequency is about $0.004\,V_*^2/\nu$. This estimate for bursting frequency is controversial. Some experts [72] think that bursting frequency does not scale with the wall units; others suggest that this result is an artifact of the measurement system [73]. Much greater detail on these near-wall structures, together with the various concepts and theories advanced to explain their formation and regeneration, can be found in Panton [74].

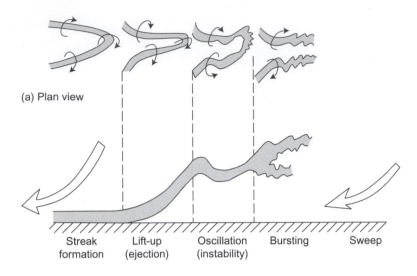

(a) Plan view

Streak   Lift-up   Oscillation   Bursting   Sweep
formation   (ejection)   (instability)

(b) Side view

FIGURE 8.39

Evolution of a hairpin vortex in the near-wall region.

Example 8.10

Determining Specifications of a MEMS Actuator for Flow Control.

Modern technology is rapidly developing the capability to build very small machines that, among many other applications, can be used for actuation and sensing in flow-control systems. Such machines are known as MEMS (micro-electro-mechanical systems), a term that usually refers to devices with overall dimensions of less than 1 mm but more than 1 μm. Such devices combine electrical and mechanical components manufactured by means of integrated-circuit batch-processing [75].

A conceptual MEMS actuator as part of a flow-control system is depicted schematically in Fig. E8.10(a). If consists of a diaphragm located at the bottom of a buried cavity that connects to the boundary layer via an exit orifice. The diaphragm is made of silicon or a suitable polymeric material and is driven by a piezoceramic driver. When a voltage is applied to the driver, depending on the sign of the electrical signal, the driver either contracts or expands, displacing the diaphragm up or down. If an alternating voltage is applied to the diaphragm, it is periodically driven up and down, which in turn alternately reduces and increases the volume of the cavity, raising and then lowering the air pressure in it. An elevated cavity pressure drives air through the exit orifice into the boundary layer; air returns to the cavity when the cavity pressure falls. This periodic outflow and inflow creates a *synthetic jet*. Even though, over a cycle, there is no net air leaving the cavity, vortical structures propagate into the boundary layer, much as they do from a steady micro-jet. It is also possible to drive the diaphragm with

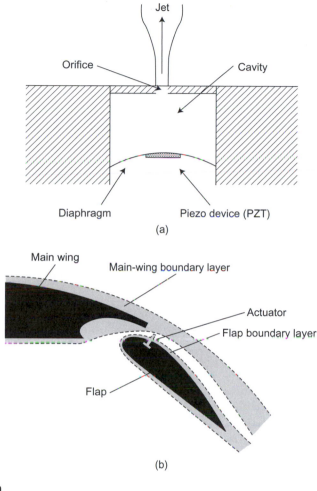

FIGURE E8.10

(a) Jet actuator. (b) Sample application of a synthetic jet actuator.

short-duration steady voltage, thereby displacing the diaphragm suddenly upward and driving a "puff" of air into the boundary layer [76].

It has been proposed that synthetic-jet actuators be used to control the near-wall streak-like structures in the turbulent boundary layer over the flap of a large airliner. In particular, the aim is to increase so-called bursting frequency to increase turbulence. Increased turbulence would lead to a delay of boundary-layer separation, allowing the flap to be deployed at a larger angle of incidence, and so increase its performance. The basic concept is illustrated in Fig. E8.10. The question is what dimensions and specifications should be chosen for the MEMS actuator in this application.

Consider an aircraft similar to the Airbus A340 with a mean wing chord of 6 m and a flap chord of 1.2 m. Assume that the approach speed is about 100 m s^{-1} and assume standard sea-level conditions so that the kinematic viscosity of the air is around $15 \times 10^{-6} \text{m}^2 \text{s}^{-1}$. The array of MEMS actuators is located near the point of minimum pressure 200 mm from the leading edge of the flap. A new boundary layer develops on the flap underneath the separated boundary layer from the main wing, which ensures that the flap boundary layer is strongly disturbed, provoking early transition to turbulence.

$$Re_x = \frac{xU_e}{\nu} \simeq \frac{0.2 \times 100}{15 \times 10^{-6}} = 1.33 \times 10^6$$

In a low-disturbance environment, we normally expect the boundary layer to be laminar at this Reynolds number (see Section 8.8), but the flap boundary layer is highly disturbed by the separated boundary layer from the main wing, and so ensures early transition. Using Eq. (8.104), we can estimate the local skin-friction coefficient:

$$c_f = (2\log_{10}(1.33 \times 10^6) - 0.65)^{-2.3} = 0.00356$$

From this we can estimate the wall shear stress and friction velocity:

$$\tau_w = \frac{1}{2}\rho U_e^2 c_f \simeq \frac{1}{2} \times 1.2 \times 100^2 \times 0.00356 = 21.36\,\text{Pa}, \quad V_* = \sqrt{\frac{\tau_w}{\rho}} = \sqrt{\frac{21.36}{1.2}} = 4.22\,\text{m s}^{-1}$$

The boundary-layer thickness is not strictly needed for determining the actuator specifications, but it is instructive. In Eq. (8.55) it was shown that $C_{Df} = 2\Theta(L)/L$, that is, there is a relationship between the coefficient of skin-friction drag and momentum thickness at the trailing edge. This relationship can be exploited to estimate the momentum thickness in the present application. We merely assume for this purpose that the flap boundary layer terminates at the point in question ($x = 200$ mm), so that

$$\Theta(x) = \frac{x}{2}C_{Df}(x)$$

Using Eq. (8.81)

$$C_{Df}(x) = \frac{0.455}{(\log_{10} Re_x)^{2.58}} = \frac{0.455}{\{\log_{10}(1.33 \times 10^6)\}^{2.58}} = 0.00424$$

Thus

$$\Theta(x) = \frac{x}{2}C_{Df}(x) = \frac{0.2}{2} \times 0.00424 = 425\,\mu\text{m}$$

From Eq. (8.67)

$$\Theta = 0.0973\delta \quad \text{giving} \quad \delta = \frac{0.424}{0.0973} = 4.36\,\text{mm}$$

This illustrates yet again just how thin the boundary layer is in aeronautical applications.

We can now calculate the wall unit and thus the other dimensions of interest. The wall unit is

$$\ell^+ \equiv \frac{\nu}{V_*} = \frac{15 \times 10^{-6}}{4.22} = 3.5\,\mu\text{m}$$

The viscous sublayer thickness is

$$5\ell^+ \simeq 17.5\,\mu\text{m}$$

The average spanwise spacing of streaks is

$$100\ell^+ \simeq 350\,\mu\text{m}$$

The bursting frequency is

$$0.004\frac{V_*^2}{\nu} = 0.004 \times \frac{4.22}{3.5 \times 10^{-6}} \simeq 4.8\,\text{kHz}$$

This suggests that the spanwise dimensions of the MEMS actuators should not exceed about $100\,\mu\text{m}$—about 30%—of average spanwise streak spacing. They also need to effect control at frequencies of at least ten times the bursting frequency, say 50 kHz.

8.10 COMPUTATIONAL METHODS

In this section we deal with methods of computational fluid dynamics (CFD). We concentrate on methods to solve boundary-layer equations. We first examine methods based on the momentum-integral equation. Subsequently, we deal with finite-difference methods and models required to solve by CFD turbulent boundary layers.

8.10.1 Methods Based on the Momentum-Integral Equation

In the general case with an external pressure gradient, the momentum-integral equation must be solved numerically. There are a number of ways this can be done. One, for laminar boundary layers, is the approximate expressions (8.48a′, b′, c′), with Eq. (8.43) or Eq. (8.82) rewritten as

$$\frac{d\theta}{dx} = \frac{C_f}{2} - \frac{V_s}{U_e} - \frac{\theta}{U_e}\frac{dU_e}{dx}(H+2) \tag{8.110}$$

$d\theta/dx$ can be related to $d\delta/dx$:

$$\frac{d\theta}{dx} = \frac{d(\delta I)}{dx} = I\frac{d\delta}{dssx} + \delta\frac{dI}{dx} = I\frac{d\delta}{dx} + \delta\frac{dI}{d\Lambda}\frac{d\Lambda}{d\delta}\frac{d\delta}{dx}$$

$$= \left(I + \delta\frac{dI}{d\Lambda}\frac{d\Lambda}{d\delta}\right)\frac{d\delta}{dx}$$

It follows from Eqs. (8.46d) and (8.48b′), respectively, that

$$\delta\frac{d\Lambda}{d\delta} = \frac{2\delta^2}{v}\frac{dU_e}{dx} = 2\Lambda$$

$$\frac{dI}{d\Lambda} = -\frac{1}{63}\left(\frac{1}{15} + \frac{\Lambda}{72}\right)$$

So

$$\frac{d\theta}{dx} = F_1(\Lambda)\frac{d\delta}{dx} \tag{8.111}$$

where

$$F_1(\Lambda) = I - \frac{2\Lambda}{63}\left(\frac{1}{15} + \frac{\Lambda}{72}\right)$$

Thus, Eq. (8.110) can be readily converted into an equation for $d\delta/dx$ by dividing both sides by $F_1(\Lambda)$. The problem is that it follows from Eq. (8.48c′) that

$$C_f \propto \frac{1}{U_e\delta}$$

Thus, in cases where either $\delta = 0$ or $U_e = 0$ at $x = 0$, the initial value of C_f, and therefore the right-hand side of Eq. (8.110) is infinite. These two cases are in fact the two most common in practice. The former corresponds to sharp leading edges; the latter, to blunt ones. To deal with the problem identified, both sides of Eq. (8.110) are multiplied by $2U_e\delta/v$, whereby, using Eq. (8.111), it becomes

$$F_1(\Lambda)\frac{U_e}{v}\frac{d\delta^2}{dx} = F_2(\Lambda) - \frac{2V_s\delta}{v}$$

where Eqs. (8.48a′, b′, c′) give

$$F_2(\Lambda) = 4 + \frac{\Lambda}{3} - \frac{3}{10}\Lambda + \frac{\Lambda^2}{60} + \frac{4}{63}\left(\frac{37}{5} - \frac{\Lambda}{15} - \frac{\Lambda^2}{144}\right)$$

To obtain the final form of Eq. (8.110) for computational purposes, $F_1(\Lambda)\delta^2 dU_e/dx$ is added to both sides, which are then divided by $F_1(\Lambda)$; thus

$$\frac{dZ}{dx} = \frac{F_2(\Lambda)}{F_1(\Lambda)} + \Lambda - \frac{2}{F_1(\Lambda)}\frac{V_s\delta}{\nu} \tag{8.112}$$

where the dependent variable changes from δ to $Z = \delta^2 U_e/\nu$. In the usual case when $V_s = 0$, the right-hand side of Eq. (8.112) is purely a function of Λ. Note that

$$\Lambda = \frac{Z}{U_e}\frac{dU_e}{dx} \quad \text{and} \quad \delta = \sqrt{\frac{Z\nu}{U_e}}$$

Since U_e is a prescribed function of x, this allows both Λ and δ to be obtained from a value of Z. The other quantities of interest can be obtained from Eq. (8.48a′, b′, c′).

With the momentum-integral equation in the form of Eq. (8.112) it is suitable for the direct application of standard methods for numerical integration of ordinary differential equations [77]. We recommend that the fourth-order Runge-Kutta method be used with an adaptive stepsize control; the advantage of this is that small steps are chosen in regions of rapid change (e.g., near the leading edge), while larger steps are taken elsewhere.

To begin the calculation, it is necessary to supply initial values for Z and Λ at, say, $x = 0$. For a sharp leading edge, $\delta = 0$, giving $Z = \Lambda = 0$ at $x = 0$, whereas for a round leading edge, $U_e = 0$, and $Z = 0$, but $\Lambda = 7.052$ (see Example 8.8). This value of Λ should be used to evaluate the right-hand side of Eq. (8.112) at $x = 0$.

Boundary-layer computations using the method just described were carried out for a circular cylinder of radius 1 m in air flowing at 20 m/s ($\nu = 1.5 \times 10^{-5}$ m²/s). In Fig. 8.40, the computed values of Λ and momentum thickness are plotted against the angle around the cylinder's surface measured from the fore stagnation point. A fourth-order Runge-Kutta integration scheme was used with 200 fixed steps between $x = 0$ and $x = 3$, which gave acceptable accuracy. According to this approximate calculation, the separation point, corresponding to $\Lambda = -12$, occurs at 106.7 degrees. This should be compared with the accurately computed value of 104.5 degrees obtained with the differential form of the equations of motion. It can be seen that the Pohlhausen [78] method gives reasonably acceptable results when compared with more accurate methods. In point of fact, neither value given for the separation point is close to the actual value found experimentally for a laminar boundary layer. Experimentally, separation occurs just ahead of the apex of the cylinder. The reason for the large discrepancy between theoretical and observed separation points is that

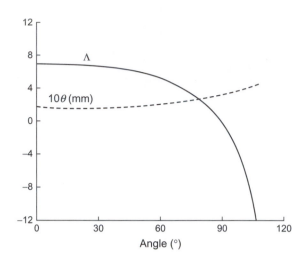

FIGURE 8.40

Computed values of Λ and momentum thickness plotted against an angle around the cylinder's surface measured from the fore stagnation point.

the large wake substantially alters the flow outside the boundary layer. This main-stream flow, accordingly, departs markedly from the potential-flow solution assumed for the boundary-layer calculations. Boundary-layer theory predicts the separation point accurately only in the case of streamlined bodies with relatively small wakes. Nevertheless, the circular cylinder is a good test case for checking the accuracy of boundary-layer computations.

Numerical solutions of the momentum-integral equation can also be found via Thwaites's method [79], which does not use the Pohlhausen approximate velocity profile Eq. (8.47) or Eq. (8.48a). It is very simple to use and, for some applications, is more accurate than the Pohlhausen method. A suitable FORTRAN program for Thwaites's method is given by Cebeci and Bradshaw (1977) [80].

Computational methods based on the momentum-integral equation are available for the turbulent boundary layer. In this case, one or more semi-empirical relationships, in addition to the momentum-integral equation, are required. For example, most methods make use of the formula for C_f due to Ludwieg and Tillmann (1949) [81]:

$$C_f = 0.246 \times 10^{-0.678H} \left(\frac{U_e \theta}{\nu} \right)^{-0.268}$$

A good method of this type, due to Head (1958) [82], is relatively simple to use but performs better than many more complex methods based on the differential equations of motion. A FORTRAN program based on Head's method is also given by Cebeci and Bradshaw.

To begin computation of the turbulent boundary layer by Head's method, it is necessary to locate the transition point and supply initial values of θ and $H \equiv \delta^*/\theta$. In Section 8.6.7, we saw that, for the boundary layer on a flat plate, Eq. (8.73) holds at the transition point—that is, there is no discontinuous change in momentum thickness. This applies equally well to the more general case. Once the transition point is located, then the starting value for θ in the turbulent boundary layer is given by the final value in the laminar part. However, since transition is assumed to occur instantaneously at a specific location along the surface, there is a discontinuous change in velocity profile shape at the transition point, which implies a discontinuous change in the shape factor H. To a reasonable approximation,

$$H_{Tt} = H_{Lt} - \Delta H$$

where

$$\Delta H = 0.821 + 0.114 \log_{10}(Re_{\theta t}) \quad Re_{\theta t} < 5 \times 10^4$$

$$\Delta H = 1.357 \quad Re_{\theta t} > 5 \times 10^4$$

8.10.2 Transition Prediction

The usual methods of determining the transition point are based on the so-called e^n method developed by A.M.O. Smith and H. Gamberoni (1956) at Douglas Aircraft Co. [83]. These methods are complex and involve specialized computational techniques. Fortunately, Smith and his colleagues devised a simple but highly satisfactory alternative [84], having found, on the basis of predictions using the e^n method, that the transition Reynolds number Re_{xt} (where $Re_x = U_e x/\nu$) is related to the shape factor H by the following semi-empirical formula:

$$\log_{10}(Re_{xt}) = -40.4557 + 64.8066H - 26.7538H^2 + 3.3819H^3 \quad 2.1 < H < 2.8$$

$$(8.113)$$

In this method, $\log_{10}(Re_x)$ is plotted against x (the distance along the surface from the forward stagnation point on the leading edge). A laminar boundary layer is calculated, and the right-hand side of Eq. (8.113) is plotted against x. Initially, the former curve lies below the latter; the transition point is located at the value of x (or Re_x) where the two curves *first* cross. This is illustrated in Fig. 8.41, where the left-hand side (LHS) and the right-hand side (RHS) of Eq. (8.113), calculated for the circular cylinder illustrated in Fig. 8.40, are plotted against Re_x. In this case, the two curves cross at $Re_x \simeq 6.2 \times 10^6$, and this is taken to correspond to the transition point.

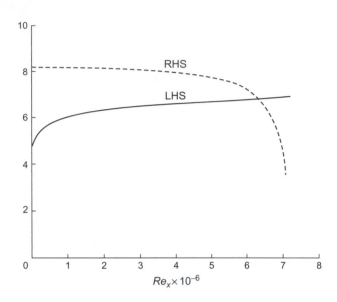

FIGURE 8.41

Computed value of the Pohlhausen parameter and momentum thickness versus angle around the surface of a cylinder as measured from the fore stagnation point.

8.10.3 Computational Solution for the Laminar Boundary-Layer Equations

Nowadays, with the availability of powerful desktop computers, it is perfectly feasible to solve the boundary-layer equations computationally in their original form as partial differential equations. For this reason, methods based on the momentum-integral equation are less widely used. Computational solution of the boundary-layer equations is now a routine matter for industry and university researchers. Nevertheless, specialized numerical techniques are necessary, and difficulties are common. Good treatments of the required techniques and pitfalls to be avoided are to be found in several textbooks [85]. All-purpose commercial software packages for computational fluid dynamics are also widely available. As a general rule, such software does not handle boundary layers well owing to the fine resolution required near the wall [86]. We offer only a brief introduction to the computational methods for boundary layers here. For full details, consult the recommended texts.

It is clear from the examples given that boundary layers in aeronautical applications are typically very thin compared with the streamwise dimensions of a body such as a wing. This in itself poses difficulties for computational solution, which is why Eqs. (8.11) and (8.10) are usually employed in the following nondimensional form:

$$\frac{\partial U}{\partial X} + \frac{\partial V}{\partial Y} = 0 \qquad (8.114)$$

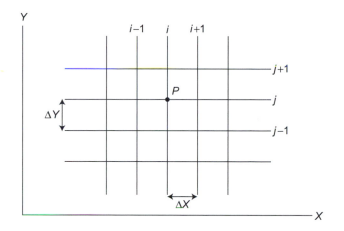

FIGURE 8.42

Illustration of finite-difference method grid.

$$U\frac{\partial U}{\partial X} + V\frac{\partial U}{\partial Y} = -\frac{dP}{dX} + \frac{\partial^2 U}{\partial Y^2} \tag{8.115}$$

where $U = u/U_\infty$, $V = v/(Re_L U_\infty)$, $P = p/(\rho U_\infty^2)$, $X = x/L$, and $Y = y/(LRe_L)$. In many respects, this form was suggested by the method of derivation given in Section 2.12.1. What it achieves is to make the effective range of both independent variables X and Y similar in size (O(1)). Both dependent variables U and V are also O(1). Thus the grid to discretize the equations for computational solution can be taken as rectangular and rectilinear, as depicted in Fig. 8.42.

Mathematically, the boundary-layer equations are *parabolic*. This means that, for computational purposes, is we starts at some initial point $X = X_0$, say, where U is known as a function of Y. For example, the stagnation-point solution (see Section 2.10.3) should be used in the vicinity of the fore stagnation point. The object of the computational scheme is to compute the solution at $X_0 + \Delta X$, where ΔX is a small step around the body surface. We repeat this procedure until the trailing edge or the boundary-layer separation point is reached. For obvious reasons, such a procedure is called a *marching* scheme. The fact that the equations are parabolic allows us to use this scheme, which would not work for the Laplace equation, which is elliptic.

The simplest approach is based on a so-called *explicit* finite-difference scheme, briefly explained next.

We assume that the values of U are known along the line $X = X_i$—that is, the discrete values at the grid points (e.g., P) are known. The objective is a scheme for calculating the values of U at the grid points along $X = X_{i+1}$. To do this we rewrite

Eq. (8.115) in a form for determining $\partial U / \partial X$ at point P in Fig. 8.42. Thus

$$\left(\frac{\partial U}{\partial X}\right)_{i,j} = \frac{1}{U_{i,j}}\left\{-V_{i,j}\left(\frac{\partial U}{\partial Y}\right)_{i,j} - \left(\frac{dP}{dX}\right)_i + \left(\frac{\partial^2 U}{\partial Y^2}\right)_{i,j}\right\} \qquad (8.116)$$

We can then estimate values at $X = X_{i+1}$ by writing

$$U_{i+1,j} = U_{i,j} + \left(\frac{\partial U}{\partial X}\right)_{i,j}\Delta X + O\left((\Delta X)^2\right) \qquad (8.117)$$

The last term on the right-hand side indicates the size of the error involved in this approximation. The pressure gradient term in Eq. (8.116) is a known function of X obtained from either experimental data or the solution for potential flow around the body. But the other terms have to be estimated using finite differences. For estimates of the first and second derivatives with respect to Y, we start with Taylor expansions about point P in the positive and negative Y directions. Thus we obtain

$$U_{i,j+1} = U_{i,j} + \left(\frac{\partial U}{\partial Y}\right)_{i,j}\Delta Y + \left(\frac{\partial^2 U}{\partial Y^2}\right)_{i,j}\frac{(\Delta Y)^2}{2} + O\left((\Delta Y)^3\right) \qquad (8.118a)$$

$$U_{i,j-1} = U_{i,j} - \left(\frac{\partial U}{\partial Y}\right)_{i,j}\Delta Y + \left(\frac{\partial^2 U}{\partial Y^2}\right)_{i,j}\frac{(\Delta Y)^2}{2} - O\left((\Delta Y)^3\right) \qquad (8.118b)$$

First we subtract Eq. (8.118a) from (8.118b) and rearrange:

$$\left(\frac{\partial U}{\partial Y}\right)_{i,j} = \frac{U_{i,j+1} - U_{i,j-1}}{2\Delta Y} + O\left((\Delta Y)^2\right) \qquad (8.119)$$

Then we add Eqs. (8.118a) and (8.118b) and rearrange:

$$\left(\frac{\partial^2 U}{\partial Y^2}\right)_{i,j} = \frac{U_{i,j+1} - 2U_{i,j} + U_{i,j-1}}{(\Delta Y)^2} + O\left((\Delta Y)^2\right) \qquad (8.120)$$

The results in Eqs. (8.119) and (8.120) are usually referred as *centered* finite differences.

Finally, the value of $V_{i,j}$ has to be estimated. We obtain this from Eq. (8.114), which we rewrite as

$$\left(\frac{\partial V}{\partial Y}\right)_{i,j} = -\left(\frac{\partial U}{\partial X}\right)_{i,j}$$

Using a result analogous to Eq. (8.119), we get

$$\frac{V_{i,j+1} - V_{i,j-1}}{2\Delta Y} = -\left(\frac{\partial U}{\partial X}\right)_{i,j} \quad \text{i.e.} \quad V_{i,j+1} = V_{i,j-1} - 2\Delta Y \left(\frac{\partial U}{\partial X}\right)_{i,j}$$

There are two problems with this result. First, it gives $V_{i,j+1}$ rather than $V_{i,j}$; this is easily remedied by replacing j by $j-1$ to obtain

$$V_{i,j} = V_{i,j-2} - 2\Delta Y \left(\frac{\partial U}{\partial X}\right)_{i,j-1} \tag{8.121}$$

Second, it requires a value of $(\partial U / \partial X)_{i,j-1}$. However, the very reason we needed to estimate $V_{i,j}$ was to obtain an estimate for $(\partial U / \partial X)_{i,j}$! Fortunately, this is not a problem if the calculations are done "in the right order."

At the wall, say $j = 0$, $U = V = 0$ (assuming the wall to be impermeable); also, from the continuity Eq. (8.114), $\partial V / \partial Y = 0$. Thus, using an equation analogous to Eq. (8.118a),

$$V_{i,1} = \underbrace{V_{i,0}}_{=0} + \underbrace{\frac{\partial V}{\partial Y}}_{=0} + O\left((\Delta Y)^2\right) \simeq 0$$

With this estimate of $V_{i,1}$, we can now estimate $(\partial U / \partial X)_{i,1}$ from Eq. (8.116), the first term on the right-hand side being equivalently zero. We then estimate $U_{i+1,1}$ from Eq. (8.117). All of the values are now known for $j = 1$. The next step is to set $j = 2$ in Eq. (8.121), so that

$$V_{i,2} = \underbrace{V_{i,0}}_{=0} - 2\Delta Y \left(\frac{\partial U}{\partial X}\right)_{i,1}$$

Next we estimate $(\partial U / \partial X)_{i,2}$ from Eq. (8.116) and so on right across the boundary layer. Of course, it is necessary to choose an upper boundary for the computational domain at a finite, but suitably large, height, $y = Y_J$, corresponding to, say, $j = J$.

This explicit finite-difference scheme is relatively simple. However, explicit schemes are far from ideal for boundary-layer calculations. Their main drawback is that using them leads to numerical instability characterized by increasingly large oscillations in the solution as the calculation marches downstream, leading to unacceptably large errors. To ensure that this does not happen, it is necessary to impose the following condition on the streamwise step length to ensure numerical stability:

$$\Delta X \leq \frac{1}{2} \left| \min(U_{i,j})(\Delta Y)^2 \right| \tag{8.122}$$

Since U is very small at the first grid point near the wall, very small values of ΔX are required.

Because of problems with numerical instability, so-called *implicit* schemes are much preferred because they permit step size to be determined by accuracy rather than numerical stability. Here a scheme based on the Crank-Nicholson method is briefly described. The essential idea is to rewrite Eq. (8.117) in the form

$$U_{i+1,j} = U_{i,j} + \frac{1}{2}\left[\left(\frac{\partial U}{\partial X}\right)_{i,j} + \left(\frac{\partial U}{\partial X}\right)_{i+1,j}\right]\Delta X + O\left((\Delta X)^3\right) \qquad (8.123)$$

so that the derivative at $x = x_i$ is replaced by the average of that and the derivative at $x = x_{i+1}$. Formally, this is more accurate and much more stable numerically. The problem is that Eq. (8.116) implies that

$$\left(\frac{\partial U}{\partial X}\right)_{i+1,j} = \frac{1}{U_{i+1,j}}\left\{-V_{i+1,j}\left(\frac{\partial U}{\partial Y}\right)_{i+1,j} - \left(\frac{dP}{dX}\right)_{i+1} + \left(\frac{\partial^2 U}{\partial Y^2}\right)_{i+1,j}\right\} \qquad (8.124)$$

and Eqs. (8.119) and (8.120) imply that

$$\left(\frac{\partial U}{\partial Y}\right)_{i+1,j} = \frac{U_{i+1,j+1} - U_{i+1,j-1}}{2\Delta Y} + O\left((\Delta Y)^2\right) \qquad (8.125)$$

$$\left(\frac{\partial^2 U}{\partial Y^2}\right)_{i+1,j} = \frac{U_{i+1,j+1} - 2U_{i+1,j} + U_{i+1,j-1}}{(\Delta Y)^2} + O\left((\Delta Y)^2\right) \qquad (8.126)$$

Thus the unknown values of U at X_{i+1}—that is, $U_{i+1,j}(j = 1,\ldots,J)$—appear on both sides of Eq. (8.123), which is what is meant by *implicit*. To solve Eq. (8.123) for these unknowns, it must be rearranged as a matrix equation of the form

$$\mathbf{AU} = \mathbf{R}$$

where $\mathbf{A}$ is the coefficient matrix, $\mathbf{U}$ is a column matrix containing the unknowns $U_{i+1,j}(j = 1,\ldots,J)$, and $\mathbf{R}$ is a column matrix of known quantities. Fortunately, $\mathbf{A}$ has a tridiagonal form: only the main diagonal and the two diagonals on either side of it are nonzero. Tridiagonal matrix equations can be solved very efficiently using the Thomas (or tridiagonal) algorithm, versions of which can be found in Press *et al.* [88] and on the website associated with Ferziger (1998) [89].

One of the most popular and widely used implicit schemes for computational solution of the boundary-layer equations is the *Keller box* [90], which is slightly more accurate than the Crank-Nicholson method.

8.10.4 **Computational Solution for Turbulent Boundary Layers**

The simplest approach is based on the turbulent boundary-layer Eqs. (8.92a, b) written in the form

$$\frac{\partial \bar{u}}{\partial x} + \frac{\partial \bar{v}}{\partial y} = 0 \tag{8.127}$$

$$\bar{u}\frac{\partial \bar{u}}{\partial x} + \bar{v}\frac{\partial \bar{u}}{\partial y} = -\frac{d\bar{p}}{dx} + \frac{\partial}{\partial y}\left(\mu\frac{\partial \bar{u}}{\partial y} - \rho\overline{u'v'}\right) \tag{8.128}$$

We do not attempt to write these equations in terms of nondimensional variables, as we did for Eqs. (8.114) and (8.115). A similar procedure would be advantageous for computational solutions, but is not necessary for the account given here.

The primary problem with solving Eqs. (8.12) and (8.11) is not computational. Rather, it is that there are only two equations but three dependent variables to determine by calculation: $\bar{u}, \bar{v}$, and $\overline{u'v'}$. The "solution" described in Section 8.9 is an eddy-viscosity model (see Eq. 8.93) whereby

$$-\rho\overline{u'v'} = \rho\varepsilon_T\frac{\partial \bar{u}}{\partial y} \tag{8.129}$$

To solve Eqs. (8.127) and (8.128), ε_T must be expressed in terms of $\bar{u}$ (and, possibly, v), which can only be done semi-empirically. In Section 8.10.5, we will be explain how a suitable semi-empirical model for eddy viscosity can be developed for computational calculation of turbulent boundary layers. First, we introduce the wider aspects of this so-called *turbulence modeling* approach.

Since 1950, when engineers first began using computers, the goal has been to develop more effective methods for computational calculation of turbulent flows. For the past two decades, it has even been possible to carry out *direct numerical simulations* (DNS) of the full unsteady, three-dimensional Navier-Stokes equations for relatively simple turbulent flows at comparatively low Reynolds numbers [92]. Despite the enormous advances in computing power, however, it is unlikely that DNS will be feasible, or even possible, for most engineering applications within the foreseeable future [93].

All alternative computational methods rely on semi-empirical approaches known collectively as *turbulent modeling*. These modern methods are based on deriving additional transport equations from the Navier-Stokes equations for quantities such as various components of the Reynolds stress tensor (see Eq. 8.91), turbulence kinetic energy, and viscous dissipation rate. In a sense, these approaches are based on an unattainable goal because each new equation contains ever more unknown quantities so that the number of dependent variables always grows faster than the number of equations. As a consequence, an increasing number of semi-empirical formulae are required. Nevertheless, despite their evident drawbacks, computational methods based on turbulence modeling have become an indispensable tool in modern engineering. We give a brief account of one of the most widely used in Section 8.10.6.

An alternative approach to this type of turbulence modeling becoming a viable computational tool for engineering applications is *large-eddy simulation* (LES), which was first developed by meteorologists. LES still relies on semi-empirical turbulence modeling; see Section 8.10.7 for a brief discussion on LES.

8.10.5 Zero-Equation Methods

Computational methods based on Eqs. (8.127) and (8.128), plus semi-empirical formulas for eddy viscosity, are often referred to as *zero-equation* methods. This name reflects the fact that no additional partial differential equations derived from the Navier-Stokes equations are used. Here we describe the Cebeci–Smith method [94], one of the most successful zero-equation methods developed in the 1970s.

Most of the zero-equation models are based on extensions of Prandtl's mixing-length concept (see Sections 8.9.4 and 8.9.5)

$$-\overline{u'v'} = \varepsilon_T \frac{\partial \overline{u}}{\partial y} \quad \varepsilon_T = \kappa y^2 \left| \frac{\partial \overline{u}}{\partial y} \right| \quad (\text{i.e. } \ell_m = \kappa y)$$

The constant κ is known as the von Kármán constant.

Three key modifications to the models were introduced in the mid-1950s:

Damping near the wall (due to Van Driest [95]): An exponential damping function comes into play as $y \to 0$. This reflects the reduction in turbulence level as the wall is approached and extends the mixing-length model into the buffer layer and viscous sublayer:

$$\ell_m = \kappa y[1 - \exp(-y^+ / A_0^+)] \quad A_0^+ = 26$$

Outer wake-like flow (due to Clauser [96]): We saw that the outer part of a boundary layer is like a free shear layer (specifically, like a wake flow), so there the Prandtl-Görtler eddy-viscosity model (Eqs. (8.94) and (8.95)), is more appropriate:

$$\varepsilon = \underbrace{\alpha}_{\text{const.}} \times U_e \delta^*$$

where U_e is the flow speed at the edge of the boundary layer and δ^* is the boundary-layer displacement thickness.

Intermittency (due to Corrsin and Kistler, and Klebannoff [97]): The outer part of the boundary layer is only intermittently turbulent (see Section 8.9.7 and Fig. 8.37). To allow for this, ε_T is multiplied by the following semi-empirical intermittency factor:

$$\gamma_{\text{tr}} = \left[1 + \left(\frac{y}{\delta} \right)^6 \right]^{-1}$$

Cebeci-Smith Method

The Cebeci-Smith method incorporates versions of the three key model modifications. For the *inner region* of the turbulent boundary layer,

$$(\varepsilon_T)_i = \ell^2 \left| \frac{\partial \bar{u}}{\partial y} \right| \gamma_{tr} \quad 0 \leq y \leq y_c \tag{8.130}$$

where the mixing length

$$\ell = \kappa y \underbrace{\left[1 - \exp\left(-\frac{y}{A} \right) \right]}_{i} \tag{8.131}$$

Term (i) is a semi-empirical modification of Van Driest's damping model that takes into account the effects of the streamwise pressure gradient. $\kappa = 0.4$, and the damping length is

$$A = \frac{26\nu}{V_* \sqrt{1 - 11.8(\nu U_e / V_*^3) dU_e/dx}} \tag{8.132}$$

For the *outer region* of the turbulent boundary layer,

$$(\varepsilon_T)_o = \alpha U_e \delta * \gamma_{tr} \quad y_c \leq y \leq \delta \tag{8.133}$$

where $\alpha = 0.0168$ when $Re_\theta \geq 5000$. y_c is determined by requiring

$$(\varepsilon_T)_i = (\varepsilon_T)_o \quad \text{at} \quad y = y_c \tag{8.134}$$

See Cebeci and Bradshaw [80] for further details of the Cebeci–Smith method, which does a reasonably good job calculating conventional turbulent boundary-layer flows. For applications involving separated flows, it is less successful, and one-equation methods like that due to Baldwin and Lomax [98] are preferred. For details on the Baldwin-Lomax and other one-equation methods, including computer codes, Wilcox (1993) and other specialist texts should be consulted.

8.10.6 $k - \varepsilon$: A Typical Two-Equation Method

Probably the most widely used method for calculating turbulent flows is the $k - \varepsilon$ model, which is incorporated into most commercial CFD software. It was independently developed at Los Alamos [99] and at Imperial College London [100].

The basis of the $k - \varepsilon$ and most other two-equation models is an eddy-viscosity formula based on dimensional reasoning and taking the form

$$\varepsilon_T = C_\mu k^{1/2} \ell \quad (C_\mu \text{ is an empirical const.}) \tag{8.135}$$

Note that the kinetic energy per unit mass $k \equiv (\overline{u'^2} + \overline{v'^2} + \overline{w'^2})/2$. Some previous two-equation models derived a transport equation for the length scale ℓ. This seemed rather unphysical so, based on dimensional reasoning, the $k - \varepsilon$ model took

$$\ell = k^{3/2}/\varepsilon \qquad (8.136)$$

where ε is the viscous dissipation rate per unit mass and should not be confused with eddy viscosity ε_T. A transport equation for ε was then derived from the Navier-Stokes equations.

The equations for both ε and turbulence kinetic energy k contain terms involving additional unknown dependent variables. These terms must be modeled semi-empirically. For flows at high Reynolds numbers, the transport equations for k and ε are modeled as follows.

For turbulence energy,

$$\rho \frac{Dk}{Dt} = \frac{\partial}{\partial y}\left(\frac{\rho \varepsilon_T}{\sigma_k}\frac{\partial k}{\partial y}\right) + \rho \varepsilon_T \left(\frac{\partial \overline{u}}{\partial y}\right)^2 - \rho \varepsilon \qquad (8.137)$$

and for energy dissipation,

$$\rho \frac{D\varepsilon}{Dt} = \frac{\partial}{\partial y}\left(\frac{\rho \varepsilon_T}{\sigma_\varepsilon}\frac{\partial \varepsilon}{\partial y}\right) + C_1 \frac{\varepsilon}{k}\rho \varepsilon_T \left(\frac{\partial \overline{u}}{\partial y}\right)^2 - C_2 \frac{\rho \varepsilon^2}{k} \qquad (8.138)$$

These equations contain five empirical constants that are usually assigned the following values:

C_μ	C_1	C_2	σ_k	σ_ε
0.09	1.55	2.0	1.0	1.3

where σ_k and σ_ε are often termed *effective turbulence Prandtl numbers*. Further modification of Eqs. (8.137) and (8.138) is required to deal with relatively low Reynolds numbers. (See Wilcox [87] for details of this and the choice of wall boundary conditions.)

The $k - \varepsilon$ model is intended for computational calculations of general turbulent flows. It is questionable whether it performs any better than, or even as well as, the zero-equation models described in Section 8.10.5 for boundary layers. However, it can be used for more complex flows, although the results should be viewed with caution. A common misconception among practicing engineers who use commercial CFD packages containing the $k - \varepsilon$ model is that they solve the exact Navier-Stokes equations. In fact, they solve a system of equations that contains several approximate semi-empirical formulae, including the eddy-viscosity model described previously. Real turbulent flows are highly unsteady and three-dimensional. The best we can expect when using $k - \varepsilon$ or any other similar turbulence model is an approximate result that gives guidance to some of the features of the real turbulent flow. At worst, the results can be completely misleading. For an example, see the discussion in Wilcox [87] of the round-jet/plane-jet anomaly.

For a full description and discussion of two-equation and other more advanced turbulence models, see Wilcox [87], Pope [91], and other specialized books.

8.10.7 Large-Eddy Simulation

Large-eddy simulation to calculate turbulent flow originated with meteorologists. In a sense, LES is halfway between the turbulence modeling based on Reynolds averaging and direct numerical simulations. It is motivated by the view that larger-scale motions are likely to vary profoundly between one type of turbulent flow and another, but that small-scale turbulence is likely to be much more universal. Accordingly, any semi-empirical modeling should be confined to small-scale turbulence. With this in mind, we partition the flow variables into

$$\{u,v,w,\ p\} = \underbrace{\{\tilde{u},\tilde{v},\tilde{w},\ \tilde{p}\}}_{\text{Resolved-field}} + \underbrace{\{u',v',w',\ p'\}}_{\text{Subgrid-scale field}} \tag{8.139}$$

The resolved or large-scale field is computed directly while the subgrid-scale field is modeled semi-empirically.

The resolved field is obtained by applying a *filter* to the flow variables—for example,

$$\tilde{u}(\vec{x}) = \int \underbrace{G(\vec{x}-\vec{\xi})}_{\text{Filter function}} u(\vec{\xi})d\vec{\xi} \tag{8.140}$$

If the filter function is chosen appropriately, this has the effect of "averaging" over the subgrid scales.

Two Common Choices of Filter Function
In the box filter [101],

$$G(\vec{x}-\vec{\xi}) = \begin{cases} 1 & |\vec{x}-\vec{\xi}| < \Delta/2 \\ 0 & \text{Otherwise} \end{cases} \tag{8.141}$$

In the Gaussian filter [102],

$$G(\vec{x}-\vec{\xi}) = \left[\left(\frac{6}{\pi}\right)^{1/2}\frac{1}{\Delta_a}\right]\exp\left[-6(\vec{x}-\vec{\xi})^2/\Delta_a^2\right] \tag{8.142}$$

The choice of size for Δ or Δ_a in Eqs. (8.141) or (8.142) determines the subgrid scale. Filtering the Navier-Stokes equations gives

$$\frac{\partial \tilde{u}_i}{\partial t} + \frac{\partial}{\partial x_1} \widetilde{u_i u_1} + \frac{\partial}{\partial x_2} \widetilde{u_i u_2} + \frac{\partial}{\partial x_3} \widetilde{u_i u_3} = -\frac{1}{\rho} \frac{\partial \tilde{p}}{\partial x_i} + \nu \nabla^2 \tilde{u}_i \quad i = 1, 2, 3 \qquad (8.143)$$

$$\frac{\partial \tilde{u}_1}{\partial x_1} + \frac{\partial \tilde{u}_2}{\partial x_2} + \frac{\partial \tilde{u}_3}{\partial x_3} = 0 \qquad (8.144)$$

where u_1, u_2, and u_3 denote u, v, and w; x_1, x_2, and x_3 denote x, y and z; and

$$\widetilde{u_i u_j} = \widetilde{\tilde{u}_i \tilde{u}_j} + \widetilde{u_i' u_j} + \widetilde{u_j' u_i} + + \widetilde{u_i' u_j'} \qquad (8.145)$$

Usually the following approximation suggested by Leonard [102] is made:

$$\widetilde{u_i u_j} \simeq \tilde{u}_i \tilde{u}_j + \underbrace{\frac{\Delta^2}{12} \frac{\partial \tilde{u}_i}{\partial u_\ell} \frac{\partial \tilde{u}_j}{\partial u_\ell}}_{\text{Leonard stress}} + \underbrace{\widetilde{u_i' u_j'}}_{\text{Modeled semi-empirically}} \qquad (8.146)$$

Subgrid Scale Modeling

A common approach, originating with Smagorinsky [103], is to use an eddy viscosity so that

$$\widetilde{u_i' u_j'} \simeq -\varepsilon_T \frac{1}{2} \left(\frac{\partial \tilde{u}_i}{\partial x_j} + \frac{\partial \tilde{u}_j}{\partial x_i} \right) \qquad (8.147)$$

A common way of modeling ε_T is also due to Smagorinsky [103]:

$$\varepsilon_T = \frac{1}{2} (c\Delta)^2 \left[\sum_{i=1}^{3} \left(\frac{\partial \tilde{u}_i}{\partial x_j} + \frac{\partial \tilde{u}_j}{\partial x_i} \right) \left(\frac{\partial \tilde{u}_i}{\partial x_j} + \frac{\partial \tilde{u}_j}{\partial x_i} \right) \right]^{1/2} \qquad (8.148)$$

where c is a semi-empirically determined constant.

For more information on LES, see Wilcox [87] and Pope [91]. LES is very demanding in terms of computational resources but with rapid increases in computing power, it is becoming more and more feasible for engineering calculations. A less demanding alternative is conventional turbulence modeling based on Reynolds averaging but including the time derivatives of the mean velocity components in the Reynolds-averaged Navier-Stokes equations. This approach is sometimes known as very-large-eddy simulation (VLES). See Tucker [104] for a specialized treatment.

8.11 ESTIMATION OF PROFILE DRAG FROM THE VELOCITY PROFILE IN A WAKE

At the trailing edge of a body immersed in a fluid flow are the boundary layers from the surfaces on either side. These boundary layers join up and move downstream in the form of a wake of retarded velocity. The velocity profile changes with distance downstream, the wake cross-section increasing in size as the magnitude of its mean velocity defect, relative to the free stream, decreases. At a sufficient distance downstream, the streamlines are all parallel and the static pressure across the wake is constant and equal to the free-stream value. If we compare conditions at this station with those in the undisturbed stream ahead of the body, the rate at which momentum has been lost while passing the body equates to the drag force on the body. The drag force so obtained includes both skin-friction and form-drag components, since these together produce the overall momentum change. A method of calculating the drag of a two-dimensional body using the momentum loss in the wake is given next. This method depends on conditions remaining steady with time.

8.11.1 Momentum-Integral Expression for the Drag of a Two-Dimensional Body

Consider a two-dimensional control volume fixed in space (see Fig. 8.43) of unit width, with two faces (planes 0 and 2) perpendicular to the free stream, far ahead of and far behind the body, respectively, the other two lying parallel to the undisturbed flow direction and situated, respectively, far above and far below the body.

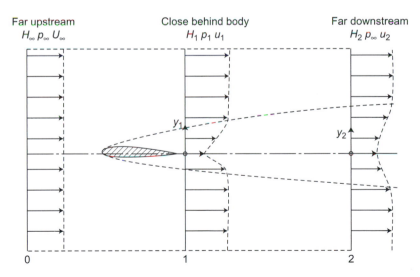

FIGURE 8.43

Control volume for drag analysis.

For any stream tube (of vertical height δy) in the wake at the downstream boundary, the mass flow per unit time is $\rho u_2 \delta y_2$ and the velocity reduction between upstream and downstream is $U_\infty - u_2$. The loss of momentum per unit time in the stream tube $= \rho u_2 (U_\infty - u_2)\delta y_2$ and, for the whole field of flow:

$$\text{Total loss of momentum per unit time} = \int_{-\infty}^{\infty} \rho u_2 (U_\infty - u_2) dy_2$$

In fact, the limits of this integration need only extend across the wake because the term $U_\infty - u_2$ becomes zero outside.

The rate of loss of momentum in the wake is brought about by the reaction on the fluid of the profile-drag force per unit span D acting on the body. Thus

$$D = \int_{w} \rho u_2 (U_\infty - u_2) dy_2 \tag{8.149}$$

This expression enables the drag to be calculated from an experiment arranged to determine the velocity profile at some considerable distance downstream of the body (i.e., where $p = p_\infty$).

For practical use, it is often inconvenient, or impossible, to arrange for measurement so far away from the body, and methods that allow measurements to be made close behind it (plane 1 in Fig. 8.43) have been developed by Betz [105] and B. M. Jones [106]. The latter's method is considerably simpler and is reasonably accurate for most purposes.

8.11.2 Jones's Wake Traverse Method for Determining Profile Drag

In the wake close behind a body static pressure, as well as velocity, varies from the value in the free stream outside the wake. B. Melville Jones allowed for this fact by assuming that, in any given stream tube between plane 1 (close to the body) and plane 2 (far downstream), the stagnation pressure can be considered constant. This is very nearly the case in practice, even in turbulent wakes.

Let H_∞ be the stagnation pressure in any stream tube at plane 0, and let $H_1 = H_2$ be its value in the same stream tube at planes 1 and 2. Then

$$H_\infty = p_\infty + \frac{1}{2}\rho U_\infty^2$$

$$H_1 = p_1 + \frac{1}{2}\rho u_1^2 = p_\infty + \frac{1}{2}\rho u_2^2$$

The velocities are given by

$$U_\infty = \sqrt{\frac{2}{\rho}(H_\infty - p_\infty)}, \quad u_1 = \sqrt{\frac{2}{\rho}(H_1 - p_1)} \quad \text{and} \quad u_2 = \sqrt{\frac{2}{\rho}(H_1 - p_\infty)}$$

Substituting the values for U_∞ and u_2 into Eq. (8.149) gives

$$D = 2 \int_{w_2} \sqrt{H_1 - p_\infty}\left(\sqrt{H_\infty - p_\infty} - \sqrt{H_1 - p_\infty}\right) dy_2 \qquad (8.150)$$

To refer this to plane 1, the equation of continuity in the stream tube must be used:

$$u_1\,\delta y_1 = u_2\,\delta y_2$$

or

$$\delta y_2 = \frac{u_1}{u_2}\delta y_1 \sqrt{\frac{H_1 - p_1}{H_1 - p_\infty}}\delta y_1 \qquad (8.151)$$

Referring to the wake at plane 1, Eq. (8.150) then becomes

$$D = 2 \int_{w_1} \sqrt{H_1 - p_1}\left(\sqrt{H_\infty - p_\infty} - \sqrt{H_1 - p_\infty}\right) dy_1 \qquad (8.152)$$

To express Eq. (8.152) nondimensionally, the profile-drag coefficient C_{Dp} is used. For unit span,

$$C_{Dp} = \frac{D}{\frac{1}{2}\rho U_\infty^2 c} = \frac{D}{c(H_\infty - p_\infty)}$$

so Eq. (8.152) becomes

$$C_{Dp} = 2 \int_{w_1} \sqrt{\frac{H_1 - p_1}{H_\infty - p_\infty}}\left[1 - \sqrt{\frac{H_1 - p_\infty}{H_\infty - p_\infty}}\right] d\left(\frac{y_1}{c}\right) \qquad (8.153)$$

Notice again that this integral needs to be evaluated only across the wake because, beyond the wake boundary, the stagnation pressure H_1 becomes equal to H_∞. Thus the second term in the bracket becomes unity and the integrand becomes zero.

Equation (8.153) may be conveniently used in experimental determination of the profile drag of a two-dimensional body when it is inconvenient, or impracticable, to use a wind-tunnel balance to obtain direct measurement. In fact, it can be used to determine the drag of aircraft in free flight. All that is required is a traversing mechanism for a Pitot-static tube to enable the stagnation and static pressures H_1 and p_1 to be recorded at a series of positions across the wake, ensuring that measurements are taken as far as the undisturbed stream on either side. Preferably an additional measurement is taken of the dynamic pressure, $H_\infty - p_\infty$, in the incoming stream ahead of the body. In the absence of the latter, it can be assumed, with reasonable accuracy, that $H_\infty - p_\infty$ is the same as the value of $H_1 - p_1$ outside the wake.

Using the recordings obtained from the traverse values of $H_1 - p_1$ and $H_1 - p_\infty$ we evaluate for a series of values of y_1/c across the wake; hence a corresponding series of values of quantity is

$$\sqrt{\frac{H_1 - p_1}{H_\infty - p_\infty}} \left(1 - \sqrt{\frac{H_1 - p_\infty}{H_\infty - p_\infty}} \right)$$

By plotting a curve of this function against the variable y_1/c, we obtain a closed area will be obtained (because the integral becomes zero at each edge of the wake). The magnitude of this area is the value of the integral, so the coefficient C_{D_P} is given directly by twice the area under the curve.

To facilitate the actual experimental procedure, it is often convenient to construct a comb or rake of Pitot and static tubes, set up at suitable spacings. The comb is positioned across the wake (it must be wide enough to read into the free stream on either side) and the Pitot and static readings are recorded.

The method can be extended to measure the drag of three-dimensional bodies by making a series of traverses at suitable lateral (or spanwise) displacements. Each individual traverse gives the drag force per unit span, so the summation of these in a spanwise direction gives the total three-dimensional drag.

8.11.3 Growth Rate of a Two-Dimensional Wake Using the General Momentum-Integral Equation

As explained, the two boundary layers at the trailing edge of a body join up and form a wake of retarded flow. The velocity profile across this wake varies appreciably with distance behind the trailing edge. Simple calculations can be made that relate the rate of growth of the wake thickness to distance downstream, provided that the wake profile shape and external mainstream conditions can be specified.

The momentum-integral equation for steady incompressible flow, Eq. (8.82), may be reduced to

$$\frac{d\theta}{dx} = \frac{C_f}{2} - \frac{\theta}{U_e} \frac{dU_e}{dx} (2 + H) \tag{8.154}$$

Now C_f is the local surface shear-stress coefficient at the base of the boundary layer, and at the wake center, where the two boundary layers join, there is no relative velocity and therefore no shearing traction. For each half of the wake, then, C_f is zero and Eq. (8.154) becomes

$$\frac{d\theta}{dx} = -\frac{\theta}{U_e}\frac{dU_e}{dx}(2+H) \tag{8.155}$$

It is clear from this that, if the mainstream velocity outside the wake is constant, $dU_e/dx = 0$ and the right-hand side becomes zero—that is, the momentum thickness of the wake is constant. We expect this from the direct physical argument that there are no overall shearing tractions at the wake edges under these conditions, so the total wake momentum remains unaltered with distance downstream. θ may represent the momentum thickness for each half of the wake, considered separately if it is unsymmetrical, or of the whole wake, if it is symmetrical.

The general thickness δ of the wake is then obtainable from the relationship

$$\theta = \delta \int_0^1 \bar{u}(1-\bar{u})d\bar{y} = I\delta$$

so that

$$\frac{\delta_b}{\delta_a} = \frac{\theta_b}{\theta_a}\frac{I_2}{I_b} \tag{8.156}$$

where subscripts a and b refer to two streamwise stations in the wake. Knowledge of the velocity profiles at stations a and b is necessary before the integrals I_a and I_b can be evaluated and used in this equation.

Example 8.11

A two-dimensional symmetrical airfoil model of 0.3-m chord with a roughened surface is immersed at zero incidence in a uniform airstream flowing at 30 m s^{-1} (Fig. E8.11).

The minimum velocity in the wake at a station 2.4 m downstream from the trailing edge is 27 m s^{-1}. Estimate the general thickness of the entire wake at this station. Assume that each boundary layer at the trailing edge has a "seventh root" profile and a thickness corresponding to a turbulent flat-plate growth from a point at 10% chord, and that each half-wake profile at the downstream station may be represented by a cubic curve of the form

$$\bar{u} = a\bar{y}^3 + b\bar{y}^2 + c\bar{y} + d$$

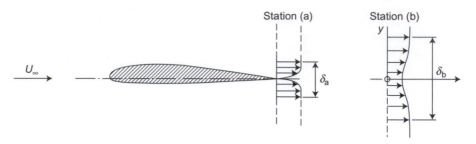

FIGURE E8.11

Illustration of velocity profiles in the wake of an airfoil.

At the trailing edge, where $x = 0.3$ m,

$$\frac{1}{2}\delta_a = \frac{0.383x}{Re_x^{1/5}} \qquad \text{(Eq. (8.65))}$$

$$Re_x = \frac{30 \times 0.3}{14.6} \times 10^6 = 6.16 \times 10^5$$

$$Re_x^{1/5} = 14.39$$

$$\frac{1}{2}\delta_a = \frac{0.383 \times 0.27}{14.39} = 0.007\,19\,\text{m}$$

Also,

$$I_a = 0.0973 \quad \text{Eq. (a), Section 8.6.5}$$

At the wake station,

$$\bar{u} = a\bar{y}^3 + b\bar{y}^2 + c\bar{y} + d, \qquad \frac{\partial \bar{u}}{\partial \bar{y}} = 3a\bar{y}^2 + 2b\bar{y} + c$$

The conditions to be satisfied are that (1) $\bar{u} = 0.9$; (2) $\partial\bar{u}/\partial\bar{y} = 0 \frac{x-\mu}{\sigma}$ when $\bar{y} = 0$; (3) $\bar{u} = 1.0$; and (4) $\partial\bar{u}/\partial\bar{y} = 0$ when $\bar{y} = 1$. (Condition 3 follows because, once the wake is established a short distance behind the trailing edge, the profile discontinuity at the centerline disappears.) Thus

$$d = 0.9, \quad c = 0, \quad 1 = a + b + 0.9, \quad \text{and} \quad 0 = 3a + 2b$$

That is,

$$a = \frac{2}{3}b$$

$$1 = \left(1 - \frac{2}{3}\right)b + 0.09 \quad \text{or} \quad b = 0.3 \quad \text{and} \quad a = -0.2$$

$$\bar{u} = -0.2\bar{y}^3 + 0.3\bar{y}^2 + 0.9$$

$$I_b = \int_0^1 \bar{u}(1-\bar{u})d\bar{y} = \int_0^1 (-0.2\bar{y}^3 + 0.3\bar{y}^2 + 0.9)(1 + 0.2\bar{y}^3 - 0.3\bar{y}^2 - 0.9)d\bar{y} = 0.0463$$

$$\frac{\delta_b}{\delta_a} = \frac{I_a}{I_b} = \frac{0.0973}{0.0463} = 2.1$$

That is,

$$\delta_b = 2.1 \times 2 \times 0.007\,19 = 0.0302\,\text{m}\,(30.2\,\text{mm})$$

8.12 SOME BOUNDARY-LAYER EFFECTS IN SUPERSONIC FLOW

We can now offer a few comments about the qualitative effects on boundary-layer flow of shock waves generated in the mainstream adjacent to the body surface. A normal shock in a supersonic stream invariably reduces the Mach number to a subsonic value, and this speed reduction is associated with a very rapid increase in pressure, density, and temperature.

For an airfoil operating in a transonic regime, the mainstream flow just outside the boundary layer accelerates from subsonic speed near the leading edge to sonic speed at some point near the subsonic-peak-suction position. At this point, the streamlines in the local mainstream are parallel and the effect of the airfoil surface curvature is to cause the streamlines to diverge downstream. Now the characteristics of a supersonic stream are such that this divergence is accompanied by an increase in Mach number with a consequent decrease in pressure. Clearly, this state of affairs cannot be maintained, because the local mainstream flow must become subsonic again at a higher pressure by the time it reaches the undisturbed free-stream conditions downstream of the trailing edge. The only mechanism available for producing the necessary retardation of the flow is a shock wave, which sets itself up approximately normal to the flow in the supersonic region of the mainstream; the streamwise position and intensity of the shock (which vary with distance from the surface) must be such that just the right conditions are established behind it so that the resulting mainstream approaches ambient conditions far downstream.

This simple picture of a near-normal shock requirement is complicated by the presence of the airfoil boundary layer, an appreciable thickness of which must be flowing at subsonic speed regardless of mainstream flow speed. Because of this, the rapid pressure rise at the shock, which cannot propagate upstream in the supersonic regions of flow, can so propagate in the subsonic region of the boundary layer. As a result, the rapid pressure rise associated with the shock becomes diffused near the base of the boundary layer and appears in the form of a progressive pressure rise starting at some appreciable distance upstream of the incident shock.

The length of the upstream diffusion depends on whether the boundary layer is laminar or turbulent. In a laminar boundary layer, the length may be as much as one hundred times nominal general thickness (δ) at the shock; for a turbulent layer, however, it is usually nearer ten times the boundary-layer thickness. This difference can be explained by the fact that, compared with a turbulent boundary layer, a larger part of the laminar boundary-layer flow near the surface is at relatively low speed so that the pressure disturbance can propagate upstream more rapidly and over a greater depth.

It was pointed out in Section 8.3 that an adverse pressure gradient in the boundary layer causes at least a thickening of the layer and may well cause separation. The latter effect is more probable in the laminar boundary layer, and an additional possibility in this type of layer is that transition to turbulence may be provoked. There are thus several possibilities, each of which may affect the external flow in different ways.

8.12.1 Near-Normal Shock Interaction with the Laminar Boundary Layer

There appear to be three general possibilities when a near-normal shock interacts with a laminar boundary layer. With a relatively weak shock, corresponding to an upstream Mach number just greater than unity, the diffused pressure rise may simply cause a gradual thickening of the boundary layer ahead of the shock with no transition and no separation. This thickening causes a family of weak compression waves to develop ahead of the main shock (these are required to produce supersonic mainstream curvature), and the latter is set at an angle between itself and the upstream surface of rather less than 90 degrees (see Fig. 8.44). The compression waves join the main shock at a small distance from the surface, giving the shock a diffused base.

Immediately behind the shock, the boundary layer tends to thin out again, and a local expansion takes place bringing a small region of the mainstream back up to slightly supersonic speed. This is followed by another weak near-normal shock that develops in the same way as the initial shock. The process may be repeated several times before the mainstream flow settles to become entirely subsonic. Generally speaking, this condition is not associated with boundary-layer separation, although there may possibly be a very limited region of separation near the base of the main shock wave.

As the mainstream speed increases so that the supersonic region is at a higher Mach number, the pattern just described tends to change, the first shock becoming much stronger than subsequent shocks; all but one of the latter may not occur at all for local upstream Mach numbers much above 1.3. This is expected because a strong first shock produces a lower Mach number in the mainstream behind it, which means that there is less likelihood of the stream regaining supersonic speed. Concurrently with this pattern change, the rate of thickening of the boundary layer, upstream of the first and major shock, accelerates and the boundary layer at the base of the normal part

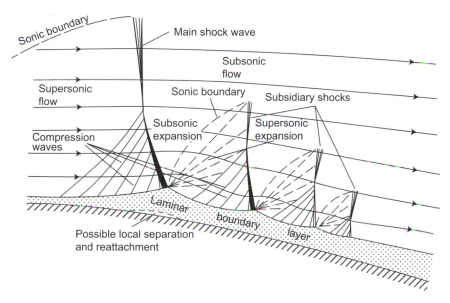

FIGURE 8.44

Illustration of shock and boundary layer interaction along a curved surface.

of the shock generally separates locally before reattaching. There is a high possibility that transition to turbulence will occur behind the single subsidiary shock. This type of flow is indicated in Fig. 8.45.

With still greater local supersonic Mach numbers, the pressure rise at the shock may be sufficient to cause separation of the laminar boundary layer well ahead of the main shock position. This results in a sharp change in direction of the mainstream flow just outside the boundary layer, which is accompanied by a well-defined oblique shock that joins the main shock at some distance from the surface. This configuration is called a lambda shock for obvious reasons. It is unlikely that the boundary layer will reattach under these conditions, and the secondary shock, which normally appears as the result of reattachment or boundary-layer thinning, will not develop. This type of flow illustrated in Fig. 8.46. The sudden separation of the upper-surface boundary layer on an airfoil, as Mach number increases, is usually associated with a sudden decrease in lift coefficient, a phenomenon known as shock stall.

8.12.2 Near-Normal Shock Interaction with the Turbulent Boundary Layer

Because the turbulent boundary layer is far less susceptible than the laminar boundary layer to disturbance by an adverse pressure gradient, separation is not likely to occur for local mainstream Mach numbers, ahead of the shock, of less than about

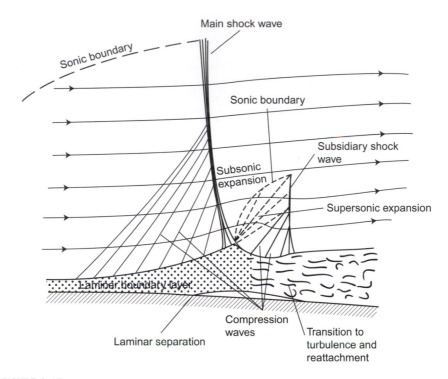

FIGURE 8.45

Illustration of shock and boundary layer interaction.

1.3 (corresponding to a downstream-to-upstream pressure ratio of about 1.8). When no separation occurs, thickening of the boundary layer ahead of the shock is rapid and the compression wavelets near the base of the main shock are very localized so that the base of the shock appears slightly diffused, although no lambda formation is apparent. Behind the shock, no subsequent thinning of the boundary layer appears to occur and the secondary shocks, typical of laminar boundary-layer interaction, do not develop.

If separation of the turbulent layer occurs ahead of the main shock, a lambda shock develops and the mainstream flow looks much like that for a fully separated laminar boundary layer.

8.12.3 Shock-Wave/Boundary-Layer Interaction in Supersonic Flow

One of the main differences between subsonic and supersonic flows, as far as boundary-layer behavior is concerned, is that the pressure gradient along the flow is of opposite sign with respect to cross-sectional area change. Thus, in a converging

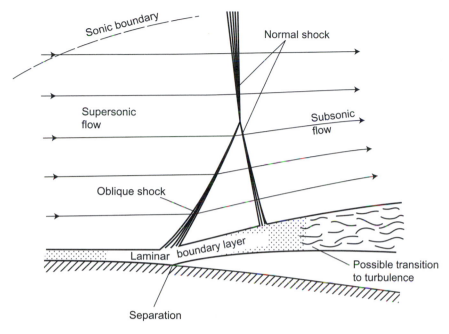

FIGURE 8.46

Illustration of shock and laminar boundary layer interaction.

supersonic flow, the pressure rises; in a diverging flow, it falls in the stream direction. As a result, the pressure gradient at a convex corner is negative and the boundary layer generally negotiates the corner without separating, causing the effect of the boundary layer on the external or mainstream flow to be negligible (Fig. 8.47(a)). Conversely, at a concave corner an oblique shock wave is generated and the corresponding pressure rise causes boundary-layer thickening ahead of the shock; in the case of a laminar boundary layer, it most likely causes local separation at the corner (see Figs. 8.47(b) and 8.48). The resultant curvature of the flow just outside the boundary layer causes a wedge of compression wavelets to develop which, in effect, diffuses the base of the shock wave, as shown in Fig. 8.47(b).

   At the nose of a wedge, the oblique nose shock is affected by growth of the boundary layer; the presence of the rapidly thickening boundary layer near the leading edge produces an effective curvature of the nose of the wedge, and a small region of expansive (Prandtl-Meyer) flow develops locally behind the nose shock, which is now curved and slightly detached from the nose (Fig. 8.49(a)). A similar effect occurs at the leading edge of a flat plate, where a small detached curved local shock develops. This shock rapidly degenerates into a very weak shock approximating a Mach wave at a small distance from the leading edge (Fig. 8.49(b)).

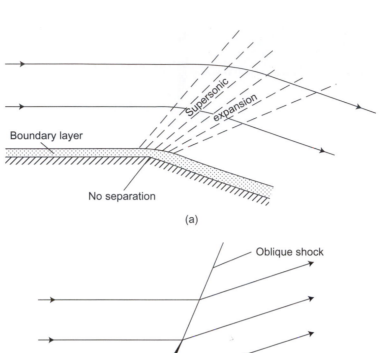

FIGURE 8.47

Illustration of (a) expansion turning and (b) compression turning.

In some cases, an oblique shock generated at some other point in the mainstream may be incident on the surface and boundary layer. Such a shock is at an angle, between the upstream surface and itself, of considerably less than 90 degrees. The general reaction of the boundary layer to this condition is similar to that discussed in the transonic case, except that the oblique shock does not, in general, reduce the mainstream flow to subsonic speed.

If the boundary layer is turbulent, it appears to reflect the shock wave as another shock wave, in much the same way as does the solid surface in the absence of the boundary layer. However, some thickening of the boundary layer occurs. There may also be local separation and reattachment, in which case the reflected shock originates

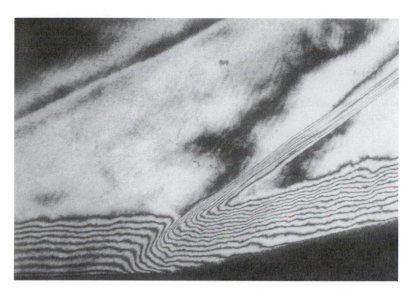

FIGURE 8.48

Supersonic flow through a sharp concave corner. The flow is from left to right at a downstream Mach number of 2.5. The holographic interferogram shows flow turning through an angle of 11 degrees forming an oblique shock wave that interacts with the turbulent boundary layer on the wall. Each fringe corresponds to constant density. The boundary layer transmits the effect of the shock wave a short distance upstream, but there is no flow separation (compare with Fig. 8.47(b)). *Source: Photograph by P.J. Bryanston-Cross, Engineering Department, University of Warwick, United Kingdom.*

just ahead of the point of incidence. A laminar boundary layer thickens gradually up to the point of incidence and may separate locally in this region, rapidly thinning again. The shock then reflects as a fan of expansion waves, followed by a diffused shock a little farther downstream. A set of weak compression waves also set up ahead of the incident shock, owing to the boundary-layer thickening, but these waves do not usually form into a lambda configuration as with a near-normal incident shock.

Approximate representations of the cases just described are shown in Fig. 8.50.

One other condition of interest occurs in a closed uniform duct (two-dimensional or circular) when a supersonic stream is retarded by back pressure set up in the duct. In the absence of boundary layers, the retardation normally occurs through a plane-normal shock across the duct, reducing flow, in one jump, from supersonic to subsonic speed. However, the boundary layer, which thickens ahead of the shock, causes the shock base to thicken or bifurcate (lambda shock), depending on the nature and thickness of the boundary layer in much the same way as for the transonic single-surface case.

Because of considerable boundary-layer thickening, the net flow area is reduced and may reaccelerate the subsonic flow to supersonic, causing the setup of another

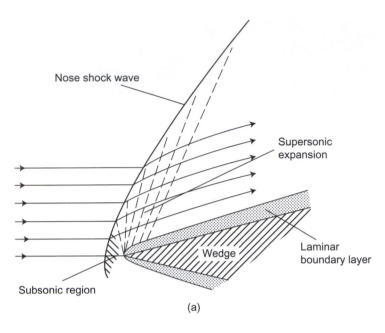

Nose shock wave

Supersonic expansion

Laminar boundary layer

Wedge

Subsonic region

(a)

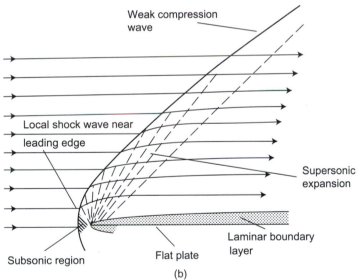

Weak compression wave

Local shock wave near leading edge

Supersonic expansion

Laminar boundary layer

Subsonic region

Flat plate

(b)

FIGURE 8.49

Illustration of bow shock.

normal shock to reestablish the subsonic condition. This situation may be repeated several times until the flow reduces to what is effectively fully developed subsonic boundary-layer flow. If the boundary layers are initially thick, the first shock may

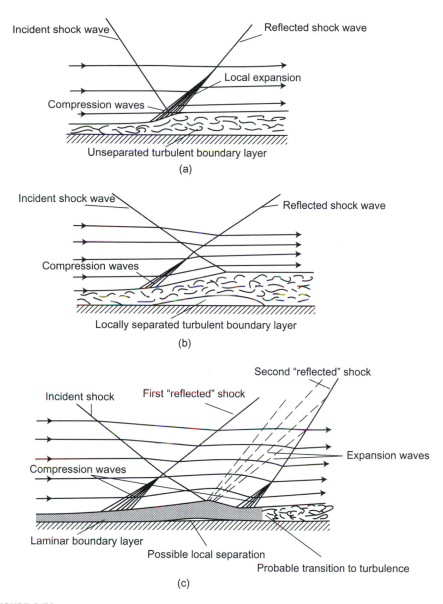

FIGURE 8.50

Illustration of shock boundary-layer interaction phenomena.

show a large degree of bifurcation owing to the large change in flow direction well ahead of the normal part of the shock. In some cases, the extent of the normal shock may reduce almost to zero and a diamond pattern of shocks develops in the duct. Several typical configurations of this sort are depicted in Fig. 8.51.

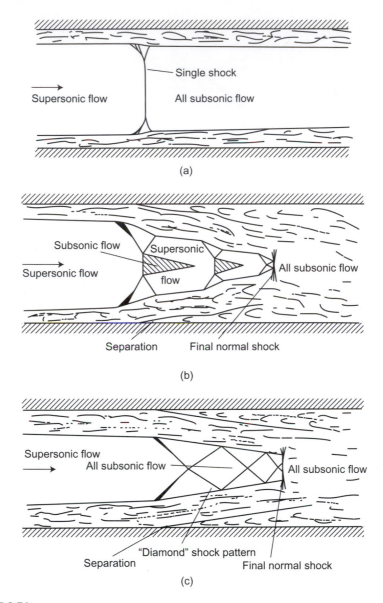

FIGURE 8.51

(a) Low upstream Mach number; thin boundary layer; relatively small pressure rise, no separation. (b) Higher upstream Mach number; thin boundary layer; larger overall pressure rise; separation at first shock. (c) Moderately high upstream Mach number; thick boundary layer; larger overall pressure rise; separation at first shock.

To sum up the last two sections, we can state that, in contrast to the case of most subsonic mainstream flows, interaction is likely to be appreciable between the viscous boundary layer and the effectively inviscid, supersonic mainstream flow. In the subsonic case, unless there is complete separation, the effect of boundary-layer development on the mainstream can usually be neglected. This means that an inviscid mainstream-flow theory can be developed independently of conditions in the boundary layer. The growth of the latter can then be investigated in terms of velocities and streamwise pressure gradients in the previously determined mainstream flow.

In the supersonic (and transonic) case, the very large pressure gradients across an incident shock wave are propagated both up- and downstream in the boundary layer. The rapid thickening and possible local separation that result frequently considerably affect how the shock is reflected by the boundary layer (see Fig. 8.52). In this way,

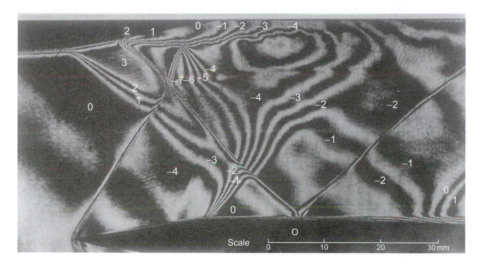

FIGURE 8.52

Complex wave interactions in supersonic flow: The flow is from left to right for this holographic interferogram. Complex interactions occur between shock waves, expansion waves, and boundary layers on the upper and lower walls. An oblique shock wave runs up and to the right from the leading edge of the wedge, interacting with a fan of expansion waves running downward and to the right from a sharp turn in the upper surface.
A subsequent compression turn in the upper surface located at the top of the photograph generates a second shock wave running downward and to the right, which interacts first with the leading-edge shock wave, then with an expansion wave emanating from the lower surface, and finally with the boundary layer on the lower surface. The pressure rise associated with this second shock wave has led to boundary-layer separation on the upper surface close to 2 (see Fig. 8.50). Interferograms can supply quantitative data in the form of density or Mach number values. The Mach numbers corresponding to the numerical labels (given in parentheses) are as follows: 0.92 (−7), 0.98 (−6), 1.05 (−5), 1.13 (−4), 1.21 (−3), 1.28 (−2), 1.36 (−1), 1.44 (0), 1.53 (1), 1.64 (2), 1.75 (3). *Source: Photograph by P.J. Bryanston-Cross, Engineering Department, University of Warwick, United Kingdom.*

the whole character of the mainstream flow may change frequently. It follows from this that a supersonic mainstream flow is much more dependent on Reynolds number than a subsonic flow because of the appreciably different effects of an incident shock on laminar and turbulent boundary layers. The Reynolds number, of course, strongly influences the type of boundary layer that occurs. The theoretical quantitative prediction of supersonic stream behavior in the presence of boundary layers is, consequently, extremely difficult.

8.13 EXERCISES

8.1 A thin plate of length 50 cm is held in uniform water flow such that its length is parallel to the flow direction. The flow speed is 10 m/s; the viscosity, $\mu = 1.0 \times 10^{-3}$ Pa/s; and the water density, 998 kg/m^3.

(a) What is the Reynolds number based on plate length? What can be deduced from its value?

(b) On the assumption that the boundary layer is laminar over the whole surface, use the approximate theory based on the momentum-integral equation to find
- The boundary-layer thickness at the trailing edge of the plate
- The skin-friction drag coefficient

(c) Repeat (b), but now assume that the boundary layer is turbulent over the whole surface. (Use the formulae derived from the one-seventh-power-law velocity profile.)
(*Answer*: $Re_L = 500,000; \delta \simeq 3.3$ mm, $C_{Df} \simeq 0.0018; \delta \simeq 13.4$ mm, $C_{Df} \simeq 0.0052$)

8.2 A thin plate of length 1.0 m is held in a uniform air flow such that its length is parallel to the flow direction. The flow speed is 25 m/s, the viscosity, $\mu = 14.96 \times 10^{-6}$ Pa/s, and the air density, 1.203 kg/m^3.

(a) If it is known that transition from laminar to turbulent flow occurs when the Reynolds number based on x reaches 500,000, find the transition point.

(b) Calculate the equivalent plate length for an all-turbulent boundary layer with the same momentum thickness at the trailing edge as at the actual boundary layer.

(c) Calculate the coefficient of skin-friction drag per unit breadth for that part of the plate with
- A laminar boundary layer
- A turbulent boundary layer
  (Use the formulas based on the one-seventh-power-law velocity profile.)

(d) Calculate the total drag per unit breadth.

(e) Estimate the percentage of drag due to the turbulent boundary layer alone.
(*Answer*: 300 mm from the leading edge; 782 mm; 0.00214, 0.00431; 1.44 N/m per side; 88%)

8.3 The geometric and aerodynamic data for a wing of a large white butterfly is as follows: Flight speed, $U_\infty = 1.35$ m/s; average chord, $c = 25$ mm; average span, $s = 50$ mm; air density $= 1.2$ kg/m^3; air viscosity, $\mu = 18 \times 10^{-6}$ Pa/s; drag at zero lift $= 120\,\mu\text{N}$ (measured on a miniature wind-tunnel balance). Estimate the boundary-layer thickness at the trailing edge. Also compare the measured drag with the estimated skin-friction drag. How do you account for any difference in value?
(*Answer*: 2.5 mm; 75 μN)

8.4 A submarine is 130 m long and has a mean perimeter of 50 m. Assume its wetted surface area is hydraulically smooth and equivalent to a flat plate measuring 130 m × 50 m. Calculate the power required to maintain a cruising speed of 16 m/s when submerged in a polar sea at 0°C. If the engines develop the same power as before, at what speed is the submarine able to cruise in a tropical sea at 20°C?

Take the water density to be 1000 kg/m^3 and its kinematic viscosity to be 1.79×10^{-6} m^2/s at 0°C and 1.01×10^{-6} m^2/s at 20°C.
(*Answer*: 20.5 MW; 16.37 m/s)

8.5 A sailing vessel is 64 m long, and its hull has a wetted surface area of 560 m^2. Its top speed is about 9 m/s. Assume that the equivalent sand-grain roughness k_s of the hull is normally about 0.2 mm.

The total resistance of the hull is composed of wave drag plus skin-friction drag. Assume that the latter can be estimated by taking it to be the same as the equivalent flat plate. The skin-friction drag is exactly half the total drag when sailing at top speed under normal conditions. If the water density and kinematic viscosity are 1000 kg/m^3 and 1.2×10^{-6} m^2/s, respectively, estimate
(a) The admissible roughness for the vessel.
(b) The power required to maintain the vessel at its top speed when the hull is unfouled (having its original sand-grain roughness).
(c) The amount by which the vessel's top speed is reduced if barnacles and seaweed adhere to the hull, thereby raising the equivalent sand-grain roughness k_s to about 5 mm.
(*Answer*: 2.2 μm; 1.06 MW; 8.04 m/s [i.e., a reduction of 10.7%])

8.6 The top surface of a light-aircraft wing traveling at an air speed of 55 m/s is assumed to be equivalent to a flat plate of length 2 m. Laminar-turbulent transition is known to occur at a distance of 0.75 m from the leading edge. Given that the kinematic viscosity of air is 15×10^{-6} m^2/s, estimate the coefficient of skin-friction drag.
(*Answer*: 0.00223)

8.7 Dolphins have been observed swimming at sustained speeds of up to 11 m/s. According to the distinguished zoologist Sir James Gray, this speed can be achieved, assuming normal hydrodynamic conditions prevail, only if the power produced per unit mass of muscle far exceeds that produced by other mammalian muscles. This result is known as Gray's paradox. The object of this exercise is to revise estimates of the power required to check the soundness of Gray's calculations.

Assume that the dolphin's body is hydrodynamically equivalent to a prolate spheroid (formed by an ellipse rotated about its major axis) of 2 m in length with a maximum thickness-to-length ratio of 1:6.

$$\text{Volume of a prolate spheroid} = \frac{4}{3}\pi ab^2$$

$$\text{Surface/area} = 2\pi b^2 + \frac{2\pi a^2 b}{\sqrt{a^2 - b^2}} \arcsin\left(\frac{\sqrt{a^2 - b^2}}{a}\right)$$

where $2a$ is the length and $2b$ is the maximum thickness.

Calculate the dimensions of the equivalent flat plate and estimate the power required to overcome the hydrodynamic drag (assuming it to be solely due to skin friction) at 11 m/s for the following cases:

(a) Assuming that the transitional Reynolds number takes the same value as the maximum found for a flat plate, say 2×10^6.

(b) Assuming that transition occurs at the point of maximum thickness (i.e., at the point of minimum pressure), which is located halfway along the body.

The propulsive power is supplied by a large group of muscles arranged around the spine. Typically their total mass is about 36 kg, the total mass of the dolphin being typically about 90 kg. Assuming that the propulsive efficiency of the dolphin's tail unit is about 75%, estimate the power required per unit mass of muscle for the two cases (a) and (b). Compare the results with the values given here:

• Running man, 40 W/kg.
• Hovering hummingbird, 65 W/kg

(*Answer*: 2 m × 0.832 m; 2.87 kW, 1.75 kW; 106 W/kg, 65 W/kg)

8.8 Many years ago *Scientific American* published a letter about the aerodynamics of pollen spores. A photograph accompanied the letter showing a spore having a diameter of about 20 μm and looking remarkably like a golf ball. The gist of the letter was that nature had discovered the principle of golf-ball aerodynamics millions of years before man. Explain why the letter writer's logic is faulty.

Flow Control and Wing Design

- Investigate the history of solutions to undesirable boundary-layer behavior on wings. What are the undesirable behaviors and what, historically, has been found to be successful in mitigating the impact of such flows?

- Learn about advantageous and adverse pressure distributions for low- and high-speed flight, how these gradients can affect flow, and how they can be manipulated to improve performace.

- Explore lift augmentation methods, such as trailing-edge flaps, at low speed.

- Introduce yourself to drag reduction methods.

9.1 INTRODUCTION

Wing design is an exceedingly complex and multifaceted subject. It is not possible to do justice to all that it involves in the present text, but it is possible to cover some of the fundamental principles that underlie design for high lift and low drag.

For fixed air properties and free-stream speed, lift can be augmented in four main ways:

- Increased wing area
- Increased angle of attack
- Increased camber
- Increased circulation by judicious application of high-momentum fluid

The extent to which the second and third approaches can be exploited is governed by the behavior of the boundary layer. A wing can only continue to generate lift successfully if boundary-layer separation is either avoided or closely controlled. Lift augmentation is usually accomplished by deploying various high-lift devices, such as flaps and multi-element airfoils. These lead to increased drag, so they are generally used only at the low speeds encountered during takeoff and landing. Nevertheless, it is instructive to examine the factors governing the maximum lift achievable with an unmodified single-element airfoil before considering the various high-lift devices. Accordingly, the maximization of lift for single-element airfoils is considered in

Section 9.2, followed by Section 9.3 on multi-element airfoils and various types of flap, and Section 9.4 on other methods of boundary-layer control. Finally, the various methods of drag reduction are described in Sections 9.5 to 9.8.

9.2 MAXIMIZING LIFT FOR SINGLE-ELEMENT AIRFOILS

This section addresses choosing the pressure distribution, particularly that on the upper wing surface, to maximize lift. Even when a completely satisfactory answer is found to this rather difficult issue, it still remains to determine the appropriate shape of the airfoil to produce the specified pressure distribution. This second step in the process is the so-called inverse problem of airfoil design. It is much more demanding than the direct problem, discussed in Chapter 4, of determining the pressure distribution for a given airfoil shape. Nevertheless, satisfactory inverse design methods are available, although they will not be discussed any further here. Only the more fundamental question of choosing the pressure distribution will be considered.

In broad terms the maximum lift achievable is limited by two factors:

• Boundary-layer separation
• The onset of supersonic flow

In both cases it is usually the upper wing surface that is the more critical. Boundary-layer separation is the more fundamental of the two factors, since supercritical wings are routinely used even for subsonic aircraft, despite the substantial drag penalty in the form of wave drag that results if there are regions of supersonic flow over the wing. However, no conventional wing can operate at peak efficiency with significant boundary-layer separation.

In two-dimensional flow, boundary-layer separation is governed by

• The severity and quality of the adverse pressure gradient
• The kinetic-energy defect in the boundary layer at the start of the adverse pressure gradient

The latter can be measured by the kinetic-energy thickness δ^{**} introduced in Section 8.2.2. The former is more vague. Precisely how is the severity of an adverse pressure gradient assessed? What is the optimum variation of adverse pressure distribution along the wing? Plainly, when seeking an answer to the first of these questions, a suitable nondimensional local pressure must be used to remove, as far as possible, the effects of scale. What soon becomes clear is that the conventional definition of coefficient of pressure is not at all satisfactory:

$$C_p = \frac{p - p_\infty}{\frac{1}{2}\rho_\infty V_\infty^2}$$

Use of this nondimensional quantity invariably results in pressure distributions with high negative values of C_p that appear to be the most severe. It is difficult to tell from the variation of C_p along an airfoil whether or not the boundary layer has a satisfactory margin of safety against separation. Yet it is known from elementary dimensional analysis that if the Reynolds number is the same for two airfoils of the same shape but of different size and free-stream speed, the boundary layers will behave identically. Furthermore, Reynolds-number effects, although very important, are relatively weak.

There is a more satisfactory definition of pressure coefficient for characterizing the adverse pressure gradient. This is the *canonical* pressure coefficient $\overline{C}_P$, introduced by A.M.O. Smith [107]. The definition of $\overline{C}_P$ is illustrated in Fig. 9.1. Note that local pressure is measured as a departure from the value of pressure p_m (the corresponding local velocity at the edge of the boundary layer is U_m) at the start of the pressure rise. Also note that the local dynamic pressure at the start of the pressure rise is now used to make the pressure difference nondimensional. When the canonical representation is used, $\overline{C}_P = 0$ at the start of the adverse pressure gradient and $\overline{C}_P = 1$, corresponding to the stagnation point where $U = 0$, is the maximum possible value. Furthermore, if two pressure distributions have the same shape, a boundary layer experiencing a deceleration of $(U/U_\infty)^2$ from 20 to 10 is no more or less likely to separate than one experiencing a deceleration of $(U/U_\infty)^2$ from 0.2 to 0.1. With the pressure-magnitude effects scaled out, it is much easier to assess the effect of the adverse pressure gradient by simple inspection than when a conventional C_p distribution is used.

How are the two forms of pressure coefficient related? From Bernoulli's equation, it follows that

$$C_p = 1 - \left(\frac{U}{U_\infty} \right)^2 \quad \text{and} \quad \overline{C}_p = 1 - \left(\frac{U}{U_m} \right)^2$$

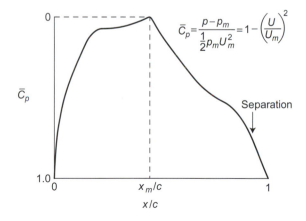

FIGURE 9.1

Smith's canonical pressure distribution around an airfoil.

Therefore, it follows that

$$C_p = 1 - \left(\frac{U}{U_\infty}\right)^2 = 1 - \left(\frac{U}{U_m}\right)^2 \left(\frac{U_m}{U_\infty}\right)^2$$

$$= 1 - (1 - \overline{C}_p)\left(\frac{U_m}{U_\infty}\right)^2$$

The factor $(U_m/U_\infty)^2$ is just a constant for a given pressure distribution or airfoil shape.

Figure 9.2 gives some idea of how the quality of the adverse pressure distribution affects boundary-layer separation. For this figure it is assumed that a length of constant pressure is followed by various types of adverse pressure gradient. Suppose that from the point $x = 0$ onward, $\overline{C}_p \propto x^m$. For the curve labeled *convex*, $m \simeq 4$; for that labeled *linear*, $m = 1$; and for that labeled *concave*, $m \simeq 1/4$. One would not normally design a wing for which the flow separates before the trailing edge is reached, so ideally the separation loci should coincide with the trailing edge. The separation loci in Fig. 9.2 depend on two additional factors:

1. The thickness of the boundary layer at the start of the adverse pressure gradient, as shown in this figure, and
2. The Reynolds number per unit length in the form of U_m/ν. This latter effect is not illustrated, but as a general rule the higher the value of U_m/ν, the higher the value of $\overline{C}_p$ that the boundary layer can sustain before separating.

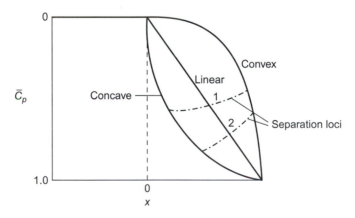

1—thick boundary layer at $x=0$; 2—thin boundary layer at $x=0$

FIGURE 9.2

Effects of different types of adverse pressure variation on separation.

It was mentioned that the separation point is affected by the energy defect in the boundary layer at the start of the adverse pressure gradient $x = 0$. Other things being equal, this implies that the thinner the boundary layer is at $x = 0$, the farther the boundary layer can develop in the adverse pressure gradient before separating. This point is illustrated in Fig. 9.3, which is based on calculations (using Head's method) of a turbulent boundary layer in an adverse pressure gradient with a preliminary constant-pressure region of variable length x_0. It is shown very clearly that the shorter x_0 is, the longer the distance Δx_s is from $x = 0$ to the separation point. It may be deduced from this that it is best to keep the boundary layer laminar, and therefore thin, up to the start of the adverse pressure gradient. Ideally, transition should occur at or shortly after $x = 0$, since turbulent boundary layers can withstand adverse pressure gradients much better than laminar ones can. Fortunately, the physics of transition (see Section 8.9) ensures that this desirable state of affairs can be easily achieved.

The canonical plot in Fig. 9.2 contains much information of practical value. For example, suppose that at typical cruise conditions the value of $(U/U_\infty)^2$ at the trailing edge is 0.8, corresponding to $C_p = 0.2$, and typically $\overline{C}_P = 0.4$ (say) there. In this case any of the $\overline{C}_p$ curves in the figure would be able to sustain the pressure rise without leading to separation. Therefore, suitable airfoils with a wide variety of pressure distributions can be designed to meet the specification. If, on the other hand, the goal is to achieve the maximum possible lift, then a highly concave pressure-rise curve with $m \simeq 1/4$ is the best choice. This is because, assuming that separation occurs at the trailing edge, the highly concave distribution not only gives the largest possible value of $(\overline{C}_p)_{TE}$, and therefore the largest possible value of U_m/U_{TE}; but also the pressure rises to its value at the trailing edge the most rapidly. This latter attribute

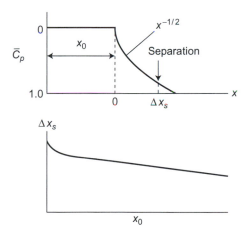

FIGURE 9.3

Variation of location of separation with length of initial flat plate for a turbulent boundary layer in a specified adverse pressure variation.

is of great advantage because it allows the region of constant pressure to be maintained over as much of the airfoil surface as possible, leading to the greatest possible average value of $|C_p|$ on the upper surface and, therefore, the greatest possible lift. For many people this conclusion is counterintuitive since it seems to violate the classic rules of streamlining that seek to make the adverse pressure gradient as gentle as possible. Nevertheless, the conclusions based on Fig. 9.2 are practically sound.

The results depicted in Fig. 9.2 naturally suggest an important practical question. Is there, for a given situation, a best choice of adverse pressure distribution? The desired goals are as above: to maximize U_m/U_{TE} and to maximize the rate of pressure rise. This question, or others very similar, has been considered by many researchers and designers. A widely quoted method of determining the optimum adverse pressure distribution is due to Stratford [108]. His theoretically derived pressure distributions lead to a turbulent boundary layer that is on the verge of separation but remains under control for much of the adverse pressure gradient. It is quite similar qualitatively to the concave distribution in Fig. 9.2. Two prominent features of Stratford's pressure distribution are

- The initial pressure gradient $d\overline{C}_P/dx$ is infinite so that small pressure rises can be accomplished in very short distances.
- It can be shown that in the early stages $C_p \propto x^{1/3}$.

If compressible effects are taken into account and it is desirable to avoid supersonic flow on the upper wing surface, the minimum pressure must correspond to sonic conditions. The consequences of this are illustrated in Fig. 9.4 where comparatively low speeds and very high values of suction pressure can be sustained before sonic conditions are reached, resulting in a pronounced peaky pressure distribution. For high subsonic Mach numbers, on the other hand, only modest maximum suction pressures are permissible before sonic conditions are reached. In this case, therefore, the pressure distribution is very flat. An example of the practical application of these ideas for low flight speeds is illustrated schematically in Fig. 9.5, which shows a Liebeck [109] airfoil of the sort used as a basis for the airfoil designed by Lissaman [110] especially for the successful man-powered aircraft *Gossamer Albatross* and *Condor*. In this application high lift and low drag were paramount. Note that there is a substantial foreportion of the airfoil with a favorable pressure gradient, rather than a very rapid initial acceleration up to a constant-pressure region. The favorable pressure gradient ensures that the boundary layer remains laminar until the onset of the adverse pressure gradient, thereby minimizing boundary-layer thickness at the start of the pressure rise. Incidentally, the maximum suction pressure in Fig. 9.5 is considerably less than that in Fig. 9.4 for the low-speed case, but, it is not, of course, suggested here that at the speeds encountered in man-powered flight the flow over the upper wing surface is close to sonic conditions.

There is a practical disadvantage with airfoils designed for concave pressure–recovery distributions. This is illustrated in Fig. 9.6, which compares the variations in lift coefficient with angle of attack for typical airfoils with convex and concave pressure distributions. It is immediately plain that the concave distribution leads to much higher values of $(C_L)_{max}$, but the trailing-edge stall is much more gentle, initially at

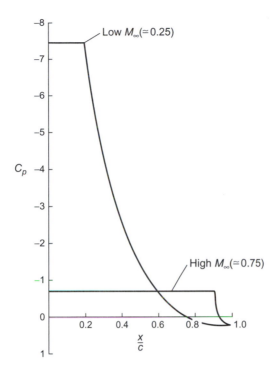

FIGURE 9.4

Upper-wing-surface pressure distributions with laminar rooftop.

least, for the airfoil with the convex distribution. This is a desirable feature from the standpoint of safety. The much sharper fall in C_L seen in the airfoil with the concave pressure distribution is explained by the fact that the boundary layer is close to separation for most of the airfoil aft of the point of minimum pressure. (Recall that the ideal Stratford distribution aims for the boundary layer to be on the verge of separation throughout pressure recovery.) Consequently, when the angle of attack that provokes separation is reached, any further rise in incidence causes the separation point to move rapidly forward.

As indicated earlier, it is not feasible to design efficient wings for aircraft cruising at high subsonic speeds without permitting a substantial region of supersonic flow to form over the upper surface. However, it is still important to minimize the wave drag as much as possible. This is achieved by tailoring the pressure distribution so as to minimize the strength of the shock-wave system that forms at the end of the supersonic-flow region. A schematic illustrating the main principles of modern supercritical airfoils is shown in Fig. 9.7. This type of airfoil would be designed for M_∞ in the range 0.75 to 0.80. The principles behind this design are not dissimilar from those behind the high-speed case in Fig. 9.4, in that a constant pressure is maintained over as much of the upper surface as possible.

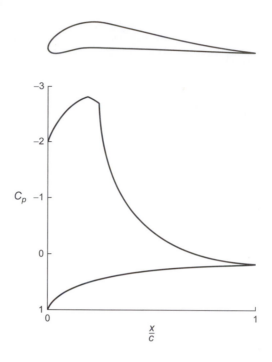

FIGURE 9.5

Typical low-speed high-lift airfoil: schematic of a Liebeck airfoil.

9.3 MULTI-ELEMENT AIRFOILS

At the low speeds encountered during landing and takeoff, lift needs to be greatly augmented and stall avoided. Lift augmentation is usually achieved by flaps [111] of various kinds (see Fig. 9.8). The plain flap shown in part (a) of the figure increases the camber and angle of attack; the Fowler flap in part (b) increases camber, angle of incidence, and wing area; the nose flap in part (g) increases camber. The flaps shown in this figure are relatively crude and are likely to lead to boundary-layer separation when deployed. Modern aircraft use combinations of these devices in the form of multi-element wings (Fig. 9.9). The slots between the wing elements effectively suppress the adverse effects of boundary-layer separation, provided that they are appropriately designed.

Multi-element airfoils are not a new idea. The basic concept dates back to the early days of aviation with the work of Handley Page in Britain and Lachmann in Germany. Nature also exploits this concept in the wings of birds. In many species a group of small feathers attached to the thumb-bone and known as the alula acts as a slat.

How do multi-element airfoils augment lift without suffering the adverse effects of boundary-layer separation? The conventional explanation is that, since a slot

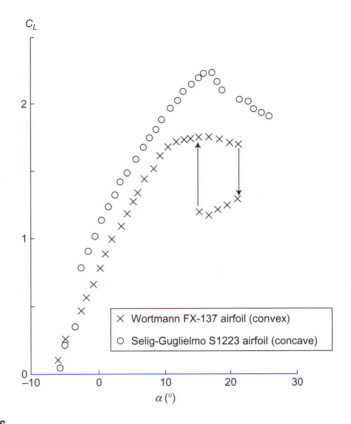

FIGURE 9.6

Comparison of the variations of lift coefficient versus angle of attack for airfoils with concave and convex pressure-recovery distributions. $Re = 2 \times 10^5$; ×, Wortmann FX-137 airfoil (convex); o, Selig-Guglielmo S1223 airfoil (concave). *Source: Based on figures in Selig J., Guglielmo J.J. High-lift low Reynolds number airfoil design. AIAA Aircraft 1997; 34:72–79.*

connects the high-pressure region on the wing's lower surface to the relatively low-pressure region on the top surface, it acts as a blowing type of boundary-layer control (see Section 9.4.2). This explanation is to be found in a large number of technical reports and textbooks, and thus is one of the most widespread misconceptions in aerodynamics. It can be traced back to no less an authority than Prandtl [112], who wrote:

> The air coming out of a slot blows into the boundary layer on the top of the wing and imparts fresh momentum to the particles in it, which have been slowed down by the action of viscosity. Owing to this help the particles are able to reach the sharp rear edge without breaking away.

This conventional view of how slots work is mistaken for two reasons. First, since the stagnation pressure in the air flowing over the lower surface of a wing is exactly

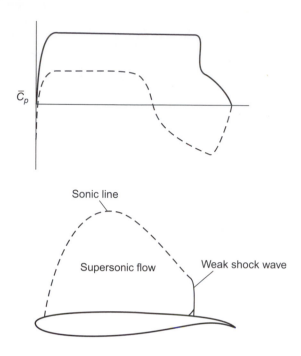

FIGURE 9.7

Schematic of a modem super-critical airfoil.

the same as that over the upper surface, the air passing through a slot cannot really be said to be high energy nor can the slot act like a kind of nozzle. Second, the slat does not give the air in the slot a high velocity compared to that over the upper surface of the unmodified single-element wing. This is readily apparent from accurate and comprehensive measurements of the flow field around a realistic multi-element airfoil reported by Nakayama et al. [113]. In fact, as will be explained, the slat and the slot usually act to reduce the flow speed over the main airfoil.

The flow field associated with a typical multi-element airfoil is highly complex. Its boundary-layer system is illustrated schematically in Fig. 9.10 based on the measurements of Nakayama et al. [113]. Note that the wake from the slot does not interact strongly with the boundary layer on the main airfoil before reaching the airfoil's trailing edge. The wake from the main airfoil and boundary layer from the flap also remain separate entities. As might be expected, given the complexity of the flow field, the true explanation of how multi-element airfoils augment lift while avoiding the detrimental effects of boundary-layer separation is multifaceted. Also the beneficial aerodynamic action of a well-designed multi-element airfoil is due to a number of primary effects that will be described in turn.[1]

----

[1]Many of the ideas described in the following passages are due to [107].

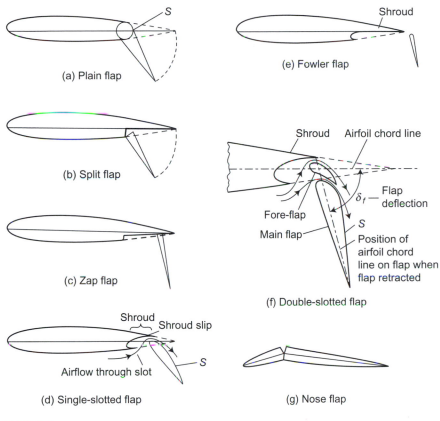

FIGURE 9.8

Flap types.

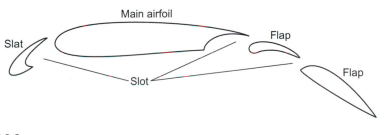

FIGURE 9.9

Schematic of a four-element airfoil.

9.3.1 The Slat Effect

To qualitatively appreciate the effect of the upstream element (the slat) on the immediate downstream element (the main airfoil) we can mode the former by a vortex.

FIGURE 9.10

Typical boundary-layer behavior for a three-element airfoil.

The effect is illustrated in Fig. 9.11. When we consider the velocity induced by the vortex in the direction of the local tangent to the airfoil contour in the vicinity of the leading edge (see inset figure), we see that the slat (vortex) reduces the velocity along the edge of the boundary layer on the upper surface and has the opposite effect on the lower surface. Thus the slat reduces the severity of the adverse pressure gradient on the main airfoil. In the case illustrated schematically in Fig. 9.11, the consequent reduction in pressure over the upper surface is counterbalanced by the rise in pressure on the lower surface. For a well-designed slat/main-wing combination, it can be arranged that the latter effect predominates, resulting in a slight rise in lift coefficient.

9.3.2 The Flap Effect

In a similar way the effect of the downstream element (the flap) on the immediate upstream element (the main airfoil) can also be modeled approximately by placing a vortex near the trailing edge of the latter, as illustrated in Fig. 9.12. This time the flap (vortex) near the trailing edge induces a velocity over the main airfoil surface that leads to a rise in velocity on both the upper and the lower surface. For the upper surface this is beneficial because it raises the velocity at the trailing edge, thereby reducing the severity of the adverse pressure gradient. The flap has a second beneficial effect, which can be understood from the figure inset. Note that, owing to the velocity induced by the flap at the trailing edge, the effective angle of attack has been increased. If matters were left unchanged, the streamline would not leave smoothly from the trailing edge of the main airfoil, violating the Kutta condition (see Section 4.1.1). What must happen is that viscous effects generate additional circulation so that the Kutta condition is satisfied once again. Thus the presence of the flap leads to enhanced circulation and therefore higher lift.

9.3.3 Off-the-Surface Recovery

With a typical multi-element airfoil, as shown in Figs. 9.9 and 9.13, the boundary layer develops in the adverse pressure gradient of the slat, reaches the trailing edge in an unseparated state, and then leaves the trailing edge, forming a wake. The slat

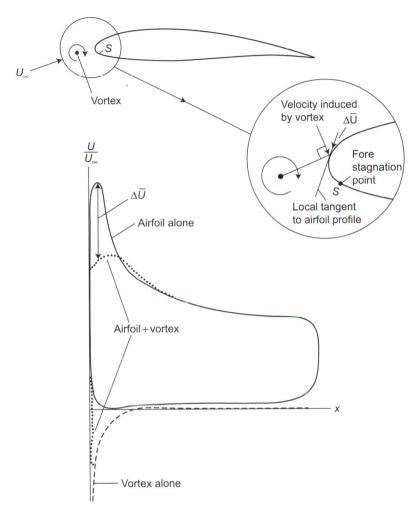

FIGURE 9.11

Slat effect (modeled by a vortex) on the velocity distribution over the main airfoil.

wake continues to develop in the adverse pressure gradient over the main airfoil. However, for well-designed multi-element airfoils, the slot is sufficiently wide for the slat wake and main-airfoil boundary layer to remain separate; likewise, the wake of the main airfoil and flap boundary layer. It is perfectly possible for the flow within the wakes to decelerate in the downstream adverse pressure gradient such that reversed flow occurs in the wake. This causes a stall, immediately destroying any beneficial effect. For well-designed cases, it appears that wake flows can withstand adverse pressure gradients to a far greater degree than attached boundary layers, so that flow reversal and wake breakdown are usually avoided. Consequently, for a multi-element airfoil the total deceleration (or recovery, as it is often called) of velocity along the edge of the boundary layer can take place in stages, as illustrated schematically in

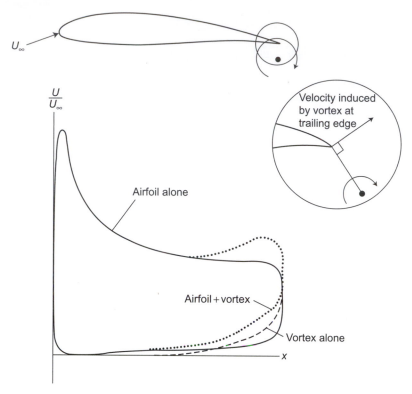

FIGURE 9.12

Flap effect (modeled by a vortex) on the velocity distribution over the main wing.

Fig. 9.13. In terms of the canonical pressure coefficient, U/U_m has approximately the same value at the trailing edge of each element; moreover, the boundary layer is on the verge of separation at the trailing edge of each element. (In fact, because of the flap effect described earlier, the value of $(U/U_m)_{TE}$ for the flap is lower than that for the main airfoil.) It is then evident that the overall reduction in (U/U_∞) from $(U_m/U_\infty)_{slat}$ to $(U_{TE}/U_\infty)_{flap}$ is much greater than the overall reduction for a single-element airfoil. In this way the multi-element airfoil can withstand a much greater overall velocity ratio or pressure difference than can a comparable single-element airfoil.

9.3.4 Fresh Boundary-Layer Effect

It is evident from Fig. 9.10 that the boundary layer on each element develops largely independently from the boundary layers on the other airfoil elements. This ensures a fresh thin boundary layer, and therefore a small kinetic-energy defect, at the start of the adverse pressure gradient on each element. The length of pressure rise that the boundary layer on each element can withstand before separating is thereby maximized (compare Fig. 9.3).

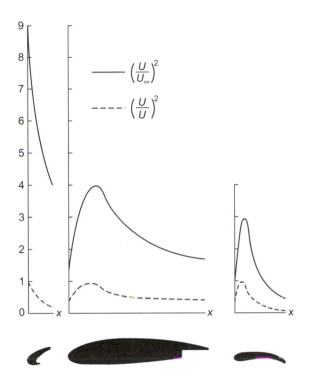

FIGURE 9.13

Typical distributions of velocity ratio over the elements of a three-element airfoil.

9.3.5 **The Gurney Flap**

The principle of the Gurney flap (see *Aerodynamics Around Us* in this chapter) was probably exploited almost by accident in aeronautics many years before its invention. Similar strips had been in use for many years, but were intended to reduce control-surface oscillations caused by patterns of flow separation changing unpredictably. It is also likely that the split and Zap flaps, Fig 9.8(b,c), which date back to the early 1930s, produced similar flow fields to those produced by the Gurney flap. Nevertheless, it is certainly fair to claim that the Gurney flap is unique as the only aerodynamic innovation made in automobile engineering that has been transferred to aeronautical engineering.

Today Gurney flaps are widely used to increase the effectiveness of helicopter stabilizers.[2] They were first used in helicopters on the trailing edge of the tail on the Sikorsky S-76B because the first flight tests had revealed insufficient maximum (upward) lift. This problem was overcome by fitting a Gurney flap to the inverted NACA 2412 airfoil used for the horizontal tail. Similar circumstances led to the use

---

[2]The information on helicopter aerodynamics used here is based on an article by R.W. Prouty in March 2000 [114].

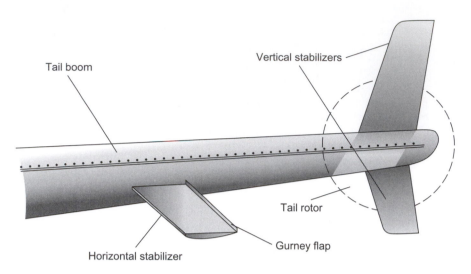

FIGURE 9.14

The Gurney flap installed on the horizontal stabilizer of a Bell 206 JetRanger.

of a Gurney flap on the horizontal stabilizer of the Bell JetRanger (Fig. 9.14). In this case, apparently the design engineers had difficulty estimating the required incidence of the stabilizer. Flight tests indicated that they had not guessed it correctly. This was remedied by adding a Gurney flap.

Another example is the double-sided Gurney flap installed on the trailing edge of the vertical stabilizer of the Eurocopter AS-355 TwinStar to cure a problem on thick surfaces with large trailing-edge angles. In such a case, lift reversal can occur for small angles of attack, as shown in Fig. 9.15, thereby making the stabilizer a "destabilizer"! The explanation for this behavior is that, at a small positive angle of attack, the boundary layer separates near the trailing edge on the upper (suction) side of the airfoil. On the lower side, the boundary layer remains attached, which means that the pressure is lower there than over the top surface. The addition of a double Gurney flap stabilizes the boundary-layer separation and eliminates the lift reversal.

AERODYNAMICS AROUND US
Gurney Flaps
As well as being a successful race-car driver, Dan Gurney became well known for his technical innovations. Most prominent is the Gurney flap, a simple small plate fixed to and perpendicular to the trailing edge of a wing. It can be seen attached to the trailing edge of the multi-element rear wing in Figs. B9.1 and B9.2.

Gurney first started fitting these "spoilers" pointing upward at the end of the rear deck of his Indy 500 cars in the late 1960s to increase the downforce. Because his innovation was contrary to the classic concepts of aerodynamics, he was able to disguise his true motives by telling his competitors that the devices were intended to prevent cut hands when the cars were pushed. It was several years before competitors realized the truth.

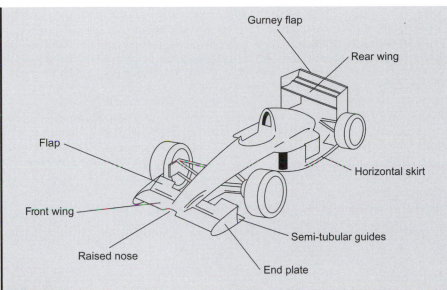

FIGURE B9.1

Main aerodynamic features of a Grand Prix car. *Source: Based on a figure in Dominy RG. Aerodynamics of Grand Prix cars. Proc. I. Mech. E. Part D. J. Automobile Eng. 1992, 206.267–274.*

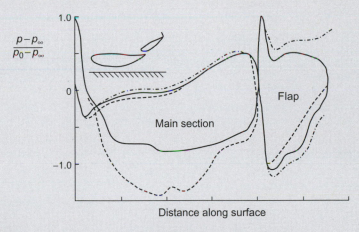

FIGURE B9.2

Effects of ground proximity and a Gurney flap on the pressure distribution over a two-element front wing—(schematic only). *Key:* solid line—wing in free flow; dashed line—wing in close proximity to the ground; dot/dash line—wing fitted with a Gurney flap and in close proximity to the ground. *Source: Based on figures in Dominy RG. Aerodynamics of Grand Prix cars, Proc. I. Mech. E. Part D, J. Automobile Eng. 1992; 206:267–274.*

Gurney flaps became known in aerodynamics after Gurney discussed his ideas with the aerodynamicist and wing designer Bob Liebeck of Douglas Aircraft. Gurney and Liebeck reasoned that the tabs should be capable of enhancing the lift generated by conventional wings, which Liebeck confirmed experimentally. The beneficial effects of a Gurney flap in generating an enhanced downforce is illustrated by the pressure distribution over the flap of the two-element airfoil shown in Fig. B9.2. The direct effects of Gurney flaps of various heights on wing lift and drag were demonstrated by other experimental studies (see Fig. B9.3). It can be seen that the maximum lift

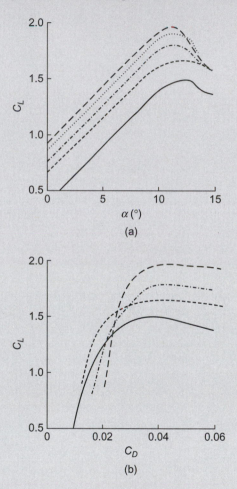

FIGURE B9.3

Effects of Gurney flaps placed at the trailing edge of a NACA 4412 wing on the variation in lift and drag with angle of attack. The flap height varies from 0.005 to 0.02 times the chord, *c. Key*: solid line—baseline without flap; dashed line—0.005*c*; dot/dash line—0.01*c*; dotted line—0.015*c*; long dashed line—0.02*c*. *Source: Based on a figure in Storms BL, Jang CS. Lift enhancement of an airfoil using a Gurney flap and vortex generators. AIAA J. Aircraft 1994; 31(3):542–547.*

rises as the height of the flap is increased from 0.005 to 0.02 chord. However, it is clear that further improvement to aerodynamic performance diminishes rapidly with increased flap height. The drag polars plotted in Fig. B9.3(b) show that, for a lift coefficient less than unity, the drag is generally greater with a Gurney flap attached. The flaps are really only an advantage for generating high lift.

Why do Gurney flaps work? The answer is to be found in the twin-vortex flow field depicted in Fig. B9.4. Something like this was hypothesized by Liebeck (1978) [115] and was confirmed by laser-Doppler measurements carried out at Southampton University (England) [116]. As can be seen in the figure, two contra-rotating vortices are created behind the flap. A trapped vortex is also immediately ahead of the flap, even though this is not clearly shown in the measurements. This vortex must be present, as was originally suggested by Liebeck.

In an important respect, Fig. B9.4 is misleading, because it cannot depict the unsteady nature of the flow field. The vortices are, in fact, shed alternately in a similar fashion to the von Kármán vortex street behind a circular cylinder (see Section 9.5). It can be also seen in the figure (showing the configuration for enhancing downforce) that the vortices behind the Gurney flap deflect the flow downstream upwards. In some respects, the vortices have a similar circulation-enhancing effect as that of the downstream flap in a multi-element airfoil (see Section 9.3.2).

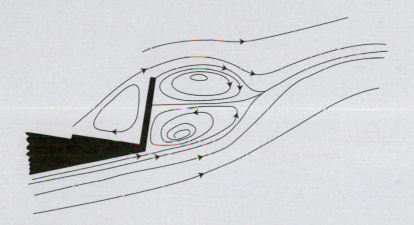

FIGURE B9.4

Flow pattern downstream of a Gurney flap. *Source: Based on figures in Jeffrey D, Zhang X, Hurst DW. Aerodynamics of Gurney flaps on a single-element high-lift wing, AIAA J. Aircraft 2000; 37(2):295–301.*

9.3.6 Movable Flaps: Artificial Bird Feathers

The concept of movable flaps is illustrated in Fig. 9.16 [117]. The basic idea is that at high angles of attack, when flow separation starts to near the trailing edge, the associated reversed flow causes the movable flap to be raised. This then acts as a barrier to the further migration of reversed flow toward the leading edge, thereby controlling flow separation.

The movable flap concept originated with Liebe [118], who was the inventor of the boundary-layer fence (see Section 9.4.3). He observed that during the landing approach or in gusty winds, the feathers on the upper surface of many bird wings

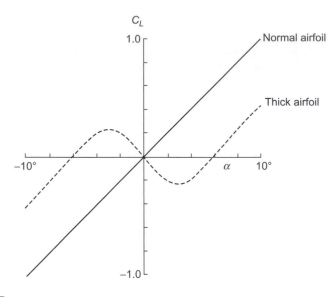

FIGURE 9.15

Lift reversal for thick airfoils.

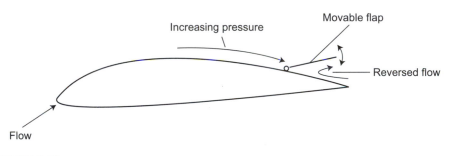

FIGURE 9.16

Schematic of the basic concept of the movable flap.

tend to rise near the trailing edge. (Photographs of the phenomenon on a skua wing can be found in Bechert et al. [117]) Liebe interpreted this behavior as a form of biological high-lift device, and his ideas led to some flight tests on a Messerchmitt Me 109 in 1938. This device led to the development of asymmetric lift distributions that made the aircraft difficult to control, and the project was abandoned. Many years later a few preliminary flight tests were carried out in Aachen on a glider [119]. In this case small movable plastic sheets were installed on the upper surface of the wing. Apparently they improved the glider's handling at high angles of attack.

There are problems with movable flaps. First, they have a tendency to flip over at high angles of attack when the reversed flow becomes too strong. Second, they tend not to lie flat at low angles of attack, leading to a deterioration in aerodynamic

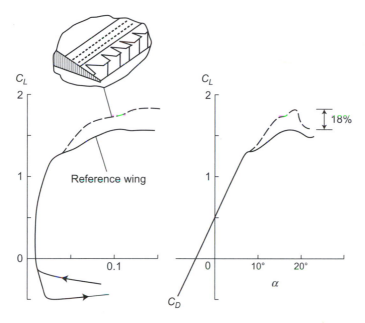

FIGURE 9.17

Improved design of the movable flap and resulting improvement in aerodynamic characteristics for a laminar glider airfoil. *Source: Based on a figure in Bechert et al. [117].*

performance. This is because, when the boundary layer is attached, the pressure rises toward the trailing edge so that the space under the flap connects with a region of slightly higher pressure that tends to lift it from the surface. These problems were largely overcome because of three features of the design depicted in Fig. 9.17, which was fitted to a laminar glider airfoil (see Bechert et al. [117]). Ties limited the maximum deflection of the flaps. Also, making the flap porous and the trailing edge jagged helped to equalize the static pressure on either side of the flap during attached-flow conditions. These last two features are also seen in birds' feathers. The improvement in the aerodynamic characteristics can also be seen in Fig. 9.17.

Successful flight tests on similar movable flaps were carried out later on a motor glider.

9.4 BOUNDARY LAYER CONTROL PREVENTION TO SEPARATION

Many of the widely used techniques were described in Section 9.3. However, there are various other methods of flow-separation control that are used on aircraft and in engineering applications. These are described here [120]. Some of the devices are active, that is, they require additional power from the propulsion units. Others are passive and require no additional power. As a general rule, however, the passive devices usually lead to increased drag at cruise when not in operation. The active techniques are discussed first.

9.4.1 Boundary-Layer Suction

The basic principle was demonstrated experimentally in Prandtl's paper introducing the boundary-layer concept to the world [121]. Prandtl showed that the suction through a slot could prevent flow separation on the surface of a cylinder. This is illustrated in Fig. 9.18. The layer of low-energy ("tired") air near the surface approaching the separation point is removed through a suction slot.

The result is a much thinner, more vigorous boundary layer that is able to progress further along the surface against the adverse pressure gradient without separating.

Suction can be used to suppress separation at high angles of incidence, thereby obtaining very high lift coefficients. In such applications the trailing edge can have an appreciable radius instead of being sharp. The circulation is then adjusted by means of a small spanwise flap, as depicted in Fig. 9.19. If sufficient boundary layer is removed by suction, a flow regime, which is virtually a potential flow, may be set up; on the basis of the Kutta–Zhukovsky hypothesis, the sharp-edged flap will locate the rear stagnation point, allowing airfoils with elliptic, or even circular, cross-sections to generate very high lift coefficients.

There are great practical disadvantages with this type of high-lift device. First, it is very vulnerable to dust blocking the suction slots. Second, it is entirely reliant on

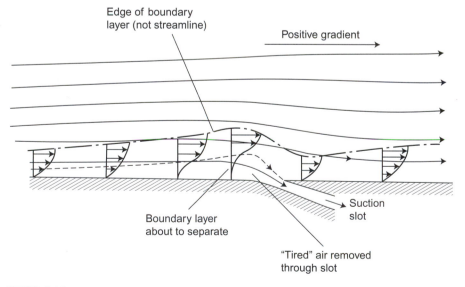

FIGURE 9.18

Detail of streamlines in the region of a boundary bleed slot.

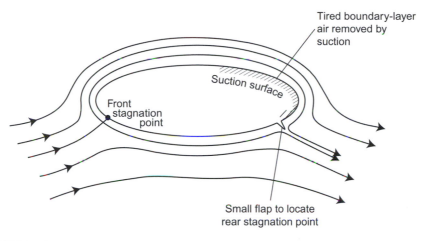

Tired boundary-layer air removed by suction

Suction surface

Front stagnation point

Small flap to locate rear stagnation point

FIGURE 9.19

Example of a blunt trailing-edge shape with boundary-layer suction and flap to control circulation.

the necessary engine power being available for suction. Either blockage or engine failure can lead to catastrophic failure. For these reasons suction has not been used for separation control in production aircraft, although it has been tested on rotors in prototype helicopters.

Many supersonic aircraft feature suction in their engines' intake to counter the effects of shock-wave/boundary-layer interaction. Without it, the boundary layers in the inlets would certainly thicken and likely separate. Also, some form of shock-wave system is indispensible because the air needs to be slowed down from supersonic flight speed to a Mach number of about 0.4 at entry to the compressor.

Two common methods of boundary-layer suction (or bleed) are porous surfaces and a throat slot bypass, both of which were used for the first time on the McDonnell Douglas F-4 Phantom. Another method is the wide slot at the throat that acts as an effective and sophisticated form of boundary-layer bleed on the Concorde, making the intake tolerant of changes in engine demand or the amount of bleed. The McDonnell Douglas (now Boeing) F-15 Eagle also incorporates a variety of such boundary-layer control methods, as illustrated in Fig. 9.20. This aircraft has porous areas on the second and third engine-inlet ramps plus a throat bypass in the form of a slot and a porous region on the sideplates in the vicinity of the terminal shock wave. All of the porous areas together account for about 30% of the boundary-layer removal, with the throat bypass accounting for the remainder.

9.4.2 Control by Tangential Blowing

Since flow separation is due to the complete loss of kinetic energy in the boundary layer immediately adjacent to the wall, another method of preventing it is to reenergize the "tired" air by blowing a thin, high-speed jet into it. This is often used

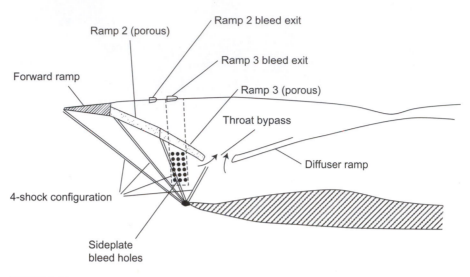

FIGURE 9.20

Features of the F-15's engine-inlet flow management.

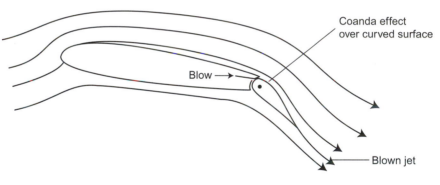

FIGURE 9.21

Blown trailing-edge flap.

with trailing-edge flaps (Fig. 9.21). To obtain reasonable results with this method, the blowing duct must be carefully designed. It is essential that good mixing take place between the blown air and the boundary layer.

Most applications of tangential blowing for flow control exploit the so-called *Coanda effect*—the tendency of a fluid jet issuing tangentially onto a curved or angled solid surface to adhere to it, as illustrated in Fig. 9.22. The name derives from the Franco-Romanian engineer, Henri Coanda, who filed a French patent in 1932 for a propulsive device exploiting the phenomenon. The Coanda effect can be understood

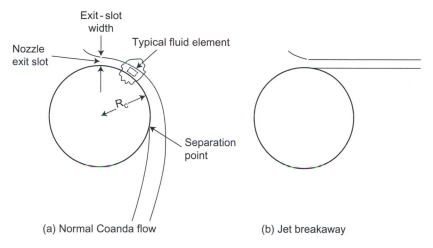

Exit-slot width

Nozzle exit slot

Typical fluid element

R_c

Separation point

(a) Normal Coanda flow   (b) Jet breakaway

FIGURE 9.22

Coanda effect: the flow of a jet around a circular cylinder. (a) shows the Coanda effect in action. (b) shows an equally valid fluid flow that fails to produse the desired effect. In (b) the flow (wall jet) detaches nearly immediately from the cylinder surface. *Source: Based on Figure 1 in Carpenter PW, Green PN. The aeroacoustics and aerodynamics of high-speed Coanda devices. J. Sound & Vibration 1997; 208(5); 777–801.*

by considering the radial equilibrium of the fluid element depicted in Fig. 9.22(a), which can be expressed in simple terms as follows:

$$\frac{\partial p}{\partial r} = \frac{\rho V^2}{r} \tag{9.1}$$

where p is the pressure in the jet boundary layer (strictly, the wall jet) issuing from the nozzle exit slot; r is the radial distance from the center of curvature of the surface; ρ is the fluid density; and V is the local flow speed. It is easy to see that the pressure field created forces the flow issuing from the nozzle to adhere to the surface. However, this does not explain why the equally valid flow solution shown in Fig. 9.22(b) is only found in practice when the Coanda effect breaks down. Presumably, the slightly enhanced viscous drag experienced by the jet on its surface side as the jet emerges from the nozzle tends to deflect the jet toward the surface. Thereafter, the pressure field created by the requirements of radial equilibrium tend to force the jet toward the surface. Another viscous effect, entrainment of the fluid between the jet and the surface, may help pull the jet toward the surface as well.

The practical limits on use of the Coanda effect can also be understood to a certain extent by considering the radial equilibrium of the fluid element in Fig. 9.22(a). Initially we assume that the flow around the curved surface is inviscid so that it obeys

Bernoulli's equation:

$$p = p_0 - \frac{1}{2}\rho V^2 \tag{9.2}$$

where p_0 is the stagnation pressure of the flow issuing from the nozzle. Equation (9.2) may be substituted into Eq. (9.1), which is then rearranged to give

$$\frac{dV}{V} = dr, \quad \text{i.e.,} \quad V = V_{\text{w}} \exp\left(\frac{r}{R_c}\right) \tag{9.3}$$

where V_{w} is the (inviscid) flow speed along the wall and R_c is the radius of curvature of the surface. When the ratio of the exit-slot width b to the radius of curvature is small, $r \simeq R_c$ and $V \simeq V_{\text{w}}$. It then follows from Eq. (9.1) that near the exit slot the pressure at the wall is given by

$$p = \int \frac{\rho V^2}{r} dr \simeq \rho \int \frac{V_{\text{w}}^2}{R_{\text{C}}} dr = p_\infty - \rho \int_r^b \frac{V_{\text{w}}^2}{R_{\text{C}}} dr = p_\infty - \frac{\rho V^2 b}{R_{\text{C}}} \tag{9.4}$$

where p_∞ is the ambient pressure outside the Coanda flow.

We see from Eq. (9.4) that the larger $\rho V^2 b/R_{\text{c}}$ is, the more the wall pressure falls below p_∞. In actual viscous flow the average flow speed tends to fall with distance around the surface, which causes the wall pressure to rise with distance around the surface, creating an adverse pressure gradient and eventual separation. This effect is intensified for large values of $\rho V^2 b/R_{\text{c}}$, so the nozzle exit-slot height b must be kept as small as possible. For small values of b/R_{c}, the Coanda effect may still break down if the exit-flow speed is high enough. But the simple analysis leading to Eq. (9.4) ignores compressible-flow effects. In fact, the blown air normally reaches supersonic speeds before the Coanda effect breaks down. At sufficiently high supersonic exit speeds, shock-wave/boundary-layer interaction will provoke flow separation and cause the breakdown of the Coanda effect [122]. This places practical limits on the strength of blowing that can be employed.

The Coanda principle may be used to delay separation over the upper surface of a trailing-edge flap. The blowing is usually powered by air ducted from the engines. Careful positioning of the flap surface relative to the blown-air jet and the main wing surface allows the Coanda effect to make the blown jet adhere to the upper surface of the flap even when it is deflected downward by as much as 60 degrees (Fig. 9.21), which greatly enhances circulation around the wing.

A more extreme version of the Coanda principle is depicted in Fig. 9.23, where only a vestigial flap is used. This arrangement is occasionally found at the trailing edge of a conventional blown flap. Although the term *jet flap* has sometimes been applied to this device, it is rather imprecise; it has even been applied to blown-flap systems in general. Here we will reserve "jet flap" for the case where air is blown so strongly as to be supersonic. This arrangement is found on fighter aircraft

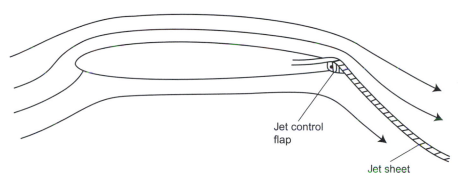

Jet control flap

Jet sheet

FIGURE 9.23

Jet flap with a vestigial control flap.

Table 9.1 Aerodynamic Performance of Some High-Lift Systems	
System	$C_{L_{max}}$
Internally blown flap	9
Upper surface blowing	8
Externally blown flap	7
Vectored thrust	3
Boeing 767 with slat + triple flap	2.8
Boeing 727 with slat + single flap	2.45

Source: Based on tables in Filippone A. 1999. Aerodynamics database–lift coefficients. Available from http://aerodyn.org/HighLift/tables.html.

with small wings, such as the Lockheed F-104 Starfighter, the Mig-21 PFM, and the McDonnell Douglas F-4 Phantom, as a way to increase lift at low speeds, thereby reducing landing speed. The air is bled from the engine compressor and blown over the trailing-edge flaps.

According to McCormick [123], prior to 1951 it was thought that, if supersonic blown air were used, it would not only fail to adhere to the flap surface but also lead to unacceptable losses due to shock waves. This view was dispelled by an undergraduate student, John Attinello, in his honors thesis at Lafayette College in the United States. Attinello's ideas were subjected to rigorous and sophisticated experimental studies before being flight-tested and ultimately used on many aircraft, including those just mentioned.

Internally blown flaps give the best performance of any high-lift system (see Table 9.1), but upper-surface blowing (Fig. 9.24) is also effective. This arrangement is used on various versions of the Antonov An 72/74 transport aircraft. A slightly less efficient system is the externally blown flap (Fig. 9.25), a version of which is used on the Boeing C-17 Globemaster heavy transport aircraft. The engine exhaust flow is directed below and through slotted flaps to produce additional lifting force,

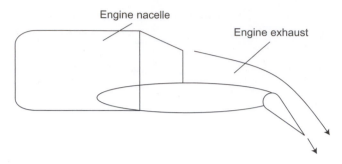

FIGURE 9.24

Upper surface blowing.

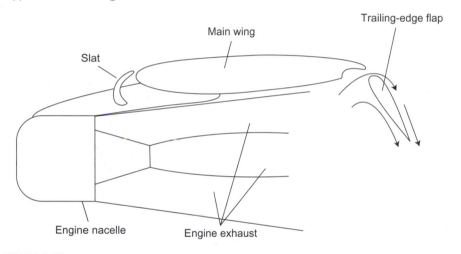

FIGURE 9.25

Externally blown flap.

allowing the aircraft to make a steep, low-speed final approach with a low landing speed for routine short-field landings. Many STOL (short takeoff and landing) and fighter aircraft use thrust vectoring that also exploits the Coanda effect. One possible arrangement is depicted in Fig. 9.26.

Blown flaps and some other high-lift systems actually generate substantial additional circulation and do not just generate the required high lift because of an increased angle of attack. For this reason, in some applications the term *circulation-control wings* is often used. It is not necessary to install a flap on a circulation-control wing (see the system depicted in Fig. 9.27). Rotors have been fitted with both suction-type circulation control (see Fig. 9.19) and the more common blown and jet flaps, and have been tested on a variety of helicopter prototypes [124]. As yet, however, circulation-control rotors have not been used on any production aircraft. A recent research development, mainly in the last ten years, is the use of periodic blowing

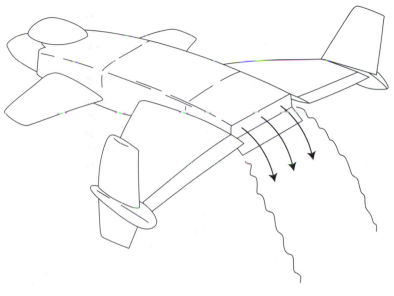

FIGURE 9.26

Use of the Coanda effect for thrust vectoring. *Source: Based on a figure in Bevilaqua PM, Lee JD. Design of supersonic Coanda jet nozzles. Proc. Circulation-Control Workshop 1986, NASA Conf. Pub. 2432.*

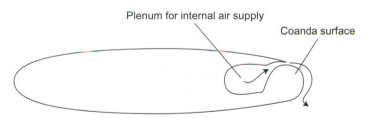

FIGURE 9.27

Circulation-control wing.

for separation control [125]. Significant lift enhancement can be achieved efficiently with the use of very low flow rates. Almost all the experimental studies are at fairly low Reynolds numbers, but Seifert and Pack [126] have carried out wind-tunnel tests at Reynolds numbers typical of flight conditions.

Tangential blowing can only be used to prevent separation, unlike suction, which can be employed for this purpose or for laminar-flow control. The flow created by blowing tends to be very vulnerable to laminar-turbulent transition, so tangential blowing almost inevitably triggers transition.

9.4.3 Other Methods of Separation Control

Passive flow control through the generation of streamwise vortices is common on aircraft and in other applications. Some of the devices in use are shown in Fig. 9.28. Part (a) of the figure shows a row of vortex generators on the upper surface of a wing. These take a variety of forms, and often two rows at two different chordwise locations are used. The basic principle is to generate an array of small streamwise vortices that promote increased mixing between high-speed air in the main stream and outer boundary layer with the relatively low-speed air nearer the surface. In this way, the boundary layer is reenergized. Vortex generators promote the reattachment of separated boundary layers within separation bubbles, which postpones fully developed stall.

Fixed vortex generators are simple, cheap, and rugged. Their disadvantages are that they cannot be used for active stall control, a technology now being used for highly maneuverable fighter aircraft; also, they generate parasitic drag at cruise conditions where stall suppression is not required. These disadvantages have led to the development of vortex-generator jets (VGJ). These are angled small jets that are blown, either in steady or in pulsatory mode, through orifices in the wing surface. The concept was first proposed by Wallis in Australia and Pearcey in the U.K. [127], primarily for control of shock-induced separation. More recently the concept was reexamined as an alternative to conventional vortex generators [128].

Wing fences (Fig. 9.28(b)) and "vortilons" act as barriers to tipward flow on swept-back wings. They also generate powerful streamwise vortices. The sawtooth

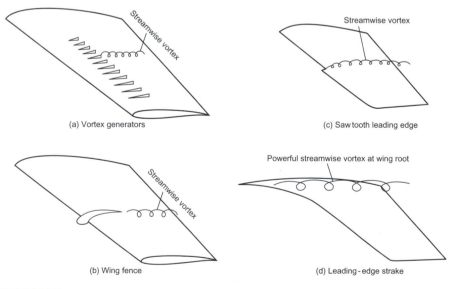

(a) Vortex generators

(c) Saw tooth leading edge

(b) Wing fence

(d) Leading-edge strake

FIGURE 9.28

Illustration of four common vortex-generating devices for controlling flow over swept wings.

leading edge (Fig. 9.28(c)) is another common device for generating a powerful streamwise vortex, as is the leading-edge strake (Fig. 9.28(d)). In this last case the vortex reenergizes the complex, three-dimensional boundary-layer flow that develops along the wing-body junction.

9.5 REDUCTION OF SKIN-FRICTION DRAG

Four main types of drag are found in aerodynamics (see Section 1.6.5): skin-friction, form, induced, and wave. The methods used to reduce each type are discussed in turn in the sections that follow. A more recent detailed account of drag reduction is given by Gad-el-Hak [129].

In broad terms, skin-friction drag [131] can be reduced in one of two ways. Either laminar flow can be maintained by postponing laminar-turbulent transition—so-called laminar-flow technology—or ways can be found to reduce the surface shear stress generated by the turbulent boundary layer. The maintenance of laminar flow by prolonging a favorable or constant-pressure region over the wing surface is discussed briefly in Section 8.2.5. Active laminar-flow control requires the use of boundary-layer suction, which is described in Section 9.5.1. Another laminar-flow technique based on compliant walls (artificial dolphin skin) is described in Section 9.5.2. Riblets are the main technique for reducing turbulent skin friction, and their use is described in Section 9.5.3.

9.5.1 Laminar Flow Control by Boundary-Layer Suction

Distributed suction acts in two main ways to suppress laminar-turbulent transition. First, it reduces the boundary-layer thickness. (Recall from Section 8.1 that for a fixed pressure gradient a critical Reynolds number based on boundary-layer thickness must be reached before transition is possible.) Second, it creates a much fuller velocity profile within the boundary layer, somewhat similar to the effect of a favorable pressure gradient. This makes the boundary layer much more stable with respect to the growth of small disturbances (e.g., Tollmien-Schlichting waves). In effect, it also greatly increases the critical Reynolds number. The earliest work on laminar-flow control (LFC), including the use of suction, was carried out in wind tunnels in Germany and Switzerland during the late 1930s [132]. The first flight tests took place in the United States in 1941 using a B-18 bomber fitted with a wing glove. The maximum flight speed available and the chord of the wing glove limited the transitional Reynolds number achieved to a lower value than that obtained in wind-tunnel tests.

Research on suction-type LFC continued until the 1960s in Great Britain and the United States. This work included several flight tests using wing gloves on aircraft like the F-94 and the Vampire. In such tests full-chord laminar flow was maintained on the wing's upper surface at Reynolds numbers up to 30×10^6. Achieving this transition delay required exceptionally well-made smooth wings. Even very small surface roughness due to insect impact, for example, caused wedges of turbulent flow to form behind each individual roughness element. Further flight tests in the United States and Great Britain (British researchers used a vertically mounted test wing on a Lancaster bomber) revealed that it was much more difficult to maintain laminar flow over swept wings because swept leading edges bring into play more powerful routes to transition than the amplification of Tollmien-Schlichting waves. First, turbulence propagates along the leading edge from the wing roots; this is termed *leading-edge contamination*. Second, completely different and more powerful disturbances form in the boundary layer over the swept wing's leading-edge region. These are called *cross-flow vortices*.

Because of the practical difficulties and the relatively low price of aviation fuel, LFC research was discontinued at the end of the 1960s, but it has been revived in response to the growing awareness of environmental requirements for fuel economy and lower engine emissions. LFC is really the only technology currently available with the potential for substantial improvement in fuel economy. For transport aircraft, the reduction in fuel burnt could exceed 30%. Recent technical advances have also helped to overcome some of the practical difficulties. The principal advances are

- Krueger flaps (Fig. 9.29) at the leading edge that increase lift and act to protect the leading-edge region from insect impact during takeoff and climb-out.

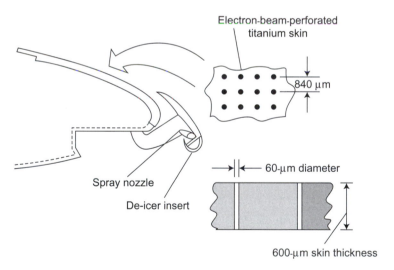

FIGURE 9.29

Leading-edge arrangement for 1983–1987 flight tests conducted on a JetStar aircraft at NASA's Dryden Flight Research Center. Important features: (1) suction on upper surface only; (2) suction through electron-beam-perforated skin; (3) leading-edge shield extended for insect protection; (4) de-icer insert on shield for ice protection; (5) supplementary spray nozzles for protection from insects and ice. *Source: Based on Braslow AL. A history of suction-type laminar flow control with emphasis on flight research. NASA History Division, Monographs in Aerospace History, Number 13; 1999.*

- Improved manufacturing techniques, such as laser drilling and electron-beam technology, that permit leading edges to be smooth perforated titanium skins.
- The use of hybrid LFC.

The first two innovations are illustrated in Fig. 9.29. Perforated skins give distributed suction, which is more effective than discrete suction slots. Hybrid LFC would be particularly useful for swept-back wings because it is not possible to maintain laminar flow over these wings by means of natural LFC alone. This depends on shaping the wing section so as to postpone the onset of an adverse pressure gradient to as far aft as possible. Tollmien-Schlichting waves can be suppressed in this way, but not the more powerful transition mechanisms of leading-edge contamination and cross-flow vortices found in the leading-edge region. With hybrid LFC, suction is used only in the leading-edge region to suppress the cross-flow vortices and leading-edge contamination. Over the remainder of the wing, where amplification of Tollmien-Schlichting waves is the main route to transition, wing-profile shaping can reduce the effects of an adverse pressure gradient. In practice, this is easier to achieve for the upper surface only. Because of the higher flow speeds there, the upper surface produces most of the skin-friction drag. Hybrid LFC wings were extensively and successfully flight-tested by Boeing on a modified 757 airliner during the early 1990s. Although LFC based on boundary-layer suction has yet to be used in any operational aircraft, the less risky hybrid LFC is now practically realizable. Based on proven current technology, a 10% to 20% improvement in fuel consumption can be achieved for moderate-sized subsonic commercial aircraft.

A detailed account of LFC technology and its history is given by Braslow [133].

9.5.2 Compliant Walls: Artificial Dolphin Skins

It is widely thought that some dolphin species possess an extraordinary laminar-flow capability. Certainly mankind has long admired the swimming skills of these fleet creatures. Scientific interest in dolphin hydrodynamics dates back at least as far as 1936, when Gray [134] published his analysis of dolphin energetics. It is widely accepted that the bottle-nosed dolphin (*Tursiops truncatus*) can maintain a sustained swimming speed of up to 9 m/s. Gray followed the usual practice of marine engineers in modeling the dolphin's body as a one-sided flat plate 2 m in length. The corresponding Reynolds number based on overall body length was about 20×10^6. Even in a very-low-noise flow environment, the Reynolds number Re_{xt} for transition from laminar to turbulent flow does not exceed 2 to 3×10^6 for flow over a flat plate. Accordingly, Gray assumed that if conventional hydrodynamics were involved, the flow would be mostly turbulent and the dolphin would experience a large drag force—so large, in fact, that at 9 m/s its muscles would have to deliver about seven times more power per unit mass than any other mammalian muscles. This led Gray and others to argue that the dolphin must be capable of maintaining laminar flow by some extraordinary means, a hypothesis that has come to be known as *Gray's paradox*.

Little was known about laminar-turbulent transition in 1936, and Gray would have been unaware of the effects of the streamwise pressure gradient along the boundary layer (see Section 8.3). We now know that transition is delayed in favorable pressure gradients and promoted in adverse ones. For the dolphin, this means that the transition point occurs near the point of minimum pressure, which for *Tursiops truncatus* is about halfway along the body, corresponding to $Re_{xt} = 10 \times 10^6$. Taking this into account, the estimated drag is much less and the required power output from the muscles exceeds the mammalian norm by no more than a factor of two. There is recent evidence that dolphin muscle is capable of a higher output. On reexamination of Gray's paradox, then, there is now much less of an anomaly to explain. Nevertheless, the dolphin may still benefit from in a laminar-flow capability. Moreover, there is ample evidence, which will be briefly reviewed momentarily, that the use of properly designed, passive artificial dolphin skins (i.e., compliant walls) can maintain laminar flow at much higher Reynolds numbers than found for rigid surfaces.

In the late 1950s, Max Kramer [135], a German aeronautical engineer working in the United States, carried out a careful study of the dolphin epidermis and designed compliant coatings closely based on what he considered to be the skin's key properties (see Fig. 9.30). Certainly, his coatings bore a considerable resemblance to dolphin skin, particularly with respect to dimensions (see also Fig. 9.31). They were manufactured from soft natural rubber and Kramer mimicked the effects of the fatty, more hydrated tissue with a layer of highly viscous silicone oil in the voids created by the short stubs. He achieved drag reductions of up to 60% for his best compliant coating compared with rigid-walled control in seawater at a maximum speed of 18 m/s. Three grades of rubber and various silicone oils with a range of viscosities were tested to obtain the largest drag reduction. The optimum viscosity was found to be about 200 times that of water.

Although no evidence existed beyond drag reduction, Kramer believed that his compliant coatings acted as a form of laminar-flow control, reducing or suppressing the small-amplitude Tollmien-Schlichting waves and thereby postponing transition to a much higher Reynolds number—or even eliminating it. He believed that the fatty tissue in the upper dermal layer of the dolphin skin, and by analogy the silicone oil in his coatings, acted as damping to suppress the growth of the waves. This must have seemed eminently reasonable at the time. Surprisingly, however, the early theoretical work by Benjamin [136], while showing that wall compliance can indeed suppress the growth of Tollmien-Schlichting waves, also showed that wall damping in itself promoted wave growth (i.e., the waves grew faster for a high level of damping than for a low level). This led to considerable skepticism about Kramer's claims. But the early theories, including Benjamin's, were rather general and made no attempt to model Kramer's coatings theoretically. A detailed theoretical assessment of the laminar-flow capabilities of his coatings was carried out much later by Carpenter and Garrad [137], who modeled the coatings as elastic plates supported on spring foundations with the effects of visco-elastic damping and the viscous damping

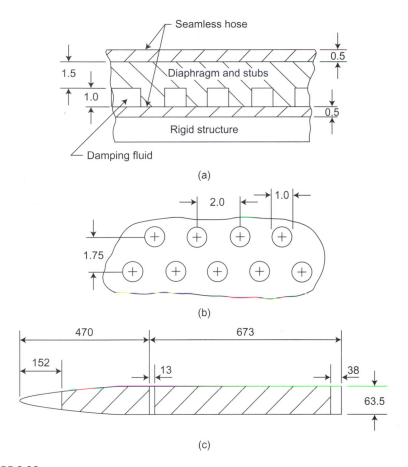

FIGURE 9.30

Kramer's compliant coating and model. All dimensions are in millimeters. (a) Cross-section; (b) cut-through stubs; (c) model: shaded regions were coated. *Source: Based on a figure in Carpenter PW, Davies C, Lucey AD. Hydrodynamics and compliant walls: Does the dolphin have a secret? Current Science 2000; 79(6):758–765.*

fluid included. Their results broadly confirmed that Kramer's coatings substantially reduced Tollmien-Schlichting wave growth.

Experimental confirmation of the stabilizing effects of wall compliance on Tollmien-Schlichting waves was provided by Gaster [138], who found close agreement between the measured growth and the theory's predictions. Many authors have used versions of this theory to show how suitably designed compliant walls can achieve a fivefold or greater increase in the transitional Reynolds number Re_{xt} as compared with the corresponding rigid surfaces. Although compliant walls have yet

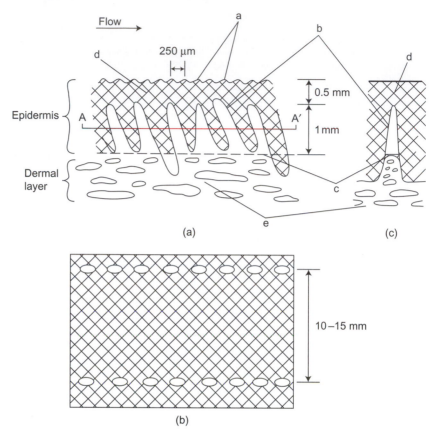

FIGURE 9.31

Structure of dolphin skin: (a) cross-section; (b) cut through the dermal papillae at AA′; (c) front view. *Key*: a—cutaneous ridges or microscales; b—dermal papillae; c—dermal ridge; d—upper epidermal layer; e—fatty tissue. *Source: Based on a figure in Carpenter PW, Davies C, Lucey AD. Hydrodynamics and compliant walls: Does the dolphin have a secret? Current Science 2000; 79(6):758–765 (see Fig. 9.30).*

to be used for laminar-flow control, there is little doubt that they have this potential in certain marine applications. In principle, they could also be used in aeronautical applications, but in practice, owing to the need to match the inertias of the air and the wall, the wall structure would have to be impractically light and flimsy [139].

9.5.3 Riblets

A moderately effective way of reducing turbulent skin friction involves surface modification in the form of *riblets*. These may take many forms, but essentially consist of minute streamwise ridges and valleys. One possible configuration is depicted in Fig. 9.32(b). Similar triangular-shaped riblets are available in the form of polymeric film from the 3M Company. The optimum, nondimensional, spanwise spacing between the riblets is given in wall units (see Section 8.10.5) by

$$s^+ = s\sqrt{\tau_w/\nu} = 10 \text{ to } 20$$

which corresponds to an actual spacing of 25 to 75 μm for flight conditions. (Note that the thickness of a human hair is approximately 70 μm). The 3M riblet film has been flight-tested on an in-service Airbus A300–600 and on other aircraft. It is currently being used on regular commercial flights of the Airbus A340–300 aircraft by Cathay Pacific, which has been observed a reduction in skin-friction drag on the order of 5% to 8%. Skin-friction drag accounts for about 50% of the total drag for the Airbus A340–300 (a rather higher proportion than for many other types of airliner). Probably only about 70% of the surface of the aircraft can be covered with riblets, leading to about 3% reduction in total drag [140]. This is fairly modest but represents a worthwhile savings in fuel and an increase in payload. Riblets have been used on Olympic-class rowing shells in the United States and on the hull of the *Stars and Stripes*, the winner of the 1987 America's Cup.

The basic concept behind riblets has many origins, but it was probably the work at NACA Langley [141] in the United States that led to the present developments. The concept was also discovered independently in Germany through the research on the hydrodynamics of riblet-like formations on shark scales [142]. The nondimensional riblet spacings found on shark scales lie in the range $8 < s^+ < 18$— that is, almost identical to the range of values given earlier for optimum drag reduction in NASA experiments and others on man-made riblets.

Given that the surface area is increased by a factor of 1.5 to 2.0, the actual reduction in mean surface shear stress achieved with riblets is some 12% to 16%. How do riblets produce a reduction in skin-friction drag? At first sight it is astonishing that such minute modifications to the surface should have such a large effect. The phenomenon is also in conflict with the classic view in aerodynamics and hydrodynamics that surface roughness leads to a drag increase. A plausible explanation for the effect of riblets is that they interfere with the development of the near-wall structures in the turbulent boundary that are mainly responsible for generating wall shear stress. (see Section 8.10.5). These structures can be thought of as "hairpin" vortices that form near the wall, as depicted in Fig. 9.32(a). As these vortices grow and develop, they reach a point where the head of a vortex is violently ejected from the wall. Simultaneously the vortex's contra-rotating, streamwise-oriented legs move closer together, thereby inducing a powerful downwash of high-momentum fluid between them. This sequence of events is often termed a "near-wall burst." It is thought that riblets impede the close approach of the vortex legs, thereby weakening the bursting process.

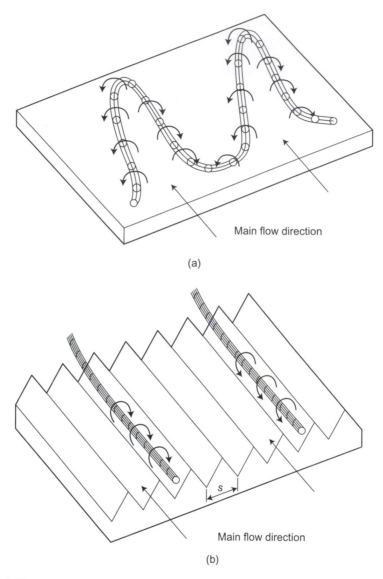

Main flow direction

(a)

Main flow direction

(b)

FIGURE 9.32

Effect of riblets on near-wall structures in a turbulent boundary layer.

9.6 REDUCTION OF FORM DRAG

Form drag is kept to a minimum by avoiding flow separation, and in this respect it was discussed in previous sections. Streamlining is vitally important for reducing form drag. It is worth noting that at high Reynolds numbers a circular cylinder has roughly the same overall drag as a classic streamlined airfoil, with a chord length equal to 100 cylinder radii. Form drag is by far the main contribution to overall drag for bluff bodies like the cylinder, whereas for streamlined bodies skin-friction drag is predominant, form drag being less than 10% of overall drag. For bluff bodies even minimal streamlining can be very effective.

9.7 REDUCTION OF INDUCED DRAG

Aspects of this topic were discussed in Chapter 5. There it was shown that, in accordance with classic wing theory, induced drag falls as the aspect ratio of the wing increases. It was also shown that, for a given aspect ratio, elliptic-shaped wings (strictly, wings with elliptic wing loading) have the lowest induced drag. Over the past 25 years, the winglet has been developed as a device for reducing induced drag without increasing the aspect ratio. A typical example is depicted in Figs. 9.33(a) and 9.36. Winglets of this and other types have been fitted to many civil aircraft, ranging from business jets to very large airliners.

The physical principle behind the winglet is illustrated in Fig. 9.33(b,c). On all subsonic wings, there is a tendency for a secondary flow to develop from the high-pressure region below the wing around the wingtip to the relatively low-pressure region on the upper surface (Fig. 9.33(b)). This is part of the formation of trailing vortices. If a winglet of the appropriate design and orientation is fitted to the wingtip, the secondary flow causes it to be at an effective angle of attack, giving rise to lift and drag components L_w and D_w relative to it, as shown in Fig. 9.33(c). Both L_w and D_w have components in the direction of flight. L_w provides a component to counter the aircraft drag, while D_w provides one that augments it. For a well-designed winglet the contribution of L_w predominates, resulting in a net reduction in overall drag, or a thrust, equal to ΔT (Fig. 9.33c). For example, data available for the Boeing 747–400 indicate that winglets reduce drag by about 2.5% corresponding to a weight savings of 9.5 tons at takeoff [143].

The winglet shown in Fig. 9.33(a) has a sharp angle where it joins the main wing. This creates the sort of corner flow seen at wing-body junctions. Over the rear part of the wing, the boundary layer in this junction is subject to an adverse streamwise pressure gradient from both the main wing and the winglet. This tends to intensify the effect of the adverse pressure gradient, leading to a risk of flow separation and increased drag, which can be avoided by the use of a blended winglet (Fig. 9.34(a)) or one that is shifted downstream (Fig. 9.34(b)). Variants of both designs are very common. The pressure distributions over the upper surface of the main wing close to the wingtip are plotted in Fig. 9.35(a) for all three winglet types and for the unmodified wing. The winglet with the sharp corner has a distribution with a narrow suction peak close to the leading edge that is followed by a steep adverse pressure gradient. This

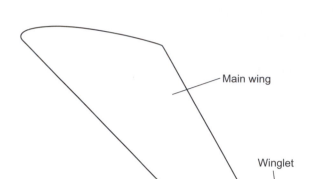

(a) Winglet attached to wingtip

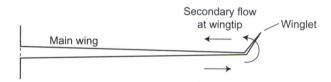

(b) Secondary flow around wingtip

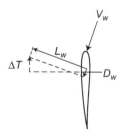

(c) Force acting on winglet

FIGURE 9.33

Using winglets to reduce induced drag.

type of pressure distribution favors early laminar-turbulent transition and also risks flow separation. In contrast, the other two designs, especially the downstream-shifted winglet, have much more benign pressure distributions. Calculations using the panel method indicate that all three types lead to a similar reduction in induced drag [144]. This suggests that the two winglet designs in Fig. 9.34 are to be preferred to the one with a sharp corner in Fig. 9.33.

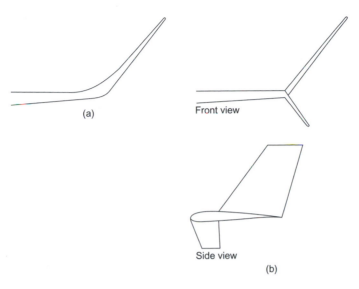

FIGURE 9.34

Alternative winglet designs: (a) blended winglet; (b) winglet shifted downstream.

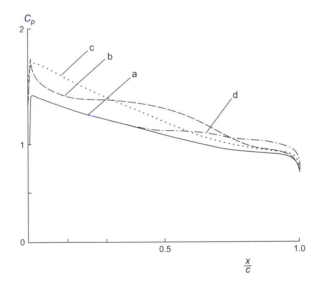

FIGURE 9.35

Streamwise pressure distributions over the upper surface of the main wing close to the wingtip for different winglet configurations (schematic only): (a) wingtip without winglet; (b) winglet with a sharp corner; (c) blended winglet; (d) winglet shifted downstream.
Source: Based on a figure from Winglets: a close look. Available from
http://www.mh-aerotools.de/airfoils/winglets.htm.

FIGURE 9.36

View of the Airbus A340 showing winglets attached to the wingtips. These devices reduce induced drag (also see Fig. 9.33). In the foreground is the wing of the Airbus A320–200 fitted with another device known as a wingtip fence. (*Source: Gert Wunderlich.*)

9.8 REDUCTION OF WAVE DRAG

Aspects of wave drag reduction were covered in the discussion of swept wings in Section 5.7 and of supercritical airfoils in Sections 7.1.1 and 9.2. In the latter case it was found that keeping the pressure uniform over the upper wing surface minimizes the shock strength, thereby reducing wave drag. A somewhat similar principle holds for the whole wing-body combination of a transonic aircraft. This was encapsulated in the *area rule* formulated in 1952 by Richard Whitcomb [145] and his team at NACA Langley. It was known that as the wing-body configuration passes through the speed of sound, the conventional straight fuselage, shown in Fig. 9.37(a), experiences a sharp rise in wave drag. Whitcomb's team proved that this rise in drag could be considerably reduced if the fuselage was waisted, as shown in Fig. 9.37(b), in such a way as to keep the total cross-sectional area of the wing-body combination as uniform as possible. Waisted fuselages of this type became common in aircraft designed for transonic operation.

The area rule was first applied to a production aircraft in the Convair F-102A, the USAF's first supersonic interceptor. Emergency application of the area rule became necessary because of a serious problem revealed during the flight tests of the prototype aircraft, the YF-102. Its transonic drag was found to exceed the thrust produced

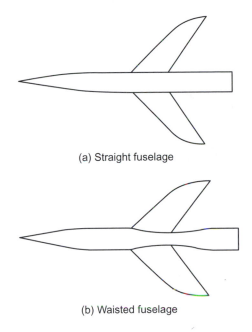

(a) Straight fuselage

(b) Waisted fuselage

FIGURE 9.37

Application of the area rule for minimizing wave drag.

by the most powerful engine then available, which threatened to jeopardize the entire program, considering that supersonic flight speed was an essential USAF specification. The area rule was used to guide a major revised design of the fuselage. This reduced the drag sufficiently for supersonic Mach numbers to be achieved.

Propulsion Devices

LEARNING OBJECTIVES

- Learn how to apply one-dimensional control-volume analysis or simple momentum theory to examine the performance of propulsion devices.

- Learn that propulsive force is obtained by increasing the momentum of the working gas in the direction opposite to that of the force.

- Learn about propellers, turbo-jets, and helicopters operating in the Earth's atmosphere.

- Learn from blade-element theory how propulsion devices using rotating lifting surfaces perform.

- Learn more about control-volume analysis by examining rocket motors and hovercraft.

The forward propulsive force, or thrust, in aeronautics is invariably obtained by increasing the rearward momentum of a quantity of gas. Aircraft propulsion systems may be divided into two classes:

- Class 1: The gas worked on is wholly or principally atmospheric air.
- Class 2: The gas does not contain atmospheric air in any appreciable quantity.

Class 1 includes turbo-jets, ram-jets, and all systems using airscrews or helicopter rotors. It also includes ornithopters (and birds, flying insects, etc.). The only class 2 system currently in use in aviation is the rocket motor.

10.1 FROUDE'S MOMENTUM THEORY OF PROPULSION

Froude's theory applies to propulsive systems of class 1. In this class, work is done on air from the atmosphere and its energy is increased. This increase is used to increase the rearward momentum of the air, the reaction to which appears as a thrust on the engine or airscrew.

The theory is based on the concept of the ideal actuator disc or pure energy supplier. This is an infinitely thin disc of area S that offers no resistance to air passing through it. The air receives energy in the form of pressure from the disc, the energy

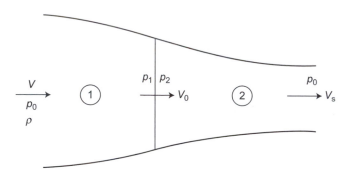

FIGURE 10.1

Ideal actuator disk and flow in the slipstream; the density, ρ, is constant in this example.

being added uniformly over the entire disc area. It is assumed that the velocity of the air through the disc is constant over the whole area and that all of the energy supplied is transferred to the air.

We consider the system shown in Fig. 10.1, which represents an actuator disc at rest in a fluid that, a long way ahead of the disc, is moving uniformly with speed V and has pressure p_0. The outer curved lines represent the streamlines that separate the fluid that passes through the disc from the fluid that does not. As the fluid between these streamlines approaches the disc, it accelerates to speed V_0, its pressure decreasing to p_1. At the disc, the pressure increases to p_2, but continuity prohibits a sudden change in speed. Behind the disc, the air expands and accelerates until, well behind the disc, its pressure returns to p_0, when its speed is V_s. The flow between the bounding streamlines behind the disc is known as the *slipstream*.

The rate of mass flowing through the disk per unit time is

$$\dot{m} = \rho S V_0 \tag{10.1}$$

The increase in the horizontal component of (or rearward) momentum of this mass flow is

$$\dot{m}(V_s - V) = \rho S V_0 (V_s - V) \tag{10.2}$$

This is the thrust on the disc. Hence

$$T = \rho S V_0 (V_s - V) \tag{10.3}$$

The thrust can also be calculated from the pressures on the two sides of the disc as

$$T = S(p_2 - p_1) \tag{10.4}$$

The flow is divided into two regions, 1 and 2, and Bernoulli's equation may be applied within each. Since the fluid receives energy at the disc, Bernoulli's equation does not apply through the disc. Thus

$$p_0 + \frac{1}{2}\rho V^2 = p_1 + \frac{1}{2}\rho V_0^2 \tag{10.5}$$

and

$$p_2 + \frac{1}{2}\rho V_0^2 = p_0 + \frac{1}{2}\rho V_s^2 \tag{10.6}$$

From Eqs. (10.5) and (10.6),

$$\left(p_2 + \frac{1}{2}\rho V_0^2\right) - \left(p_1 + \frac{1}{2}\rho V_0^2\right) = \left(p_0 + \frac{1}{2}\rho V_s^2\right) - \left(p_0 + \frac{1}{2}\rho V^2\right)$$

that is,

$$p_2 - p_1 = \frac{1}{2}\rho\left(V_s^2 - V^2\right) \tag{10.7}$$

Substituting this into Eq. (10.4) and equating the result to Eq. (10.3) (i.e., equating the two expressions for the thrust), we get

$$\frac{1}{2}\rho S\left(V_s^2 - V^2\right) = \rho S V_0 \left(V_s - V\right)$$

Dividing this by $\rho S(V_s - V)$ and rearranging terms, we get

$$V_0 = \frac{1}{2}\left(V_s + V\right) \tag{10.8}$$

This shows that the velocity through the disc is the arithmetic mean of the velocities well upstream and in the fully developed slipstream. Furthermore, if the velocity through the disc V_0 is written as

$$V_0 = V\left(1 + a\right) \tag{10.9}$$

it follows from Eq. (10.8) that

$$V_s + V = 2V_0 = 2V\left(1 + a\right)$$

whence

$$V_s = V\left(1 + 2a\right) \tag{10.10}$$

The quantity a is the *inflow factor*.

A unit mass of the fluid upstream of the disc has kinetic energy of $\frac{1}{2}V^2$ and pressure energy appropriate to pressure p_0, whereas the same mass well behind the disc has, after passing through the disc, kinetic energy of $\frac{1}{2}V_s^2$ and pressure energy appropriate to pressure p_0. Thus the unit mass of the fluid receives an energy increase of $\frac{1}{2}(V_s^2 - V^2)$ on passing. The rate of increase in fluid energy in the system, dE/dt, is given by

$$
\begin{aligned}
\frac{dE}{dt} &= \rho S V_0 \frac{1}{2}\left(V_s^2 - V^2\right) \\
&= \frac{1}{2}\rho S V_0 \left(V_s^2 - V^2\right)
\end{aligned}
\tag{10.11}
$$

This rate of increase is, in fact, the power supplied to the actuator disc.

If we now imagine that the disc is moving from right to left at speed V into initially stationary fluid, we see that useful work is done at the rate TV. Thus the efficiency of the disc as a propulsive system is

$$
\eta_i = \frac{TV}{\frac{1}{2}\rho S V_0 \left(V_s^2 - V^2\right)}
$$

Substituting for T from Eq. (10.3) gives

$$
\begin{aligned}
\eta_i &= \frac{\rho S V_0 \left(V_s - V\right) V}{\frac{1}{2}\rho S V_0 \left(V_s^2 - V^2\right)} \\
&= \frac{V}{\frac{1}{2}\left(V_s + V\right)}
\end{aligned}
\tag{10.12}
$$

This is the ideal propulsive efficiency, or the Froude efficiency of the propulsive system.

In practice, the function of the ideal actuator disc is carried out by the airscrew or jet engine, which violates some or all of the assumptions made. Each departure from the ideal leads to a reduction in efficiency; thus the efficiency of a practical propulsive system is always less than the Froude efficiency as calculated for an ideal disc of the same area producing the same thrust under the same conditions.

Equation (10.12) may be treated to give different expression for efficiency, each of which has its own merit and use. Thus

$$
\begin{aligned}
\eta_i &= \frac{V}{\frac{1}{2}\left(V_s + V\right)} \\
&= \frac{2}{\left[1 + \left(V_s/V\right)\right]}
\end{aligned}
\tag{10.12a}
$$

$$= \frac{V}{V_0} \tag{10.12b}$$

$$= \frac{1}{(1+a)} \tag{10.12c}$$

Also, since useful power is equal to TV and the efficiency is V/V_0, the power supplied is

$$P = \frac{TV}{V/V_0} = TV_0 \tag{10.13}$$

Of particular interest is Eq. (10.12a), which shows that, for a given flight speed V, efficiency decreases with increasing V_s. Now the thrust is obtained by accelerating a mass of air. Consider two extreme cases. In the first, a large mass of air is affected— that is, the disc diameter is large. Then the required increase in air speed is small, so V_s/V differs little from unity and efficiency is relatively high. In the second case, a small-diameter disc affects a small mass of air, requiring a large increase in speed to give the same thrust. Thus V_s/V is large, leading to low efficiency. Therefore, to achieve a given thrust at a high efficiency, it is necessary to use the largest practicable actuator disc.

In fact, an airscrew affects a relatively large mass of air and therefore has high propulsive efficiency. A simple turbo-jet or ram-jet, on the other hand, is closer to the second extreme considered previously and so has poor propulsive efficiency. However, at high forward speeds, compressibility marked by reduces the efficiency of a practical airscrew; the advantage then shifts to the jet engine. It was to improve the propulsive efficiency of the turbo-jet engine that the bypass or turbo-fan type of engine was introduced. In this engine type, only part of the air taken is fully compressed and passed through the combustion chambers and turbines. The remainder is slightly compressed and ducted around the combustion chambers and then exhausted at a relatively low speed, producing thrust at fairly high propulsive efficiency. The air passed through the combustion chambers is ejected at high speed, producing thrust at a comparatively low efficiency. Overall propulsive efficiency is thus slightly greater than that of a simple turbo-jet engine giving the same thrust. The turbo-prop engine is, in effect, an extreme form of bypass engine in which nearly all of the thrust is obtained at high efficiency.

Another very useful equation in Froude's theory may be obtained by expressing Eq. (10.3) in a different form. Since

$$V_0 = V(1+a) \quad \text{and} \quad V_s = V(1+2a)$$

$$T = \rho S V_0 (V_s - V) = \rho S V (1+a)[V(1+2a) - V] \tag{10.14}$$

$$= 2\rho S V^2 a (1+a)$$

Example 10.1

An airscrew is required to produce a thrust of 4000 N at a flight speed of $120\,\mathrm{m\,s^{-1}}$ at sea level. If the diameter is 2.5 m, estimate the minimum power that must be supplied on the basis of Froude's theory:

$$T = 2\rho S V^2 a (1+a)$$

that is,

$$a + a^2 = \frac{T}{2\rho S V^2}$$

Now $T = 4000\,\mathrm{N}$, $V = 120\,\mathrm{m\,s^{-1}}$, and $S = 4.90\,\mathrm{m^2}$. Thus

$$a + a^2 = \frac{4000}{2 \times 1.226 \times 14400 \times 4.90} = 0.0232$$

whence

$$a = 0.0227$$

Then the ideal efficiency is

$$\eta_i = \frac{1}{1.0227}$$

$$\text{Useful power} = TV = 480000\ \mathrm{W}$$

Therefore, the minimum power supplied P is given by

$$P = 480000 \times 1.0227 = 491\ \mathrm{kW}$$

The actual power required by a practical airscrew is probably about 15% greater than this (approximately 560 kW).

Example 10.2

A pair of airscrews is placed in tandem (Fig. E10.2) at a streamwise spacing sufficient to eliminate mutual interference. The rear airscrew is of such a diameter that it just fills the slipstream of the front airscrew. Using the simple momentum theory, calculate: (1) the efficiency of the combination and (2) the efficiency of the rear airscrew if the front airscrew has a Froude efficiency of 90% and if both airscrews deliver the same thrust. (University of London [U of L])

For the front airscrew, $\eta_i = 0.90 = \frac{9}{10}$. Therefore,

$$1/(1+a) = \frac{9}{10}; \quad 1+a = \frac{10}{9}; \quad a = \frac{1}{9}$$

Thus

$$V_0 = V(1+a) = \frac{10}{9}V$$

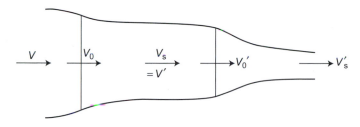

FIGURE E10.2

Actuator discs in tandem.

and

$$V_s = V(1 + 2a) = \frac{11}{9}V$$

The thrust of the front airscrew is

$$T = \rho S_1 V_0 (V_s - V) = \rho S_1 \left(\frac{10}{9}V\right)\left(\frac{2}{9}V\right) = \frac{20}{81}\rho S_1 V^2$$

The second airscrew is working entirely in the slipstream of the first, so the speed of the approaching flow is V_s (i.e., $\frac{11}{9}V$). The thrust is

$$T = \rho S_2 V_0' (V_s' - V') = \rho S_2 V_0' (V_s' - V_s)$$

Now, by continuity,

$$\rho S_2 V_0' = \rho S_1 V_0$$

Also, the thrusts from the two airscrews are equal, so

$$T = \rho S_1 V_0 (V_s - V) = \rho S_2 V_0' (V_s' - V') = \rho S_1 V_0 (V_s' - V_s)$$

whence

$$V_s - V = V_s' - V_s$$

that is,

$$V_s' = 2V_s - V = \left(\frac{22}{9} - 1\right)V = \frac{13}{9}V$$

Then, if the rate of mass flow through the discs is $\dot{m}$,

$$\text{Thrust of rear airscrew} = \dot{m}(V_s' - V_s) = \dot{m}\left(\frac{13}{9} - \frac{11}{9}\right)V = \frac{2}{9}\dot{m}V$$

The useful power given by the second airscrew is TV, not TV_s; therefore,

$$\text{Useful power from 2nd airscrew} = \frac{2}{9}\dot{m}V^2$$

Kinetic energy added per second by the second airscrew, which is the power supplied by (and to) the second disc, is

$$\frac{dE}{dt} = P = \frac{1}{2}\dot{m}\left(V_s'^2 - V_s^2\right) = \frac{1}{2}\dot{m}\left(\frac{169}{81} - \frac{121}{81}\right)V^2 = \frac{8}{27}\dot{m}V^2$$

Thus the efficiency of the rear components is

$$\frac{\dfrac{2}{9}\dot{m}V^2}{\dfrac{8}{27}\dot{m}V^2} = 0.75 \quad \text{or} \quad 75\%$$

$$\text{Power input to front airscrew} = \frac{TV}{0.90}$$

$$\text{Power input to rear airscrew} = \frac{TV}{0.75}$$

Therefore,

$$\text{Total power input} = \left(\frac{10}{9} + \frac{4}{3}\right)TV = \frac{22}{9}TV$$

$$\text{Total useful power output} = 2TV$$

Therefore,

$$\text{Efficiency of combination} = \frac{2TV}{\dfrac{22}{9}TV} = \frac{9}{11} = 0.818 \quad \text{or} \quad 81.8\%$$

10.2 AIRSCREW COEFFICIENTS

The performance of an airscrew may be determined by model tests. As is the case with all such tests, it is necessary to find some way of relating them to full-scale performance. Dimensional analysis is used for this purpose, which leads to a number of coefficients, analogous to the lift and drag coefficients of a body. These coefficients also serve as a convenient way of presenting airscrew performance data, which may be calculated by blade-element theory (Section 10.4) for use in aircraft design.

10.2.1 Thrust Coefficient

Consider an airscrew of diameter D revolving at n revolutions per second, driven by a torque Q and giving a thrust T. The characteristics of the fluid are defined by its density ρ, its kinematic viscosity ν, and its modulus of bulk elasticity K. The forward speed of the airscrew is V. We assume that

$$T = h(D, n, \rho, v, K, V)$$
$$= CD^a n^b \rho^c v^d K^e V^f \qquad (10.15)$$

Then, putting this in dimensional form, we get

$$\left[\text{MLT}^{-2} \right] = \left[(\text{L})^a (\text{T})^{-b} (\text{ML}^{-3})^c (\text{L}^2 \text{T}^{-1})^d (\text{ML}^{-1} \text{T}^{-2})^e (\text{LT}^{-1})^f \right]$$

Separating this into the three fundamental equations, we get

$$(\text{M}) \quad 1 = c + e$$
$$(\text{L}) \quad 1 = a - 3c + 2d - e + f$$
$$(\text{T}) \quad 2 = b + d + 2e + f$$

Solving these three equations for a, b, and c in terms of d, e, and f, we get

$$a = 4 - 2e - 2d - f$$
$$b = 2 - d - 2e - f$$
$$c = 1 - e$$

Substituting these in Eq. (10.15), we get

$$T = CD^{4-2e-2d-f} n^{2-d-2e-f} \rho^{1-e} v^d K^e V^f$$
$$= C\rho n^2 D^4 f \left[\left(\frac{v}{D^2 n} \right)^d \left(\frac{K}{\rho D^2 n^2} \right)^e \left(\frac{V}{nD} \right)^f \right] \qquad (10.16)$$

Consider the three factors within the square brackets.

- $v/D^2 n$: The product Dn is a multiple of the rotational component of the blade-tip speed; thus the complete factor is of the form $v/(\text{length} \times \text{velocity})$ and is therefore of the form of the reciprocal of a Reynolds number. Ensuring equality of Reynolds numbers as between model and full scales takes care of this term.
- $K/\rho D^2 n^2$: $K/\rho = a^2$, where a is the speed of sound in the fluid. As just noted, Dn is related to blade-tip speed and therefore the complete factor is related to (speed of sound/velocity)^2—that is, to the tip Mach number. Care in matching the tip Mach number in model test and full-scale flight allows for this factor.
- V/nD : V is the forward speed of the airscrew; V/n is therefore the distance advanced per revolution. V/nD is this advance per revolution expressed as a multiple of the airscrew diameter and is known as the advance ratio, denoted J.

Equation (10.16) may be thus written as

$$T = C\rho n^2 D^4 h\,(Re, M, J) \tag{10.17}$$

The constant C and the function $h(Re, M, J)$ are usually collected together and denoted k_T, the thrust coefficient. Finally, we have

$$T = k_T \rho n^2 D^4 \tag{10.18}$$

where k_T is a dimensionless quantity dependent on airscrew design; it is also dependent on Re, M, and J. This dependence may be found experimentally or by theoretical means—for example, the blade-element theory, the lifting-line/lifting-surface theory, or other advanced computational methodology.

10.2.2 Torque Coefficient

Torque Q is a force multiplied by a length, and it follows that a rational expression for Q is

$$Q = k_Q \rho n^2 D^5 \tag{10.19}$$

where k_Q is the torque coefficient, which, like k_T, depends on airscrew design and on Re, M, and J.

10.2.3 Efficiency

The power supplied to an airscrew is P_{in}, where

$$P_{\text{in}} = 2\pi n Q$$

whereas the useful power output is P_{out}, where

$$P_{\text{out}} = TV$$

Therefore, the airscrew efficiency η is given by

$$
\begin{aligned}
\eta &= \frac{TV}{2\pi n Q} = \frac{k_T \rho n^2 D^4 V}{k_Q \rho n^2 D^5 2\pi n} \\
&= \frac{1}{2\pi} \frac{k_T}{k_Q} \frac{V}{nD} = \frac{1}{2\pi} \frac{k_T}{k_Q} J
\end{aligned}
\tag{10.20}
$$

10.2.4 Power Coefficient

The power required to drive an airscrew is

$$P = 2\pi n Q = 2\pi n \left(k_Q \rho n^2 D^5 \right) = 2\pi k_Q \rho n^3 D^5 \tag{10.21}$$

The power coefficient C_P is then defined by

$$P = C_P \rho n^3 D^5 \tag{10.22}$$

that is,

$$C_P = \frac{P}{\rho n^3 D^5} \tag{10.22a}$$

By comparing Eqs. (10.21) and (10.22), we see that

$$C_P = 2\pi k_Q \tag{10.22b}$$

Then, from Eq. (10.20), the efficiency of the airscrew is

$$\eta = J\left(\frac{k_T}{C_P}\right) \tag{10.23}$$

10.2.5 Activity Factor

The activity factor is a measure of the power-absorbing capacity of the airscrew, which, for optimum performance, must be accurately matched to the power produced by the engine.

Consider an airscrew of diameter D rotating at n with zero forward speed. Consider in particular an element of the blade at radius r; the chord of the element is c. The airscrew generally produces a thrust, and therefore there is flow of finite speed through the disc. We ignore this inflow, however, so the motion and forces on the element are as shown in Fig. 10.2.

$$\frac{\delta Q}{r} = C_D \frac{1}{2} \rho \, (2\pi rn)^2 \, c \delta r$$

The torque associated with the element is thus

$$\delta Q = 2\pi^2 \rho C_D n^2 \left(cr^3\right) \delta r$$

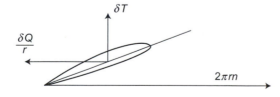

FIGURE 10.2

Relationship between thrust, torque, and rotation speed.

We further assume that C_D is constant for all blade sections, although this is not normally true since much of the blade is stalled. However, within the accuracy required by the activity factor, this assumption is acceptable. The total torque required to drive an airscrew with B blades is then

$$Q = 2\pi^2 \rho C_D B n^2 \int_{\text{root}}^{\text{tip}} cr^3 dr$$

Thus the power absorbed by the airscrew under static conditions is approximately

$$P = 2\pi n Q = 4\pi^3 \rho C_D B n^3 \int_{\text{root}}^{\text{tip}} cr^3 dr$$

In a practical airscrew, the blade roots are usually shielded by a spinner, and the lower limit of the integral, by convention, changes from zero (the root) to $0.1D$. Thus

$$P = 4\pi^3 \rho C_D B n^3 \int_{0.1D}^{0.5D} cr^3 dr$$

Defining the activity factor (AF) as

$$AF = \frac{10^5}{D^5} \int_{0.1D}^{0.5D} cr^3 dr$$

leads to

$$P = 4\pi^3 \rho C_D B n^3 \left(\frac{D}{10}\right)^5 \times (AF)$$

Further work on the airscrew coefficients is most conveniently carried out through examples.

Example 10.3

An airscrew of 3.4-m diameter has the following characteristics:

J	1.06	1.19	1.34	1.44
k_Q	0.0410	0.0400	0.0378	0.0355
η	0.76	0.80	0.84	0.86

Calculate the forward speed at which it absorbs 750 kW at 1250 rpm at 3660 m ($\sigma = 0.693$) and the thrust under these conditions. Compare the efficiency of the airscrew with that of the ideal actuator disc of the same area, giving the same thrust under the same conditions.

$$\text{Power} = 2\pi n Q$$

Therefore,

$$\text{torque } Q = \frac{750000 \times 60}{2\pi \times 1250} = 5730 \text{ N m}$$

$$n = \frac{1250}{60} = 20.83 \text{ rps} \quad n^2 = 435 \,(\text{rps})^2$$

so

$$k_Q = \frac{Q}{\rho n^2 D^5} = \frac{5730}{0.639 \times 435 \times (3.4)^5 \times 1.226} = 0.0368$$

Plotting the given values of k_Q and η against J shows that, for $k_Q = 0.0368$, $J = 1.39$ and $\eta = 0.848$. Now $J = V/nD$, so

$$V = JnD = 1.39 \times 20.83 \times 3.4 = 98.4 \,\text{m s}^{-1}$$

Since the efficiency is 0.848 (or 84.8%), the thrust power is

$$750 \times 0.848 = 635 \text{ kW}$$

Therefore, the thrust is

$$T = \frac{\text{Power}}{\text{Speed}} = \frac{635000}{98.4} = 6460 \text{ N}$$

For the ideal actuator disc,

$$a(1+a) = \frac{T}{2\rho SV^2} = \frac{6460}{2 \times 0.693 \times \dfrac{\pi}{4}(3.4)^2 \times (98.4)^2 \times 1.226} = 0.0434$$

whence

$$a = 0.0417$$

Thus the ideal efficiency is

$$\eta_1 = \frac{1}{1.0417} = 0.958 \quad \text{or} \quad 95.8\%$$

The efficiency of the practical airscrew is (0.848/0.958) of that of the ideal actuator disc, so the relative efficiency of the practical airscrew is 0.885, or 88.5%.

Example 10.4

An airplane is powered by a single engine with speed-power characteristics as follows:

Speed (rpm)	1800	1900	2000	2100
Power (kW)	1072	1113	1156	1189

The fixed-pitch airscrew of 3.05-m diameter has the following characteristics:

J	0.40	0.42	0.44	0.46	0.48	0.50
K_T	0.118	0.115	0.112	0.109	0.106	0.103
k_Q	0.0157	0.0154	0.0150	0.0145	0.0139	0.0132

The airscrew is directly coupled to the engine crankshaft. What is the airscrew thrust and efficiency during the initial climb at sea level, when the aircraft speed is 45 m s^{-1}?

Preliminary calculations required are

$$Q = k_Q \rho n^2 D^5 = 324.2 k_Q n^2$$

after using the appropriate values for ρ and D, and

$$J = V/nD = 14.75/n$$

The power required to drive the airscrew P_r is

$$P_r = 2\pi n Q$$

With these expressions, the following tabulated results may be calculated. In this table, P_a is the brake power available from the engine, as given in the data, whereas the values of k_Q for the calculated values of J are read from a graph.

rpm	1800	1900	2000	2100
P_a (kW)	1072	1113	1156	1189
n (rps)	30.00	31.67	33.33	35.00
n^2 (rps)^2	900	1003	1115	1225
J	0.492	0.465	0.442	0.421
k_Q	0.01352	0.01436	0.01464	0.01538
Q (N m)	3950	4675	5405	6100
P_r (kW)	745	930	1132	1340

A graph is now plotted of P_a and P_r against rpm, the intersection of the two curves giving the equilibrium condition. This is found to be at a rotational speed of 2010 rpm (i.e., $n = 33.5$ rps). For this value of n, $J = 0.440$, giving $k_T = 0.112$ and $k_Q = 0.0150$. Then

$$T = 0.112 \times 1.226 \times (33.5)^2 \times (3.05)^4 = 13330 \text{ N}$$

and

$$\eta = \frac{1}{2\pi}\frac{k_T}{k_Q}J = \frac{1}{2\pi}\frac{0.112}{0.0150} \times 0.440 = 0.523 \quad \text{or} \quad 52.3\%$$

As a check on the correctness and accuracy of this result, note that

$$\text{Thrust power} = TV = 13300 \times 45 = 599 \text{ kW}$$

At 2010 rpm, the engine produces 1158 kW (from engine data), and therefore the efficiency is $599 \times 100/1158 = 51.6\%$, which is in satisfactory agreement with the earlier result.

10.3 AIRSCREW PITCH

By analogy with screw threads, the pitch of an airscrew is the advance per revolution. This definition, as it stands, is of little use for airscrews. Consider two extreme cases. If the airscrew is turning at, say, 2000 rpm while the aircraft is stationary, the advance per revolution is zero. If, on the other hand, the aircraft is gliding with the engine stopped, the advance per revolution is infinite. Thus the pitch can take any value and so is useless as a term to describe the airscrew. To overcome this difficulty, two more definite measures of airscrew pitch are acceptable.

10.3.1 Geometric Pitch

Consider the blade section shown in Fig. 10.3, at radius r from the airscrew axis. The broken line is the zero-lift line of the section, or the direction relative to the section of the undisturbed stream when the section gives no lift. Then the geometric pitch of the element is $2\pi r \tan\theta$—the pitch of a screw of radius r and helix angle $(90 - \theta)$ degrees. This is frequently constant for all sections of a given airscrew. In some cases, however, it varies from blade section to blade section. In such cases, the geometric pitch of each section at 70% of the airscrew radius is taken and is called the geometric mean pitch.

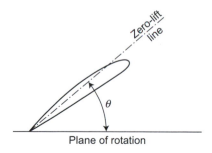

FIGURE 10.3

Location of zero-lift angle with respect to the plane of rotation.

The geometric pitch depends solely on blade geometry. It is thus a definite length for a given airscrew and does not depend on the precise conditions of operation at any instant, although many airscrews are mechanically variable in pitch.

10.3.2 Effect of Geometric Pitch on Airscrew Performance

Consider two airscrews differing only in the helix angles of the blades, and let the blade sections at, say, 70% radius be as drawn in Fig. 10.4. The blade in part (a) of the figure has a fine pitch, whereas the blade in part (b) has a coarse pitch. When the aircraft is at rest, say at the start of the takeoff run, the air velocity relative to the blade section is the resultant V_R of the velocity due to rotation $2\pi nr$ and the inflow velocity V_{in}. The blade section of the fine-pitch airscrew is working at a reasonable incidence, so the lift δL is large and the drag δD is small. Thus the thrust δT is large and the torque δQ is small, and so the airscrew is efficient. The section of the coarse-pitch airscrew, on the other hand, is stalled and therefore gives little lift and much drag. Thus the thrust is small and the torque is large, and so the airscrew is inefficient. At high flight speeds, the situation is much changed, as shown in Figs. 10.4(c) and 10.4(d). Here the section of the coarse-pitch airscrew is efficient, whereas the fine-pitch airscrew gives a negative thrust, a situation that might cause a steep dive. Thus an airscrew with a pitch suitable for low-speed flight and takeoff is liable to perform poorly at high forward speeds and vice versa. This is the one factor that limited aircraft performance in the early days of powered flight.

A great advance was achieved by the development of the two-pitch airscrew, in which each blade may be rotated bodily and set in either of two positions at will. One position gives a fine pitch for takeoff and climb; the other gives a coarse pitch for cruising and high-speed flight. Figure 10.5 shows typical variations of efficiency η with J for (a) a fine-pitch and (b) a coarse-pitch airscrew.

For low advance ratios, corresponding to takeoff and low-speed flight, the fine pitch is obviously better; for higher speeds, the coarse pitch is preferable. If the pitch is varied at will between these two values, the overall performance attainable is as given by the hatched line in Fig. 10.6, which is clearly better than that attainable from either pitch separately.

Subsequent research led to the development of the constant-speed airscrew in which the blade pitch is infinitely variable between predetermined limits. A mechanism in the airscrew hub varies the pitch to keep the engine speed constant, permitting the engine to work at its most efficient speed. The pitch variations also result in the airscrew working close to its maximum efficiency at all times. Figure 10.6 shows the variation in efficiency with J for a number of possible settings. Since the blade pitch may take any value between the curves drawn, airscrew efficiency varies with J, as shown by the dashed curve, which is the envelope of all separate η, J curves. The requirement that the airscrew always work at its optimum efficiency, while absorbing the power produced by the engine at the predetermined constant speed, calls for very skillful design in matching the airscrew with the engine.

The constant-speed airscrew, in turn, led to the innovations of feathering and reverse thrust. In feathering, the geometric pitch is made so large that the blade

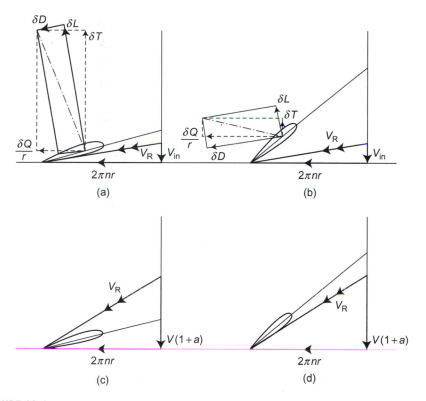

FIGURE 10.4

Effect of geometric pitch on airscrew performance.

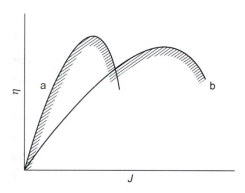

FIGURE 10.5

Efficiency for a two-pitch airscrew.

sections are almost parallel to the direction of flight. This reduces drag and prevents the airscrew from turning the engine ("windmilling") in the event of engine failure. For reverse thrust, the geometric pitch is made negative, enabling the airscrew to give

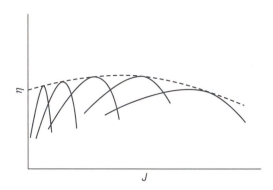

FIGURE 10.6

Efficiency for a constant-speed airscrew.

a negative thrust to supplement the brakes during the landing ground run and to assist in maneuvering the aircraft on the ground.

10.3.3 Experimental Mean Pitch

The experimental mean pitch is defined as the advance per revolution when the airscrew is producing zero net thrust. It is thus a suitable parameter for experimental measurement on an existing airscrew. Like geometric pitch, experimental mean pitch has a definite value for any given airscrew, provided that the conditions of test approximate reasonably well to practical flight conditions.

10.4 BLADE-ELEMENT THEORY

Blade-element theory permits direct calculation of the performance of an airscrew and its design to achieve a given performance.

10.4.1 Vortex System of an Airscrew

An airscrew blade is a form of lifting airfoil and as such may be replaced by a hypothetical bound vortex. In addition, a trailing vortex is shed from the tip of each blade. Since the tip traces out a helix as the airscrew advances and rotates, the trailing vortex is itself helical. Thus a two-bladed airscrew may be replaced by the vortex system of Fig. 10.7. Photographs taken of aircraft taking off in humid air show very clearly the helical trailing vortices behind the airscrew.

Rotational Interference

The slipstream behind an airscrew rotates, in the same sense as the blades rotate, about the airscrew axis. This rotation is due in part to circulation around the blades (the hypothetical bound vortex) and in part to the helical trailing vortices. Consider three planes: plane (1) immediately ahead of the airscrew blades; plane (2)

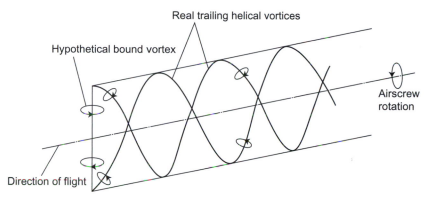

FIGURE 10.7

Simplified vortex system for a two-bladed airscrew.

the plane of the airscrew blades; and plane (3) immediately behind the blades. Ahead of the airscrew, in plane (1), the angular velocity of the flow is zero. Thus in this plane the effects of the bound and trailing vortices exactly cancel each other. In plane (2), the angular velocity of the flow is due entirely to the trailing vortices since the bound vortices cannot produce an angular velocity in their own plane. In plane (3), the angular velocity due to the bound vortices is equal in magnitude and opposite in sense to that in plane (1), and the effects of the trailing and bound vortices are now additive.

Let the angular velocity of the airscrew blades be Ω, the angular velocity of the flow in the plane of the blades be $b\Omega$, and the angular velocity induced by the bound vortices in planes ahead of and behind the disc be $\pm\beta\Omega$. This assumes that these planes are equidistant from the airscrew disc. It also assumes that the distance between these planes is small so that the effect of the trailing vortices at the three planes is practically constant. Then, ahead of the airscrew (plane 1),

$$(b - \beta)\,\Omega = 0$$

that is,

$$b = \beta$$

Behind the airscrew (plane 3), if ω is the angular velocity of the flow,

$$\omega = (b + \beta)\,\Omega = 2b\Omega$$

Thus the angular velocity of the flow behind the airscrew is twice the angular velocity in the airscrew plane. Note the similarity between this result and that for axial velocity in the simple momentum theory.

10.4.2 Performance of a Blade Element

Consider an element of length δr and chord c at radius r of an airscrew blade. This element has a speed in the plane of rotation of Ωr. The flow is itself rotating in the same plane and sense at $b\Omega$, and thus the speed of the element relative to the air in this plane is $\Omega r(1-b)$. If the airscrew is advancing at a speed of V, the velocity through the disc is $V(1+a)$, a being the inflow at radius r. Note that, in this theory, it is not necessary for a and b to be constant over the disc. The total velocity of the flow relative to the blade is V_R, as shown in Fig. 10.8.

If the line CC′ represents the zero-lift line of the blade section, then θ is, by definition, the geometric helix angle of the element, related to the geometric pitch, and α is the absolute angle of incidence of the section. The element therefore experiences lift and drag forces, respectively perpendicular and parallel to the relative velocity V_R, appropriate to the absolute incidence α. The values of C_L and C_D are those for a two-dimensional airfoil of the appropriate section at absolute incidence α, since three-dimensional effects have been allowed for in the rotational interference term $b\Omega$. This lift and drag may be resolved into components of thrust and "torque-force," as in Fig. 10.8. Here δL is the lift and δD is the drag on the element. δR is the resultant aerodynamic force, making the angle γ with the lift vector. δR is resolved into components of thrust δT and torque force $\delta Q/r$, where δQ is the torque required to rotate the element about the airscrew axis. Then

$$\tan\gamma = \delta D/\delta L = C_D/C_L \tag{10.24}$$

$$V_R = V(1+a)\mathrm{cosec}\phi = \Omega r(1-b)\sec\phi \tag{10.25}$$

$$\delta T = \delta R\cos(\phi+\gamma) \tag{10.26}$$

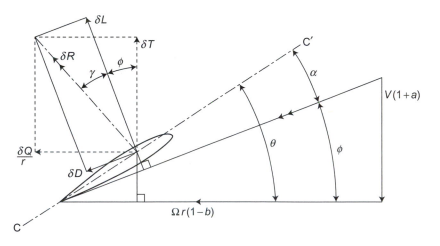

FIGURE 10.8

General blade element.

$$\frac{\delta Q}{r} = \delta R \sin(\phi + \gamma) \tag{10.27}$$

$$\tan \phi = \frac{V(1+a)}{\Omega r(1-b)} \tag{10.28}$$

The efficiency of the element η_1 is the ratio of useful power out to power in:

$$\eta_1 = \frac{V}{\Omega} \frac{\delta T}{\delta Q} = \frac{V \cos(\phi + \gamma)}{\Omega r \sin(\phi + \gamma)} \tag{10.29}$$

Now, from the triangle of velocities and Eq. (10.28),

$$\frac{V}{\Omega r} = \frac{1-b}{1+a} \tan \phi$$

whence, by Eq. (10.29),

$$\eta_1 = \frac{1-b}{1+a} \frac{\tan \phi}{\tan(\phi + \gamma)} \tag{10.30}$$

Let the solidity of the annulus σ be defined as the ratio of the total area of blade in the annulus to the total area of annulus. Then

$$\sigma = \frac{Bc}{2\pi r} \frac{\delta r}{\delta r} = \frac{Bc}{2\pi r} \tag{10.31}$$

where B is the number of blades.

Now

$$\delta L = Bc \delta r \frac{1}{2} \rho V_R^2 C_L \tag{10.32a}$$

$$\delta D = Bc \delta r \frac{1}{2} \rho V_R^2 C_D \tag{10.32b}$$

From Fig. 10.8,

$$\delta T = \delta L \cos \phi - \delta D \sin \phi$$

$$= Bc \delta r \frac{1}{2} \rho V_R^2 (C_L \cos \phi - C_D \sin \phi)$$

Therefore,

$$\frac{dT}{dr} = Bc \frac{1}{2} \rho V_R^2 (C_L \cos \phi - C_D \sin \phi)$$

$$= 2\pi r \sigma \frac{1}{2} \rho V_R^2 (C_L \cos \phi - C_D \sin \phi) \tag{10.33}$$

Bearing in mind Eq. (10.24), Eq. (10.33) may be written as

$$\frac{dT}{dr} = \pi r \sigma \rho V_R^2 C_L \left(\cos\phi - \tan\gamma \sin\phi \right)$$

$$= \pi r \sigma \rho V_R^2 C_L \sec\gamma \left(\cos\phi \cos\gamma - \sin\phi \sin\gamma \right)$$

Now, for moderate incidences of the blade section, $\tan\gamma$ is small, about 0.02. That is, $L/D \approx 50$, and therefore $\sec\gamma \approx 1$. In this case the previous equation may be written as

$$\frac{dT}{dr} = \pi r \sigma \rho V_R^2 C_L \cos(\phi + \gamma)$$

Let

$$t = C_L \cos(\phi + \gamma) \tag{10.34}$$

Then

$$\frac{dT}{dr} = \pi r \sigma t \rho V_R^2 \quad \text{for the airscrew} \tag{10.35a}$$

$$= Bc\frac{1}{2}\rho V_R^2 t \quad \text{for the airscrew} \tag{10.35b}$$

$$= c\frac{1}{2}\rho V_R^2 t \quad \text{per blade} \tag{10.35c}$$

Similarly,

$$\frac{\delta Q}{r} = \delta L \sin\phi + \delta D \cos\phi$$

whence, using Eq. (10.32a,b),

$$\frac{dQ}{dr} = 2\pi r^2 \sigma \frac{1}{2}\rho V_R^2 \left(C_L \sin\phi + C_D \cos\phi \right)$$

Let

$$q = C_L \sin(\phi + \gamma) \tag{10.36}$$

which leads to

$$\frac{dQ}{dr} = \pi r^2 \sigma q \rho V_R^2 \quad \text{total} \tag{10.37a}$$

$$= Bcr\frac{1}{2}\rho V_R^2 q \quad \text{total} \tag{10.37b}$$

$$= cr\frac{1}{2}\rho V_R^2 q \quad \text{per blade} \tag{10.37c}$$

The quantities dT/dr and dQ/dr are known as thrust grading and torque grading, respectively.

Consider now the axial momentum of the flow through the annulus. Thrust δT is equal to the product of the rate of mass flow through the element and the change in the axial velocity—$\delta T = \dot{m}\delta V$. Now

$$\dot{m} = \text{area of annulus} \times \text{velocity through annulus} \times \text{density}$$

$$= (2\pi r\delta r)[V(1+a)]\rho$$

$$= 2\pi r\rho\delta rV(1+a)$$

$$\Delta V = V_s - V = V(1+2a) - V = 2aV$$

whence

$$\delta T = 2\pi r\rho\delta rV^2 2a(1+a)$$

giving

$$\frac{dT}{dr} = 4\pi\rho rV^2 a(1+a) \tag{10.38}$$

Equating Eqs. (10.38) and (10.35a) and also using Eq. (10.25) leads to

$$4\pi\rho rV^2 a(1+a) = \pi r\sigma t\rho V^2(1+a)^2\cosec^2\phi$$

whence

$$\frac{a}{1+a} = \frac{1}{4}\sigma t\cosec^2\phi \tag{10.39}$$

In the same way, by considering the angular momentum,

$$\delta Q = \dot{m}\Delta\omega r^2$$

where $\Delta\omega$ is the change in angular velocity of the air on passing through the airscrew. Then

$$\delta Q = (2\pi r\delta r)[\rho V(1+a)](2b\Omega)r^2$$
$$= 4\pi r^3\rho Vb(1+a)\Omega\delta r \tag{10.40}$$

whence

$$\frac{dQ}{dr} = 4\pi r^3\rho Vb(1+a)\Omega \tag{10.41}$$

Substituting for V_R, as given in Eq. (10.25), this becomes

$$\frac{dQ}{dr} = \pi r^2\sigma\rho[V(1+a)\cosec\phi][\Omega r(1-b)\sec\phi]q$$

Equating the expressions for dQ/dr, we get

$$\frac{b}{1-b} = \frac{1}{4}\sigma q \operatorname{cosec}\phi \sec\phi$$

$$= \frac{1}{2}\sigma q \operatorname{cosec}2\phi \qquad (10.42)$$

The local efficiency of the blade at element η_1 is found as follows:

$$\text{Useful power output} = V\delta T = V\frac{dT}{dr}\delta r$$

$$\text{Power input} = 2\pi n\delta Q = 2\pi n\frac{dQ}{dr}\delta r$$

Therefore,

$$\eta_1 = \frac{V}{2\pi n}\frac{dT/dr}{dQ/dr}$$

$$= \frac{V}{2\pi n}\frac{2\pi r\sigma}{2\pi r^2\sigma}\frac{\frac{1}{2}\rho V_R^2 t}{\frac{1}{2}\rho V_R^2 q} \qquad (10.43)$$

$$= \frac{V}{2\pi nr}\frac{t}{q}$$

which is an alternative to Eq. (10.30).

With these expressions, we can evaluate dT/dr and dQ/dr at several radii of an airscrew blade given blade geometry and section characteristics, forward and rotational speeds, and air density. Then, by plotting dT/dr and dQ/dr against radius r and measuring the areas under the curves, we can estimate total thrust and torque for each blade and for the entire airscrew. In the design of a blade, this is the usual first step. With both thrust and torque grading known, the blades, deflection, and twist under load can be calculated. This furnishes new values of θ along the blade that we use to repeat the process until we reach our desired accuracy.

A further point to note is that portions of the blade toward the tip may attain appreciable Mach numbers—large enough for the effects of compressibility to become important. The principal effect of compressibility in this connection is its effect on the lift-curve slope of the airfoil section. Provided that the Mach number of the relative flow does not exceed about 0.75, the effect on the lift-curve slope may be approximated by the Prandtl-Glauert correction, which states that, if the lift-curve slope at zero Mach number (i.e., in incompressible flow) is a_0, the lift-curve slope at a subsonic Mach number M is a_M, where

$$a_M = \frac{a_0}{\sqrt{1 - M^2}}$$

If the Mach number does not exceed about 0.75, as stated, the effect of compressibility on section drag is very small. If the Mach number of any part of the blade exceeds the value just given, although the exact value depends on the profile and thickness-to-chord ratio of the blade section, that part of the blade loses lift and its drag rises sharply, leading to a very marked loss in overall efficiency and increased noise.

Example 10.5

At 1.25-m radius on a four-bladed airscrew of 3.5-m diameter, the local chord of each blade is 250 mm and the geometric pitch is 4.4 m. The lift-curve slope of the blade section in incompressible flow is 0.1 per degree, and the lift/drag ratio may, as an approximation, be taken to be constant at 50. Estimate thrust grading, torque grading, and local efficiency in flight at 4600 m ($\sigma = 0.629$, temperature $= -14.7°C$) at a flight speed of 67 m s^{-1} TAS and a rotational speed of 1500 rpm.

The solution to this problem is essentially a process of successive approximation to the values of a and b:

$$\text{Solidity } \sigma = \frac{Bc}{2\pi r} = \frac{4 \times 0.25}{2\pi \times 1.25} = 0.1273$$

$$1500 \text{ rpm} = 25\text{rps} = n$$

$$\tan \gamma = \frac{1}{50} \quad \text{whence} \quad \gamma = 1.15°$$

$$\text{Geometric pitch} = 2\pi r \tan\theta = 4.4$$

whence

$$\tan\theta = 0.560, \quad \theta = 29.3°$$

$$\tan\phi = \frac{V(1+a)}{\Omega r(1-b)} = \frac{67(1+a)}{62.5\pi (1-b)} = 0.3418\frac{1+a}{1-b}s$$

$$\text{Speed of sound in atmosphere} = 20.05 (273 - 14.7)^{1/2} = 325 \text{ m s}^{-1}$$

Suitable values for initial guesses for a and b are $a = 0.1$, and $b = 0.02$. Then

$$\tan\phi = 0.3418\frac{1.1}{0.98} = 0.383$$

$$\phi = 20.93°, \quad \alpha = 29.3 - 20.93 = 8.37°$$

$$V_R = V(1+a)\text{cosec}\phi$$

$$= \frac{V(1+a)}{\sin\phi} = \frac{67 \times 1.1}{0.357} = 206\,\mathrm{m\,s^{-1}}$$

$$M = \frac{206}{325} = 0.635, \quad \sqrt{1 - M^2} = 0.773$$

$$\frac{\mathrm{d}C_L}{\mathrm{d}\alpha} = \frac{0.1}{0.773} = 0.1295 \text{ per degree}$$

Since α is the absolute incidence (i.e., the incidence from zero lift),

$$C_L = \alpha\frac{\mathrm{d}C_L}{\mathrm{d}\alpha} = 0.1295 \times 8.37 = 1.083$$

Then

$$q = C_L\sin(\phi + \gamma) = 1.083\sin(20.93 + 1.15)^\circ = 0.408$$

and

$$t = C_L\cos(\phi + \gamma) = 1.083\cos 22.08^\circ = 1.004$$

$$\frac{b}{1-b} = \frac{1}{2}\sigma q\,\mathrm{cosec}\,2\phi = \frac{\sigma q}{2\sin 2\phi} = \frac{0.1274 \times 0.408}{2 \times 0.675} = 0.0384$$

giving

$$b = \frac{0.0384}{1.0384} = 0.0371$$

$$\frac{a}{1+a} = \frac{1}{4}\sigma t\,\mathrm{cosec}^2\phi = \frac{0.1274 \times 1.004}{4 \times 0.357 \times 0.357} = 0.2515$$

so

$$a = \frac{0.2515}{0.7485} = 0.336$$

Thus the assumed values $a = 0.1$ and $b = 0.02$ lead to the better approximations $a = 0.336$ and $b = 0.0371$. A further iteration may be carried out using these values we can more quickly find the final values of a and b by using, as the initial values for an iteration, the arithmetic mean of the input and output values of the previous iteration. Thus, in the present example, the values for the next iteration are $a = 0.218$ and $b = 0.0286$. The arithmetic mean is particularly convenient when giving instructions to computers (whether human or electronic).

We continue the iteration process until we obtain agreement with the desired accuracy between the assumed and derived values of a and b. The results of the iterations are

$$a = 0.1950, \quad b = 0.0296$$

to four significant figures. With these values substituted in the appropriate equations, we have the following results:

$$\phi = 22^\circ 48'$$

$$\alpha = 6^\circ 28'$$

$$V_R = 207\,\mathrm{m\,s^{-1}} \quad M = 0.640$$

giving

$$\frac{dT}{dr} = \frac{1}{2}\rho V_R^2 ct = 3167 \, \text{N m}^{-1} \quad \text{per blade}$$

and

$$\frac{dQ}{dr} = \frac{1}{2}\rho V_R^2 crq = 1758 \, \text{N m m}^{-1} \quad \text{per blade}$$

Thus the thrust grading for the whole airscrew is $12670 \, \text{N m}^{-1}$ and the torque grading is 7032 N m m^{-1}. The local efficiency is

$$\eta_1 = \frac{V}{2\pi nr}\frac{t}{q} = 0.768 \quad \text{or} \quad 76.8\%$$

10.5 THE MOMENTUM THEORY APPLIED TO THE HELICOPTER ROTOR

In most, but not all, states of helicopter flight, the effect of the rotor may be approximated by replacing it with an ideal actuator disc to which the simple momentum theory applies. More specifically, momentum theory may be used for translational—forward, sideways, or rearward—flight, climb, and slow descent under power and hovering.

10.5.1 Actuator Disc in Hovering Flight

In steady hovering flight, the speed of the oncoming stream well ahead of (i.e., above) the disc is zero, while the thrust equals the helicopter weight, ignoring any downward force by down flow from the rotor acting on the fuselage or other part of the craft. If the weight is W and the rotor area is A, and using the normal notation of the momentum theory, with ρ as the air density,

$$W = \rho A V_0 (V_s - V) = \rho A V_0 V_s \tag{10.44}$$

since $V = 0$. V_s is the slipstream velocity and V_0 is the velocity at the disc.

The general momentum theory shows that

$$V_0 = \frac{1}{2}(V_s + V) \quad \text{(Eq. (10.8))}$$

$$= \frac{1}{2}V_s \quad \text{in this case} \tag{10.45}$$

or

$$V_s = 2V_0$$

which, substituted in Eq. (10.44), gives

$$W = 2\rho A V_0^2 \qquad (10.46)$$

that is,

$$V_0 = \sqrt{W/2\rho A} \qquad (10.47)$$

Defining the effective disc loading l_{de} as

$$l_{de} = W/A\sigma \qquad (10.48)$$

where σ is the relative density of the atmosphere,

$$\frac{W}{2\rho A} = \frac{W}{A\sigma}\frac{1}{2}\frac{\sigma}{\rho} = \frac{1}{2\rho_0}l_{de}$$

ρ_0 being sea-level standard density. Then

$$V_0 = \sqrt{l_{de}/2\rho_0} \qquad (10.49)$$

The power supplied is equal to the rate of increase of kinetic energy of the air:

$$
\begin{aligned}
P &= \frac{1}{2}\rho A V_0 \left(V_s^2 - V^2\right) \\
&= \frac{1}{2}\rho V_0 V_s^2 A = 2\rho A V_0^3
\end{aligned}
\qquad (10.50)
$$

Substituting for V_0 from Eq. (10.47) leads to

$$P = 2\rho A \left(\frac{W}{2\rho A}\right)^{3/2} = \sqrt{\frac{W^3}{2\rho A}} \qquad (10.51a)$$

$$= W\sqrt{\frac{l_{de}}{2\rho_0}} \qquad (10.51b)$$

This is the power that must be supplied to the ideal actuator disc. A real rotor requires considerably greater power.

10.5.2 Vertical Climbing Flight

The problem of vertical climbing flight is identical to the problem in Section 10.1. The thrust is equal to the helicopter weight plus the air resistance of the fuselage or other parts of the craft to the vertical motion, with the oncoming stream speed V equal to the rate of climb of the helicopter.

10.5.3 Slow, Powered Descending Flight

In this case, the air approaches the rotor from below and its momentum decreases on passing through the disc. The associated loss of its kinetic energy appears as a power input to the ideal actuator, which therefore acts as a windmill. A real rotor, however, must be driven by the engine unless the rate of descent is high. This case, for the ideal actuator disc, may be treated by the methods of Section 10.1 with the appropriate changes in sign: V positive, $V_s < V_0 < V$, $p_1 > p_2$, and thrust $T = -W$.

10.5.4 Translational Helicopter Flight

We assume that the effect of the actuator disc in approximating the rotor is to add incremental velocities v_v and v_h vertically and horizontally, respectively, at the disc. We further assume, in accordance with the simple axial momentum theory of Section 10.1, that, in the slipstream well behind the disc, these incremental velocities increase to $2v_v$ and $2v_h$, respectively. The resultant speed through the disc is denoted U; the resultant speed in the fully developed slipstream, U_1. Then, by considering vertical momentum,

$$W = \rho A U (2v_v) = 2\rho A U v_v \tag{10.52}$$

Also, from the vector addition of velocities,

$$U^2 = (V + v_h)^2 + (v_v)^2 \tag{10.53}$$

where V is the speed of horizontal flight. By consideration of horizontal momentum,

$$\frac{1}{2}\rho V^2 A C_D = 2\rho A U v_h \tag{10.54}$$

where C_D is the drag coefficient of the fuselage, based on the rotor area A.

Power input equals rate of increase of kinetic energy:

$$P = \frac{1}{2}\rho A U \left(U_1^2 - V^2 \right) \tag{10.55}$$

and, from the vector addition of velocities,

$$U_1^2 = (V + 2v_h)^2 + (2v_v)^2 \tag{10.56}$$

The most useful solution of Eqs. (10.52) through (10.56) is obtained by eliminating U_1, v_h, and v_v as follows. Rearranging (10.52) and (10.54), we get

$$v_v = \frac{W}{2\rho A U} \tag{10.52a}$$

$$v_h = \frac{\frac{1}{2}\rho V^2 A C_D}{2\rho A U} = \frac{C_D}{4U}V^2 \tag{10.54a}$$

Then, from Eq. (10.53),

$$U^2 = V^2 + 2V v_h + v_h^2 + v_v^2$$

Substituting for v_v and v_h and multiplying by U^2 gives

$$U^4 - U^2 V^2 = \frac{1}{2}C_D U V^3 + \frac{1}{16}C_D^2 V^4 + \left(\frac{W}{2\rho A}\right)^2$$

Introducing the effective disc loading l_{de} from Eq. (10.48) leads to

$$U^4 - U^2 V^2 - \frac{1}{2}C_D V^3 U = \frac{1}{16}C_D^2 V^4 + \left(\frac{l_{de}}{2\rho_0}\right)^2 \tag{10.57}$$

which is a quartic equation for U in terms of given quantities. Since, from Eq. (10.56),

$$U_1^2 = V^2 + 4V v_h + 4v_h^2 + 4v_v^2$$

then

$$
\begin{aligned}
P &= \frac{1}{2}\rho A U \left(U_1^2 - V^2\right) = \frac{1}{2}\rho A U \left[4V v_h + 4v_h^2 + 4v_v^2\right] \\
&= 2\rho A \left[\frac{1}{4}C_D V^3 + \frac{1}{16}C_D^2 \frac{V^4}{U} + \frac{1}{U}\left(\frac{l_{de}}{2\rho_0}\right)^2\right]
\end{aligned}
\tag{10.58}
$$

which, with the value of U calculated from Eq. (10.57) and the given quantities, may be used to calculate the power required.

Example 10.6

A helicopter weighs 24,000 N and has a single rotor of 15-m diameter. Using momentum theory, estimate the power required for level flight at a speed of 15 m s^{-1} at sea level. The drag coefficient, based on the rotor area, is 0.006:

$$A = \frac{\pi}{4}(15)^2 = 176.7 \text{ m}^2$$

$$l_{de} = \frac{W}{A\sigma} = \frac{24000}{176.7 \times 1} = 136 \text{ N m}^{-2}$$

$$\frac{l_{de}}{2\rho_0} = \frac{136}{2 \times 1.226} = 55.6 \text{ m}^2\text{s}^{-2}$$

With these values, and with $V = 15\,\mathrm{m\,s^{-1}}$, Eq. (10.57) is

$$U^4 - 225U^2 - \frac{1}{2}U(0.006)(3375) = (55.6)^2 + \frac{(0.006)^2}{16}(15)^4$$

that is,

$$U^4 - 225U^2 - 10.125U = 3091$$

This quartic equation in U may be solved by any of the standard methods (e.g., Newton-Raphson), with the solution being $U = 15.45\,\mathrm{m\,s^{-1}}$ to four significant figures. Then

$$P = 2 \times 1.226 \times 176.7 \left[\frac{0.006 \times (15)^3}{4} + \frac{0.006^2 \times (15)^4}{16 \times 15.45} + \frac{(55.6)^2}{15.45} \right]$$

$$= 88.9\,\mathrm{kW}$$

This is the power required if the rotor behaves as an ideal actuator disc. A practical rotor requires considerably more power than this.

10.6 THE ROCKET MOTOR

As noted at the beginning of this chapter, the rocket motor is the only current example of aeronautical interest in class 2 of propulsive systems. Since it does not work by accelerating atmospheric air, it cannot be treated by Froude's momentum theory. The rocket motor is unique among current aircraft power plants in that it can operate independently of air from the atmosphere. The consequences of this are these:

- It can operate in a rarefied atmosphere or in an atmosphere of inert gas.
- Its maximum speed is not limited by the thermal barrier set up by the high ram-compression of the air in all air-breathing engines.

In a rocket, some form of chemical is converted in the combustion chamber into gas at high temperature and pressure. This gas is then exhausted at supersonic speed through a nozzle. Suppose that a rocket is traveling at speed V, and let the gas leave the nozzle with speed v relative to the rocket. Let the rate of mass flow of the gas be $\dot{m}$.[1] The gas is produced by the consumption, at the same rate, of the chemicals in the rocket fuel tanks (or solid charge). While in the tank, the mean m of fuel has a forward momentum of mV. After discharge from the nozzle, the gas has a rearward momentum of $m(v - V)$. Thus the rate of increase of rearward momentum of the fuel/gas is

$$\dot{m}(v - V) - (-\dot{m}V) = \dot{m}v \tag{10.59}$$

---

[1] Some authors denote mass flow by m in rocketry, using the mass discharged (per second, understood) as the parameter.

and this rate is equal to the rocket's thrust. Thus the thrust depends only on the rate of fuel consumption and the velocity of discharge relative to the rocket. It does not depend on the speed of the rocket itself. In particular, it is possible that the speed of the rocket V can exceed the speed of the gas relative to the rocket v and relative to the axes of reference $v - V$.

When in the form of fuel in the rocket, mass m has a kinetic energy of $\frac{1}{2}mV^2$. After discharge, it has a kinetic energy of $\frac{1}{2}m(v - V)^2$. Thus the rate of change of kinetic energy is (in Watts)

$$\frac{dE}{dt} = \frac{1}{2}\dot{m}[(v - V)^2 - V^2] = \frac{1}{2}\dot{m}(v^2 - 2vV) \qquad (10.60)$$

Useful work is done at the rate TV, where the thrust is $T = \dot{m}vs$. The propulsive efficiency of the rocket is thus

$$\eta_P = \frac{\text{rate of useful work}}{\text{rate of increase of KE of fuel}}$$
$$= \frac{2vV}{v^2 - 2vV} = \frac{2}{(v/V) - 2}s \qquad (10.61)$$

Now, if $v/V = 4$, then

$$\eta_P = \frac{2}{4 - 2} = 1 \quad \text{or} \quad 100\%$$

If $v/V < 4$ (i.e., $V > v/4$), the propulsive efficiency exceeds 100%.

This derivation of efficiency, while theoretically sound, is not normally accepted because engineers are unaccustomed to such efficiencies. Accordingly, an alternative measure of efficiency is used, in which the energy input is that liberated in the jet, plus the initial kinetic energy of the fuel while in the tanks. The total energy input is then

$$\frac{dE}{dt} = \frac{1}{2}\dot{m}v^2 + \frac{1}{2}\dot{m}V^2$$

giving for the efficiency,

$$\eta_P = \frac{2(v/V)}{(v/V)^2 + 1} \qquad (10.62)$$

By differentiating with respect to v/V, this is a maximum when $v/V = 1$, the propulsive efficiency then being 100%. In this way, the definition of efficiency leads to a maximum efficiency of 100% when the speed of the rocket equals the speed of the exhaust gas relative to the rocket—that is, when the exhaust is at rest relative to an observer past whom the rocket has speed V.

If the speed of the rocket V is small compared with the exhaust speed v, as is the case for most aircraft applications, V^2 may be ignored compared with v^2, giving

$$\eta_P \simeq \frac{2vV}{v^2} = 2\frac{V}{v} \tag{10.63}$$

10.6.1 Free Motion of a Rocket-Propelled Body

Imagine a rocket-propelled body moving in a region where aerodynamic drag and lift and gravitational force may be neglected, say, in space remote from any planets. At time t, let the mass of the body plus unburned fuel be denoted M and the speed of the body relative to some axes be denoted V. Let the fuel be consumed at a rate $\dot{m}$, the resultant gas being ejected at a speed v relative to the body. Further, let the total rearward momentum of the rocket exhaust, produced from the instant of firing to time t, be I relative to the axes. Then, at time t, the total forward momentum is

$$H_1 = MV - I \tag{10.64}$$

At time $(t + \delta t)$, the mass of the body plus unburned fuel is $(M - \dot{m}\delta t)$, and its speed is $(V + \delta V)$. A mass of fuel $\dot{m}\delta t$ has been ejected rearward with a mean speed, relative to the axes, of $(v - V - \frac{1}{2}\delta V)$. Thus, the total forward momentum is

$$H_2(M - \dot{m}\delta t)(V + \delta V) - \dot{m}\delta t\left(v - V - \frac{1}{2}\delta V\right)s - I$$

Now, by the conservation of momentum of a closed system,

$$H_1 = H_2$$

that is,

$$MV - I = MV + M\delta V - \dot{m}V\delta t - \dot{m}\delta t\delta V - \dot{m}v\delta t + \dot{m}V\delta t + \frac{1}{2}\dot{m}\delta t\delta V - I$$

which reduces to

$$M\delta V - \frac{1}{2}\dot{m}\delta t\delta V - \dot{m}v\delta t = 0$$

Dividing by δt and taking the limit as $\delta t \to 0$, this becomes

$$M\frac{dV}{dt} - \dot{m}v = 0 \tag{10.65}$$

Note that this equation can be derived directly from Newton's second law: force = mass × acceleration. However, it is not always immediately clear how to apply this law to bodies of variable mass. The fundamental appeal to momentum made previously removes any doubts as to the legitimacy of such an application. Equation (10.65) may now be rearranged as

$$\frac{dV}{dt} = \frac{\dot{m}}{M}v$$

that is,

$$\frac{1}{v}dV = \frac{\dot{m}}{M}dt$$

Now $\dot{m} = dM/dt$, since $\dot{m}$ is the rate at which fuel is burnt; therefore,

$$\frac{1}{v}dV = -\frac{1}{M}\frac{dM}{dt}dt = -\frac{dM}{M}$$

so

$$V/v = -M + \text{constant}$$

where v, but not necessarily $\dot{m}$, is assumed to be constant. If the rate of fuel injection into the combustion chamber is constant, and if the pressure into which the nozzle exhausts is also constant (e.g., the near-vacuum implicit in the initial assumptions), both $\dot{m}$ and v are closely constant. If the initial conditions are $M = M_0$, $V = 0$ when $t = 0$, then

$$0 = -\ln M_0 + \text{constant}$$

that is, the constant of integration is $\ln M_0$. With this,

$$\frac{V}{v} = \ln M_0 - \ln M = \ln\left(\frac{M_0}{M}\right)$$

or, finally,

$$V = v\ln(M_0/M) \tag{10.66}$$

The maximum speed of a rocket in free space is reached when all fuel is burnt, or at the instant the motor ceases to produce thrust. Let the mass with all fuel burnt be M_1. Then, from Eq. (10.66),

$$V_{\max} = v\ln(M_0/M_1) = v\ln R \tag{10.66a}$$

where R is the mass ratio M_0/M_1. Note that, if the mass ratio exceeds $e = 2.718\ldots$, the base of natural logarithms, the speed of the rocket exceeds the speed of ejection of the exhaust relative to the rocket.

Distance Traveled During Firing
From Eq. (10.66),

$$V = v \ln(M_0/M) = v \ln M_0 - v \ln(M_0 - \dot{m}t)$$

Now, if the distance traveled from the instant of firing is x in time t,

$$x = \int_0^t V \, dt$$

$$= v \int_0^t [\ln M_0 - \ln(M_0 - \dot{m}t)] \, dt \qquad (10.67)$$

$$= vt \ln M_0 - v \int_0^t \ln(M_0 - \dot{m}t) \, dt$$

To solve the integral (G, say) in Eq. (10.67), let

$$y = \ln(M_0 - \dot{m}t)$$

Then

$$\exp(y) = M_0 - \dot{m}t \quad \text{and} \quad t = \frac{1}{\dot{m}}(M_0 - e^y)$$

whence

$$dt = -\frac{1}{\dot{m}}e^y \, dy$$

so

$$G = \int_0^t \ln(M_0 - \dot{m}t) \, dt$$

$$= \int_{y_0}^{y_1} y \left(-\frac{1}{\dot{m}} e^y \, dy \right)$$

where

$$y_0 = \ln M_0 \quad \text{and} \quad y_1 = \ln(M_0 - \dot{m}t)$$

Therefore,

$$G = -\frac{1}{\dot{m}} \int_{y_0}^{y_1} y e^y \, dy = \frac{1}{\dot{m}} \int_{y_2}^{y_1} y \, d(e^y)$$

which, on integrating by parts, gives

$$G = -\frac{1}{\dot{m}} \left[e^y(y-1) \right]_{y_0}^{y_1} = \frac{1}{\dot{m}} \left[e^y(1-y) \right]_{y_0}^{y_1}$$

Substituting back for y in terms of M_0, $\dot{m}$, and t gives

$$G = \frac{1}{\dot{m}} [(M_0 - \dot{m}t)(1 - \ln\{M_0 - \dot{m}t\})]_0^t$$

$$= \frac{1}{\dot{m}} [M(1 - \ln M) - M_0(1 - \ln M_0)]$$

where $M = M_0 - \dot{m}t$. Thus, finally,

$$G = \frac{1}{\dot{m}} [(M - M_0) - M \ln M + M_0 \ln M_0]$$

Substituting this value of the integral back into Eq. (10.67) gives, for the distance traveled,

$$x = vt \ln M_0 - \frac{v}{\dot{m}} \{ (M - M_0) - M \ln M + M_0 \ln M_0 \} \tag{10.68}$$

Now, if $\dot{m}$ is constant,

$$t = \frac{1}{\dot{m}} (M_0 - M) \tag{10.69}$$

which, substituted into Eq. (10.68), gives

$$x = \frac{v}{\dot{m}} \{ (M_0 - M) \ln M_0 - (M - M_0) + M \ln M - M_0 \ln M_0 \}$$

$$= v \frac{M_0}{\dot{m}} \left[\left(1 - \frac{M}{M_0} \right) - \frac{M}{M_0} \ln \left(\frac{M_0}{M} \right) \right] \tag{10.70}$$

For the distance at all-burnt fuel, when $x = X$ and $M = M_1 = M_0/R$,

$$X = v\frac{M_0}{\dot{m}}\left(1 - \frac{1}{R}(1 + \ln R)\right) \qquad (10.71)$$

Alternatively, this may be written as

$$X = \frac{v}{\dot{m}}\frac{M_0}{R}[(R - 1) - \ln R]$$

that is,

$$X = v\frac{M_1}{\dot{m}}[(R - 1) - \ln R] \qquad (10.72)$$

Example 10.7

A rocket-propelled missile has an initial total mass of 11,000 kg. Of this, 10,000 kg is fuel, which is completely consumed in five minutes burning time. The exhaust is $1500\ \mathrm{m\,s^{-1}}$ relative to the rocket. Determine the variation of acceleration, speed, and distance with time during the burning period by calculating these quantities at each half-minute.

For the acceleration,

$$\frac{dV}{dt} = \frac{\dot{m}}{M}v \qquad (a)$$

Now

$$\dot{m} = \frac{10000}{5} = 2000\ \mathrm{kg\ min^{-1}} = 33.\dot{3}\ \mathrm{kg\ s^{-1}}$$

$$M = M_0 - \dot{m}t = 11000 - 33.\dot{3}t\ \mathrm{kg}$$

where t is the time from firing in seconds, or

$$M = 11000 - 1000\,N\ \mathrm{kg}$$

where N is the number of half-minute periods elapsed since firing, so

$$\frac{M_0}{\dot{m}} = \frac{11000}{100/3} = 330\ \mathrm{seconds}$$

Substituting these values into the appropriate equations leads to the final results in Table E10.7. The student should plot the curves defined by the values given there. Note that, in the five minutes of burning time, the missile travels only 342 kilometers but at the end of this time is traveling at $3600\ \mathrm{m\,s^{-1}}$, or $13000\ \mathrm{km\,h^{-1}}$. Another point to note is the rapid increase in acceleration toward the end of the burning time, consequent on the rapid percentage

Table E10.7 Prediction of Flight Data for Missile in Example 10.7

t (min)	M (1000 kg)	Acceleration (m s^{-2})	V (m s^{-1})	x (km)
0	11	4.55	0	0
0.5	10	5.00	143	2.18
1	9	5.55	300	9.05
1.5	8	6.25	478	19.8
2	7	7.15	679	37.6
2.5	6	8.33	919	61.4
3	5	10.0	1180	92.0
3.5	4	12.5	1520	133
4	3	16.7	1950	185
4.5	2	25.0	2560	256
5	1	50.0	3600	342
5.5	1	0	3600	450

decrease of total mass. In Table E10.1, results are given as well for the first half-minute after all-burnt.

10.7 THE HOVERCRAFT

In conventional winged aircraft, lift, associated with circulation around the wings, is used to balance the weight. For helicopters, the "wings" rotate but the lift generation is the same. A radically different principle is used for hovercraft. In machines of this type, a more or less static region of air, at slightly more than atmospheric pressure, is formed and maintained below the craft. The difference between the pressure of the air on the lower side and the atmospheric pressure on the upper side produces a force that provides lift. The trapped mass of air under the craft is formed by the effect of an annular jet of air, directed inward and downward from near the periphery of the underside. The downward ejection of the annular jet produces an upward reaction on the craft, lifting it. In steady hovering, the weight is balanced by the jet thrust and the force due to the cushion of air below. The difference between a hovercraft and a normal jet-lift machine lies in the air cushion effect, which amplifies the vertical force available, permitting the direct jet thrust to be only a small fraction of the craft's weight. The cushion effect requires that the craft's hovering height/diameter ratio be small (e.g., 1/50), and this severely limits the attainable altitude.

Consider the simplified system of Fig. 10.9, which shows a hovercraft with a circular planform of radius r hovering a height h above a flat, rigid horizontal surface. An annular jet of radius r, thickness t, velocity V, and density ρ is ejected at an angle θ to the horizontal surface. It is directed inward but, in a steady equilibrium state,

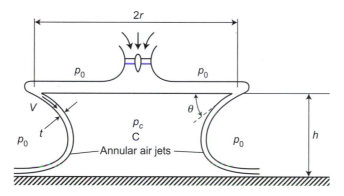

FIGURE 10.9

Simplified hovercraft system.

must turn to flow outward as shown. If it did not, there would be a continuous increase in mass within the region C, which is impossible. Note that such an increase of mass occurs for a short time immediately after starting, while the air cushion builds up. The curvature of the path of the air jet shows that it possesses a centripetal acceleration, which is produced by a difference between the pressure p_c within the air cushion and the atmospheric pressure p_0.

Consider a short peripheral length δs of the annular jet and assume

1. That the pressure p_c is constant over the depth h of the air cushion.
2. That the speed V of the annular jet is unchanged throughout the motion.

Then the rate of mass flow within the element of peripheral length δs is $\rho V t \delta s$ kg s^{-1}. This mass has an initial momentum parallel to the rigid surface (or ground) of $\rho V t \delta s V \cos\theta = \rho V^2 t \cos\delta s$ inward.

After turning to flow radially outward, the air has a momentum parallel to the ground of $\rho V t \delta s V = \rho V^2 t \delta s$. Therefore, the rate of change in momentum parallel to the ground is $\rho V^2 t (1 + \cos\theta)\delta s$. This is due to the pressure difference $(p_c - p_0)$ and must, indeed, be equal to the force exerted on the jet by this pressure difference, parallel to the ground, which is $(p_c - p_0)h\delta s$. Thus

$$(p_c - p_0)h\delta s = \rho V^2 t (1 + \cos\theta)\delta s$$

or

$$(p_c - p_0) = \frac{\rho V^2 t}{h}(1 + \cos\theta) \tag{10.73}$$

The lift L_c due to the cushion of air on a circular body of radius r is then

$$L_c = \pi r^2 (p_c - p_0) = \frac{\pi \rho r^2 V^2 t}{h}(1 + \cos\theta) \tag{10.74}$$

The direct lift due to the downward ejection of the jet is

$$L_j = \rho V t 2\pi r V \sin\theta = 2\pi r \rho V^2 t \sin\theta \qquad (10.75)$$

and thus the total lift is

$$L = \pi r \rho V^2 t \left\{ 2\sin\theta + \frac{r}{h}(1 + \cos\theta) \right\} \qquad (10.76)$$

If the craft is remote from any horizontal surface, such as the ground or sea, so that the air cushion has negligible effect, the lift derives only from the direct jet thrust, with the maximum value $L_{jo} = 2\pi r \rho V^2 t$ when $\theta = 90$ degrees. Thus the lift amplification factor L/L_{jo} is

$$\frac{L}{L_{jo}} = \sin\theta + \frac{r}{2h}(1 + \cos\theta) \qquad (10.77)$$

Differentiation with respect to θ shows that this has a maximum value when

$$\tan\theta = \frac{2h}{r} \qquad (10.78)$$

Since machines of this type operate under conditions such that h is very small compared to r, it follows that maximum amplification is achieved when θ is close to zero—that is, the jet is directed radially inward. Then, with the approximations $\sin\theta = 0$, $\cos\theta = 1$,

$$\frac{L}{L_{jo}} = \frac{h}{r} \qquad (10.79)$$

and

$$L = \frac{r}{h}L_{jo} = \frac{r}{h}2\pi r \rho V^2 t = \frac{2\pi r^2 \rho V^2 t}{h} \qquad (10.80)$$

Note that the direct jet lift is now negligible.

The power supplied is equal to the kinetic energy contained in the jet per unit time,[2] which is

$$\frac{1}{2}2\pi r \rho V t V^2 = \pi r \rho V^3 t \qquad (10.81)$$

---

[2] The power supplied to the jet also contains a term relating to the increase in potential (pressure) energy, since the jet static pressure is slightly greater than atmospheric. Since the jet pressure is approximately equal to p_c, which is typically about 750 N m^{-2} above atmospheric, the increase in pressure energy is very small and is neglected in this simplified analysis.

Denoting this P, combining Eqs. (10.80) and (10.81), and setting lift L equal to weight W, we get

$$\frac{P}{W} = \frac{Vh}{2r}$$

as the minimum power necessary for sustentation, while, if $\theta \neq 0$, then

$$\frac{P}{W} \simeq \frac{Vh}{r(1 + \cos\theta)}$$

ignoring a term involving $\sin\theta$. Thus, if V is small and if h is small compared to r, it becomes possible to lift the craft with comparatively low power.

The foregoing analysis for hovering flight involves a number of simplifying assumptions. The first is that of a level, rigid surface below the machine. This is reasonably accurate for operation over land but is not justified over water, when a depression is formed in the water below the craft. Remember that the weight of the craft is reacted by a pressure distributed over the surface below it, which leads to deformation of a nonrigid surface.

Another assumption is that the pressure p_c is constant throughout the air cushion. In fact, mixing between the annular jet and the air cushion produces eddies, which cause nonuniformity of the pressure within the cushion. The mixing referred to, together with friction between the air jet and the ground (or water) leads to a loss of kinetic energy and speed in the air jet, whereas it is assumed that the speed of the jet remains constant throughout the motion. These effects produce only small corrections to the results of the previous analysis.

If the power available is greater than necessary to sustain the craft at the selected height h, the excess may be used either to raise the machine to a greater height or to propel it forward.

10.8 EXERCISES

10.1   If an aircraft of wing area S and drag coefficient C_D is flying at speed V in air of density ρ, and if its single airscrew, of disc area A, produces a thrust equal to the aircraft drag, show that the speed in the slipstream V_s is, on the basis of Froude's momentum theory,

$$V_s = V C_D \sqrt{1 + \frac{S}{A}}$$

10.2   A cooling fan is required to produce a stream of air 0.5 m in diameter with a speed of 3 $\mathrm{m\,s^{-1}}$ when operating in a region of otherwise stationary air of standard density. Assuming the stream of air to be the fully developed

slipstream behind an ideal actuator disc, and ignoring mixing between the jet and the surrounding air, estimate the fan diameter and power input required.
(*Answer:* 0.707-m diameter; 3.24 W)

10.3 Repeat Example 10.2 in the text for the case where two airscrews absorb equal power, and find (1) the thrust of the second airscrew as a percentage of the thrust of the first; (2) the efficiency of the second; and (3) the efficiency of the combination.
(*Answer:* 84%; 75.5%; 82.75%)

10.4 Calculate the flight speed at which the airscrew in Example 10.3 produces a thrust of 7500 N. Also calculate the power absorbed at the same rotational speed.
(*Answer:* 93 m s^{-1}; 840 kW)

10.5 At 1.5-m radius, the thrust and torque gradings on each blade of a three-bladed airscrew revolve at 1200 rpm at a flight speed of 90 m s^{-1} TAS at an altitude where $\sigma = 0.725$ are 300 N m^{-1} and 1800 N m m^{-1}, respectively. If the blade angle is 28 degrees, find the blade section absolute incidence. Ignore compressibility.
(*Answer:* 1°48′)

10.6 At 1.25-m radius on a three-bladed airscrew, the airfoil section has the following characteristics:

$$\text{Solidity} = 0.1; \quad \theta = 29°7'; \quad \alpha = 4°7'; \quad C_L = 0.49; \quad L/D = 50$$

Allowing for both axial and rotational interference, find the local efficiency of the element.
(*Answer:* 0.885)

10.7 The thrust and torque gradings at 1.22-m radius on each blade of a two-bladed airscrew are 2120 N m^{-1} and 778 N m m^{-1}, respectively. Find the speed of rotation (in rad s^{-1}) of the airstream immediately behind the disc at 1.22-m radius.
(*Answer:* 735 rad s^{-1})

10.8 A four-bladed airscrew is required to propel an aircraft at 125 m s^{-1} at sea level, with a rotational speed of 1200 rpm. The blade element at 1.25-m radius has an absolute incidence of 6 degrees, and the thrust grading is 2800 N m^{-1} per blade. Assuming a reasonable value for the sectional lift-curve slope, calculate the blade chord at 1.25-m radius. Neglect rotational interference, sectional drag, and compressibility.
(*Answer:* 240 mm)

10.9 A three-bladed airscrew is driven at 1560 rpm at a flight speed of 110 m s^{-1} at sea level. At 1.25-m radius, the local efficiency is estimated to be 87%,

while the lift/drag ratio of the blade section is 57.3. Calculate the local thrust grading, ignoring rotational interference.
(*Answer:* 9000 N m^{-1} per blade)

10.10 Using simple momentum theory, develop an expression for the thrust of a propeller in terms of disc area, air density, and axial velocities of the air a long way ahead, and in the plane, of the propeller disc. A helicopter has an engine developing 600 kW and a rotor of 16-m diameter with a disc loading of 170 N m^{-2}. When ascending vertically with constant speed at low altitude, the product of lift and axial velocity of the air through the rotor disc is 53% of the power available. Estimate the velocity of ascent.
(*Answer:* 110 m min^{-1})

Appendix A: Symbols and Notation

A	Moment of inertia about OX. Aspect ratio (also AR). With subscripts, the coefficients in a Fourier series, or a polynomial series in z.
AF	Activity factor of a propeller.
AR	Aspect ratio (also A).
a	Speed of sound. Axial inflow factor in airscrew theory. Lift curve slope. dCL/da (subscripts denote particular values). Radius of vortex core. Acceleration or deceleration.
B	Number of blades on a propellor.
b	Rotational interference factor in airscrew theory. Wing span. Hinge moment coefficient slope.
CG	Center of gravity.
C_D	Total drag coefficient.
C_{D_o}	Zero-lift drag coefficient.
C_{D_V}	Trailing vortex drag coefficient.
C_{D_L}	Lift-dependent drag coefficient. (Other suffices are used in particular cases.)
C_H	Hinge moment coefficient.
C_L	Lift coefficient.
C_M	Pitching moment coefficient.
C_p	Pressure coefficient.
C_p	Power coefficient for propellers.
C_R	Resultant force coefficient.
c	Wing chord. A distance. Ultimate velocity.
$\bar{c}$	Standard or geometric mean chord.
$\bar{c}$ or $\bar{c}_A$	Aerodynamic mean chord
c_0	Root chord.
c_T	Tip chord.
c_p	Specific heat at constant pressure.
c_v	Specific heat at constant volume.
CP	Center of pressure.
D	Drag (subscripts denote particular values). Propellor diameter. A length.

d	Diameter, occasionally a length.
d_v	Spanwise trailing vortex drag grading ($=\rho w \Gamma$).
E	Internal energy per unit mass. Kinetic energy.
F	Fractional flap chord. Force.
$f(\ ), fn(\ )$	Function of the stated variables.
g	Acceleration due to gravity.
$g(\ )$	Function of the stated variables.
H	Hinge moment. Total pressure. Momentum. Shape factor, δ^*/θ.
h	Fractional camber of a flapped plate airfoil. Distance between plates in Newton's definition of viscosity. Enthalpy per unit mass.
$h(\ )$	Function of the stated variables.
h_0	Fractional position of the aerodynamic center.
I	Momentum of rocket exhaust.
i	The imaginary unit which is $\sqrt{-1}$.
J	Advance ratio of a propellor.
K	Modulus of bulk elasticity.
k	Chordwise variation of vorticity. Lift-dependent drag coefficient factor.
k_{CP}	Coefficient of the location of the center of pressure.
K_T, k_Q	Thrust and torque coefficients for propellers.
L	Lift. Dimension of length. Temperature lapse rate in the atmosphere.
l	Length. Lift per unit span.
l_{de}	Effective disc loading of a helicopter.
M	Dimension of mass. Mach number. Pitching moment.
m	Mass. Strength of a source (or sink). An index.
$\dot{m}$	Rate of mass flow.
N	Rpm of a propellor. Normal influence coefficient. Number of panel Points.
n	Revolutions per second of a propellor. Frequency. An index.
$\hat{n}$	Unit normal vector.
O	Origin of coordinates.
P	Power. The general point in space.
p	Static pressure in a fluid.
Q	Torque, or a general moment. Total velocity of a uniform stream.
q	Angular velocity in pitch about Oy. Local resultant velocity. A coefficient in airscrew theory.

q_n, q_t	Radial and tangential velocity components.
Re	Real part of a complex number.
Re	Reynolds number.
R	Resultant force. Characteristic gas constant. Radius of a circle.
r	Radius vector, or radius generally.
S	Projected wing area. Vortex tube area. Area of actuator disc. Entropy.
S'	Tail plane area.
s	Distance. Specific entropy.
s'	Separation of trailing vortex pair.
T	Dimension of time. Thrust. Temperature (suffices denote particular values). Tangential influence coefficient.
t	Time. Airfoil section thickness. A coefficient in airscrew theory.
$\hat{t}$	Tangential unit vector.
U	Velocity. Steady velocity parallel to Ox.
U_∞	Free stream flow speed. Flight speed.
U_e	Mainstream flow speed.
u	Velocity component parallel to Ox.
u'	Disturbance velocity parallel to Ox.
V	Velocity. Volume. Steady velocity parallel to Oy.
V_s	Stalling speed.
V_E	Equivalent air speed.
V_R	Resultant speed.
v	Velocity component parallel to Oy. Velocity.
v'	Disturbance velocity parallel to Oy.
W	Weight. Steady velocity parallel to Oz.
w	Wing loading. Downwash velocity. Velocity parallel to Oz.
X, Y, Z	Components of aerodynamic or external force.
x, y, z	Coordinates of the general point P.
x, X	Distance.
z	Distance. Spanwise coordinate.
α	Angle of attack (or incidence). An angle, generally.
β	A factor in propellor theory. An angle generally.
Γ	Circulation. Half the dihedral angle; the angle between each wing and the Oxy plane.
γ	Ratio of specific heats, c_p/c_V. Shear strain.
δ	Boundary layer thickness. A factor. Camber of an airfoil section.

ε	Downwash angle. Surface slope. Strain.
ζ	Vorticity. Complex variable in transformed plane ($= \xi + i\eta$).
η	Efficiency. Ordinate in ζ-plane.
θ	Dimension of temperature. Polar angular coordinate. Blade helix angle. Momentum thickness.
Λ	Angle of sweepback or sweep-forward. Pohlhausen pressure-gradient parameter.
λ	Taper ratio ($= c_T/c_0$). A constant.
μ	Strength of a doublet. Dynamic viscosity. Airfoil parameter in lifting line theory.
ν	Kinematic viscosity. Prandtl-Meyer angle.
ξ	Abscissa in ζ-plane.
ρ	Density. Radius of curvature.
Σ	Summation sign.
σ	Relative density. Blade or annular solidity (propellor theory). Stress.
τ	Shear stress.
ϕ	Sweepback angle. Velocity potential. A polar coordinate. Angle of relative wind to plane of airscrew disc.
ψ	The stream function.
Ω	Angular velocity of propellor.
ω	Angular velocity in general.
∇^2	Laplace's operator.

SUBSCRIPTS

0	No lift. Standard sea level. Straight and level flight. Undisturbed stream.
$c/4$	Quarter chord point.
1	A particular value.
2	A particular value.
3	A particular value.
∞	Infinity or two-dimensional conditions.
AC	Aerodynamic center.
a	available.
c	Chord from Ox axis. Compressible.
CP	Center of pressure.
f	Full scale or flight.
g	Ground.

h	Horizontal.
i	Ideal, computation numbering sequence. Incompressible.
in	Input. Computation numbering sequence. Length.
L	Lower surface.
LE	Leading edge.
l	Local.
m	Model.
max	Maximum.
min	Minimum.
md	Minimum drag.
n	normal.
n	Denotes general term.
Opt	Optimum.
Out	Output.
P	Prandtl-Meyer. Propulsive parallel.
r	Required.
s	Stagnation or reservoir conditions. Slipstream. Stratosphere. Surface.
TE	Trailing edge.
t	Thickness (airfoil), panel identification in computation. Tangential.
U	Upper surface.
V	Vertical.
W	Wall.

PRIMES AND SUPERSCRIPTS

$'$	Perturbation or disturbance.
$*$	Throat (locally sonic) conditions. Boundary-layer displacement thickness.
$**$	Boundary-layer energy thickness.
$\hat{}$	Unit vector.
$\rightarrow$	Vector.
$\cdot$	The dot notation is frequently used for derivatives, e.g., $\dot{y} = dy/dx$, the rate of change of y with x.

Appendix B

Altitude feet	Temperature Rankine	Temperature Fahrenheit	Pressure psi	Density slug/ft^3
−3,000	529.4	69.69	16.36	2.594E-03
2,000	525.8	66.11	15.79	2.520E-03
−1,000	522.2	62.54	15.23	2.448E-03
0	518.7	58.98	14.70	2.378E-03
1,000	515.1	55.42	14.17	2.309E-03
2,000	511.5	51.85	13.66	2.242E-03
3,000	508.0	48.29	13.17	2.176E-03
4,000	504.4	44.71	12.69	2.112E-03
5,000	500.8	41.14	12.23	2.049E-03
6,000	497.2	37.58	11.78	1.987E-03
7,000	493.7	34.01	11.34	1.927E-03
8,000	490.1	30.45	10.92	1.869E-03
9,000	486.6	26.89	10.51	1.812E-03
10,000	483.0	23.32	10.11	1.756E-03
11,000	479.4	19.76	9.722	1.702E-03
12,000	475.9	16.19	9.348	1.648E-03
13,000	472.3	12.63	8.986	1.596E-03
14,000	468.8	9.066	8.635	1.546E-03
15,000	465.2	5.520	8.296	1.496E-03
16,000	461.6	1.956	7.967	1.448E-03
17,000	458.1	−1.608	7.650	1.401E-03
18,000	454.5	−5.172	7.342	1.356E-03
19,000	451.0	−8.736	7.045	1.311E-03
20,000	447.4	−12.30	6.757	1.267E-03
21,000	443.8	−15.85	6.480	1.225E-03
22,000	440.3	−19.41	6.211	1.184E-03
23,000	436.7	−22.97	5.952	1.144E-03
24,000	433.2	−26.54	5.701	1.104E-03
25,000	429.6	−30.08	5.459	1.066E-03
26,000	426.0	−33.65	5.226	1.029E-03
27,000	422.5	−37.21	5.000	9.932E-04
28,000	418.9	−40.76	4.783	9.581E-04
29,000	415.4	−44.32	4.573	9.239E-04

(Continued)

Altitude feet	Temperature Rankine	Temperature Fahrenheit	Pressure psi	Density slug/ft^3
30,000	411.8	−47.89	4.371	8.907E-04
31,000	408.3	−51.43	4.176	8.584E-04
32,000	404.7	−55.00	3.989	8.271E-04
33,000	401.1	−58.54	3.808	7.966E-04
34,000	397.6	−62.11	3.634	7.670E-04
35,000	394.0	−65.65	3.466	7.382E-04
36,000	390.5	−69.22	3.305	7.103E-04
37,000	390.0	−69.72	3.150	6.779E-04
38,000	390.0	−69.72	3.003	6.462E-04
39,000	390.0	−69.72	2.863	6.160E-04
40,000	390.0	−69.72	2.729	5.872E-04
41,000	390.0	−69.72	2.601	5.597E-04
42,000	390.0	−69.72	2.480	5.336E-04
43,000	390.0	−69.72	2.364	5.086E-04
44,000	390.0	−69.72	2.253	4.848E-04
45,000	390.0	−69.72	2.148	4.622E-04
46,000	390.0	−69.72	2.047	4.406E-04
47,000	390.0	−69.72	1.952	4.200E-04
48,000	390.0	−69.72	1.861	4.004E-04
49,000	390.0	−69.72	1.774	3.816E-04
50,000	390.0	−69.72	1.691	3.638E-04
51,000	390.0	−69.72	1.612	3.468E-04
52,000	390.0	−69.72	1.536	3.306E-04
53,000	390.0	−69.72	1.465	3.152E-04
54,000	390.0	−69.72	1.396	3.005E-04
55,000	390.0	−69.72	1.331	2.864E-04
56,000	390.0	−69.72	1.269	2.731E-04
57,000	390.0	−69.72	1.210	2.603E-04
58,000	390.0	−69.72	1.153	2.481E-04
59,000	390.0	−69.72	1.099	2.366E-04
60,000	390.0	−69.72	1.048	2.255E-04
61,000	390.0	−69.72	0.9991	2.150E-04
62,000	390.0	−69.72	0.9525	2.050E-04
63,000	390.0	−69.72	0.9080	1.954E-04
64,000	390.0	−69.72	0.8657	1.863E-04
65,000	390.0	−69.72	0.8253	1.776E-04
66,000	390.1	−69.61	0.7868	1.693E-04

(Continued)

Altitude feet	Temperature Rankine	Temperature Fahrenheit	Pressure psi	Density slug/ft^3
67,000	390.6	−69.07	0.7501	1.611E-04
68,000	391.2	−68.53	0.7152	1.534E-04
69,000	391.7	−67.97	0.6819	1.461E-04
70,000	392.3	−67.43	0.6503	1.391E-04
71,000	392.8	−66.89	0.6202	1.325E-04
72,000	393.4	−66.34	0.5915	1.262E-04
73,000	393.9	−65.80	0.5642	1.202E-04
74,000	394.4	−65.26	0.5381	1.145E-04
75,000	395.0	−64.72	0.5134	1.091E-04
76,000	395.5	−64.16	0.4897	1.039E-04
77,000	396.1	−63.62	0.4673	9.900E-05
78,000	396.6	−63.08	0.4458	9.433E-05
79,000	397.2	−62.54	0.4254	8.988E-05
80,000	397.7	−61.98	0.4059	8.565E-05
81,000	398.3	−61.44	0.3874	8.163E-05
82,000	398.8	−60.90	0.3697	7.780E-05
83,000	399.3	−60.36	0.3529	7.416E-05
84,000	399.9	−59.80	0.3369	7.069E-05
85,000	400.4	−59.26	0.3215	6.739E-05
86,000	401.0	−58.72	0.3070	6.424E-05
87,000	401.5	−58.18	0.2931	6.125E-05
88,000	402.1	−57.62	0.2798	5.840E-05
89,000	402.6	−57.08	0.2672	5.569E-05
90,000	403.1	−56.54	0.2551	5.311E-05
91,000	403.7	−56.00	0.2436	5.065E-05
92,000	404.2	−55.45	0.2327	4.831E-05
93,000	404.8	−54.91	0.2222	4.607E-05
94,000	405.3	−54.37	0.2123	4.395E-05
95,000	405.9	−53.83	0.2028	4.193E-05
96,000	406.4	−53.29	0.1937	4.000E-05
97,000	407.0	−52.73	0.1851	3.816E-05
98,000	407.5	−52.19	0.1768	3.641E-05
99,000	408.0	−51.65	0.1689	3.474E-05
100,000	408.6	−51.11	0.1614	3.315E-05

Altitude meters	Temperature Kelvin	Temperature Celcius	Pressure Pascals	Density kg/m^3
−1,000	294.7	21.50	1.139E+05	1.347
−500	291.4	18.25	1.075E+05	1.285
0	288.2	15.00	1.013E+05	1.225
500	284.9	11.75	9.546E+05	1.167
1,000	281.7	8.500	8.988E+05	1.112
1,500	278.4	5.250	8.456E+05	1.058
2,000	275.2	2.000	7.950E+05	1.007
2,500	271.9	−1.240	7.469E+05	9.570E-01
3,000	268.7	−4.490	7.012E+05	9.093E-01
3,500	265.4	−7.740	6.578E+05	8.634E-01
4,000	262.2	−10.98	6.166E+05	8.194E-01
4,500	258.9	−14.23	5.775E+05	7.770E-01
5,000	255.7	−17.47	5.405E+05	7.364E-01
5,500	252.4	−20.72	5.054E+05	6.975E-01
6,000	249.2	−23.96	4.722E+05	6.601E-01
6,500	245.9	−27.21	4.408E+05	6.243E-01
7,000	242.7	−30.45	4.111E+05	5.900E-01
7,500	239.5	−33.69	3.830E+05	5.572E-01
8,000	236.2	−36.93	3.565E+05	5.258E-01
8,500	233.0	−40.18	3.315E+05	4.958E-01
9,000	229.7	−43.42	3.080E+05	4.671E-01
9,500	226.5	−46.66	2.859E+05	4.397E-01
10,000	223.3	−49.90	2.650E+05	4.135E-01
10,500	220.0	−53.14	2.454E+05	3.886E-01
11,000	216.8	−56.38	2.270E+05	3.648E-01
11,500	216.7	−56.50	2.099E+05	3.374E-01
12,000	216.7	−56.50	1.940E+05	3.119E-01
12,500	216.7	−56.50	1.793E+05	2.884E-01
13,000	216.7	−56.50	1.658E+05	2.666E-01
13,500	216.7	−56.50	1.533E+05	2.465E-01
14,000	216.7	−56.50	1.417E+05	2.279E-01
14,500	216.7	−56.50	1.310E+05	2.107E-01
15,000	216.7	−56.50	1.211E+05	1.948E-01
15,500	216.7	−56.50	1.120E+05	1.801E-01
16,000	216.7	−56.50	1.035E+05	1.665E-01
16,500	216.7	−56.50	9.572E+05	1.539E-01
17,000	216.7	−56.50	8.850E+05	1.423E-01
17,500	216.7	−56.50	8.182E+05	1.316E-01

(Continued)

Altitude meters	Temperature Kelvin	Temperature Celcius	Pressure Pascals	Density kg/m³
18,000	216.7	−56.50	7.565E+05	1.217E-01
18,500	216.7	−56.50	6.995E+05	1.125E-01
19,000	216.7	−56.50	6.468E+05	1.040E-01
19,500	216.7	−56.50	5.980E+05	9.616E-02
20,000	216.7	−56.50	5.529E+05	8.891E-02
20,500	217.1	−56.07	5.113E+05	8.205E-02
21,000	217.6	−55.57	4.729E+05	7.572E-02
21,500	218.1	−55.07	4.375E+05	6.988E-02
22,000	218.6	−54.58	4.048E+05	6.451E-02
22,500	219.1	−54.08	3.746E+05	5.956E-02
23,000	219.6	−53.58	3.467E+05	5.501E-02
23,500	220.1	−53.09	3.210E+05	5.081E-02
24,000	220.6	−52.59	2.972E+05	4.694E-02
24,500	221.1	−52.09	2.752E+05	4.337E-02
25,000	221.6	−51.60	2.549E+05	4.008E-02
25,500	222.1	−51.10	2.362E+05	3.705E-02
26,000	222.5	−50.61	2.188E+05	3.426E-02
26,500	223.0	−50.11	2.028E+05	3.168E-02
27,000	223.5	−49.61	1.880E+05	2.930E-02
27,500	224.0	−49.12	1.743E+05	2.710E-02
28,000	224.5	−48.62	1.616E+05	2.508E-02
28,500	225.0	−48.13	1.499E+05	2.321E-02
29,000	225.5	−47.63	1.390E+05	2.148E-02
29,500	226.0	−47.14	1.290E+05	1.988E-02
30,000	226.5	−46.64	1.197E+05	1.841E-02

Appendix C: A Solution of Integrals of the Type of Glauert's Integral*

In Chapters 4 and 5 much use is made of the integral

$$G_n = \int_0^\pi \frac{\cos n\theta}{\cos\theta - \cos\theta_1}\, d\theta$$

which is known as Glauert's Integral. The result quoted in the chapters above is

$$\int_0^\pi \frac{\cos n\theta}{\cos\theta - \cos\theta_1}\, d\theta = \pi \frac{\sin n\theta_1}{\sin\theta_1}$$

This may be proved, by contour integration, as follows. In the complex plane, integrate the function

$$f(z) = \frac{z''}{z^2 - 2z\cos\theta_1 + 1}$$

with respect to z around the circle of unit radius centered at the origin. On this circle $z = e^{i\theta}$ and therefore

$$\int_c \frac{z''\, dz}{z^2 - 2z\cos\theta_1 + 1} = \int_{-\pi}^\pi \frac{e^{2in\theta}\, ie^{i\theta}\, d\theta}{z^2 e^{2i\theta} - 2e^{i\theta}\cos\theta_1 + 1}$$

which, cancelling $e^{i\theta}$ from numerator and denominator, applying $e^{i\theta} = \cos\theta + i\sin\theta$ and using De Moivre's theorem, reduces to

$$\frac{i}{2}\int_{-\pi}^\pi \frac{\cos n\theta + i\sin n\theta}{\cos\theta - \cos\theta_1}\, d\theta \qquad (C.1)$$

The *poles* or *singularities* of the function $f(z)$ are those points where $f(z)$ is infinite, i.e. in this case where

$$z^2 - 2z\cos\theta_1 + 1 = 0$$

---

*This section may be omitted at a first reading.

Which when solved provides

$$z = \cos\theta_1 \pm \sin\theta_1 = e^{\pm i\theta_1} \tag{C.2}$$

In general if a function $f(z)$ has a simple pole at the point $z = c$, then

$$\lim_{z \to c}(z - c)f(z)$$

is finite and its value is called the *residue* at the pole. In this case

$$\lim_{z \to c}(z - c)f(z) = \lim_{z \to c}\left(\frac{(z - c)z''}{z^2 - 2z\cos\theta_1 + 1}\right)$$

This limit is found by L'Hôpital's rule,

$$\lim_{z \to c}\left(\frac{(z - c)z''}{z^2 - 2z\cos\theta_1 + 1}\right) = \lim_{z \to c}\left[\frac{\frac{d}{dz}((z - c)z'')}{\frac{d}{dz}(z^2 - 2z\cos\theta_1 + 1)}\right]_{z=c}$$

Thus differentiating and reducing, and for this case using $c = e^{\pm i\theta_1}$, from Eqn (C.2) the residues at the two poles are $(\sin n\theta_l \pm \cos n\theta)/2\sin\theta_l$ and the sum of the residues is

$$\frac{\sin n\theta_1}{\sin\theta_1} \tag{C.3}$$

Now for this case the poles (at the points $c = e^{\pm i\theta_1}$) are on the contour of integration and by Cauchy's residue theorem the value of the integral (Eqn (C.1)) is equal to the product of $i\pi$ and the sum of the residues on the contour. Thus

$$\frac{i}{2}\int_{-\pi}^{\pi}\frac{\cos n\theta + i\sin n\theta}{\cos\theta - \cos\theta_1}d\theta = i\pi\frac{\sin n\theta_1}{\sin\theta_1} \tag{C.4}$$

Equating the imaginary parts of this equation,

$$\frac{1}{2}\int_{-\pi}^{\pi}\frac{\cos n\theta}{\cos\theta - \cos\theta_1}d\theta = \pi\frac{\sin n\theta_1}{\sin\theta_1}$$

and by the symmetry of the integrand: i.e.

$$\int_{0}^{\pi}\frac{\cos n\theta}{\cos\theta - \cos\theta_1}d\theta = \frac{1}{2}\int_{-\pi}^{\pi}\frac{\cos n\theta}{\cos\theta - \cos\theta_1}d\theta$$

Then

$$\int_0^\pi \frac{\cos n\theta}{\cos\theta - \cos\theta_1} d\theta = \pi \frac{\sin n\theta_1}{\sin\theta_1} \tag{C.5}$$

This equation is the solution for Glauert's integral. A useful extension of this work can be performed rapidly, as below.

Using the trigonometric identity,

$$\sin n\theta \sin\theta = \frac{1}{2}\cos(n-1)\theta - \frac{1}{2}\cos(n+1)\theta$$

it follows that

$$\int_0^\pi \frac{\sin n\theta \sin\theta}{\cos\theta - \cos\theta_1} d\theta = \frac{1}{2}\int_0^\pi \frac{\cos(n-1)\theta}{\cos\theta - \cos\theta_1} d\theta - \frac{1}{2}\int_0^\pi \frac{\cos(n+1)\theta}{\cos\theta - \cos\theta_1} d\theta$$

Then from Eqn (C.5) it follows that

$$\int_0^\pi \frac{\sin n\theta \sin\theta}{\cos\theta - \cos\theta_1} d\theta = \frac{\pi}{2}\frac{\sin(n-1)\theta_1 - \sin(n+1)\theta_1}{\sin\theta_1}$$

and since

$$\sin(n-1)\theta_1 - \sin(n+1)\theta_1 = -2\cos n\theta_1 \sin\theta_1$$

the integral becomes

$$\int_0^\pi \frac{\sin n\theta \sin\theta}{\cos\theta - \cos\theta_1} d\theta = -\pi\cos n\theta_1 \text{ for } n = 0, 1, 2, 3... \tag{C.6}$$

Appendix D: Conversion of Imperial Units to Systéme International (SI) Units

The conversion between Imperial units and SI units is based on the fact that the fundamental units (slug, foot, second and degree Centigrade) of the Imperial (English) system have been defined in terms of the corresponding units of the SI:

$$1\,\text{slug} = 14.59388\,\text{kg}$$

$$1\,\text{foot} = 0.3048\,\text{m}$$

$$T_F = (9/5)T_C + 32$$

Where T_F is temperature in degrees Fahrenheit and T_C is temperature in degrees Celsius. Conversion of absolute temperatures is even simpler as 5 Kelvins $= 9°$ Rankine. Note too that the mass unit of *pound mass* is also used, and a slug is equal to 32.174 pound mass. The unit of time, second, is identical in the two systems.

Working from these definitions, the conversion factors given in Table D.1 are calculated. This table covers the more common quantities encountered in aerodynamics. The conversion factors have been rounded to five significant figures where appropriate.

Table D.1 Conversion factors between Imperial units and SI units.

One of these	is equal to this number	of these
ft	0.3048	m
in	25.4	mm
statute mile	1,609.3	m
nautical mile	1,853.2	m
ft^2	0.0929	m^2
in^2	6.4516×10^{-4}	m^2
in^2	645.16	mm^2
ft^3	0.02832	m^3
in^3	1.6387×10^{-5}	m^3
in^3	16,387	mm^3
slug	14.594	kg
$slug\ ft^{-3}$	515.38	$kg\ m^{-3}$

(Continued)

Table D.1 *(Continued)*

One of these	is equal to this number	of these
lbf	4.4482	N
lbf ft^{-2}	47.880	Nm^{-2}
lbf in^{-2}	68,948	Nm^{-2}
ft lbf	1.3558	J
hp	745.70	W
lbf ft	1.3558	Nm
ft s^{-1}	0.3048	m s^{-1}
mile hr^{-1}	0.44704	m s^{-1}
knot	0.51477	m s^{-1}

The knot: The knot continues to be used as a preferred non-metric unit in practical aeronautics. The knot is a unit of speed, and is defined as one nautical mile per hour, where one nautical mile is equal to 1853 meters $\approx$ 6080 feet $\approx$ 1.15 mile.

Bibliography

[1] Mair WA, et al. Definitions to be used in the description and analysis of drag. Aeronautical Research Committee Current Paper No. 369 which was also published in the Journal of the Royal Aeronautical Society, vol. 62. November 1958.

[2] Shapiro AH. The dynamics and thermodynamics of compressible fluid flow, vol.1. New York: Ronald Press; 1953.

[3] Batchelor GK. An introduction to fluid dynamics. New York: Cambridge University Press; 1967.

[4] Bird RB, Stewart WE, Lightfoot EN. Transport phenomena. 2nd ed. New York: J. Wiley & Sons; 2002.

[5] Fuhrmann G. Drag and pressure measurements on balloon models, Z. Flugtech 1911;11:165.

[6] Munk MM. The aerodynamic forces on airship hulls. NACA Report 184; 1924.

[7] Munk MM. Fluid mechanics, Part VI, Section Q. In: Durand W, editor. Aerodynamic theory, vol. 1. Berlin: Springer; New York: Dover; 1934.

[8] Thwaites B, editor. Incompressible aerodynamics. Oxford, UK: Oxford University Press; 1960.

[9] Karamcheti K. Principles of ideal-fluid aerodynamics. New York: J. Wiley & Sons; 1966.

[10] Hess JL, Smith AMO. Calculation of potential flow about arbitrary bodies. Prog Aero Sci 1967;8:1–138.

[11] Press WH, et al. Numerical recipes: the art of scientific computing. 2nd ed. Cambridge: Cambridge University Press; 1992.

[12] Katz J, Plotkin A. Low-speed aerodynamics. 2nd ed. Cambridge: Cambridge University Press; 2000.

[13] Anderson Jr JD. Ludwig Prandtl's boundary layer. Phys Today 2005;42–8.

[14] Kutta W. Lift forces in flowing fluids. Ill Aeronaut Mitt 1902;6:133 [in German].

[15] Joukowsky N. On the shape of the lifting surfaces of kites. Z Flugtech Motorluftschiffahrt 1910;1:281 and 1912;3:81 [in German].

[16] von Kármán T, Trefftz E. Potential flow round aerofoil profiles. Z Flugtech Motorluftschiffahrt 1918;9:111.

[17] Theodorsen T. Theory of wing sections of arbitrary shape. NACA Report No. 411; 1931.

[18] Halsey ND. Potential flow analysis of multi-element airfoils using conformal mapping. AIAA J 1979;12:1281.

[19] Munk MM. General theory of thin wing sections. NACA TR 142; 1922.

[20] Glauert H. A theory of thin aerofoils. Aeronautical Research Council Reports and Memoranda No. 910; 1924.

[21] Glauert H. Theoretical relationships for an aerofoil with hinged flap. Aeronautical Research Council Reports and Memoranda No. 1095; 1927.

[22] Spence DA. The lift coefficient of a thin jet flapped wing. Proc Roy Soc Series A 1956;238,(1212):46–68.

[23] Spence DA. Some simple results for two-dimensional jet-flapped aerofoils. Aero Quart 1958.

[24] Glauert H. The elements of aerofoil and airscrew theory. 2nd ed. Cambridge Science Classics. Cambridge: Cambridge University Press; 1947. First published in 1926.

[25] Abbott IH, von Doenhoff AE. Theory of wing sections. New York: Dover; 1959.

[26] Lighthill MJ. A new approach to thin airfoil theory. Aero Quart 1951;3:193.

[27] Weber J. Calculation of the pressure distribution over the surface of two-dimensional and swept wings with symmetrical sections. Aeronautical Research Council Reports and Memoranda No. 2918;1953.

[28] Hess JL, Smith AMO. Calculation of Potential Flow about arbitrary bodies. Prog Aero Sci 1967;8.

[29] Rubbert PE. Theoretical characteristics of arbitrary wings by a nonplanar vortex lattice method, report D6-9244, The Boeing Co.; 1964.

[30] Hess JL. Calculation of potential flow about arbitrary three-dimensional lifting bodies. Douglas Aircraft Co. Report. MDC J5679/01; 1972.

[31] Brockett TE. Steady two-dimensional pressure distribution on arbitrary profiles. David Taylor Naval Ship Research and Development Center Report 1821; 1965.

[32] Pinkerton RM. Calculated and measured pressure distributions over the midspan section of the NACA 4412 airfoil. NACA Report 563; 1936.

[33] A brief history of Prandtl's contributions can be found inthe article by John D. Anderson Jr. (2005): "Ludwig Prandtl's Boundary Layer", Physics Today, December, pp. 42-48, American Institute of Physics, New York.

[34] Lancaster, FW. Aerodynamics. London: Archibald Constable & Co., Ltd.; 1907.

[35] Lancaster, FW. Aerodynamics. London: Archibald Constable & Co., Ltd.; 1908.

[36] Prandtl L. Tragflügeltheorie. Nachr Ges Wiss 1918;107:451.

[37] Weis-Fogh T. Quick estimates of flight fitness in hovering animals, including novel mechanisms for lift production. J Expl Biol 1973;59:169–230.

[38] Lighthill MJ. On the Weis-Fogh mechanism of lift generation. J Fluid Mech 1973; 60: 1–17.

[39] Saffman PG. Vortex dynamics. Cambridge: Cambridge University Press; 1992.

[40] Vachon MJ, Ray R, Walsh K, Ennix K. F/A-18 aircraft performance benefits measured during the Autonomous Formation Flight Project, AIAA-2002-4491. In: AIAA Atmospheric Flight Mechanics Conference and Exhibit, Monterey, California; August 5–8, 2002.

[41] Ning A, Kroo I. Tip extensions, winglets, and C-wings: conceptual design and optimization. Paper AIAA-2008-7052, presented at the 26th AIAA Applied Aerodynamics Conference, Honolulu, HI; August 2008.

[42] Deighton L. Fighter. Jonathan Cape Ltd.; 1977.

[43] Pohlhamus EC. A concept of the vortex lift of sharp-edge delta wings based on a leading-edge-suction analogy. NASA TN D-3767; See also Applying slender wing benefits to military aircraft. AIAA J Aircraft 1984;21:545–59.

[44] Hunt B. The panel method for subsonic aerodynamic flows: a survey of mathematical formulations and numerical models and an outline of the new British Aerospace scheme. In: Kollmann W, editor. Computational fluid dynamics. Hemisphere Pub. Corp. 1978. p. 100–165; and a review by Hess JL. Panel methods in computational fluid dynamics. Ann Rev Fluid Mech 1990;22:255–74.

[45] Morgenstern J, Arslan A, Lyman V, Vadyak J. F-5 shaped sonic boom demonstrator's persistence of boom shaping reduction through turbulence. Paper AIAA-2005-12, 43rd AIAA Aerospace Sciences Meeting and Exhibit, Reno, Nevada; January 2005.

[46] Holder DW, North RJ, Chinneck A. Experiments with static tubes in a supersonic airstream. ARC R&M-2782; 1950.

[47] Prandtl L, Busemann A. Naherungsverfahren zur Zeichnerischen Ermittlung von Ebenen Strömungen mit Überschallgeschwindigkeit. Stodola Festschrift (Zurich). Füssli Verlag; 1929;S. 499–509.

[48] Glauert H. The effect of compressibility on airfoil. ARC R&M No. 1135; 1927.

[49] Ackeret J. Zlugtech motorluftschiff [DM Trans.]. NAC & TM 1925:72–4.

[50] Houghton EL, Brock AE. Equations, tables, and charts for compressible flow. NACA Technical Report 1135; 1953.

[51] Ferri A. Experimental results with airfoils tested in the high-speed tunnel at Guidornia, Atti Guidornia 1939; 17.

[52] Stanton TE. A high-speed wind channel for tests on airfoils. ARC R&M 1130; 1928.

[53] Houghton EL, Brock AE. Tables for the compressible flow of dry air. 3rd ed. London: Edward Arnold; 1975.

[54] Jameson A. Full-potential Euler and Navier-Stokes schemes. In: Henne PA, editor. Applied computational aerodynamics. Progress in astronautics and aeronautics, vol. 125. New York: AIAA; 1990. p. 39–88.

[55] Schlichting, H. Boundary-layer theory. New York: McGraw-Hill; 1955.

[56] Owen, PR, Klanfer L. On the laminar boundary layer separation from the leading edge of a thin aerofoil. R.A.E. Report 2508; 1953.

[57] Tollmien W. Über die Entstehung der Turbulenz. 1. Mitt. Nachr. Ges. Wiss. Göttingen, Math Phys Klasse 1929; 21–44.

[58] Schlichting H. Zur Entstehung der Turbulenz bei der Plattenströmung. Z angew Math Mech 1933;13:171–4.

[59] Prandtl L. Bermerkungen über die Enstehung der Turbulenz. Z angew Math Mech 1921;1:431–6.

[60] Schubauer GB, Skramstadt HK. Laminar boundary layer oscillations and transition on a flat plate. NACA Rep. 909; 1948.

[61] Reynolds O. On the dynamical theory of incompressible viscous fluids and the determination of the criterion. Philos Trans R Soc Lond Ser A 1895;186:123.

[62] Boussinesq J. Essai sur la théorie des ea ux courantes. Mémoires Acad des Science 1872;23(1).

[63] Laufer J. The structure of turbulence in fully developed pipe flow. NACA report 1174;1954.

[64] Prandtl L. Bemerkungen zur Theorie der freien Turbulenz. ZAMM 1942;22:241–243.

[65] Reichardt H. Gesetzmässigkeiten der freien Turbulenz. VDI-Forschungsheft 1942;414. Berlin.

[66] Prandtl L. Bericht über Untersuchunger zur ausgebildeten Turbulenz. ZAMM 1925;5:136–139.

[67] von Kármán T. Mechanische Ähnlichkeit und Turbulenz, Nachrichten der Akademie der Wissenschaften Göttingen. Math Phys Klasse 1930;68:58–76.

[68] Barenblatt GI, Prostokishin VM. Scaling laws for fully-developed turbulent shear flows. J Fluid Mech 1993;248:513–29.

[69] Carpenter PW. The right sort of roughness. Nature 1997;388:713–4.

[70] Nikuradze J, vDI Forschungsheft. 1933;361 [in English, in Technical Memorandum 1292, National Advisory Committee for Aeronautics (1950)].

[71] Kline SJ, Reynolds WJ, Schraub FA, Runstadler PW. The structure of turbulent boundary layers. J Fluid Mech 1967;30:741–73.

[72] Narahari Rao K, Narasimha R, Badri Narayanan MA. The 'bursting' phenomenon in a turbulent boundary layer. J Fluid Mech 1971;48:339.

[73] Blackwelder RF, Haritonidis JH. Scaling of the bursting frequency in turbulent boundary layers. J Fluid Mech 1983;132:87–103.

[74] Panton RL, editor. Self-sustaining mechanisms of wall turbulence. Southampton, UK: Computational Mechanics Publications; 1997.

[75] For a general description of MEMS technology see Maluf N. An introduction to microelectro-mechanical systems engineering. Boston/London: Artech House; 2000. For a description of their potential use for flow control, see Gad-el-Hak M. 2000. Flow control: passive, active, and reactive flow management. Cambridge: Cambridge University Press; 2000.

[76] Smith BL, Glezer A. The formation and evolution of synthetic jets. Phys. Fluids 1998;1:2281–97. See also Crook A, Sadri AM, Wood NJ. The development and implementation of synthetic jets for the control of separated flow. AIAA Paper 99–3176; Amitay M, et al. Aerodynamic flow control over an unconventional airfoil using synthetic jet actuators. AIAA J 2001;39:361–70; Lockerby DA, Carpenter PW, Davies C. Numerical simulation of the interaction of MEMS actuators and boundary layers. AIAA J 2001;39:67–73.

[77] Press WH. Numerical recipes in Fortran: The art of scientific computing. 2nd ed. Cambridge, UK: Press Syndicate of the University of Cambridge; 1992.

[78] Pohlhausen K. Zur niiherungsweisen integration der DilTerentiaJgJeichung der jaminaren grenzschicht. Z. Angew. Math. M echo 1921;1:252–68. Trans. Anderson RC. Dept. of Eng. Sci., Univ. Fla., Gainesville, 1935.

[79] Thwaites B. Approximate calculation of the laminar boundary layer. Aero Quart 1949;1:245.

[80] Cebeci T, Bradshaw P. Momentum transfer in boundary layers. New York: Hemisphere Publishing Corporation; 1977.

[81] Ludwieg H, Tillmann W. Untersuchungen fiber die wandschubspannung in turbulenten reibungsschichten. Ing Arch 1949;1:228–99. (Investigation of the wall-shearing stress in turbulent boundary layers, NACA TM 1285, 1950).

[82] Head MR. Entrainment in the turbulent boundary layers. ARCR & M, 3152.

[83] Jaffe NA, Okamura TT, Smith AMO. Determination of spatial amplification factors and their application to predicting transition. AIAA J 1970;1:301–8.

[84] Wazzan AR, Gazley C. Smith AMO. $H - R_x$ method for predicting transition. AIAA J. 1981;19:810–2.

[85] For example, see Ferziger JH. Numerical methods for engineering application. 2nd ed. New York: Wiley; 1998. Fersiger JH, Peric M. Computational methods for fluid dynamics. 2nd ed. Berlin: Springer; 2001. Schlichting H, Gersten K. Boundary layer theory, 8th ed. New York: McGraw-Hill; 2000 [Chapter 23].

[86] A guide on commercial CFD is given by Shaw CT. Using computational fluid mechanics. Englewood Cliffs: NJ Prentice Hall; 1992.

[87] Wilcox DC. Turbulence modeling for CFD. 3rd ed. La Canada, CA: DCW Industries, Inc.; 2006.

[88] Press WH, Teukolsky SA, Vetterling WT, Flannery BP. Numerical recipes 3rd edition: the art of scientific computing. Cambridge: Cambridge University Press; 2007.

[89] Ferziger, JH. Numerical methods for engineering applications. 2nd ed. New York: Wiley-Interscience; 1998.

[90] Keller HB. Numerical methods in boundary-layer theory. Ann Rev Fluid Mech 1978;10:417–33.

[91] Stephen BP. Turbulent flows. UK: Cambridge University Press; 2000.

[92] Spalart PR, Watmuff JH. Direct simulation of a turbulent boundary layer up to $R_\theta =$ 1410. J. Fluid Mech 1993;249:337–71; Moin P, Mahesh K. Direct numerical simulation: a tool in turbulence research. Ann Rev Fluid Mech 1998;30:539–78; Friedrich R, et al. Direct numerical simulation of incompressible turbulent flows. Comput Fluid 2001;30:555–79.

[93] Agarwal R. Computational fluid dynamics of whole-body aircraft. Ann Rev Fluid Mech 1999;31:125–70.

[94] Cebeci T, Smith AMO. Analysis of turbulent boundary layers. New York: Academic Press; 1974.

[95] Van Driest ER. On the turbulent flow near a wall. J. Aeronautical Sciences 1956;23:1007–11.

[96] Clauser FH. The turbulent boundary layer. Adv. in Applied Mech 1956;4:1–51.

[97] Corrsin S, Kistler AL. The free-stream boundaries of turbulent flows. NACA Tech. Note 3133; 1954. Klebannoff PS. Characteristics of influence in a boundary layer with zero pressure gradient. NACA Tech. Note 3178 and NACA Rep.1247; 1954.

[98] Baldwin BS, Lomax H. Thin layer approximation and algebraic model for separated turbulent flows. AIAA Paper; 1978:78–257.

[99] Harlow FH, Nakayama PI. Transport of turbulence energy decay rate. Univ. of California, Los Alamos Science Lab. Rep. LA-3854; 1968.

[100] Jones WP, Launder BE. The prediction of laminarization with a two-equation model of turbulence. Int. J. Heat Mass Transfer 1972;15:301–14.

[101] Deardorff JW. A numerical study of three-dimensional turbulent channel flow at large Reynolds numbers. J. Fluid Mech 1970;41:453–480.

[102] Leonard A. Energy cascade in large eddy simulations of turbulent fluid flow. Adv. Geophys 1974;18A:237–48.

[103] Smagorinsky J. General circulation experiments with the primitive equations: I. The basic equations. Mon. Weather Rev 1963;91:99–164.

[104] Tucker PG. Computation of unsteady internal flows. Norwell, MA: Kluwer Academic Publishers; 2001.

[105] Betz A. A method for the direct determination of profile drag. ZFM 1925;16:42–4.

[106] Jones B. Melville: the measurement of profile drag by the Pitot-Traverse method. ARC R. & M. No. 1688; 1936.

[107] Smith AMO. High-lift aerodynamics. J. Aircraft 1975;12:501–530. Many of the topics discussed in Sections 9.1 and 9.2 are covered in greater depth by Smith.

[108] Stratford BS. The prediction of separation of the turbulent boundary layer. J. Fluid Mech 1975;5:1–16.

[109] Liebeck RH. A class of airfoils designed for high lift in incompressible flow. J. Aircraft 1973;10:610–7.

[110] Lissaman PBS. Low-Reynolds-number airfoils. Ann Rev Fluid Mech 1983;15:223–39.

[111] Young AD. The aerodynamic characteristics of flaps. ARC R&M 2622; 1947.

[112] Prandtl L, Tietjens OG. Applied hydro- and aeromechanics. New York: Dover; 1957:227.

[113] Nakayama A, Kreplin HP, Morgan HL. Experimental investigation of flowfield about a multielement airfoil. AIAA J 1990;26:14–21.

[114] Prouty RW. The Gurney flap, Part 2 in the March 2000 issue of Rotor & Wing. Available from: http://www.aviationtoday.com/reports/rotorwing/.

[115] Liebeck RH. Design of subsonic airfoils for high lift. AIAA J. Aircraft 1978;15(9): 547–61.

[116] Jeffrey D, Zhang X, Hurst DW. Aerodynamics of Gurney flaps on a single-element high-lift wing. AIAA J. Aircraft 2000;37(2):295–301; Jeffrey D, Zhang X, Hurst DW. Some aspects of the aerodynamics of Gurney flaps on a double-element wing. Trans. ASME, J. Fluids Eng. 2001;123:99–104.

[117] Bechert DW, Bruse M, Hage W, Meyer R. Biological surfaces and their technological application—laboratory and flight experiments on drag reduction and separation control. AIAA-1997-1960 Fluid Dynamics Conference, 28th, Snowmass Village, CO, June 29-July 2, 1997.

[118] Liebe W. Der Auftrieb am Tragflügel: Enstehung and Zusammenbruch. Aerokurier 1975;12:1520–3.

[119] Malzbender B. Projekte der FV Aachen, Erfolge im Motor- und Segelflug. Aerokurier 1984;1:4.

[120] A more complete recent account is to be found in Gad-el-Hak M. Flow control: passive, active and reactive flow management. Cambridge: Cambridge University Press; 2000.

[121] Prandtl L. Über Flüssigkeitsbewegung bei sehr kleiner Reibung. In: Proceedings of the Third International Mathematical Congress. Heidelberg: 1904;5:p. 484–91.

[122] For a recent review on the aerodynamics of the Coanda effect, see Carpenter PW, Green PN. The aeroacoustics and aerodynamics of high-speed Coanda devices. J. Sound & Vibration 1997;208(5):777–801.

[123] McCormick BW. Aerodynamics, aeronautics and flight mechanics. 2nd ed. New York: Wiley, 1995.

[124] See Prouty's articles on aerodynamics in Aviation Today: Rotor & Wing. Published in May, June, and July 2000. Available from: http://www.aviationtoday.com/reports/rotorwing/.

[125] See the recent review by Greenblatt D, Wygnanski I. The control of flow separation by periodic excitation. Prog. Aero Sci 2000;36:487–545.

[126] Seifert A, Pack LG. Oscillatory control of separation at high Reynolds number. AIAA J 1999;37(9):1062–71.

[127] Wallis RA. The use of air jets for boundary layer control. Aerodynamic Research Laboratories, Australia, Aero. Note 110 (N-34736); 1952; Pearcey HH. Shock-induced separation and its prevention. In: Lachmann V, editor. Boundary layer and flow control, vol. 2. Pergamon; 1961.

[128] For example, see Johnston JP, Nishi M. Vortex generator jets—means for flow separation control. AIAA J 1920;28(6):989–994; see also the recent reviews by Greenblatt and Wygnanski (2000) referenced in Section 9.4.2 and Gad-el-Hak (2000) referenced at the beginning of Section 9.4., and Magill JC. McManus KR. Exploring the feasibility of pulsed jet separation control for aircraft configurations. AIAA J. Aircraft 2001;38(1): 48–56.

[129] Gad-el-Hak M. Flow control: passive, active and reactive flow management. Cambridge: Cambridge University Press; 2000.

[130] Braslow AL. A history of suction-type laminar flow control with emphasis on flight research. NASA History Division, Monographs in Aerospace History, Number 13; 1999.

[131] Reviews of many aspects of this subject are to be found in Bushneil DM, Hefner JN, editors. Viscous drag reduction in boundary layers. Washington, DC: AIAA; 1990.

[132] Holstein H. Messungen zur Laminarhaltung der Grenzschicht an einem Flügel. Lilienthal Bericht 1940;S10:17–27. Ackeret J, Ras M, Pfenninger W. Verhinderung des Turbulentwerdens einer Grenzschicht durch Absaugung. Naturwissenschaften, 1941;29:622–623; and Ras M, Ackeret J. Über Verhinderung der Grenzschicht-Turbulenz durch Absaugung. Helv. Phys. Acta 1941;14:323.

[133] Braslow AL. Laminar-flow control. Available from: http://www.nasa.gov/centers/dryden/history/Publications/index.html, 2000.

[134] Gray J. Studies in animal locomotion. VI: the propulsive powers of the dolphin. J. Experimental Biol 1936;13:192–199.

[135] Kramer MO. Boundary layer stabilization by distributed damping. J. Aeronautical Sciences 1960;24:459; and J. Ame Soc Naval Engineers;74:341–348.

[136] Benjamin TB. Effects of a flexible boundary on hydrodynamic stability. J. Fluid Mech 1960;9:513–532.

[137] Carpenter PW, Garrad AD. The hydrodynamic stability of flows over Kramer-type compliant surfaces. Pt. 1: Tollmien-Schlichting instabilities. J. Fluid Mech 1985;155: 465–510.

[138] Gaster M. Is the dolphin a red herring? In: Liepmann HW, Narasimha R, editors. Proceedings of the IUTAM symposium on turbulence management & relaminarisation. New York: Springer; 2006. p. 285–204, 1987. See also Lucey AD, Carpenter PW. Boundary layer instability over compliant walls: comparison between theory and experiment. Physics of Fluids 1995;7(11):2355–2363.

[139] Detailed reviews of recent progress can be found in Carpenter PW. Status of transition delay using compliant walls, viscous drag reduction in boundary layers. In: Bushnell DM, Hefner JN, editors. Progress in astronautics and aeronautics, vol. 123. Washington, DC: AIAA; 1990. p. 79–113; Gad-el-Hak M. Flow control. Cambridge: Cambridge University Press; 2000 and Carpenter PW, Lucey AD, Davies C. Progress on the use of compliant walls for laminar-flow control. AIAA J. Aircraft 2001;38(3):504–12.

[140] Bechert DW, Bruse M, Hage W, Meyer R. Biological surfaces and their technological application – laboratory and flight experiments and separation control. In: AIAA-1997-1960 fluid dynamics conference; 1997.

[141] Walsh MJ, Weinstein LM. Drag and heat transfer on surfaces with longitudinal fins. AIAA Paper No. 1161; 1978.

[142] Reif W-E, Dinkelacker A. Hydrodynamics of the squamation in fast swimming sharks. Neues Jahrbuch für Geologie und Paleontologie 1982;164:184–187; and Bechert DW, Hoppe G, Reif W-E. On the drag reduction of the shark skin. In: 23rd aerospace sciences meeting. Reno, NV: AIAA; 1985.

[143] Filippone A. Wing-tip devices. Available from: http://aerodyn.org/Drag/tip_devices .html.

[144] Hepperle M. A close look at Winglets. Available from: http://www.mh-aerotools.de/airfoils/index.htm.

[145] Whitcomb RT. A study of the zero-lift drag-rise characteristics of wing-body combinations near the speed of sound. NACA Rep. 1273: January 1956.

Index

Page numbers followed by "*f*" indicates figures and "*t*" indicates tables.

715